COLLEGE MATHEMATICS
WITH APPLICATIONS TO MANAGEMENT, ECONOMICS, AND THE SOCIAL AND NATURAL SCIENCES

CHESTER PIASCIK
Bryant College

Charles E. Merrill Publishing Company
A Bell & Howell Company
Columbus   Toronto   London   Sydney

To my wife, Francine, and sister, Joan, who gave countless hours towards the preparation of early versions of this manuscript.

Published by Charles E. Merrill Publishing Co.
A Bell & Howell Company
Columbus, Ohio 43216

This book was set in Univers
Production Editor: Rex E. Davidson
Developmental Editor: Annamaria Doney
Cover Design: Tony Faiola
Cover Photo: Melvin L. Prueitt,
   Los Alamos National Laboratory
Text Designer: Cynthia Brunk

Material from the Uniform CPA Examinations and Unofficial Answers, copyright © 1970, 1971, 1974, 1975, 1976, 1977, 1978, by the American Institute of Certified Public Accountants, Inc., is reprinted with permission.

Copyright © 1984, by Bell & Howell Company. All rights reserved. No part of this book may be reproduced in any form, electronic or mechanical, including photocopy, recording, or any information storage and retrieval system, without permission in writing from the publisher.

Library of Congress Catalog Card Number: 83-61872
International Standard Book Number: 0-675-20094-6
Printed in the United States of America
1 2 3 4 5 6 7 8 9 10—88 87 86 85 84

PREFACE

This text is designed to provide the mathematical concepts required in today's undergraduate business administration curriculum. Typically the algebra backgrounds of the students taking these courses are varied. Thus, one of my goals has been to make the text as readable as possible without sacrificing its mathematical content.

Another goal of this text is to train the student to think graphically. Accordingly, I have included sections in Chapter 2 on graphing polynomial and rational functions and in Chapter 3 on graphing exponential functions. These sections are also designed to prepare the student for calculus. Since a calculus student is often sketching and working with higher degree polynomial functions, rational functions, and exponentials and logarithmic functions, a prior study of these topics will help to reduce their "shock effect" when encountered in calculus.

This text contains a more thorough coverage of mathematics of finance in response to managements' express needs. Feedback from management professionals indicates that this is the mathematical topic requiring more extensive coverage in undergraduate studies. Many business decisions involve an analysis of cash flows under various conditions. This text includes, in addition to the usual mathematics of finance topics, equations of value, deferred annuities, and complex annuities.

Since we live in an age of inexpensive computing, greater emphasis is placed upon problem formulation. In this regard, I have included one section (Section 6–8) at the end of Chapter 6 (Linear Programming) on formulating linear programming problems from various fields of application.

Chapter 7 (Probability) provides the instructor with many options. The section on Bayes' formula may be omitted if desired. Similarly, Section 7–7 (Counting, Permutations, Combinations, and Probability) may be omitted, as well as the sections on Markov chains, if time does not permit their treatment.

The calculus chapters are designed to provide the student with essential concepts and viable applications. Chapter 8 presents an intuitive approach to the concepts of average rate of change and instantaneous rate of change. These serve as the structural foundation for the "rate of change" and "slope of tangent line" concepts of the derivative. The "Limits" and "Differentiability and Continuity" topics have been placed in separate sections so that the instructor has the flexibility to choose the level of treatment of these topics.

The chapter on optimization stresses both graphics and problem-solving. Applications to real-world problems accompany the calculus concepts.

Applications appear at numerous places in the calculus chapters as they do throughout the text. Section 10–8 (Continuous Cash Flows) attempts to integrate these financial models with those presented in the Mathematics of Finance chapter. In this regard, the Improper Integrals section contains an application to the present value of a perpetual cash flow. Finally, continuous probability distributions are presented as an application of integral calculus. The intent is to introduce the concept of a continuous probability distribution for germination in future courses.

We have six business cases in this text. Each is intended to show how math techniques are actually applied to business problems to solve such practical problems as forecasting the price per share for a stock, calculating the present value, or determining the annual inventory cost for a company. By working through these problems, it is hoped the student will see how mathematical techniques are applied to solve real-world business problems.

Acknowledgments

I wish to thank the many reviewers of this manuscript for their valuable suggestions. These include especially Robert A. Moreland, Texas Tech University; Franklin Sheehan, San Francisco State University; Laurence Maher, North Texas State University; Joseph M. Mutter, Northern Arizona University; and Merlin M. Ohmer, Nichols State University.

A special note of thanks goes to my colleague, Alan Olinsky of Bryant College, who provided assistance with the computer generation of tables for this text. Also, I thank my colleague, Robert Muksian of Bryant College, for his valuable suggestions.

I thank Bryant College for its support with the secretarial and typing aspects of this project. A special thanks goes to Jackie David who coordinated the typing of this project. Her efficiency and willingness to help were invaluable. I am indebted to the typists who worked faithfully and diligently on this project. These include Mary Afeltra, Sherry Maddison, Pamela Eddleston, and Susan LaRosa.

My sincere thanks goes to the staff of Charles E. Merrill Publishing Company for their support and expertise in the production of this text. Specifically, I thank Christopher R. Conty for his support and patience during the earlier phases of this project. I thank Marianne Taflinger for her editorial support during the final stages of this manuscript. I especially thank Annamaria Doney for her dedication and tireless efforts towards the goal of producing an excellent text.

A very special thanks goes to Ho Key Min for valuable suggestions and comments regarding the applications in this text. Another very special thanks goes to Patricia A. Fox for invaluable suggestions and perceptive criticisms for both the text and Instructor's Manual.

Finally, I thank my family for their support and patience during the countless hours I have spent writing this manuscript.

Chester Piascik

CONTENTS

1 LINEAR FUNCTIONS — 1

- **1-1** The Real Numbers 2
- **1-2** Functions 7
- **1-3** Slope of a Straight Line 22
- **1-4** Slope and Equations of Straight Lines 30
- **1-5** Graphing Linear Functions 41
- **1-6** Applications of Linear Functions 47

2 POLYNOMIAL AND RATIONAL FUNCTIONS — 59

- **2-1** Exponents 60
- **2-2** Quadratic Functions 67
- **2-3** Applications of Quadratic Functions 78
- **2-4** Polynomial Functions 92
- **2-5** Applications of Polynomial Functions 99
- **2-6** Rational Functions 106

3 EXPONENTIAL AND LOGARITHMIC FUNCTIONS — 119

- **3-1** Exponential Functions 120
- **3-2** Applications of Exponential Functions 132
- **3-3** Logarithms 139
- **3-4** Applications of Logarithms 149
- Case A Stock Price Forecasting—Microchips, Inc. 156

4 MATHEMATICS OF FINANCE — 159

- 4-1 Simple Interest 160
- 4-2 Compound Interest 169
- 4-3 Continuous Compounding 180
- 4-4 Geometric Series and Annuities 185
- 4-5 Present Value of an Annuity 193
- 4-6 Sinking Funds and Amortization 200
- 4-7 Equations of Value 207
- 4-8 Deferred Annuities 214
- 4-9 Complex Annuities 219
- 4-10 Complex Annuities Due 227
- Chapter Exercises 232
- Case B Capital Investment Decisions—Write Graphics, Inc. 236

5 LINEAR SYSTEMS AND MATRICES — 239

- 5-1 Two Equations in Two Variables 240
- 5-2 Matrices 248
- 5-3 Multiplying Matrices 255
- 5-4 Gauss-Jordan Method of Solving Linear Systems 269
- 5-5 Inverse of a Square Matrix 283
- 5-6 Solving Square Linear Systems by Matrix Inverses 290
- 5-7 Applications 294
- Case C Oil Refinery Scheduling—Merco Oil Refinery 302

6 LINEAR PROGRAMMING — 303

- 6-1 Algebra Refresher 304
- 6-2 Linear Inequalities in Two Variables 307
- 6-3 Linear Programming 317
- 6-4 Simplex Method Formulation 326
- 6-5 Simplex Method 335
- 6-6 Shadow Prices 349
- 6-7 Simplex Method: Minimization 352

CONTENTS

6-8 Linear Programming: Formulation and Applications 362
Case D Profit Maximization—Comfort House Furniture Company 368

7 PROBABILITY 369

- **7-1** Sets 370
- **7-2** Basic Probability Formula 382
- **7-3** Laws of Addition 390
- **7-4** Conditional Probability 397
- **7-5** Laws of Multiplication 402
- **7-6** Bayes' Formula 409
- **7-7** Counting, Permutations, Combinations, and Probability 416
- **7-8** Probability Distributions and Random Variables 429
- **7-9** Binomial Experiments 435
- **7-10** Markov Chains 445
- **7-11** Markov Chains in Equilibrium 451
  Case E Probability Distributions—Next Day Courier, Inc. 457

8 DIFFERENTIAL CALCULUS 459

- **8-1** The Derivative 460
- **8-2** Limits 467
- **8-3** Differentiability and Continuity 482
- **8-4** Rules for Finding Derivatives 487
- **8-5** Chain Rule 502
- **8-6** Applications 507
- **8-7** Derivatives of Exponential and Logarithmic Functions 518

9 DIFFERENTIAL CALCULUS: OPTIMIZATION 531

- **9-1** Critical Values and the First Derivative 532
- **9-2** The Second Derivative and Curve Sketching 541
- **9-3** Applications 550

x CONTENTS

9–4 Partial Derivatives 560
9–5 Maxima and Minima 566
Case F Back-order Inventory Model—BFI, Inc. 572

10 INTEGRATION 575

10–1 Antidifferentiation 576
10–2 The Definite Integral and Area Under a Curve 584
10–3 Area Between Two Curves 599
10–4 Application: Consumers' and Producers' Surpluses 604
10–5 Integration by Substitution 607
10–6 Integrals Involving Exponential and Logarithmic Functions 615
10–7 Using Tables of Integrals 620
10–8 Application: Continuous Cash Flows 624
10–9 Improper Integrals 626
10–10 Continuous Probability Distributions 630

APPENDIX A-1
ANSWERS TO SELECTED EXERCISES A-33
INDEX I-1

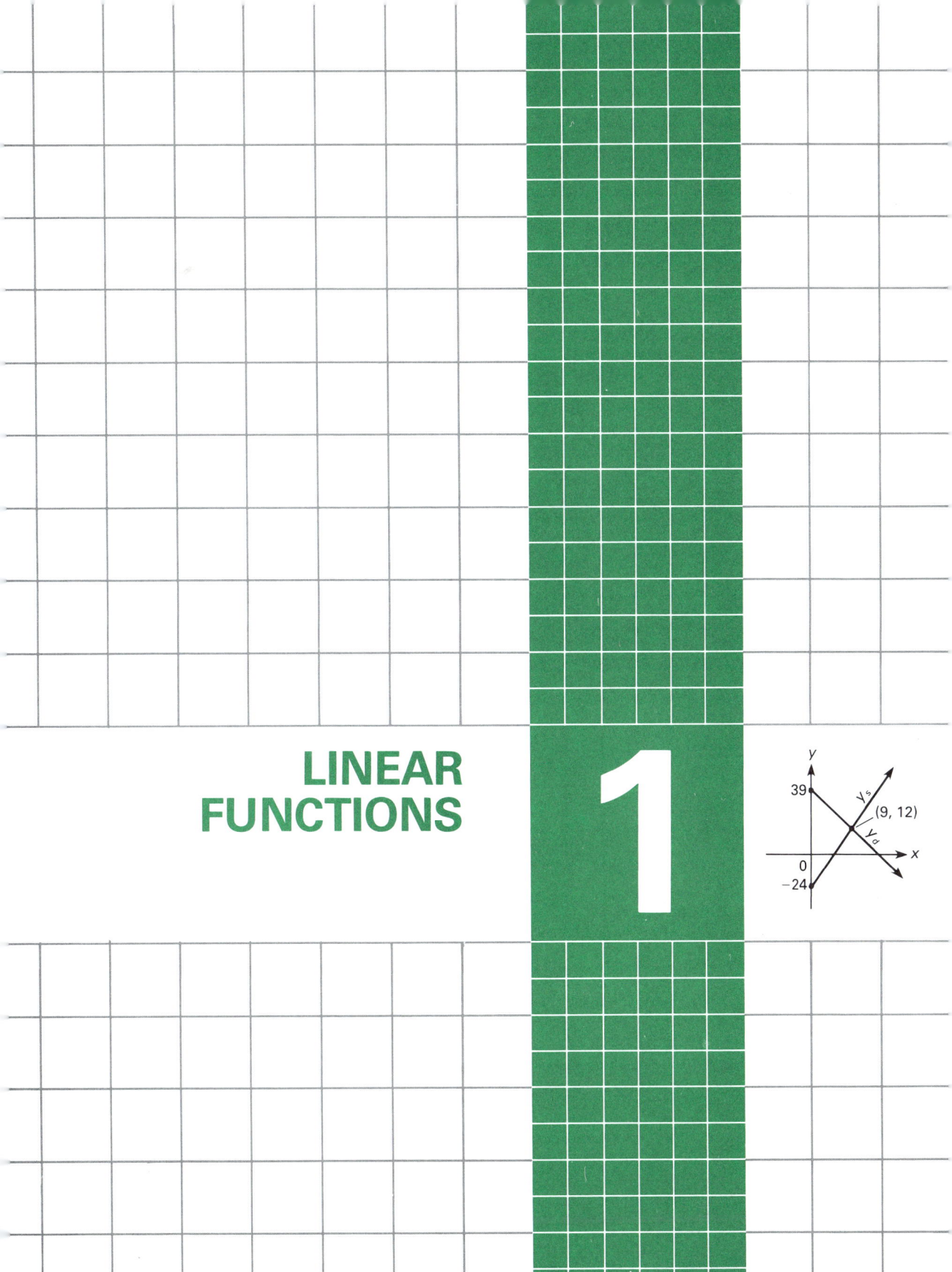
LINEAR FUNCTIONS
1

1-1

THE REAL NUMBERS

All the numbers we will use in this text can be represented as points on a straight line. Such a representation can be constructed as follows. Begin with a straight line. Choose an arbitrary point, called the **origin,** on the line and label it 0. Then choose another point to the right of 0 and label it 1. Let the distance between 0 and 1 represent 1 unit of measure (see Figure 1–1). The point on the straight line 1 unit to the right of 1 is labeled 2, the point 1 unit to the right of 2 is labeled 3, etc. (see Figure 1–2). Also, the point on the straight line 1 unit to the left of 0 is labeled −1, the point 1 unit to the left of −1 is labeled −2, the point 1 unit to the left of −2 is labeled −3, etc. (see Figure 1–2). The straight line of Figure 1–2 is called the **real number line.** There is a one-to-one correspondence between the points on the real number line and the set of real numbers. In other words, any real number is associated with some point on this number line. Also, any point on this number line is associated with some real number. For example, the fraction 1/2 is associated with the point midway between 0 and 1; the number $1\frac{3}{4}$ is associated with the point three-quarters of the distance between 1 and 2; and the number −1/3 is associated with the point one-third the distance between 0 and −1 (see Figure 1–3).

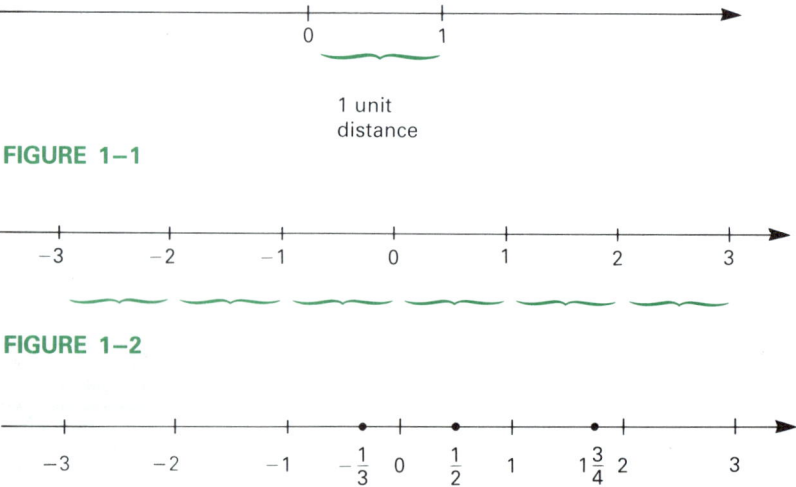

FIGURE 1–1

FIGURE 1–2

FIGURE 1–3

There are several types of real numbers:

1. **Counting Numbers** (also called **natural numbers**):

$$1, 2, 3, 4, 5, \ldots *$$

3 LINEAR FUNCTIONS

2. **Whole Numbers:**

$$0, 1, 2, 3, 4, \ldots *$$

3. **Integers:**

$$\ldots, -4, -3, -2, -1, 0, 1, 2, 3, 4, \ldots$$

4. **Rational Numbers**—All numbers that can be expressed as a quotient of two integers, where the denominator is not equal to 0. Examples are numbers such as 1/2, 3/4, 4/1, 3, 1.23, and nonterminating repeating decimals such as 5.838383 . . ., which can be written as $5.\overline{83}$.

5. **Irrational Numbers**—All real numbers that are not rational. Irrational numbers have decimal representations that are nonterminating and nonrepeating. Some examples are:

$$\sqrt{2} = 1.4142135\ldots **$$
$$\pi = 3.1415926\ldots$$
$$e = 2.718281\ldots$$
$$-\sqrt{5} = -2.2360679\ldots$$

Thus, a real number is either rational or irrational. The rational numbers include the integers, the integers include the whole numbers, and the whole numbers include the counting or natural numbers.

Inequality If a number a lies to the left of a number b on the real number line, then "a is less than b." This is written $a < b$ (see Figure 1–4). Also, if a number b lies to the right of a number a on the real number line, then "b is greater than a." This is written $b > a$ (see Figure 1–5). Thus, the statement "5 is less than 6" is written $5 < 6$.

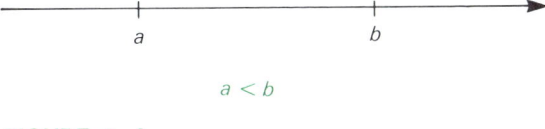

FIGURE 1–4

FIGURE 1–5

---

*Here, the three dots indicate that the numbers continue indefinitely in the same manner.
**These dots indicate that the decimal representations are nonterminating.

4 CHAPTER ONE

The inequality phrases and their respective symbols are summarized as follows:

Inequality Phrase	Symbol
"Is less than"	<
"Is greater than"	>
"Is less than or equal to"	≤
"Is greater than or equal to"	≥
"Is not equal to"	≠

Intervals   Sometimes, it is necessary to refer to all real numbers located between two numbers a and b on the real number line (see Figure 1–6). Such a set of numbers is called an **interval** and is expressed as all real numbers x such that

$$a < x < b$$

Observe that the endpoints, a and b, are not included in the above interval. This situation is graphically expressed by using an open circle at each endpoint (see Figure 1–6). If the endpoints are to be included, then the set must be written as

$$a \leq x \leq b$$

and graphically expressed by using a solid circle at each endpoint (see Figure 1–7).

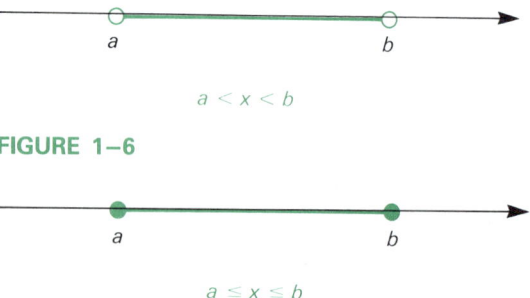

$a < x < b$

FIGURE 1–6

$a \leq x \leq b$

FIGURE 1–7

EXAMPLE 1–1   Graph all real numbers x such that $5 \leq x \leq 10$.

5 LINEAR FUNCTIONS

Solution This interval includes all real numbers between 5 and 10. The endpoints are included. The graph appears in Figure 1–8.

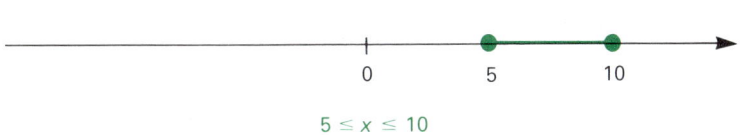

FIGURE 1–8

EXAMPLE 1–2 Express the interval of Figure 1–9 by using the variable x.

FIGURE 1–9

Solution This interval includes all real numbers between -7 and -3. The endpoints are not included. Hence, the interval is written as all real numbers x such that $-7 < x < -3$.

EXAMPLE 1–3 Graph all real numbers x such that $x \leq 9$.
Solution This interval includes all real numbers less than or equal to 9. The endpoint, 9, is included. The graph appears in Figure 1–10.

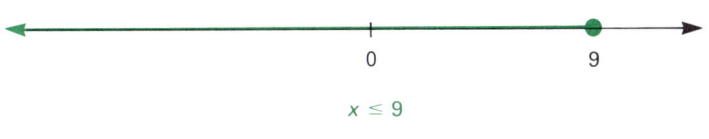

FIGURE 1–10

Absolute Value   The **absolute value** of a number x, written $|x|$, is the distance on the real number line from 0 to x. Thus, the absolute value of -3, written $|-3|$, is 3 since the distance from 0 to -3 is 3 units (see Figure 1–11). Also, the absolute value of 3, written $|3|$, is 3 since the distance from 0 to 3 is 3 units (see Figure 1–12). Note that the absolute value of a number is always nonnegative.

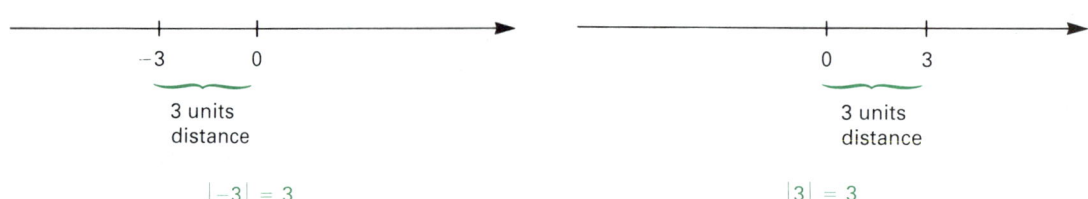

|−3| = 3 |3| = 3

FIGURE 1–11   **FIGURE 1–12**

EXAMPLE 1–4   Solve for $|-7|$.
Solution $|-7| = 7$ since the distance between 0 and −7 on the real number line is 7 units.

EXAMPLE 1–5   Solve for $|8|$.
Solution $|8| = 8$ since the distance between 0 and 8 on the real number line is 8 units.

EXAMPLE 1–6   Graph all real numbers x such that $|x| \leq 5$ on the real number line.
Solution   This interval includes all numbers x on the real number line such that the distance between 0 and x is less than or equal to 5. Thus, any real number x within 5 units distance of 0 has an absolute value less than or equal to 5. Hence, we write

$$-5 \leq x \leq 5$$

This interval is graphed in Figure 1–13.

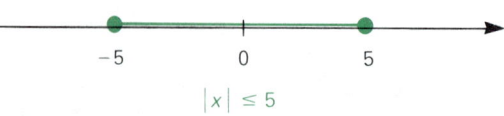

$|x| \leq 5$

FIGURE 1–13

EXERCISES   **1.** State whether each of the following is true or false:
(A) $3 < 7$ (B) $-3 < -7$ (C) $-2 < -5$
(D) $2 < 5$ (E) $-6 < -2$ (F) $-3 > -7$
(G) $-2 > -5$ (H) $0 < 5$ (I) $0 > -3$
(J) $9 > 6$ (K) $8 > 10$ (L) $-6 < -1$

7 LINEAR FUNCTIONS

2. State whether each of the following is true or false:
   - (A) Every counting number is a whole number.
   - (B) Every whole number is an integer.
   - (C) Every counting number is an integer.
   - (D) Every integer is a rational number.
   - (E) Every rational number is a real number.
   - (F) Every integer is a whole number.
   - (G) Every whole number is a counting number.
   - (H) Every irrational number is a real number.

3. State whether each of the following is true or false:
   - (A) 7 is a rational number.
   - (B) $\frac{3}{5}$ is a rational number.
   - (C) $-\frac{2}{3}$ is a rational number.
   - (D) $\sqrt{11}$ is a rational number.
   - (E) $\sqrt{11}$ is an irrational number.
   - (F) 3.56345 . . . is an irrational number.
   - (G) 4.7065 is an irrational number.
   - (H) 2.767676 . . . is a rational number.

4. Sketch each of the following on the real number line:
   - (A) $-5 \leq x \leq -1$
   - (B) $7 \leq x \leq 11$
   - (C) $-4 < x < -2$
   - (D) $9 < x < 15$
   - (E) $-3 < x \leq 2$
   - (F) $2 \leq x < 9$
   - (G) $5 \leq x$
   - (H) $x \geq 5$
   - (I) $x \leq -3$
   - (J) $x < 10$
   - (K) $x > -2$
   - (L) $x > 4$
   - (M) $2 < x$
   - (N) $x \geq -1$
   - (O) $x \neq 2$
   - (P) $x = -3, x \neq 5$

5. Solve for each of the following:
   - (A) $|0|$
   - (B) $|-1|$
   - (C) $|1|$
   - (D) $|-21|$
   - (E) $|-2|$
   - (F) $|15|$
   - (G) $|-15|$
   - (H) $|-20|$
   - (I) $|20|$

6. Sketch each of the following intervals on the real number line:
   - (A) $|x| \leq 4$
   - (B) $|x| < 8$
   - (C) $|x| \leq 10$
   - (D) $|x| < 10$
   - (E) $|x| \geq 6$
   - (F) $|x| > 6$
   - (G) $|x| \neq 5$
   - (H) $|x| \neq 3$

1–2

FUNCTIONS A **function** is a rule that associates a unique **output value** with each element in the set of possible **input values**. Consider, for example, the conversion of temperature from degrees Fahrenheit to degrees Celsius. Given a temperature in degrees Fahrenheit (input value), we can find the corresponding value in degrees Celsius (output value) by the following rule:

8 CHAPTER ONE

$$\underbrace{\text{Celsius temperature}}_{\substack{\text{output}\\\text{value}}} = \frac{5}{9} \underbrace{(\text{Fahrenheit temperature} - 32)}_{\substack{\text{input}\\\text{value}}}$$

If C is temperature in degrees Celsius and F is temperature in degrees Fahrenheit, then this rule may be expressed by the equation

$$C = \frac{5}{9}(F - 32)$$

To determine the Celsius temperature (output value) associated with 50 degrees Fahrenheit, we substitute $F = 50$ (input value) into the equation and obtain

$$C = \frac{5}{9}(50 - 32)$$
$$= \frac{5}{9}(18)$$
$$= 10$$

Thus, 10 degrees Celsius is associated with 50 degrees Fahrenheit. Since only one value of C is associated with a value of F, then this equation defines C as a function of F.

Observing the equation

$$C = \frac{5}{9}(F - 32)$$
$$\underset{\substack{\text{output}\\\text{value}}}{\uparrow} \quad \underset{\substack{\text{input}\\\text{value}}}{\uparrow}$$

note that the output value, C, is dependent upon the input value, F. Thus, C is called the **dependent variable,** and F is called the **independent variable.** This relationship is usually indicated by saying that C is a *function* of F.

Functional Notation Often, a letter is used to represent a function. Specifically, if the letter f is used to name the function defined by the equation

$$y = 5x^2 + 2x + 7$$

then the dependent variable, y, is represented by the symbol $f(x)$, read "f of x." Thus, the preceding equation is written as

$$f(x) = 5x^2 + 2x + 7$$

To find the output value associated with $x = 3$, we replace x with 3 to obtain

9 LINEAR FUNCTIONS

$$f(3) = 5(3)^2 + 2(3) + 7$$
$$= 45 + 6 + 7$$
$$= 58$$

Rectangular Coordinate System

It is often useful to graph functions on a plane called the **rectangular coordinate system.** Such a system consists of two perpendicular real number lines in the plane, as shown in Figure 1–14. The horizontal number line is called the **x-axis,** and the vertical number line is called the **y-axis.** The point where the lines intersect is the zero point of both lines and is called the **origin.**

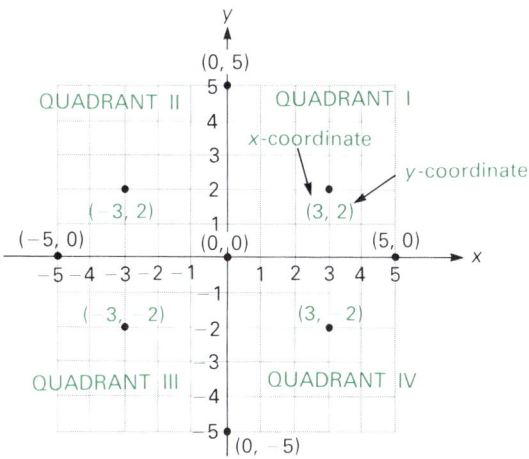

FIGURE 1–14

The plane consists of infinitely many **points.** Each point is assigned an **ordered pair** of numbers which locates its position relative to both axes. For example, looking at Figure 1–14, the ordered pair (3, 2) is associated with the point that may be plotted by starting at the origin and moving 3 units to the right horizontally, and then 2 units upward vertically. The numbers 3 and 2 of the ordered pair (3, 2) are called the **x-coordinate** and **y-coordinate,** respectively.

Similarly, the ordered pair (−3, 2) of Figure 1–14 is associated with the point that may be plotted by starting at the origin, moving 3 units to the left horizontally, then 2 units upward vertically. The ordered pair (−3, −2) is associated with the point that may be plotted by starting at the origin, moving 3 units to the left horizontally, then 2 units downward vertically. The ordered pair (3, −2) is associated with the point that may be plotted by starting at the origin, moving 3 units to the right horizontally, then 2 units downward vertically. The ordered pair (0, 0) is associated with the origin.

Further studying Figure 1–14, we note the following:

1. The x and y axes partition the plane into four quadrants.
2. For any point in Quadrant I, the x and y coordinates are both positive, i.e., (+, +).
3. For any point in Quadrant II, the x-coordinate is negative and the y-coordinate is positive, i.e., (−, +).
4. For any point in Quadrant III, the x and y coordinates are both negative, i.e., (−, −).
5. For any point in Quadrant IV, the x-coordinate is positive and the y-coordinate is negative, i.e., (+, −).
6. Points on the axes belong to no quadrant. Points on the x-axis have y-coordinates of 0, and points on the y-axis have x-coordinates of 0.

Functions as Sets of Ordered Pairs

It is possible to express a function as a set of ordered pairs (x, y) such that each value of y is the number associated with its corresponding value of x in accordance with the rule defined by the equation. For example, consider the situation of a young entrepreneur manufacturing sneakers. She initially invests $1000 to pay for overhead items such as heat, electricity, etc. Additionally, each pair of sneakers costs her $5 to manufacture. Thus, the total cost of producing x pairs of sneakers is given by the equation

$$y = C(x) = 5x + 1000$$

During the first week of operation, the entrepreneur plans to manufacture either 50, 100, 150, or 200 pairs of sneakers. Table 1–1 shows the output value associated with each input value. The set of input values is called the **domain** of the function, and the set of output values is called the **range** of the function. Observing Table 1–1, note that the domain consists of the set of x-values (in the left column) and that the range consists of the set of y- or C(x)-values (in the right column). The function C consists of the set of ordered pairs (x, y) graphed in Figure 1–15.

TABLE 1–1

x	y	$y = C(x) = 5x + 1000$
50	1250	$C(50) = 5(50) + 1000 = 1250$
100	1500	$C(100) = 5(100) + 1000 = 1500$
150	1750	$C(150) = 5(150) + 1000 = 1750$
200	2000	$C(200) = 5(200) + 1000 = 2000$

11 LINEAR FUNCTIONS

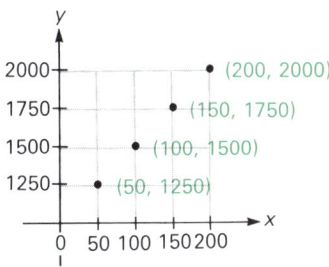

FIGURE 1-15

We are now ready for a more formal definition of a function.

> A **function** is a set of ordered pairs (x, y) such that no two ordered pairs have the same first element x and different 2nd elements y. The set of first elements, x, is the **domain** of the function, and the set of second elements, y, is the **range** of the function.

Consider the equation

$$y^2 = x$$

The ordered pairs defined by this equation include $(4, 2)$, $(4, -2)$, $(9, 3)$, and $(9, -3)$. Note that the two numbers $+2$ and -2 are associated with $x = 4$. Since this equation associates more than one y-value for each x-value, it is not a function.

To graph the equation $y^2 = x$, we plot the points corresponding to its ordered pairs, some of which are $(4, 2)$, $(4, -2)$, $(9, 3)$, and $(9, -3)$. Figure 1-16 illustrates the graph of this equation. Referring to this illustration, note that since $x = 4$ is associated with the two y-values of $+2$ and -2, a vertical line cuts the graph at two points. Generalizing, we have the following **vertical line test** for a function.

> **Vertical Line Test**
>
> If a vertical line cuts a graph at more than one point, then that graph does not represent a function.

12 CHAPTER ONE

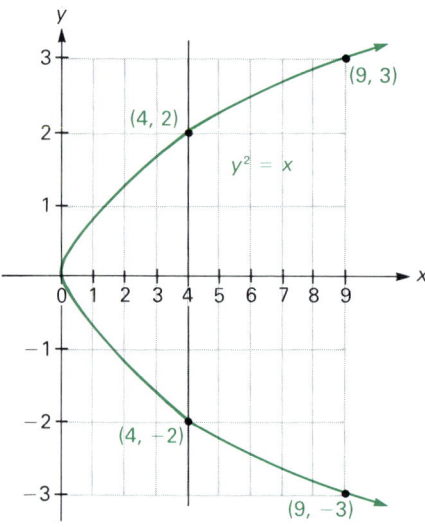

FIGURE 1–16

EXAMPLE 1–7 Use the vertical line test to determine if the graph of Figure 1–17 represents a function.

FIGURE 1–17

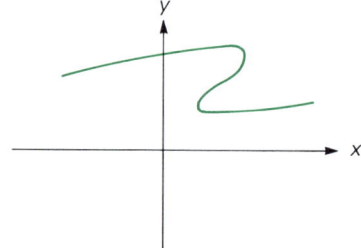

Solution Since a vertical line can be drawn to cut the graph at more than one point (see Figure 1–18), the graph does not represent a function.

FIGURE 1–18

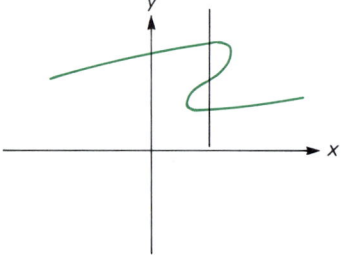

13 LINEAR FUNCTIONS

EXAMPLE 1-8 Use the vertical line test to determine if the graph of Figure 1-19 represents a function.

FIGURE 1-19

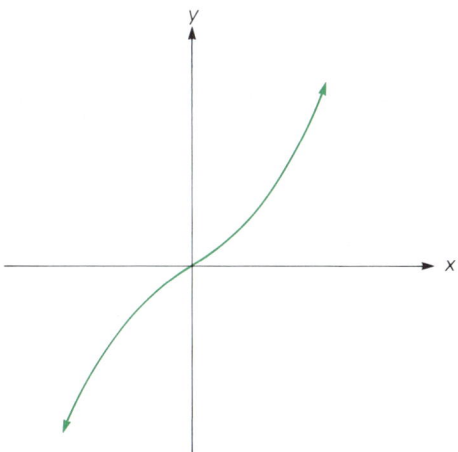

Solution Since it is not possible for a vertical line to cut the graph at more than one point (see Figure 1-20), the graph does represent a function.

FIGURE 1-20

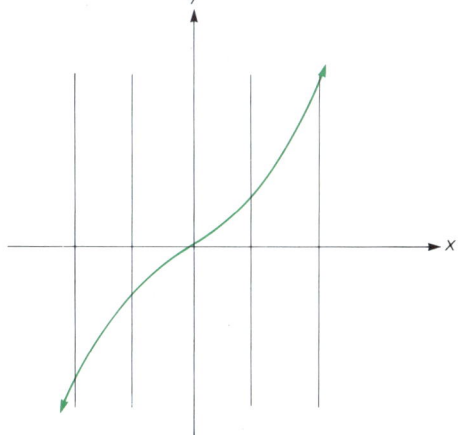

Often the domain of a function is not specified. If this is the case, it is assumed that the domain is the set of all real numbers for which the function is defined. Consider the function defined by the equation

$$g(x) = \frac{1}{x-2}$$

14 CHAPTER ONE

Since the domain of g is not specified, it is the set of all real numbers x for which $g(x)$ is defined. Note that $g(x)$ is defined for all real numbers x except $x = 2$, since

$$g(2) = \frac{1}{2-2} = \frac{1}{0} \quad \text{which is undefined.}$$

Thus, the domain of g is all real numbers x such that $x \neq 2$.

EXAMPLE 1–9 Specify the domain of h for the function defined by

$$h(x) = \frac{1}{(x-3)(x+5)}$$

Solution Since the domain of h is not specified, it is the set of all real numbers x for which $h(x)$ is defined. Note that $h(x)$ is defined for all real numbers x except $x = 3$ and $x = -5$, since

$$h(3) = \frac{1}{(3-3)(3+5)} = \frac{1}{0} \quad \text{which is undefined.}$$

$$h(-5) = \frac{1}{(-5-3)(-5+5)} = \frac{1}{0} \quad \text{which is undefined.}$$

Thus, the domain of h is all real numbers x such that $x \neq 3$ and $x \neq -5$.

EXAMPLE 1–10 Given that $f(x) = 3x^2 - 2x + 5$, calculate each of the following:
(A) $f(4)$
(B) $f(x + h)$
(C) $f(x + h) - f(x)$
(D) $\dfrac{f(x + h) - f(x)}{h}$

Solutions
(A) Since $f(x) = 3x^2 - 2x + 5$, then $f(4)$ is calculated by replacing x with 4. This gives us

$$f(4) = 3(4)^2 - 2(4) + 5$$
$$= 45$$

(B) Since $f(x) = 3x^2 - 2x + 5$, then $f(x + h)$ is calculated by replacing x with $x + h$. Therefore, we have

15 LINEAR FUNCTIONS

$$f(x + h) = 3(x + h)^2 - 2(x + h) + 5$$
$$= 3(x^2 + 2hx + h^2) - 2x - 2h + 5$$
$$= 3x^2 + 6hx + 3h^2 - 2x - 2h + 5$$

(C) Subtracting $f(x) = 3x^2 - 2x + 5$ from the result of part (B) yields

$$f(x + h) - f(x) = 3x^2 + 6hx + 3h^2 - 2x - 2h + 5$$
$$- (3x^2 - 2x + 5)$$
$$= 6hx + 3h^2 - 2h$$

(D) Dividing the result of part (C) by h, we obtain

$$\frac{f(x + h) - f(x)}{h} = \frac{6hx + 3h^2 - 2h}{h}$$
$$= 6x + 3h - 2$$

This result is called a **difference quotient** and will be discussed further when we study calculus.

We now illustrate more functions and their graphs.

EXAMPLE 1–11 Graph the function defined by

$$g(x) = \begin{cases} x & \text{if } x \geq 2 \\ 1 & \text{if } x < 2 \end{cases}$$

Solution The domain is all real numbers since a value of $g(x)$ is associated with any real number x. If x is at least 2, then the function g associates x with itself. Thus, the ordered pairs $(2, 2)$, $(3, 3)$, $(4, 4)$, and $(9/2, 9/2)$ are among those belonging to g. All numbers less than 2 are associated with the number 1. Thus, $(1.99, 1)$, $(1.5, 1)$, $(1, 1)$, $(0, 1)$, $(-1, 1)$, and $(-2, 1)$ are also among the ordered pairs belonging to g. The graph of g appears in Figure 1–21.

FIGURE 1–21

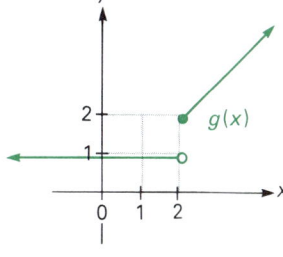

EXAMPLE 1-12 A private parcel service charges the following rates for delivering small packages:

$1.00 for a package weighing less than 4 ounces
$1.50 for a package weighing at least 4 ounces but less than 20 ounces
$2.00 for a package weighing at least 20 ounces but less than 32 ounces

The service does not deliver packages 32 ounces or heavier. Express delivery cost as a function of weight and graph the function.

Solution If x is the weight of a package, its delivery cost, $C(x)$, is given by

$$C(x) = \begin{cases} 1.00 \text{ if } 0 < x < 4 \\ 1.50 \text{ if } 4 \leq x < 20 \\ 2.00 \text{ if } 20 \leq x < 32 \end{cases}$$

Note that the domain of $C(x)$ is the interval $0 < x < 32$. The graph of $C(x)$ appears in Figure 1-22.

FIGURE 1-22

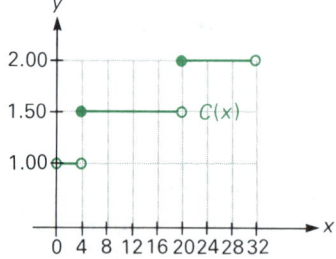

Absolute Value Function

In Section 1-1, we defined the absolute value of a number x as the distance on the real number line from 0 to x. We now define the **absolute value function,** $a(x)$, as follows:

$$a(x) = |x| = \begin{cases} x \text{ if } x \geq 0 \\ -x \text{ if } x < 0 \end{cases}$$

Note that this function associates a nonnegative number with itself. A negative number is associated with its additive inverse. Thus, (−3, 3), (−1, 1), (−1/2, 1/2), (0, 0), (1, 1), (2, 2), and (5/2, 5/2) are some of the ordered pairs belonging to the absolute value function. The graph of $a(x)$ appears in Figure 1-23.

17 LINEAR FUNCTIONS

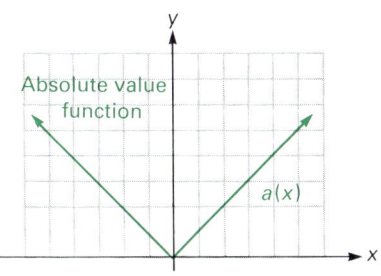

FIGURE 1–23

EXERCISES

1. A function is defined by the equation
$$y = 3x - 2$$
   (A) Which number does this equation associate with $x = 0$?
   (B) Which number does this equation associate with $x = 4$?

2. A function is defined by the equation
$$y = 3x^2 - 4x + 5$$
   (A) Which number does this equation associate with $x = 0$?
   (B) Which number does this equation associate with $x = 2$?

3. If $f(x) = -2x + 7$, calculate each of the following:
   (A) $f(0)$
   (B) $f(1)$
   (C) $f(5)$
   (D) $f(-3)$

4. If $w(r) = r^3 - 7r^2 + 8r - 5$, calculate each of the following:
   (A) $w(0)$
   (B) $w(2)$
   (C) $w(-2)$
   (D) $w(4)$

5. If $z(t) = \dfrac{5}{t + 7}$ calculate each of the following:
   (A) $z(0)$
   (B) $z(1)$
   (C) $z(-6)$
   (D) $z(8)$

6. The Fluff Detergent Company produces detergent. The overhead cost is $5000. Additionally, each box of Fluff Detergent costs $0.40 to produce. If $C(x)$ represents the total cost of producing x boxes of Fluff Detergent, then

$$C(x) = 0.40x + 5000$$

The function C is called a **cost function**.
   (A) Determine the total cost of producing 10,000 boxes of Fluff Detergent.
   (B) Calculate and interpret: $C(0)$, $C(1000)$, $C(5000)$.

The cost of producing 0 boxes, $C(0)$, is called the **fixed cost**, and the cost of producing each additional box of detergent, $0.40, is called the **variable cost per unit**, or **the unit variable cost**.

7. The Soft Soap Company has a fixed cost of $10,000 and a variable cost per unit of $0.25 per bar of soap. Assume that x represents the number of bars of soap produced and $C(x)$ represents the total cost of producing x bars of soap.
   (A) Write the equation that defines the cost function.
   (B) Calculate and interpret: $C(0)$, $C(10)$, $C(1000)$.

8. A manufacturer of wooden shoes sells each pair for $100. If $R(x)$ represents the total sales revenue gained from selling x pairs of wooden shoes, then

$$R(x) = 100x$$

   The function R is called a **sales revenue function**.
   (A) Calculate and interpret: $R(0)$, $R(20)$, $R(50)$.
   (B) Find the total sales revenue gained from selling 40 pairs of wooden shoes.

9. A manufacturer of alarm clocks sells each clock for $10. Let $R(x)$ represent the total sales revenue gained from selling x clocks.
   (A) Write the sales revenue equation relating $R(x)$ with x.
   (B) Calculate and interpret: $R(0)$, $R(10)$, $R(20)$.

10. The number of bushels of wheat demanded at a price of p dollars per bushel is given by

$$D(p) = \frac{5000}{p} \quad (p > 0)$$

   The function D is called a **demand function**. Calculate $D(10)$ and interpret its meaning.

11. The demand function for EZ Chew Peanuts is defined by

$$D(p) = -6p + 48 \quad (0.50 \le p \le 8)$$

   where p = dollar price of peanuts per pound and $D(p)$ = number of pounds of peanuts demanded at price p.
   (A) Calculate the demand for EZ Chew Peanuts at: $p = \$1$ per pound, $p = \$3$ per pound.
   (B) Calculate and interpret: $D(2)$, $D(4)$.

12. Let the variable p represent the dollar price per pound of Great Smell Tobacco and $S(p)$ represent the number of pounds the manufacturer is willing to supply at price p. These two variables are related by the equation

19 LINEAR FUNCTIONS

$$S(p) = 5p - 10 \qquad (p \geq 2)$$

The function S is called a **supply function**.

(A) Calculate the supply of Great Smell Tobacco at: $p = \$3$ per pound, $p = \$7$ per pound.

(B) Calculate and interpret: $S(5)$, $S(10)$.

13. For each of the points A through J of Figure 1–24, find the associated ordered pair.

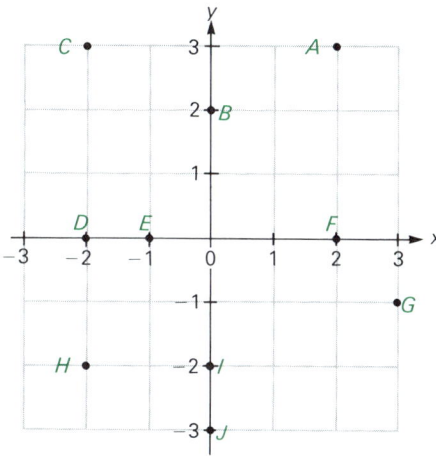

FIGURE 1–24

14. Plot each of the following points on the rectangular coordinate system: $(0, 0)$, $(2, 0)$, $(-5, 0)$, $(0, 2)$, $(0, 5)$, $(3, 1)$, $(5, 2)$, $(-7, 3)$, $(-8, -2)$, $(9, -3)$.

15. Plot the following points on the rectangular coordinate system and state the quadrant in which each is located: $(4, 2)$, $(5, 8)$, $(-9, 3)$, $(-2, 1)$, $(-3, -5)$, $(-2, -7)$, $(8, -3)$, $(9, 2)$.

16. Graph the function defined by

$$f(x) = 3x + 2$$

with domain equal to the set of x-values in Table 1–2.

TABLE 1–2

x	f(x)
0	
1	
2	
3	

17. Graph the function defined by

$$S(x) = 1/x$$

with domain equal to the set of x-values in Table 1–3.

TABLE 1–3

x	S(x)
1/4	
1/2	
1	
2	

20 CHAPTER ONE

18. If $f(x) = \dfrac{5}{(x-2)(x+7)}$ specify the domain of f.

19. If $g(x) = \dfrac{8}{(x-5)}$ specify the domain of g.

20. Which of the graphs of Figure 1–25 are graphs of functions?

(A)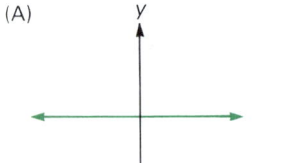

(B)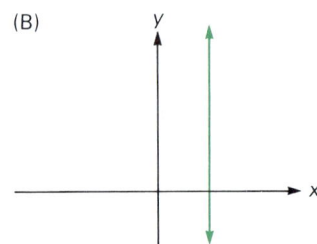

(C)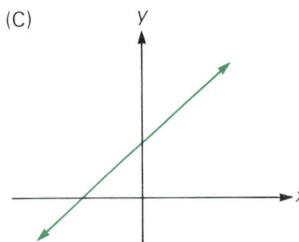

(D)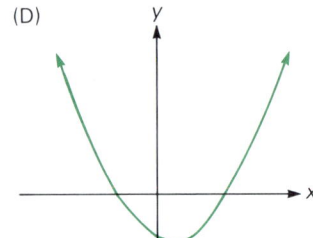

(E)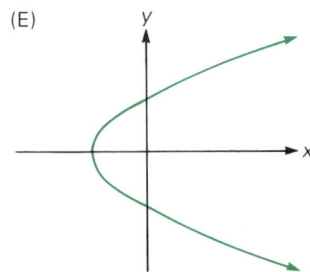

(F)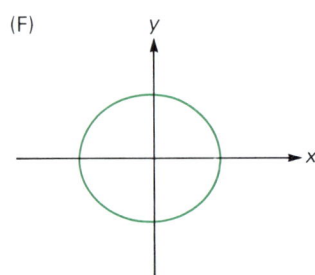

FIGURE 1–25

21. Which of the graphs of Figure 1–26 are graphs of functions?
22. Does the equation $y^2 = x + 5$ define a function? Why or why not?
23. Does the equation $y^2 = 4x + 1$ define a function? Why or why not?
24. Given that $f(x) = x^2 - 4x + 5$, calculate each of the following:
    (A) $f(x + h)$
    (B) $f(x + h) - f(x)$
    (C) $\dfrac{f(x + h) - f(x)}{h}$

21 LINEAR FUNCTIONS

(A)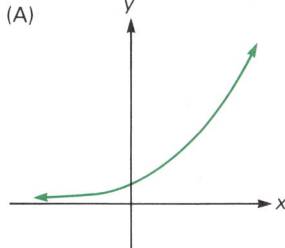

(B)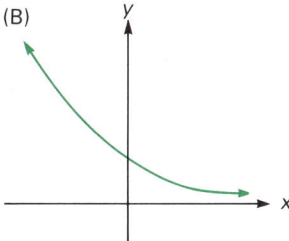

(C)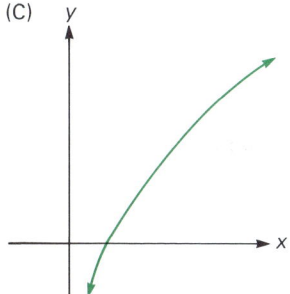

(D)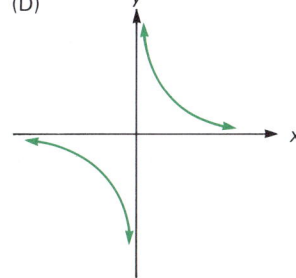

(E)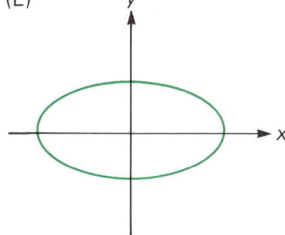

(F)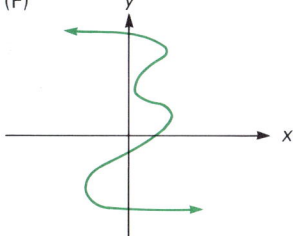

FIGURE 1–26

25. Given that $f(x) = -3x^2 + 5x - 2$, calculate each of the following:
(A) $f(x + h)$
(B) $f(x + h) - f(x)$
(C) $\dfrac{f(x + h) - f(x)}{h}$

26. Given that $f(x) = 5x^2 - 2x + 4$, calculate the difference quotient

$$\dfrac{f(x + h) - f(x)}{h}$$

27. Given that $f(x) = -2x^2 + 3x - 9$, calculate the difference quotient

$$\dfrac{f(x + h) - f(x)}{h}$$

22 CHAPTER ONE

28. Given that $g(x) = x^3 - 4x^2 + 5x - 9$, calculate the difference quotient

$$\frac{g(x + h) - g(x)}{h}$$

29. Given that $g(x) = 2x^3 - 3x + 5$, calculate the difference quotient

$$\frac{g(x + h) - g(x)}{h}$$

30. Graph the function defined by $f(x) = \begin{cases} x \text{ if } x > 5 \\ 5 \text{ if } x \leq 5 \end{cases}$

31. Graph the function defined by $g(x) = \begin{cases} -x \text{ if } x \leq 2 \\ 7 \text{ if } x > 2 \end{cases}$

32. Graph the function defined by $h(x) = \begin{cases} x \text{ if } x \geq 6 \\ 4 \text{ if } 1 \leq x < 6 \\ 2 \text{ if } x < 1 \end{cases}$

33. Graph the function defined by $k(x) = \begin{cases} 9 \text{ if } x > 9 \\ x \text{ if } 4 < x \leq 9 \\ 3 \text{ if } x \leq 4 \end{cases}$

34. A parcel service charges the following rates for delivering small packages:

    $1.25 for a package weighing less than 8 ounces
    $2.00 for a package weighing at least 8 ounces and at most 16 ounces
    $5.00 for a package weighing more than 16 ounces and at most 40 ounces

    The service delivers no packages heavier than 40 ounces.
    (A) Express delivery cost as a function of weight.
    (B) Graph the function of part (A).

35. In designing a windmill, an engineer must use the fact that power y available in wind varies with the cube of wind speed. Thus, if x represents wind speed (in miles per hour), we have

$$y = kx^3$$

where k is a constant real number.
    (A) If a 25-mile-an-hour wind produces 5000 watts, find k.
    (B) How many watts of power will a 35-mile-per-hour wind produce?

23 LINEAR FUNCTIONS

1-3

SLOPE OF A STRAIGHT LINE

Consider the straight line drawn through the two points (2, 8) and (4, 11) (see Figure 1–27). Suppose an ant begins at (2, 8) and moves along the line toward (4, 11). As the ant moves farther and farther toward (4, 11), it experiences a change in vertical distance as well as a change in horizontal distance. Once the ant reaches (4, 11), its total change in vertical distance has been $11 - 8 = 3$ units, and its total change in horizontal distance has been $4 - 2 = 2$ units (see Figure 1–28). The ratio 3/2 represents the rate of change of vertical distance with respect to horizontal distance and is called the **slope** of the straight line. The slope 3/2 implies that as the ant moves along the straight line, for every 2 units of horizontal change, it experiences 3 units of vertical change.

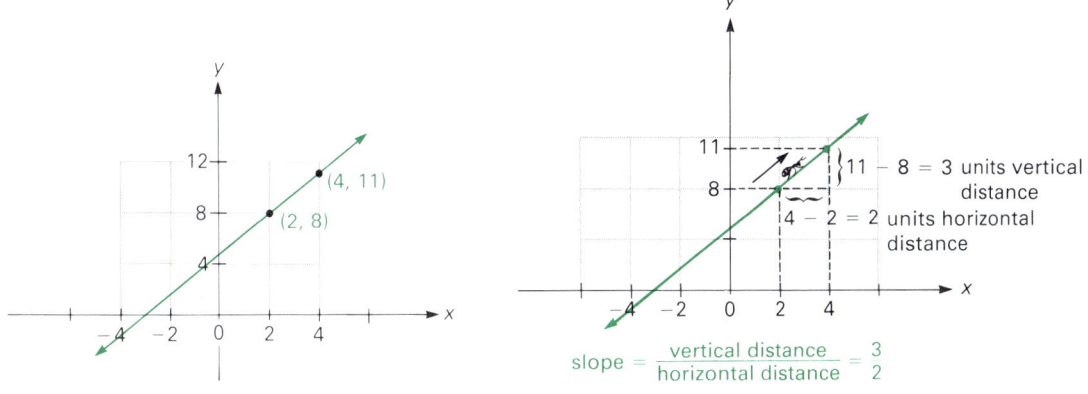

FIGURE 1–27   FIGURE 1–28

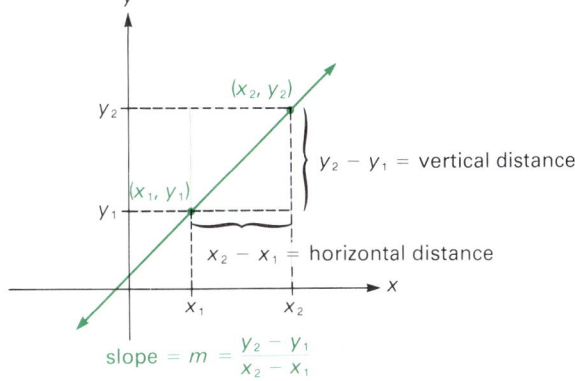

$$\text{slope} = m = \frac{y_2 - y_1}{x_2 - x_1}$$

FIGURE 1–29

24 CHAPTER ONE

In general, if (x_1, y_1) and (x_2, y_2) represent two points through which a straight line passes (see Figure 1–29), then the slope, m, of that straight line is found by the following formula.

$$\text{slope} = m = \frac{y_2 - y_1}{x_2 - x_1}$$

EXAMPLE 1–13
(A) Find the slope of the straight line passing through (5, 7) and (3, 11).
(B) Interpret the result of part (A).
Solutions
(A) If $(x_1, y_1) = (5, 7)$ and $(x_2, y_2) = (3, 11)$, then

$$\text{slope} = m = \frac{y_2 - y_1}{x_2 - x_1} = \frac{11 - 7}{3 - 5} = \frac{4}{-2} = -2$$

Note that it makes no difference which point is designated (x_1, y_1) and which is designated (x_2, y_2). Thus, if $(x_1, y_1) = (3, 11)$ and $(x_2, y_2) = (5, 7)$, we find that

$$\text{slope} = m = \frac{y_2 - y_1}{x_2 - x_1} = \frac{7 - 11}{5 - 3} = \frac{-4}{2} = -2$$

(B) Referring to Figure 1–30, as one moves along the straight line from (3, 11) to (5, 7), for each unit of hori-

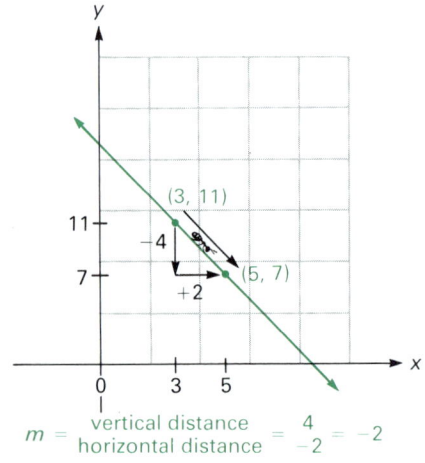

$$m = \frac{\text{vertical distance}}{\text{horizontal distance}} = \frac{4}{-2} = -2$$

FIGURE 1–30

25 LINEAR FUNCTIONS

zontal distance gained, 2 units of vertical distance are lost. Also, referring to Figure 1–31, as one moves along the straight line from (5, 7) to (3, 11), for each unit of horizontal distance lost, 2 units of vertical distance are gained.

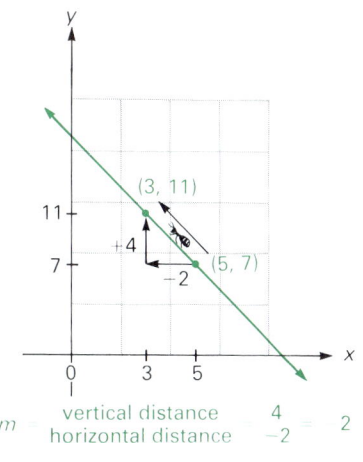

FIGURE 1–31

Compare the straight line of Figure 1–28 with that of Figures 1–30 and 1–31. Note that a straight line with positive slope slants *upward* to the right whereas a straight line with negative slope slants *downward* to the right.

CONSTANT RATE OF CHANGE

Figure 1–32 again illustrates the straight line of Figure 1–27. Referring to Figure 1–32, suppose an ant moves up along the line away from (4, 11) so that it has undergone 2 units of horizontal distance. Since the slope of the straight line is 3/2, then for each 2 units of horizontal distance gained, the ant will gain 3 units of vertical distance. Hence, the ant will be located at point (4 + 2, 11 + 3) = (6, 14).

Again referring to Figure 1–32, suppose that the ant begins at point (2, 8) and moves down along the straight line until it has undergone −2 units of horizontal distance. Since the slope of the straight line is 3/2 or −3/(−2), then for each 2 units of horizontal distance lost, the ant will lose 3 units of vertical distance. Hence, the ant will now be located at point (2 − 2, 8 − 3) = (0, 5). Note that movement along this straight line results in the constant rate of change 3/2 of vertical distance with respect to horizontal distance.

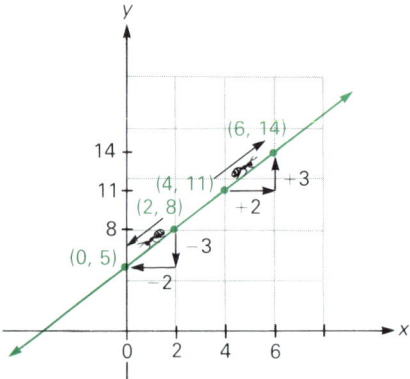

FIGURE 1–32

> In general, movement along a straight line of slope m results in m vertical units gained for each horizontal unit gained. Thus, the slope, m, is the rate of change of vertical distance (y) with respect to horizontal distance (x).

Again, refer to Figure 1–32. Note that the straight line crosses the y-axis at $(0, 5)$. This point is called the **y-intercept** of the straight line.

EXAMPLE 1–14 A manufacturer of chairs can produce 10 chairs at a total cost (fixed + variable) of $1100, while 50 such chairs will cost $3500. Assuming that x = number of chairs produced and y = total cost, use the concept of slope to calculate the variable cost per chair (i.e., the cost of each chair over and above the fixed cost).

Solution Since 10 chairs cost $1100 and 50 chairs cost $3500, let $(x_1, y_1) = (10, 1100)$ and $(x_2, y_2) = (50, 3500)$. Hence,

$$m = \frac{y_2 - y_1}{x_2 - x_1} = \frac{3500 - 1100}{50 - 10} = \frac{2400}{40} = 60$$

Note that the production of 40 additional chairs costs an additional $2400. Thus, the variable cost per chair is the slope, i.e., $60 (see Figure 1–33).

27 LINEAR FUNCTIONS

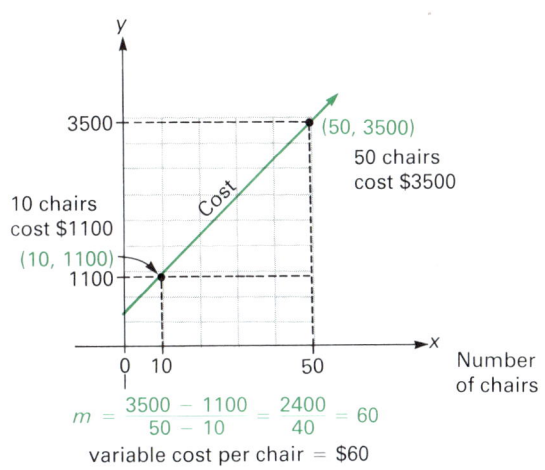

FIGURE 1–33

EXAMPLE 1–15 Find the slope of the straight line passing through (1, 3) and (5, 3).
Solution If $(x_1, y_1) = (1, 3)$ and $(x_2, y_2) = (5, 3)$, then

$$m = \frac{y_2 - y_1}{x_2 - x_1} = \frac{3 - 3}{5 - 1} = \frac{0}{4} = 0$$

Since both points have the same y-coordinate, the straight line passing through them is horizontal (see Figure 1–34). In general, the slope of a horizontal straight line is always 0.

FIGURE 1–34

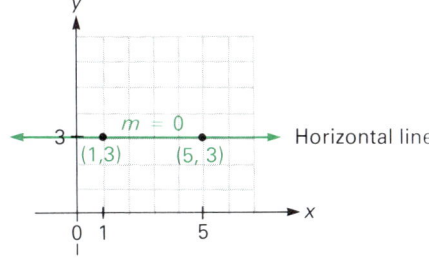

EXAMPLE 1–16 Find the slope of the straight line passing through (2, 3) and (2, 5).
Solution If $(x_1, y_1) = (2, 3)$ and $(x_2, y_2) = (2, 5)$, then

$$m = \frac{y_2 - y_1}{x_2 - x_1} = \frac{5 - 3}{2 - 2} = \frac{2}{0} \quad \text{which is undefined.}$$

EXAMPLE 1–16 (continued)

Since both points have the same x-coordinate, the straight line passing through them is vertical (see Figure 1–35). In general, the slope of a vertical line is undefined.

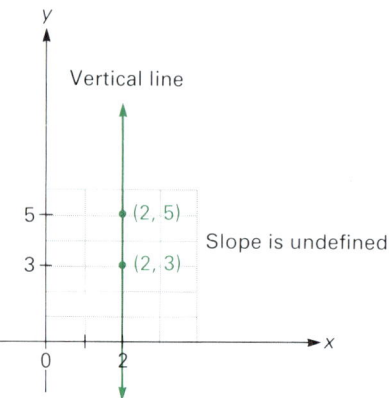

FIGURE 1–35

EXAMPLE 1–17 Table 1–4 exhibits a relationship between a nation's **disposable income,** x (in billions of dollars), and **personal consumption expenditures,** y (in billions of dollars). Find the rate of change of personal consumption expenditures with respect to disposable income.

TABLE 1–4

x	y
56	50
76	67.2

Solution We seek the slope between $(x_1, y_1) = (56, 50)$ and $(x_2, y_2) = (76, 67.2)$. Therefore,

$$m = \frac{y_2 - y_1}{x_2 - x_1} = \frac{67.2 - 50}{76 - 56} = \frac{17.2}{20} = 0.86$$

Economists call this result the **marginal propensity to consume,** abbreviated **MPC.** Since MPC = 0.86, then for each dollar increase in disposable income, consumption increases by $0.86. In other words, 86% of each additional dollar earned is spent, and 14% is saved.

29 LINEAR FUNCTIONS

EXERCISES

1. Find the slope of the straight line passing through each of the following pairs of points:
   - (A) (4, 5), (7, 16)
   - (B) (1, 4), (6, 13)
   - (C) (−1, 3), (7, 5)
   - (D) (0, 6), (9, 4)
   - (E) (−4, 3), (−5, −7)
   - (F) (5, 9), (8, 7)
   - (G) (−8, −2), (−3, −9)
   - (H) (2, 11), (5, −3)

2. Graphically interpret each of the slopes found in Exercise 1.

3. For each of the following, graph the straight line, given slope m through the indicated point:
   - (A) $m = 1/3$, (0, 0)
   - (B) $m = -1/3$, (0, 0)
   - (C) $m = 5$, (1, 8)
   - (D) $m = -2$, (2, 1)
   - (E) $m = 6/5$, (0, 1)
   - (F) $m = -3/4$, (4, −5)
   - (G) $m = 6/5$, (10, 13)
   - (H) $m = -5/3$, (0, 2)

4. Find the slope of the straight line passing through each of the following pairs of points and graph the line:
   - (A) (4, 6), (−7, 6)
   - (B) (−9, −3), (−8, −3)
   - (C) (8, 1), (11, 1)
   - (D) (4, −2), (−6, −2)
   - (E) (5, 3), (5, −1)
   - (F) (6, 4), (6, −4)
   - (G) (−2, 16), (−2, 4)
   - (H) (4, −6), (4, 9)

5. It costs the ABC Corporation $3000 to produce 20 gadgets and $5000 to produce 60 gadgets. Find the variable cost per gadget.

6. It costs the Eatco Corporation $1050 to produce 100 cases of cereal and $1250 to produce 500 cases of cereal. Find the variable cost per case of cereal.

7. If clocks are priced at $5 apiece, there will be a demand for 75 clocks. If clocks are priced at $10 apiece, the demand will decrease to 50 clocks.
   - (A) Given that x = price per clock and y = corresponding demand, use the concept of slope to find the rate of change of demand with respect to price.
   - (B) If the price of a clock decreases by $1, demand will increase by how many clocks?
   - (C) If the price of a clock increases by $1, demand will decrease by how many clocks?

8. If a gadget is priced at $9 apiece, suppliers are willing to produce 86 gadgets. If a gadget's price drops to $5, suppliers are willing to produce only 46 gadgets.
   - (A) If x = price per gadget and y = corresponding supply, use the concept of slope to find the rate of change of supply with respect to price.
   - (B) If the price of a gadget decreases by $1, the supply will decrease by how many gadgets?
   - (C) If the price of a gadget increases by $1, the supply will increase by how many gadgets?

9. Table 1–5 exhibits a relationship between a nation's disposable income, x (in billions of dollars), and personal consumption expenditures, y (in billions of dollars).
   - (A) Calculate the MPC.

TABLE 1–5

x	y
48	44
68	60

30 CHAPTER ONE

(B) According to the MPC of part (A), for each dollar increase in disposable income, consumption increases by how much?
(C) Another rate of change is the marginal propensity to save, abbreviated MPS. It is defined as

$$MPS = 1 - MPC$$

Calculate the MPS for the nation of this example.
(D) According to the MPS calculated in part (C), for each dollar increase in disposable income, how many additional dollars are saved?

1–4

SLOPE AND EQUATIONS OF STRAIGHT LINES

Study the straight line of Figure 1–36. Recall from Section 1–3 that this straight line was initially illustrated with the two points (2, 8) and (4, 11). After determining the slope to be 3/2, the rate of change concept was used to find other points on the line, specifically (6, 14) and (0, 5). The concept of slope can be used to determine other points on the line. However, there is a better way. We must realize that the x and y coordinates of any point (x, y) on a straight line are related by an equation. Such an equation is called the **equation of the straight line**. The equation of a straight line exhibits the relationships between the x and y coordinates of any point (x, y) on the line.

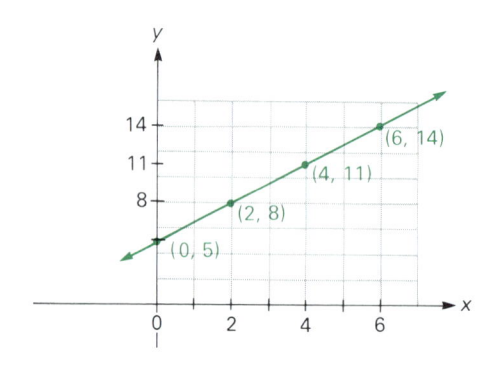

FIGURE 1–36

We will now determine the equation of the straight line of Figure 1–36. Choose one of the known points on the line, say, (2, 8). Let (x, y) represent any point on the line. Since the slope of the straight line is 3/2, the slope between (x, y) and (2, 8) must equal 3/2. Hence,

$$\frac{y - 8}{x - 2} = \frac{3}{2}$$

LINEAR FUNCTIONS

Multiplying both sides by $x - 2$, we obtain

$$y - 8 = \frac{3}{2}(x - 2)$$

This equation is called the **point–slope form** of the equation of the straight line. Observe that the coordinates of the point, (2, 8), and the slope, 3/2, appear conspicuously in the equation. Hence, the term *point–slope form*.

Solving the point–slope form for y, we obtain

$$y = \frac{3}{2}(x - 2) + 8$$

or

$$y = \frac{3}{2}x + 5$$

This equation is called the **slope–intercept form** of the equation of the straight line. Note that the slope, 3/2, and the y-intercept, 5, appear conspicuously in the equation. Hence, the term *slope–intercept form*.

Thus, the equation

$$y = \frac{3}{2}x + 5$$

exhibits the relationship between the x and y coordinates of any point (x, y) on the straight line. Observe that the points of the straight line satisfy its equation.

$(0, 5) \qquad 5 = \frac{3}{2}(0) + 5$

$(2, 8) \qquad 8 = \frac{3}{2}(2) + 5$

$(4, 11) \qquad 11 = \frac{3}{2}(4) + 5$

$(6, 14) \qquad 14 = \frac{3}{2}(6) + 5$

As previously mentioned, the equation of a straight line may be used to find other points on the line. Specifically, if $x = -4$, then

$$y = \frac{3}{2}(-4) + 5 = -1$$

Thus, $(-4, -1)$ is a point on this straight line. And if $x = 8$, then

$$y = \frac{3}{2}(8) + 5 = 17$$

Therefore, (8, 17) is also a point on the line.

We can now state the following two generalizations.

Given a point (x_1, y_1) on a straight line of slope m (see Figure 1-37), the **point–slope form** of the equation of that straight line is

$$y - y_1 = m(x - x_1)$$

FIGURE 1-37

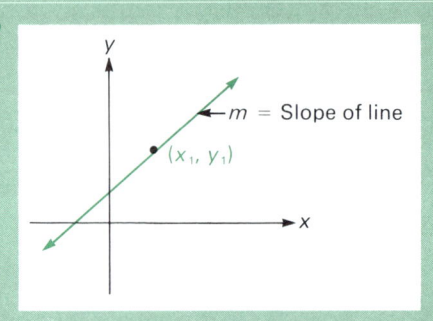

If the point–slope form is solved for y, the resulting equation is the **slope–intercept form**

$$y = mx + b$$

(see Figure 1-38).

FIGURE 1-38

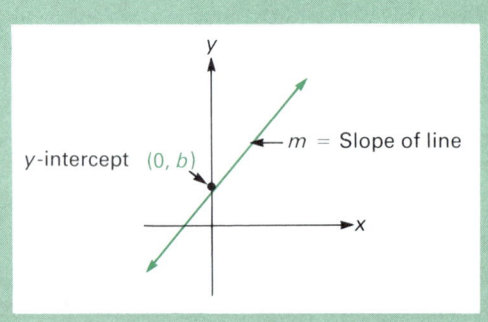

EXAMPLE 1-18 A straight line passes through (4, 7) and (9, 17).
(A) Find its equation in point–slope form.
(B) Convert the point–slope form into the slope–intercept form.
(C) Find the point on this straight line corresponding to $x = 5$.
(D) Sketch this straight line.

33 LINEAR FUNCTIONS

Solutions
(A) If $(x_1, y_1) = (4, 7)$ and $(x_2, y_2) = (9, 17)$,* then

$$m = \frac{y_2 - y_1}{x_2 - x_1} = \frac{17 - 7}{9 - 4} = \frac{10}{5} = 2$$

Substituting $(x_1, y_1) = (4, 7)$ and $m = 2$ into the point–slope form

$$y - y_1 = m(x - x_1)$$

we obtain

$$y - 7 = 2(x - 4)$$

(B) Solving for y gives us

$$y = 2(x - 4) + 7$$

or

$$y = \underset{\uparrow}{2x} \underbrace{- 1}_{\uparrow}$$
$$\text{slope} \quad y\text{-intercept}$$

(C) If $x = 5$, then $y = 2(5) - 1 = 9$. Thus, $(5, 9)$ is a point on this line.

(D) Observing the slope–intercept form, note that $(0, -1)$ is the y-intercept of this straight line. Using a straightedge, we connect the y-intercept with another point on the straight line, say, $(5, 9)$. The resulting graph is illustrated in Figure 1–39. Note that the points $(4, 7)$ and $(9, 17)$ also appear on the line.

FIGURE 1–39

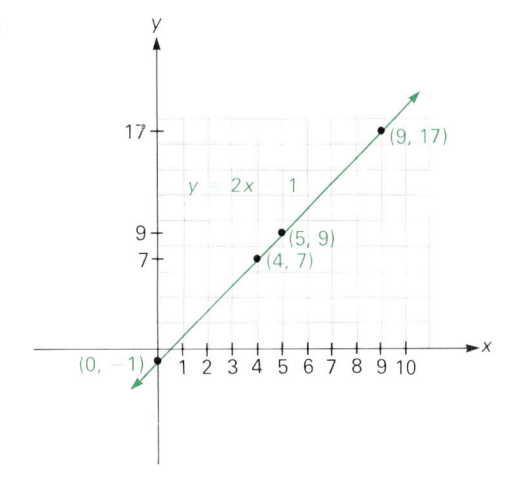

---

*Either point may be chosen as (x_1, y_1) or (x_2, y_2).

EXAMPLE 1–19 A straight line has a slope of 6 and y-intercept of -3. Find the equation of this line.
Solution Substituting $m = 6$ and $b = -3$ into the slope–intercept form

$$y = mx + b$$

yields

$$y = 6x - 3$$

EXAMPLE 1–20 A manufacturer of chairs finds that he can produce 10 chairs at a total cost of $1100, whereas 50 such chairs will cost $3500. Let x = number of chairs produced and y = total production cost.
(A) Find the equation relating these two quantities.
(B) Determine the fixed cost.
(C) Sketch the equation found in part (A).
Solutions
(A) We already calculated the variable cost per chair (i.e., the slope) in Example 1–14. Recall that $(x_1, y_1) = (10, 1100)$ and $(x_2, y_2) = (50, 3500)$. Hence,

$$m = \frac{y_2 - y_1}{x_2 - x_1} = \frac{3500 - 1100}{50 - 10} = \frac{2400}{40} = 60$$

Substituting $(x_1, y_1) = (10, 1100)$ and $m = 60$ into the point–slope form

$$y - y_1 = m(x - x_1)$$

we obtain

$$y - 1100 = 60(x - 10)$$

Solving for y, we have

$$y = 60(x - 10) + 1100$$

or

$$y = 60x + 500$$

(B) If $x = 0$, then $y = 60(0) + 500 = 500$. Since it costs $500 to produce 0 chairs, then $500 is the fixed cost.
(C) Observing the slope–intercept form $y = 60x + 500$, note that $(0, 500)$ is the y-intercept. Using a straight-edge, we connect the y-intercept with one of the other

35 LINEAR FUNCTIONS

points on the straight line, say, (10, 1100). The resulting graph is illustrated in Figure 1–40. Since $x \geq 0$, the straight line is drawn solid in the first quadrant.

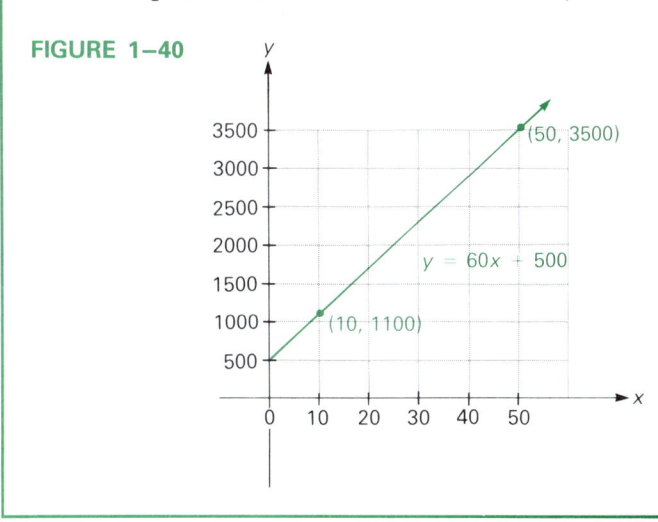

FIGURE 1–40

Horizontal Lines Consider the straight line passing through the points (2, 5) and (8, 5). Since the y-coordinates are equal, the line passing through these points is horizontal (see Figure 1–41). Thus, its slope is 0. Observing Figure 1–41, note that the y-intercept is 5. Substituting $m = 0$ and $b = 5$ into the slope–intercept form

$$y = mx + b$$

we obtain

$$y = 0x + 5$$

or

$$y = 5$$

The equation $y = 5$ expresses the fact that the y-coordinate of any point on this horizontal straight line is 5.

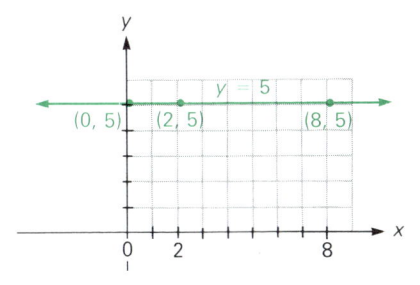

FIGURE 1–41

In general, the equation

$$y = b$$

represents a horizontal straight line with a y-intercept of b.

Vertical Lines Consider the straight line passing through the points (3, 2) and (3, 7). Since the x-coordinates are equal, the straight line is vertical (see Figure 1–42). Its x-intercept is 3. Because any point on this vertical line has an x-coordinate of 3, the equation of the line is appropriately $x = 3$.

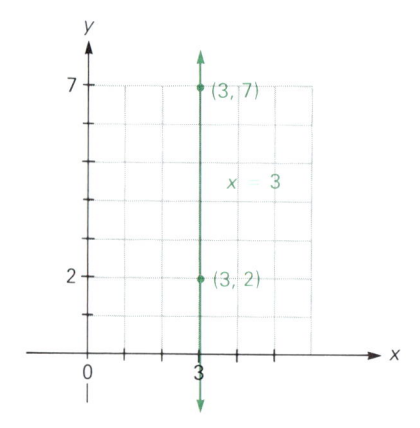

FIGURE 1–42

In general, the equation

$$x = k$$

represents a vertical straight line with an x-intercept of k.

Linear Equations Up to this point, we have encountered equations of straight lines in either the *point–slope form*

$$y - y_1 = m(x - x_1)$$

or the *slope–intercept form*

$$y = mx + b$$

37 LINEAR FUNCTIONS

These equations, which represent straight lines, are called **linear equations**. Linear equations occur in forms other than the preceding. For example,

$$3x + 5y = 16$$

is a linear equation. This is verified by converting it into slope–intercept form. Solving for y yields

$$5y = -3x + 16$$
$$y = -\frac{3}{5}x + \frac{16}{5}$$

> In general, equations of the form
>
> $$ax + by = c$$
>
> where a, b, and c are constant real numbers, with a and b not both 0, are linear equations. This form is called the **general form** of a linear equation.

EXAMPLE 1–21 Convert $4x - 7y = 18$ into slope–intercept form.
Solution Solving for y yields

$$-7y = -4x + 18$$
$$y = \frac{-4}{-7}x + \frac{18}{-7}$$

or

$$y = \underbrace{\frac{4}{7}}_{\text{slope}} x \underbrace{- \frac{18}{7}}_{\text{y-intercept}}$$

EXAMPLE 1–22 Convert $5x^2 + 2y = 16$ into slope–intercept form.
Solution Solving for y, we obtain

$$2y = -5x^2 + 16$$
$$y = -\frac{5}{2}x^2 + 8$$

38 CHAPTER ONE

> **EXAMPLE 1–22** (continued)
>
> But this equation is not the slope–intercept form $y = mx + b$ because the term, $-(5/2)x^2$, is of degree 2. Thus, the equation is not linear.

In summary, the most common forms of linear equations are as follows:

1. **Point–Slope Form**

    $y - y_1 = m(x - x_1)$ [Slope m, line passes through (x_1, y_1)]

2. **Slope–Intercept Form**

    $y = mx + b$ (Slope m, y-intercept $= b$)

3. **General Form**

    $ax + by = c$ (a, b, and c are constant real numbers with a and b not both 0)

4. **Horizontal Line**

    $y = b$ (y-intercept $= b$, slope $= 0$, no x-intercept)

5. **Vertical Line**

    $x = k$ (x-intercept $= k$, no y-intercept, slope undefined)

Note that a linear equation does not contain a variable with a non-zero exponent other than 1.

EXERCISES

1. For each of the following, find the equation, in point–slope form, of the straight line passing through the given pair of points. Then convert the equation into slope–intercept form and sketch the straight line.
    - (A) (1, 4), (3, 8)
    - (B) (4, −1), (9, −8)
    - (C) (5, −10), (7, −14)
    - (D) (3, 2), (7, 10)
    - (E) (−2, −3), (−5, 9)
    - (F) (5, 5), (7, 7)
    - (G) (4, 12), (6, 18)
    - (H) (5, −3), (4, 2)

2. For each of the following, find the equation, in point–slope form, of the straight line passing through the given point and having slope m. Then convert the equation into slope–intercept form and sketch the straight line.
    - (A) (4, 3), $m = 2$
    - (B) (5, −1), $m = -3$
    - (C) (0, −2), $m = -4$
    - (D) (−4, −9), $m = 6$
    - (E) (−7, 2), $m = -1/2$
    - (F) (0, 6), $m = 5$
    - (G) (3, 1), $m = -1/4$
    - (H) (−4, 7), $m = 1$

39 LINEAR FUNCTIONS

3. For each of the following, find the equation of the straight line passing through the given pair of points. Also sketch the straight line.
   (A) (4, 2), (4, 7)
   (B) (−3, 1), (−3, 9)
   (C) (5, −9), (8, −9)
   (D) (4, 1), (6, 1)
4. For each of the following, find the equation of the straight line passing through the given point and having slope m. Also sketch the straight line.
   (A) (4, 3), $m = 0$
   (B) (−8, 1), $m = 0$
   (C) (9, −1), m is undefined
   (D) (7, 2), m is undefined
5. Straight lines with the same slope are said to be parallel. Figure 1–43 illustrates two parallel lines of positive slope while Figure 1–44 illustrates two parallel lines of negative slope.

   Find the equation of the straight line passing through (5, 7) and parallel to the straight line $y = 4x + 6$. Sketch both lines.

Parallel lines with positive slope

Parallel lines with negative slope

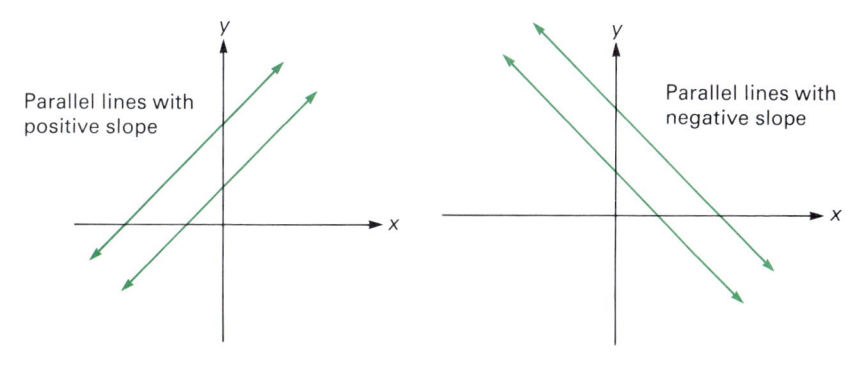

FIGURE 1–43

FIGURE 1–44

6. Find the equation of the straight line passing through (9, −5) and parallel to $6x + 2y = 18$. Sketch both lines.
7. Straight lines intersecting each other at right angles (90°) are said to be perpendicular. Figure 1–45 illustrates two perpendicular straight lines. Slopes of perpendicular straight lines are negative reciprocals of each other. In other words, if a straight line has slope m, then a straight line perpendicular to it has slope $-1/m$. This relationship

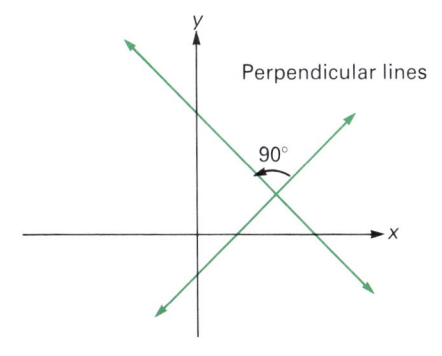

Perpendicular lines

90°

FIGURE 1–45

40 CHAPTER ONE

between slopes of perpendicular lines does not hold for vertical or horizontal lines.

Find the equation of the straight line passing through (5, 6) and perpendicular to $y = 3x - 2$. Sketch both lines.

8. Find the equation of the straight line passing through (8, −2) and perpendicular to $4x - 3y = 24$. Sketch both lines.

9. Which of the following pairs of straight lines are parallel?
   (A) $y = 6x - 14$, $y = 6x + 13$
   (B) $3x - 2y = 15$, $6x - 4y = 60$
   (C) $y = -4x + 8$, $y = 4x + 16$
   (D) $5x - 8y = 11$, $6x + y = 13$

10. Which of the following pairs of straight lines are perpendicular?
    (A) $y = 3x + 5$, $y = -\frac{1}{3}x + 16$
    (B) $y = 6x + 5$, $y = -6x + 5$
    (C) $y = 2x - 3$, $2y + x = 10$
    (D) $3x + 6y = 11$, $5x + 2y = 13$

11. It costs ABC Corporation $3000 to produce 20 gadgets and $5000 to produce 60 gadgets. Let x = number of gadgets produced and y = corresponding cost.
    (A) Find the linear cost equation.
    (B) Determine the fixed cost and the variable cost per unit.
    (C) Sketch the cost function.

12. It costs Corny Corporation $1050 to produce 100 boxes of cereal and $1250 to produce 500 boxes of cereal. If x = number of boxes and y = corresponding cost, do the following.
    (A) Find the linear cost equation.
    (B) Determine the fixed cost and the variable cost per unit.
    (C) Sketch the cost function.

13. If a widget is priced at $5 apiece, there will be a demand for 75 widgets. If a widget is priced at $10 apiece, then the demand will decrease to 50 widgets. If x = price per widget and y = corresponding demand, find the linear demand equation relating the two quantities.

14. If a gadget is priced at $9 each, the suppliers are willing to produce 86 gadgets. If a gadget's price drops to $5, then suppliers are willing to produce only 46 gadgets. If x = price per gadget and y = corresponding supply, find the linear supply equation relating the two quantities.

15. Convert each of the following linear equations into slope–intercept form:
    (A) $3x - 5y = 11$
    (B) $x = -2y + 15$
    (C) $3x - 2y = 0$
    (D) $2x + 5y + 16 = 0$

16. Which of the following equations are linear?
    (A) $y = 6x - 54$
    (B) $y = x^2 - 4x$
    (C) $y = 3x^2 - 4$
    (D) $y^2 = 5x + 7$
    (E) $3x - 4y = 5$
    (F) $y = xy + 7$
    (G) $y = x^3 - 5$
    (H) $y^3 = x + 4$

41 LINEAR FUNCTIONS

17. For each of the following linear equations, determine which of the given points lie on its corresponding straight line:
    (A) $3x - 5y = 30$; $(0, -6)$, $(5, 2)$, $(10, 0)$, $(15, 3)$
    (B) $2x + 7y = 21$; $(7, 1)$, $(3, 0)$, $(4, -3)$, $(14, -1)$
    (C) $8x + 6y = 0$; $(0, 0)$, $(1, 5)$, $(-2, 8/3)$, $(7, 13/7)$
    (D) $y = 6x + 3$; $(0, 6)$, $(1, 9)$, $(-1/2, 0)$, $(-1, -3)$
    (E) $f(x) = 2x + 4$; $(0, 4)$, $(1, 5)$, $(-1, 2)$, $(-2, 0)$
    (F) $g(x) = -3x$; $(0, 0)$, $(1, -3)$, $(2, -6)$, $(5, 14)$

18. The following problem appeared on a past Uniform CPA Examination. Maintenance expenses of a company are to be analyzed for the purpose of constructing a flexible budget. Examination of past records disclosed the following cost and volume measures:

	Highest	Lowest
Cost per month	$39,200	$32,000
Machine-hours	24,000	15,000

    (A) Determine the variable cost per machine-hour.
    (B) Determine the monthly fixed cost for maintenance expenditures.

19. Table 1–6 exhibits a relationship between a nation's disposable income, x (in billions of dollars), and personal consumption expenditures, y (in billions of dollars).
    (A) Find the linear equation relating x and y.
    (B) Calculate the personal consumption expenditures corresponding to disposable income of $85 billion.
    (C) Interpret the slope.

    TABLE 1–6

x	y
60	55
90	79

1–5 GRAPHING LINEAR FUNCTIONS

Since all straight lines, except vertical straight lines, are graphs of functions, their equations define **linear functions**. In this section, we will discuss the graphing of linear functions given their equations. In Section 1–4, we graphed a linear function by determining two points on the straight line and connecting them with a straightedge. In this section, instead of determining any two points on the straight line, we will determine the x-intercept and the y-intercept. Recall from Section 1–3 that the **y-intercept** is the point at which the straight line crosses the y-axis. Thus, the **x-intercept** is the point at which the straight line crosses the x-axis.

We now graph the linear function defined by

$$2x - 3y = 12$$

by finding its x-intercept and y-intercept.

The x-Intercept The x-intercept is a point on the x-axis. Hence, its y-coordinate is 0. Thus, we seek the point (x, 0) that satisfies the equation $2x - 3y = 12$. Setting $y = 0$ and solving for x, we obtain

$$2x - 3(0) = 12$$
$$2x = 12$$
$$x = 6$$

Thus, the x-intercept is (6, 0).

The y-Intercept The y-intercept is a point on the y-axis. Hence, its x-coordinate is 0. Thus, we seek that point (0, y) that satisfies the equation $2x - 3y = 12$. Setting $x = 0$ and solving for y, we obtain

$$2(0) - 3y = 12$$
$$-3y = 12$$
$$y = -4$$

Thus, the y-intercept is (0, −4).

The straight line defined by $2x - 3y = 12$ is graphed in Figure 1–46.

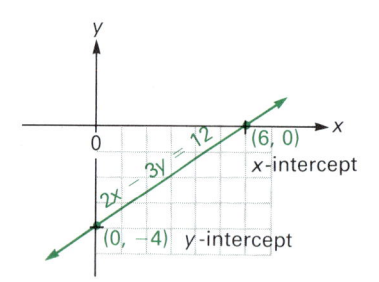

FIGURE 1–46

EXAMPLE 1–23 Graph $-7x + 2y = 28$.
Solution First, we find the x-intercept. We set $y = 0$ and solve for x. Hence,

$$-7x + 2(0) = 28$$
$$-7x = 28$$
$$x = -4$$

Thus, the x-intercept is (−4, 0). Next, we determine the y-intercept. We set $x = 0$ and solve for y. Hence,

$$-7(0) + 2y = 28$$
$$2y = 28$$
$$y = 14$$

43 LINEAR FUNCTIONS

Therefore, the y-intercept is (0, 14). The straight line is graphed in Figure 1–47.

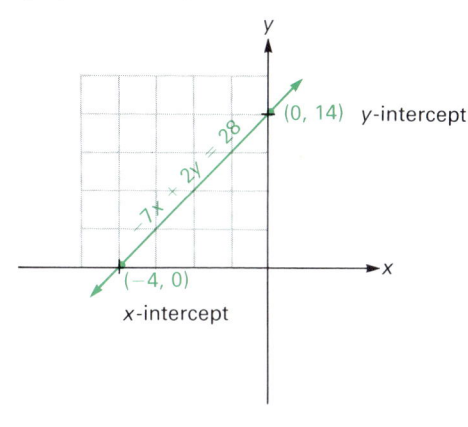

FIGURE 1–47

EXAMPLE 1–24 Sketch the graph of $y = f(x) = -3x - 2$.
Solution First, find the x-intercept. Set $y = 0$ and solve for x. Therefore,

$$0 = -3x - 2$$
$$3x = -2$$
$$x = -\frac{2}{3}$$

Thus, the x-intercept is $(-2/3, 0)$. Then, find the y-intercept. Since the equation $y = -3x - 2$ is in the slope–intercept form, by inspection we determine the y-intercept to be $(0, -2)$. The straight line is graphed in Figure 1–48.

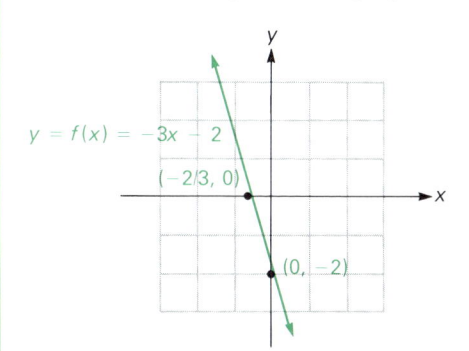

FIGURE 1–48

EXAMPLE 1–25 A toy manufacturer sells wagons for $10 each. If x = number of wagons sold and $R(x)$ = sales revenue from selling x wagons, then

$$R(x) = 10x \quad (x \geq 0)$$

Graph the sales revenue function, $R(x)$.

Solution The equation $y = R(x) = 10x$ is in the slope–intercept form $y = mx + b$, with $m = 10$ and $b = 0$. Thus, the y-intercept is the origin, $(0, 0)$. Since the origin is also the x-intercept, we do not have the two distinct points to connect with a straightedge. Therefore, we choose an arbitrary x-value and calculate its corresponding y-value. Hence, if $x = 1$, then $y = R(1) = 10(1) = 10$. Thus, $(1, 10)$ is a point on this function. The graph appears in Figure 1–49. Since $x \geq 0$, the straight line is drawn solid in the first quadrant.

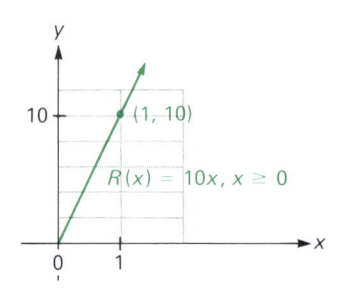

FIGURE 1–49

EXAMPLE 1–26 The equation

$$y = 0.85x + 34 \quad (x \geq 0)$$

expresses a relationship between a nation's disposable income, x (in billions of dollars), and personal consumption expenditures, y (in billions of dollars). Such an equation defines what economists call a **consumption function**. Graph the consumption function.

Solution Find the x-intercept. Set $y = 0$ and solve for x. Hence,

$$0 = 0.85x + 34$$
$$-0.85x = 34$$
$$x = -40$$

45 LINEAR FUNCTIONS

Therefore, the x-intercept is $(-40, 0)$. Then, find the y-intercept. Since the equation is in slope–intercept form, by inspection we determine the y-intercept to be $(0, 34)$. The straight line is graphed in Figure 1–50. With $x \geq 0$, the line is drawn solid in the first quadrant. Note that the slope of the consumption function is the MPC.

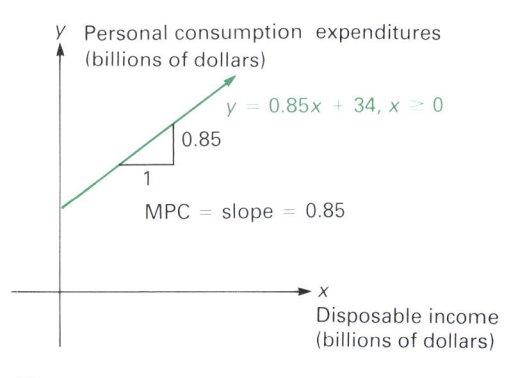

FIGURE 1–50

EXERCISES

1. Graph each of the following linear equations by finding the x and y intercepts:
   - (A) $4x - 6y = 24$
   - (B) $5x + 3y = 45$
   - (C) $-3x + 2y = 8$
   - (D) $5x - 6y = 10$
   - (E) $2x + 7y = 11$
   - (F) $-3x - 5y = 9$
   - (G) $y = -4x + 13$
   - (H) $f(x) = -2x + 15$
   - (I) $f(x) = 6x + 4$
   - (J) $g(x) = \frac{5}{2}x + 13$

2. Graph each of the following linear equations:
   - (A) $y = 2x$
   - (B) $f(x) = -3x$
   - (C) $3x + 2y = 0$
   - (D) $2x - 5y = 0$
   - (E) $g(x) = 7x$
   - (F) $f(x) = -\frac{1}{2}x$
   - (G) $f(x) = x$
   - (H) $y = -x$

3. Graph each of the following:
   - (A) $f(x) = 3x + 5, \ x \geq 0$
   - (B) $g(x) = 4x, \ x \geq 0$
   - (C) $f(x) = 2x + 3, \ x \geq 1$
   - (D) $h(x) = 3x - 15, \ x \geq 5$

4. A pizza house sells its pizzas for $5 apiece. Let x = number of pizzas produced and $R(x)$ = sales revenue from selling x pizzas.
   (A) Find the equation defining the sales revenue function.
   (B) Graph the sales revenue function.
   (C) If 1000 pizzas are sold, what is the sales revenue?

5. A firm producing hats uses the cost equation

$$C(x) = 10x + 5000 \qquad (x \geq 0)$$

   to determine the total production cost $C(x)$ of x hats.
   (A) Graph the cost function.
   (B) Determine the fixed cost and the variable cost.
   (C) Find the total cost of producing 2000 hats.
   (D) Find the total cost of producing 5000 hats.

6. The linear demand equation

$$D(x) = -6x + 1200 \qquad (0 \leq x \leq 200)$$

   relates demand, $D(x)$, with the price per widget, x.
   (A) Graph the demand function.
   (B) If a widget is priced at $5, there will be a demand for how many widgets?
   (C) If a demand for 900 widgets is desired, what should be the price per widget?
   (D) If the price per widget decreases by $1, the demand for widgets increases by how much?

7. The linear supply equation

$$S(x) = 5x - 20 \qquad (x \geq 4)$$

   relates supply, $S(x)$, with the price per widget, x.
   (A) Graph the supply function.
   (B) If a widget is priced at $20 apiece, suppliers are willing to produce how many widgets?
   (C) If suppliers are to produce 480 widgets, what should be the price per widget?
   (D) If the price per widget increases by $1, the supply of widgets increases by how much?

8. A nation's consumption function is defined by

$$y = 0.87x + 25$$

   where x is disposable income (in billions of dollars) and y is personal consumption expenditures (in billions of dollars).
   (A) Graph the consumption function.
   (B) If disposable income increases by $1 billion, do personal consumption expenditures increase or decrease? By how much?
   (C) Calculate the personal consumption corresponding to a disposable income of $26 billion.

1-6
APPLICATIONS OF LINEAR FUNCTIONS

In this section, we will consider several applications of linear functions.

Break-even Point

Henry Jackson, a recent college graduate, plans to start his own business manufacturing bicycle tires. After careful research, he concludes the following:

1. He will need $3000 for overhead (fixed cost).
2. Additionally, each tire will cost him $2 to produce (variable cost per unit).
3. To remain competitive, he must sell his tires for no more than $5 apiece.

Having compiled this information, Henry asks this crucial question:

How many tires must be produced and sold in order to break even?

This type of question is answered in the following manner. First, we must find the cost equation. If x represents the number of tires produced and sold, and $C(x)$ represents the total cost of producing x tires, then

$$C(x) = 2x + 3000 \quad \text{Cost equation}$$

where $2x$ is the variable cost per tire and 3000 is the fixed cost.

Second, we must find the sales revenue equation. The sales revenue is simply the unit selling price times the number of units sold. If $R(x)$ represents the total sales revenue gained from selling x tires at $5 each, then

$$R(x) = 5x \quad \text{Sales revenue equation}$$

Third, we sketch both cost and sales revenue functions on the same set of axes (see Figure 1–51). Observing Figure 1–51, note that the intersection of the graphs of a revenue and a cost function is the **break-even point**, (x_B, y_B). Here,

$$\text{total sales revenue} = \text{total cost}$$

or

$$\text{profit} = 0$$

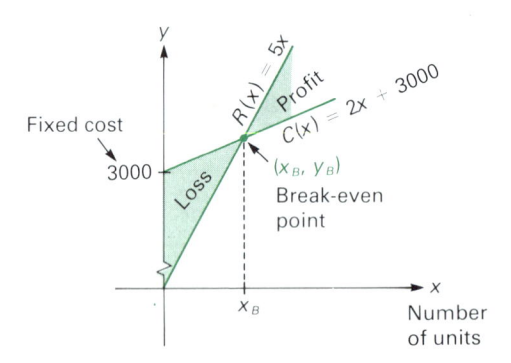

FIGURE 1-51

If $x < x_B$, then total cost is greater than total sales revenue, and the result is a loss. If $x > x_B$, then total sales revenue is greater than total cost, and the result is a profit. Thus, Henry Jackson's business will be profitable as long as x, the number of tires produced and sold, is greater than x_B, the x-coordinate of the break-even point.

To find the break-even point, (x_B, y_B), we equate $R(x)$ with $C(x)$. Thus,

$$R(x) = C(x)$$
$$5x = 2x + 3000$$

Solving for x, we obtain

$$3x = 3000$$
$$x = 1000$$

Thus, x_B, the x-coordinate of the break-even point, is 1000. The corresponding y-coordinate, y_B, is determined by calculating either $C(1000)$ or $R(1000)$. Hence,

$$C(x) = 2x + 3000 \qquad \text{or} \qquad R(x) = 5x$$
$$C(1000) = 2(1000) + 3000 \qquad\qquad R(1000) = 5(1000)$$
$$= 5000 \qquad\qquad\qquad\qquad\qquad = 5000$$

Therefore, the break-even point is (1000, 5000), and Henry must produce and sell 1000 tires to break even. If Henry produces and sells more than 1000 tires, he will make a profit.

Having determined the break-even point, Henry engages a marketing research study which indicates that he can sell at least 10,000 tires at $5 apiece. Also, there is a strong possibility that he can sell as many as 20,000 tires at this price. Since both of these demand levels are greater than $x_B = 1000$, Henry knows he will make a profit. However, he wishes to determine the amount of

49 LINEAR FUNCTIONS

profit corresponding to each of the demand levels of 10,000 and 20,000 tires.

One way of calculating profits is to find the equation relating profit, P, with x, the number of tires sold. If

$P(x)$ = total profit from selling x tires
$R(x)$ = total sales revenue from selling x tires
$C(x)$ = total cost of producing x tires

then

$$P(x) = R(x) - C(x)$$
$$= 5x - (2x + 3000)$$
$$= 3x - 3000$$

Thus

$$P(x) = 3x - 3000$$

This equation defines a **profit function**. Hence, if 10,000 tires are sold, the corresponding profit is

$$P(10,000) = 3(10,000) - 3000 = \$27,000$$

And if 20,000 tires are sold, the corresponding profit is

$$P(20,000) = 3(20,000) - 3000 = \$57,000$$

Since Henry is satisfied with either of these amounts, he decides to start his tire-manufacturing business.

EXAMPLE 1–27 (This problem appeared on a past Uniform CPA Examination.) In a recent period, the Zero Company had the following experience:

	Fixed	Variable	
Sales (10,000 units @ $200 per unit)			$2,000,000
Costs			
Direct material	$ —	$ 200,000	
Direct labor	—	400,000	
Factory overhead	160,000	600,000	
Administrative expenses	180,000	80,000	
Other expenses	200,000	120,000	1,940,000
Total costs	$540,000	$1,400,000	$ 60,000

(A) Find the sales revenue equation.
(B) Find the cost equation.

EXAMPLE 1–27 (continued)

(C) Find the break-even point.
(D) How many units must be sold in order to generate a profit of $96,000?
(E) What is the break-even point if management makes a decision that increases fixed costs by $18,000?

Solutions

(A) If $R(x)$ is the sales revenue from selling x units at $200 each, then

$$R(x) = 200x$$

(B) We have

$$\text{fixed cost} = \$540{,}000$$

$$\frac{\text{variable cost}}{\text{per unit}} = \frac{\text{total variable cost}}{\text{number of units}}$$

$$= \frac{1{,}400{,}000}{10{,}000} = \$140$$

Thus, if $C(x)$ is the total cost of producing x units, then

$$C(x) = 140x + 540{,}000$$

(C) At the break-even point,

$$R(x) = C(x)$$
$$200x = 140x + 540{,}000$$

Solving for x yields

$$60x = 540{,}000$$
$$x = 9000 \text{ units}$$

Thus, the Zero Company must produce and sell 9000 units in order to break even. Note that $R(9000) = C(9000) = \$1{,}800{,}000$.

(D) If $P(x)$ is the profit from selling x units, then

$$P(x) = R(x) - C(x)$$
$$= 200x - (140x + 540{,}000)$$
$$= 60x - 540{,}000$$

If $P(x) = 96{,}000$, then

$$96{,}000 = 60x - 540{,}000$$

51 LINEAR FUNCTIONS

> Solving for x yields
> $$-60x = -636{,}000$$
> $$x = 10{,}600 \text{ units}$$
>
> Thus, the Zero Company must sell 10,600 units in order to generate a profit of $96,000.
>
> (E) The cost equation becomes
> $$C(x) = 140x + \underbrace{540{,}000 + 18{,}000}_{\text{fixed cost}}$$
> $$= 140x + 558{,}000$$
>
> At the break-even point,
> $$R(x) = C(x)$$
>
> Thus,
> $$200x = 140x + 558{,}000$$
>
> Solving for x yields
> $$60x = 558{,}000$$
> $$x = 9300 \text{ units}$$
>
> Therefore, the Zero Company must produce 9300 units to break even. Note that $R(9300) = C(9300) = \$1{,}860{,}000$.

Equilibrium Point

Pamela Johnson sells bracelets. The demand, D, is related to p, the price per bracelet, by the demand equation

$$D(p) = -2p + 130$$

The supply, S, is related to p by the supply equation

$$S(p) = 3p - 70$$

> Pamela wishes to determine the price p at which supply equals demand.

If
$$S(p) = D(p)$$

then we have

$$3p - 70 = -2p + 130$$

Solving for p yields

$$5p = 200$$
$$p = 40$$

Thus, at a price of $40 per bracelet, supply will equal demand. The corresponding supply and demand are calculated as follows:

$$S(p) = 3p - 70 \qquad D(p) = -2p + 130$$
$$S(40) = 3(40) - 70 \qquad D(40) = -2(40) + 130$$
$$= 50 \qquad\qquad\qquad = 50$$

The point (40, 50) is called the **equilibrium point.** Its first coordinate is called the **equilibrium price,** and its second coordinate is called the **equilibrium demand.** The equilibrium point, along with the supply and demand functions, are illustrated in Figure 1–52.

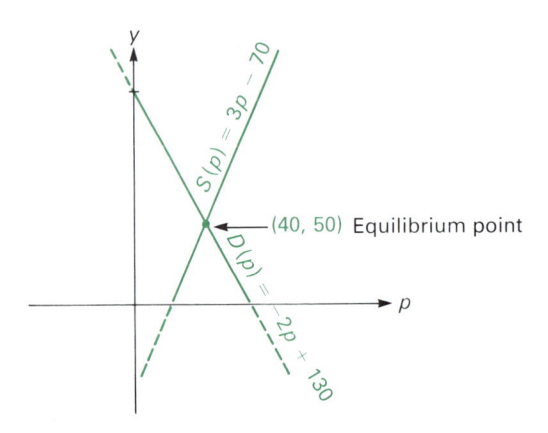

FIGURE 1–52

Depreciation (Straight-Line Method)

Most assets decrease in value over a period of time. This decrease in value is called **depreciation.** Suppose a company spends $11,000 (total cost) for a truck expected to last 5 years **(economic life),** at which time it will probably be worth $1000 **(salvage value).** For tax purposes, such an asset may be considered to depreciate each year by a fixed amount determined as follows:

$$\frac{\text{depreciable amount}}{\text{economic life}} = \frac{\text{total cost} - \text{salvage value}}{\text{economic life}} = \frac{11{,}000 - 1000}{5}$$
$$= \$2000$$
$$\uparrow$$
annual depreciation

53 LINEAR FUNCTIONS

Therefore, after 1 year, the value of the truck, y, is

$$y = 11,000 - 2000 = \$9000$$

After 2 years, the value of the truck, y, is

$$y = 11,000 - 2000(2) = \$7000$$

After 3 years, the value of the truck, y, is

$$y = 11,000 - 2000(3) = \$5000$$

After x years, the value of the truck, y, is

$$y = 11,000 - 2000x$$

Thus, the linear equation $y = 11,000 - 2000x$ relates the value of the truck, y, with its age, x, in years. This method of depreciation is called the **straight-line method**. The equation is graphed in Figure 1–53.

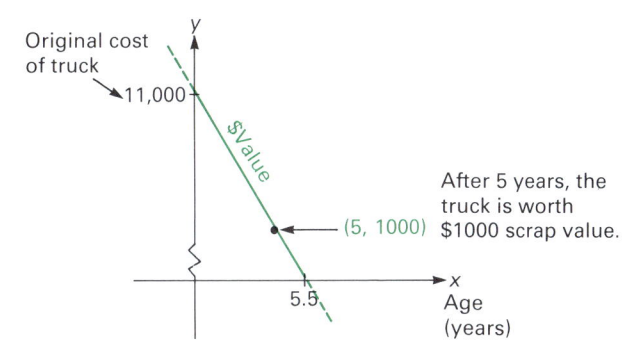

FIGURE 1–53

We will now determine formulas for the y-intercept and slope of the linear equation relating the value of an asset with its age in years. Referring to the linear equation

$$y = 11,000 - 2000x$$

y-intercept   slope (annual
(total cost)   depreciation)

note that the y-intercept, 11,000, is the total cost of the asset and that the slope, -2000, indicates that the value of the truck decreases by \$2000 per year.

54 CHAPTER ONE

Therefore, if

C = total cost of the asset
n = number of years of economic life of the asset
S = salvage value of the asset

then the y-intercept is C and the slope is

$$-\left(\frac{C-S}{n}\right)$$

Thus, the linear equation relating the value, y, of an asset with its age, x, in years is

$$y = C - \left(\frac{C-S}{n}\right)x$$

EXAMPLE 1–28 The Beefup Company buys a refrigerator for $20,000. The refrigerator has an economic life of 10 years, after which time it will probably have a salvage value of $2000. Find the linear equation relating the value, y, of the refrigerator to its age, x. Use the straight-line method of depreciation.

Solution We know that $C = \$20{,}000$, $n = 10$, and $S = \$2000$. Substituting these values into the linear equation

$$y = C - \left(\frac{C-S}{n}\right)x$$

gives us

$$y = 20{,}000 - \left(\frac{20{,}000 - 2000}{10}\right)x$$
$$= 20{,}000 - 1800x$$

EXERCISES

1. A manufacturer has fixed costs of $2000 and a variable cost of $5 per unit. She sells her product for $15 apiece.
   (A) Find the cost equation that relates cost and the number of units produced.
   (B) Find the sales revenue equation that relates sales revenue and the number of units produced.
   (C) Sketch both cost and sales revenue equations on the same set of axes.

55 LINEAR FUNCTIONS

(D) Find the break-even point.
(E) Find the profit equation that relates profit and the number of units produced.
(F) Sketch the profit equation.
(G) Find the profit that results from producing and selling 300 units of this product.
(H) Find the profit that results from producing and selling 100 units of this product.
(I) How many units must be produced and sold in order to attain a profit of $40,000?

2. The WVW Corporation has fixed costs of $5000. Its variable cost for producing 400 units is $16,000. The corporation sells its product for $60 apiece.
   (A) Find the cost equation.
   (B) Find the sales revenue equation.
   (C) Sketch both cost and sales revenue equations on the same set of axes.
   (D) Find the break-even point.
   (E) Find the profit equation.
   (F) Sketch the profit equation.
   (G) Find the profit that results from producing and selling 450 units.
   (H) How many units must be produced and sold in order to yield a profit of $25,000?

3. The VUV Corporation has presented the following information to its accountant:

   Fixed costs
      Administrative $200,000
      Building and land   50,000
      Other fixed overhead   10,000
   Variable cost for 100 units of production
      Materials $ 50,000
      Labor   70,000
      Selling expense   10,000
      Selling price per unit $ 1,800

   (A) Find the variable cost per unit.
   (B) Find the cost equation.
   (C) Find the sales revenue equation.
   (D) Sketch both the cost and sales revenue functions on the same graph.
   (E) Find the break-even point.
   (F) Find the profit equation.
   (G) Sketch the profit equation.
   (H) Find the profit that results from the production and selling of 1000 units.
   (I) How many units must be produced and sold in order to attain a profit of $104,000?

56 CHAPTER ONE

4. Susan Time can sell 150 watches at a price of $35 each. If the price drops to $20 apiece, Susan can sell 300 watches.
   (A) Find the linear demand equation.
   Susan's suppliers are willing to supply only 125 watches if the price per watch is $25. However, if the price per watch increases to $40 per watch, the suppliers are willing to provide Susan with 350 watches.
   (B) Find the linear supply equation.
   (C) Sketch both supply and demand equations on the same set of axes.
   (D) Find the equilibrium point.
   (E) For supply to equal demand, Susan's watches must be priced at how much apiece?

5. The demand, y_d, for a commodity is represented by the equation
   $$y_d = -3x + 39$$
   and the supply, y_s, is represented by the equation
   $$y_s = 4x - 24$$
   (A) Sketch both supply and demand functions on the same set of axes.
   (B) Find the equilibrium point.
   (C) Find the equilibrium price.

6. A corporation buys an automobile for $10,500. The automobile's useful life is 5 years, after which time it will have a scrap value of $500.
   (A) Find the linear equation relating the value of the automobile, y, with its age, x. Use the linear depreciation method.
   (B) Sketch this linear equation.
   (C) What is the value of the automobile after 3 years?

7. Made-Fresh Bakery buys an oven for $30,000. The useful life of the oven is 10 years, after which time it will have a scrap value of $1000.
   (A) Find the linear equation relating the value of the oven, y, with its age, x.
   (B) Graph the linear function defined by the equation of part (A).
   (C) What is the value of the oven after 6 years?

8. (This problem appeared on a past Uniform CPA Examination.)

Full Ton Company
Financial Projection For Product USA
For the Year Ended December 31, 19x7

Sales (100 units @ $100 a unit)		$10,000
Manufacturing cost of goods sold		
Direct labor	$1,500	
Direct materials used	1,400	
Variable factory overhead	1,000	
Fixed factory overhead	500	
Total manufacturing cost of goods sold		4,400
Gross profit		5,600

57 LINEAR FUNCTIONS

Selling expenses		
Variable	600	
Fixed	1,000	
Administrative expenses		
Variable	500	
Fixed	1,000	
Total selling and administrative expenses		3,100
Operating income		$ 2,500

Let x = number of units produced and sold.
- (A) Find the sales revenue equation.
- (B) Find the cost equation.
- (C) Find the break-even point.
- (D) Find the profit equation.
- (E) How many units should be sold in order to yield a profit of 4000?
- (F) What would be the operating income if sales increased by 25%?
- (G) What would be the break-even point if fixed factory overhead increased by $1700?

9. (This problem appeared on a past Uniform CPA Examination.) The Jarvis Company has fixed costs of $200,000. It has two products that it can sell: Tetra and Min. Jarvis sells these products at a rate of 2 units of Tetra to 1 unit of Min. The unit profit is $1 per unit for Tetra and $2 per unit for Min. How many units of Min would be sold at the break-even point?

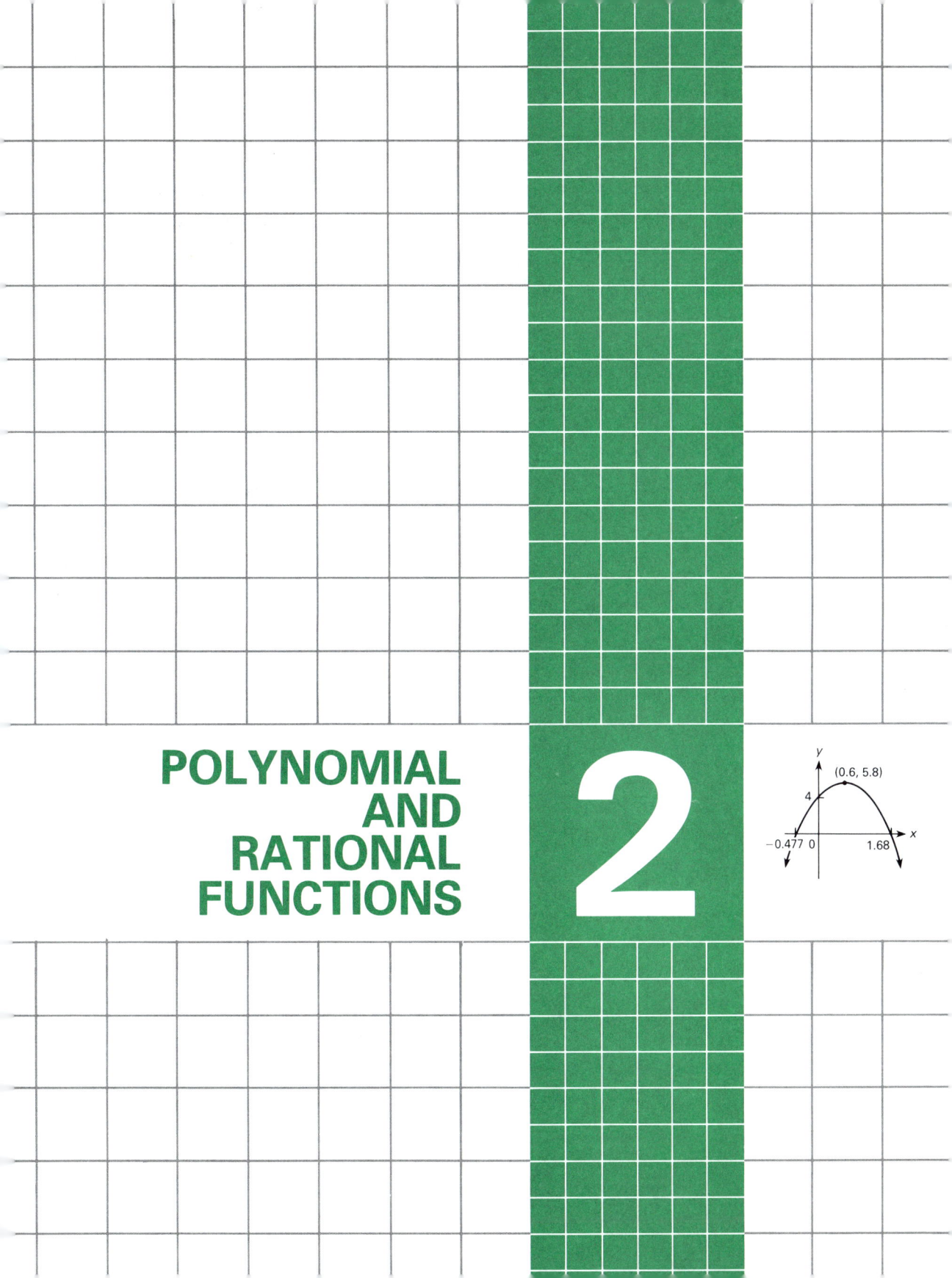

POLYNOMIAL AND RATIONAL FUNCTIONS

2

In Chapter 1, we discussed linear functions and some of their applications. In this chapter, we will consider **polynomial** and **rational functions**. The graphs of these functions are **curvilinear*** and often provide models of real-world problems. Since polynomial and rational functions involve exponents, we will first review some characteristics of exponents.

2-1

EXPONENTS The product of a number x multiplied by itself n times is denoted by x^n. That is,

$$x^n = \underbrace{x \cdot x \cdot \ldots \cdot x}_{n\ x\text{'s}}$$

The integer n, which indicates the number of times x is multiplied, is an **exponent**. The number x is called the **base**.

We define a negative exponent as follows:

$$x^{-n} = \frac{1}{x^n} = \frac{1}{\underbrace{x \cdot x \cdot \ldots \cdot x}_{n\ x\text{'s}}} \qquad x \neq 0$$

We define

$$x^0 = 1 \qquad (x \neq 0)$$

Note that 0^0 is undefined.

Laws of Exponents In this section, we will discuss exponents and the laws governing their algebraic manipulation.

> **First Law of Exponents**
>
> If x is any nonzero real number, then
>
> $$x^m \cdot x^n = x^{m+n}$$
>
> for any nonnegative integers m and n.

---

*In this chapter, we discuss polynomial functions of degree 2 or greater. The graphs of such functions are curvilinear (i.e., consist of curved lines). A polynomial function of degree 1 is a linear function. In Chapter 1, we learned that the graph of a linear function is a straight line.

61 POLYNOMIAL AND RATIONAL FUNCTIONS

To verify this law, we observe that

$$x^m \cdot x^n = \underbrace{(x \cdot x \cdot \ldots \cdot x)}_{m \text{ x's}} \underbrace{(x \cdot x \cdot \ldots \cdot x)}_{n \text{ x's}}$$

$$= \underbrace{x \cdot x \cdot \ldots \cdot x}_{(m+n) \text{ x's}}$$

$$= x^{m+n}$$

For example,

$$2^3 \cdot 2^4 = 2^{3+4} = 2^7$$
$$5^3 \cdot 5^6 = 5^{3+6} = 5^9$$

Second Law of Exponents

If x is any nonzero real number, then

$$\frac{x^m}{x^n} = x^{m-n}$$

for any nonnegative integers m and n.

To verify this law, we note that if $m = 5$ and $n = 2$, then

$$\frac{x^m}{x^n} = \frac{x^5}{x^2} = \frac{\cancel{x} \cdot \cancel{x} \cdot x \cdot x \cdot x}{\cancel{x} \cdot \cancel{x}}$$

$$= x^{5-2} = x^3$$

And, similarly, we have

$$\frac{5^7}{5^3} = 5^{7-3} = 5^4$$

$$\frac{4^3}{4^5} = 4^{3-5} = 4^{-2} = \frac{1}{4^2}$$

It should be noted that the first two laws of exponents also hold for negative integers m and n.

Third Law of Exponents

If x is any real number, then

$$(x^m)^n = x^{m \cdot n}$$

for all integers m and n.

To verify this law, we see that

$$(x^m)^n = \underbrace{x^m \cdot x^m \cdot \ldots \cdot x^m}_{n\ x^m\text{'s}}$$
$$= x^{\overbrace{m+m+\ldots+m}^{n\ m\text{'s}}}$$
$$= x^{m \cdot n}$$

As examples, we have

$$(x^3)^2 = x^{3 \cdot 2} = x^6$$
$$(7^4)^5 = 7^{4 \cdot 5} = 7^{20}$$

Roots and Radicals

We now give meaning to the symbol $\sqrt[n]{x}$, where n is a positive integer. The symbol $\sqrt[n]{x}$ is called a **radical** and is read "the nth root of x." The nth root of x, $\sqrt[n]{x}$, represents a number which when multiplied by itself n times (i.e., raised to the nth power) yields x. Thus, if y is an nth root of x, then

$$y^n = x$$

If $n = 2$, then $\sqrt[2]{x}$ is called the **square root** of x and is usually written $\sqrt{x}$. Thus, $\sqrt{16} = 4$ since $4^2 = 16$. However, the equation $y^2 = 16$ has another solution, -4, since $(-4)^2 = 16$. In such a case, we take the **positive square root**, often called the **principal square root**. Thus, 4 is the principal square root of 16, and we write $\sqrt{16} = 4$. If we want to refer to the **negative square root** of 16, we indicate it by writing $-\sqrt{16} = -4$. Note that $\sqrt{-16}$ is undefined since there exists no real number with a square of -16.

If $n = 3$, then $\sqrt[3]{x}$ is called the **cube root** of x. Note that $\sqrt[3]{8} = 2$ since $2^3 = 8$, and that $\sqrt[3]{-8} = -2$ since $(-2)^3 = -8$. Because $n = 3$ and 3 is an odd number, there is no need to define a principal cube root since the cube of a negative number is a negative number and the cube of a positive number is a positive number.

Observe that $\sqrt[4]{16} = 2$ since $2^4 = 16$. However, since $(-2)^4 = 16$, then $-\sqrt[4]{16} = -2$. Again, for even values of n, we call the positive nth root the **principal nth root**.

The preceding comments about $\sqrt[n]{x}$ are summarized as follows:

1. If $x > 0$ and n is even, then there are two solutions to the equation $y^n = x$. One is positive and the other is negative. To avoid ambiguity, we take the positive nth root, often called the principal nth root.
2. If $x < 0$ and n is even, then there are no real nth roots of x.

63 POLYNOMIAL AND RATIONAL FUNCTIONS

3. If n is odd, then there is one real nth root of x. Its sign is the same as that of x.
4. If $x = 0$, then the nth root of x is 0.

Rational Exponents

Up to this point, we have discussed only integer exponents. Thus, the first three laws of exponents were restricted to integral exponents. We now take a look at rational exponents. Consider the expression

$$x^{1/n}$$

where n is a nonzero integer. Hence, $1/n$ is a rational exponent. If $x^{1/n}$ is to be well defined, it must obey the laws of exponents. Specifically, the third law of exponents states that

$$(x^m)^n = x^{m \cdot n}$$

If this law is to hold for $x^{1/n}$, then

$$(x^{1/n})^n = x^{(1/n)n} = x$$

This implies that the product of $x^{1/n}$ multiplied by itself n times equals x. Hence, $x^{1/n}$ must equal the **nth root of x**, or

$$x^{1/n} = \sqrt[n]{x}$$

Thus, $x^{1/n}$ is now defined. Specifically,

$$x^{1/2} = \sqrt{x}$$
$$x^{1/3} = \sqrt[3]{x}$$
$$4^{1/2} = \sqrt{4} = 2$$

Note that $(-4)^{1/2} = \sqrt{-4}$, which is undefined.

Since we have defined the expression $x^{1/n}$ for integral values of n, we now discuss expressions of the form

$$x^{m/n}$$

for integral values of m and n. If the expression $x^{m/n}$ is to be well defined, it must obey the laws of exponents. Specifically, if the third law of exponents is to hold, then

$$(x^{1/n})^m = x^{(1/n)m} = x^{m/n}$$

Thus, $x^{m/n}$ may be defined as the **mth power of the nth root of x**, or

$$x^{m/n} = (\sqrt[n]{x})^m$$

As examples, we have

$$x^{5/2} = (\sqrt{x})^5$$
$$5^{2/3} = (\sqrt[3]{5})^2$$

Since the third law of exponents also indicates that

$$x^{m/n} = (x^m)^{1/n}$$

then $x^{m/n}$ may also be defined as the **nth root of x^m,** or

$$x^{m/n} = \sqrt[n]{x^m}$$

In summary, if m/n is reduced to lowest terms, we have

$$x^{m/n} = (\sqrt[n]{x})^m = \sqrt[n]{x^m}$$

Thus,

$$x^{5/2} = (\sqrt{x})^5 = \sqrt{x^5}$$
$$5^{2/3} = (\sqrt[3]{5})^2 = \sqrt[3]{5^2}$$

Since we have now defined rational exponents, it is appropriate to state that the first three laws of exponents hold for all rational exponents m and n for which x^m and x^n are defined.

Fourth Law of Exponents

If x and y are real numbers, then

$$(x \cdot y)^n = x^n \cdot y^n$$

for any integral exponent n.

To verify this law, observe that

$$(xy)^n = \underbrace{xy \cdot xy \cdot \ldots \cdot xy}_{n\,xy\text{'s}}$$
$$= \underbrace{(x \cdot x \cdot \ldots \cdot x)}_{n\,x\text{'s}}\underbrace{(y \cdot y \cdot \ldots \cdot y)}_{n\,y\text{'s}}$$
$$= x^n \cdot y^n$$

And, specifically,

$$(4 \cdot 3)^5 = 4^5 \cdot 3^5$$

65 POLYNOMIAL AND RATIONAL FUNCTIONS

$$(5x)^3 = 5^3 \cdot x^3 = 125x^3$$
$$(-2xy)^5 = (-2)^5 x^5 y^5 = -32x^5 y^5$$

It should be noted that the fourth law of exponents also holds for any rational exponent n for which x^n and y^n are defined.

> **Fifth Law of Exponents**
>
> If x and y are real numbers, with $y \neq 0$, then
>
> $$\left(\frac{x}{y}\right)^n = \frac{x^n}{y^n}$$
>
> for any integral exponent n.

To verify this law, note that

$$\left(\frac{x}{y}\right)^n = \underbrace{\frac{x}{y} \cdot \frac{x}{y} \cdot \ldots \cdot \frac{x}{y}}_{n \; \frac{x}{y}\text{'s}}$$

$$= \frac{\overbrace{x \cdot x \cdot \ldots \cdot x}^{n \; x\text{'s}}}{\underbrace{y \cdot y \cdot \ldots \cdot y}_{n \; y\text{'s}}}$$

$$= \frac{x^n}{y^n}$$

Thus,

$$\left(\frac{5}{2}\right)^3 = \frac{5^3}{2^3}$$

$$\left(\frac{-3}{y}\right)^4 = \frac{(-3)^4}{y^4} = \frac{81}{y^4}$$

Again, we note that the fifth law of exponents also holds for any rational exponent n for which x^n and y^n are defined.

Scientific Notation Any positive number can be expressed in the form

$$c \times 10^n$$

where $1 \leq c < 10$ and n is an integral exponent. When a number is expressed in this form, it is said to be in **scientific notation**. Specifically,

66 CHAPTER TWO

$$543 = 5.43 \times 10^2$$
$$84{,}000{,}000 = 8.4 \times 10^7$$
$$0.0003 = 3.0 \times 10^{-4}$$

EXERCISES

1. Simplify each of the following. No exponents should appear in the final answer.
   - (A) 3^2
   - (B) $\left(\dfrac{2}{3}\right)^4$
   - (C) $(-5)^2$
   - (D) $(-5)^3$
   - (E) $(4^2)^3$
   - (F) 5^{-3}
   - (G) 2^{-4}
   - (H) $(-3)^{-2}$
   - (I) $\dfrac{2^7}{2}$
   - (J) $\dfrac{(-3)^9}{(-3)^7}$
   - (K) $5^3 \cdot 5^4$
   - (L) $4^2 \cdot 4^3$
   - (M) $\left(\dfrac{3}{5}\right)^2$
   - (N) $\left(\dfrac{4}{3}\right)^0$
   - (O) $(2^{-3})^2$

2. Simplify each of the following. No exponents should appear in the final answer.
   - (A) $64^{1/2}$
   - (B) $64^{-1/2}$
   - (C) $216^{1/3}$
   - (D) $16^{1/2}$
   - (E) $16^{-1/2}$
   - (F) $49^{1/2}$
   - (G) $49^{-1/2}$
   - (H) $64^{5/2}$
   - (I) $64^{-5/2}$
   - (J) $49^{3/2}$
   - (K) $49^{-3/2}$
   - (L) $216^{2/3}$
   - (M) $216^{-2/3}$
   - (N) 867^0
   - (O) $(-3)^0$
   - (P) $49^{5/2}$

3. Rewrite each of the following using negative exponents:
   - (A) $\dfrac{1}{3^2}$
   - (B) $\dfrac{1}{5^6}$
   - (C) $\dfrac{1}{x^7}$
   - (D) $\dfrac{1}{(-5)^3}$
   - (E) $\dfrac{1}{x^n}$
   - (F) $\dfrac{1}{x^8}$

4. Rewrite each of the following using rational exponents:
   - (A) $\sqrt[3]{5}$
   - (B) $(\sqrt{4})^9$
   - (C) $\sqrt[3]{x^5}$
   - (D) $\sqrt[5]{2}$
   - (E) $\sqrt{8^7}$
   - (F) $\sqrt[7]{9^4}$
   - (G) $(\sqrt{5})^3$
   - (H) $(\sqrt[3]{5})^7$
   - (I) $\sqrt[4]{x}$
   - (J) $\dfrac{1}{\sqrt{5^3}}$
   - (K) $\dfrac{1}{(\sqrt[3]{5})^8}$
   - (L) $\dfrac{1}{\sqrt{x^5}}$
   - (M) $\dfrac{1}{\sqrt{x^3}}$
   - (N) $\dfrac{1}{\sqrt[3]{x^2}}$
   - (O) $\dfrac{1}{(\sqrt[3]{x})^7}$

5. Using the fourth law of exponents, $(x \cdot y)^n = x^n \cdot y^n$, simplify each of the following:
   - (A) $(-3x)^2$
   - (B) $(2y)^4$
   - (C) $(5xy)^3$
   - (D) $(xyz)^9$
   - (E) $(4 \cdot 81)^{1/2}$
   - (F) $\sqrt{9 \cdot 64}$

6. Using the fifth law of exponents, $(x/y)^n = x^n/y^n$, simplify each of the following:
   - (A) $\left(\dfrac{5}{6}\right)^3$
   - (B) $\left(\dfrac{x}{y}\right)^4$
   - (C) $\left(\dfrac{x}{3}\right)^5$
   - (D) $\left(\dfrac{x}{5}\right)^2$
   - (E) $\left(\dfrac{4}{x}\right)^3$
   - (F) $\left(\dfrac{-2}{x}\right)^3$

67 POLYNOMIAL AND RATIONAL FUNCTIONS

7. Simplify each of the following:
   - (A) $\left(\dfrac{2^{-3} \cdot 2^5}{2^{-2}}\right)^3$
   - (B) $3^{1/2} \cdot 3^{5/2}$
   - (C) $\dfrac{3^{-7/2} \cdot 3^{3/2}}{3^{1/2} \cdot 3^{-3/2}}$
   - (D) $\left(\dfrac{27^{5/3} \cdot 27^{-1/3}}{27^{1/3}}\right)^2$

8. Express each of the following in scientific notation:
   - (A) 496
   - (B) 5,870,000
   - (C) 8,000,000,000
   - (D) 0.00045
   - (E) 0.0000008
   - (F) 59.5
   - (G) 0.56
   - (H) 8730
   - (I) 0.00357

2-2

QUADRATIC FUNCTIONS

In this and the next section, we will learn to graph equations like

$$y = 3x^2 - 4x + 7$$

Such an equation defines a **quadratic function**. Note that the highest powered term of a quadratic equation is of the **second degree**.

$$y = f(x) = 3x^2 - 4x + 7$$

second-degree term   first-degree term   constant term

In general, any equation of the form

$$y = ax^2 + bx + c$$

where a, b, and c are constant real numbers and $a \neq 0$, is called a **quadratic equation**. Without the restriction $a \neq 0$, the equation becomes the linear equation $y = bx + c$.

If the quadratic equation

$$y = 3x^2 - 4x + 7$$

is compared with the general form

$$y = ax^2 + bx + c$$

then $a = 3$, $b = -4$, and $c = 7$.

As a start in learning to graph quadratic functions, we consider the simplest of all quadratic equations, which is

$$y = x^2$$

Comparing this equation with the general quadratic form $y = ax^2 + bx + c$, we have $a = 1$, $b = 0$, and $c = 0$.

A sketch of $y = x^2$ may be obtained by finding some ordered pairs (x, y) satisfying the equation and plotting their correspond-

68 CHAPTER TWO

ing points on the rectangular coordinate system. Arbitrarily choosing values of x and finding their corresponding y-values, we have the following:

x-Value	Equation $(y = x^2)$	Ordered Pair (x, y)
If $x = -3$, then	$y = (-3)^2 = 9$	$(-3, 9)$
If $x = -2$, then	$y = (-2)^2 = 4$	$(-2, 4)$
If $x = -1$, then	$y = (-1)^2 = 1$	$(-1, 1)$
If $x = 0$, then	$y = 0^2 = 0$	$(0, 0)$
If $x = 1$, then	$y = 1^2 = 1$	$(1, 1)$
If $x = 2$, then	$y = 2^2 = 4$	$(2, 4)$
If $x = 3$, then	$y = 3^2 = 9$	$(3, 9)$

Plotting the ordered pairs (x, y) and sketching the curve through them, we obtain the graph of Figure 2–1. This graph form is called a **parabola**. In fact, the graph of any quadratic function is a parabola. Note that the origin, (0, 0), is the lowest point on the parabola $y = x^2$ of Figure 2–1. The lowest point on the parabola is called the **vertex**. Thus, (0, 0) is the vertex of $y = x^2$.

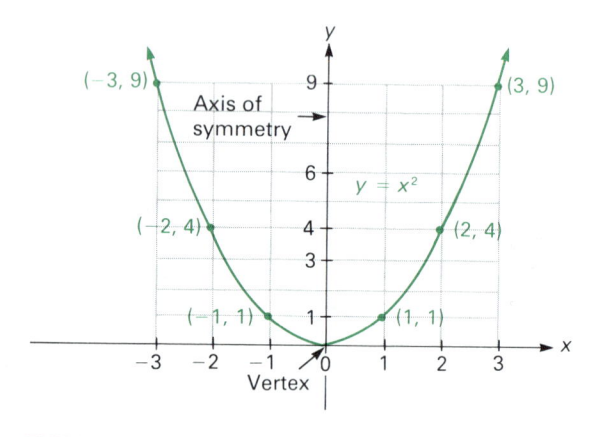

FIGURE 2–1

Still referring to the parabola $y = x^2$ of Figure 2–1, note that the y-axis separates the parabola into two symmetric parts, each the mirror image of the other. Such a vertical line passing through the vertex is called the **axis of symmetry**.

We now consider quadratic equations of the form

$$y = ax^2$$

with $a \neq 1$. If $a > 0$, then the parabola has the same general shape as $y = x^2$ of Figure 2–1, but it is pushed together if $a > 1$ and is

POLYNOMIAL AND RATIONAL FUNCTIONS

flattened if $a < 1$. If $a < 0$, then the y-coordinates are negative and the parabola appears below the x-axis. To illustrate these comments, see the graphs of $y = x^2$, $y = 3x^2$, $y = (1/2)x^2$, and $y = -3x^2$ in Figure 2–2.

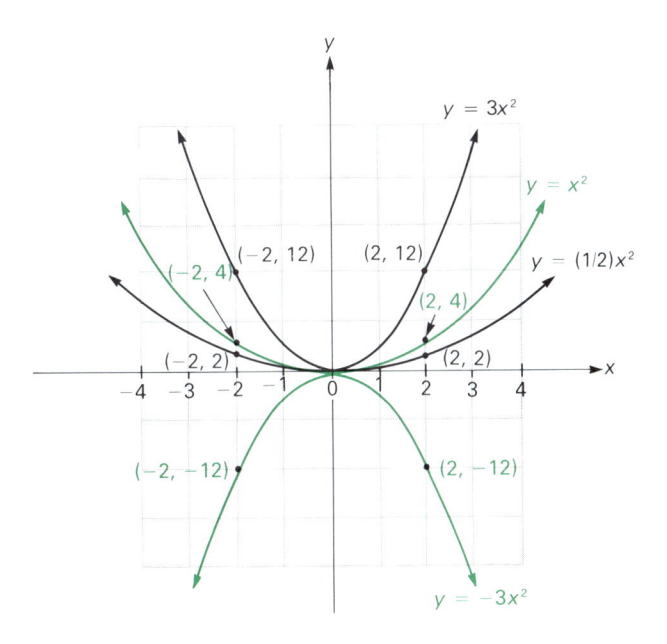

FIGURE 2–2

The graphs of quadratic functions with $a > 0$ are said to *open up*, whereas those with $a < 0$ are said to *open down*. Thus, the graphs of $y = x^2$, $y = 3x^2$, and $y = (1/2)x^2$ of Figure 2–2 are described as opening up, while the graph of $y = -3x^2$ is said to open down.

We now consider quadratic equations of the form

$$y = ax^2 + c$$

The constant, c, has the effect of *lifting* the graph of $y = ax^2$ *vertically by c units if $c > 0$*, and *lowering* it *vertically by c units if $c < 0$*. To illustrate these effects, Figure 2–3 contains the graphs of $y = 3x^2$, $y = 3x^2 + 5$, and $y = 3x^2 - 5$.

We now consider quadratic equations of the form

$$y = a(x - h)^2$$

A specific example of this case is

$$y = 3(x - 5)^2$$

[Figure 2-3: Graphs of $y = 3x^2 + 5$, $y = 3x^2$, and $y = 3x^2 - 5$]

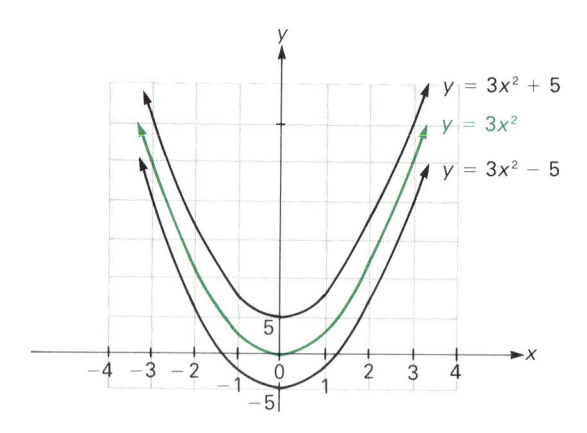

FIGURE 2-3

Note that $3(x - 5)^2 \geq 0$. Thus, the smallest value that y takes on is 0. In fact, y assumes its smallest value, 0, at $x = 5$. Hence, the vertex of the parabola is at (5, 0), as illustrated in Figure 2-4.

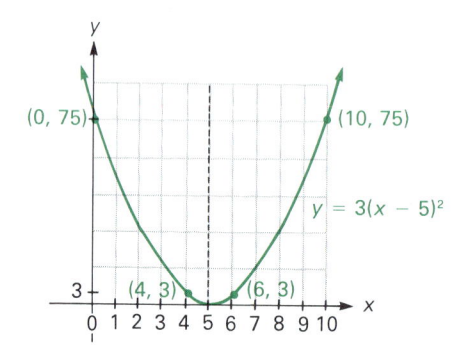

[Figure 2-4: Graph of $y = 3(x - 5)^2$ with points (0, 75), (10, 75), (4, 3), (6, 3)]

FIGURE 2-4

If we compare the graph of $y = 3(x - 5)^2$, shown in Figure 2-4, with that of $y = 3x^2$ (see Figure 2-5), we note that the graphs are the same except that we must shift the graph of $y = 3x^2$ horizontally to the right by 5 units to obtain the graph of $y = 3(x - 5)^2$. In general, the graph of $y = a(x - h)^2$ may be obtained by *shifting* the graph of $y = ax^2$ *horizontally to the right if* h > 0, and *horizontally to the left if* h < 0.

Finally, we consider quadratic functions of the form

$$y = a(x - h)^2 + k$$

A specific example is

$$y = 3(x - 5)^2 + 4$$

POLYNOMIAL AND RATIONAL FUNCTIONS

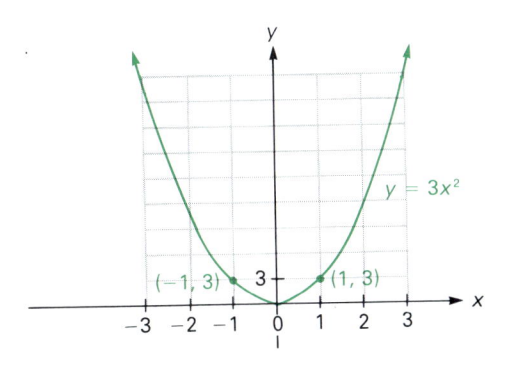

FIGURE 2–5

As was concluded when we discussed forms $y = ax^2$ and $y = ax^2 + c$, the constant c added to the function $y = ax^2$ has the effect of *lifting* the graph of $y = ax^2$ vertically by c units if $c > 0$, and *lowering* it vertically by c units if $c < 0$. Similarly, the constant k added to $y = a(x - h)^2$ has the effect of *lifting* the graph of $y = a(x - h)^2$ vertically by k units if $k > 0$, and *lowering it vertically* by k units if $k < 0$. Thus, the quadratic function defined by

$$y = 3(x - 5)^2 + 4$$

may be graphed by lifting the graph of $y = 3(x - 5)^2$ vertically by 4 units, as is illustrated in Figure 2–6. Note that the vertex becomes (5, 4).

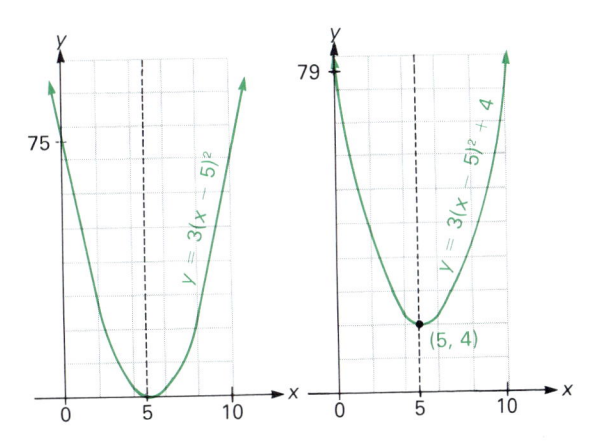

FIGURE 2–6

72 CHAPTER TWO

> In general, the graph of the quadratic function defined by
> $$y = a(x - h)^2 + k$$
>
> 1. opens up if $a > 0$ and opens down if $a < 0$;
> 2. has its vertex at (h, k);
> 3. has the line $x = h$ as its axis of symmetry.

EXAMPLE 2–1 Graph $y = f(x) = -4(x - 3)^2 + 7$.

Solution Since $a = -4$, a negative number, the parabola opens down. The vertex is located at (3, 7). The axis of symmetry is the line $x = 3$. The y-intercept is

$$y = f(0) = -4(0 - 3)^2 + 7$$
$$= -4(-3)^2 + 7$$
$$= -29$$

The x-intercepts are obtained by setting $y = 0$ and solving for x. Hence, we obtain

$$0 = -4(x - 3)^2 + 7$$
$$4(x - 3)^2 = 7$$
$$(x - 3)^2 = \frac{7}{4}$$
$$x - 3 = \pm\sqrt{\frac{7}{4}}$$
$$x = 3 \pm \sqrt{\frac{7}{4}}$$

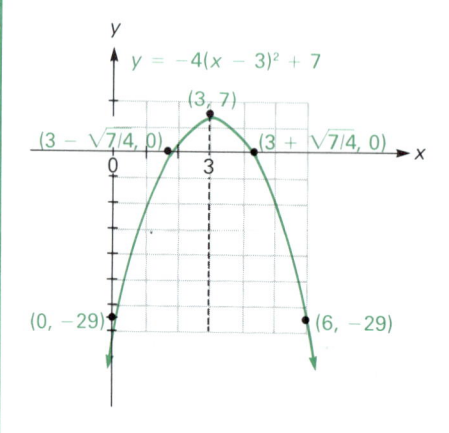

FIGURE 2–7

POLYNOMIAL AND RATIONAL FUNCTIONS

> The graph appears in Figure 2–7. Observing the parabola of Figure 2–7, note that if a parabola opens down, then the vertex is the highest point on the parabola.

Up to this point, we have graphed quadratic equations of the form

$$y = f(x) = a(x - h)^2 + k$$

Suppose we encounter a quadratic equation of the form

$$y = f(x) = ax^2 + bx + c$$

We will derive formulas for the coordinates of the vertex, (h, k), in terms of a, b, and c by expressing the equation $y = a(x - h)^2 + k$ in the form $y = ax^2 + bx + c$. We begin with $y = a(x - h)^2 + k$ and replace $(x - h)^2$ with its equivalent expression $x^2 - 2hx + h^2$ to obtain

$$y = a(x^2 - 2hx + h^2) + k$$

Simplifying this expression, we get

$$y = ax^2 - 2ahx + ah^2 + k$$

Comparing this result with the general quadratic form $y = ax^2 + bx + c$, we have

$$b = -2ah \quad \text{and} \quad c = ah^2 + k$$

Solving the equation $b = -2ah$ for h, we obtain

$$h = \frac{-b}{2a}$$

And solving the equation $c = ah^2 + k$ for k, we have

$$k = c - ah^2$$

Replacing h here with $-b/2a$ yields

$$k = c - a\left(\frac{-b}{2a}\right)^2$$
$$= c - \frac{b^2}{4a}$$
$$= \frac{4ac - b^2}{4a}$$

Thus, the x-coordinate of the vertex, h, is $-b/2a$, and its y-coordinate, k, is $(4ac - b^2)/4a$. Hence, we summarize as follows.

> The graph of a quadratic equation of the form
>
> $$y = f(x) = ax^2 + bx + c$$
>
> is a parabola with vertex (h, k), where
>
> $$h = -\frac{b}{2a} \quad \text{and} \quad k = \frac{4ac - b^2}{4a}$$

EXAMPLE 2–2 Graph $y = f(x) = 3x^2 + 5x + 2$.

Solution Comparing this equation with the form $y = f(x) = ax^2 + bx + c$, we have $a = 3$, $b = 5$, and $c = 2$. Since a is positive, the parabola opens up. The coordinates of the vertex, (h, k), are

$$h = -\frac{b}{2a} = -\frac{5}{2(3)} = -\frac{5}{6}$$

$$k = \frac{4ac - b^2}{4a} = \frac{4(3)(2) - 5^2}{4(3)} = -\frac{1}{12}$$

Note that the y-coordinate of the vertex, k, may also be obtained by

$$f\left(-\frac{b}{2a}\right) = f\left(-\frac{5}{6}\right) = 3\left(-\frac{5}{6}\right)^2 + 5\left(-\frac{5}{6}\right) + 2 = -\frac{1}{12}$$

Thus, the vertex is $(-5/6, -1/12)$. The y-intercept is

$$y = f(0) = 3(0^2) + 5(0) + 2 = 2$$

The x-intercepts are found by setting $y = 0$ and solving for x. Hence,

$$0 = 3x^2 + 5x + 2$$
$$0 = (3x + 2)(x + 1)$$

$$3x + 2 = 0 \qquad x + 1 = 0$$
$$x = -\frac{2}{3} \qquad x = -1$$

Therefore, the x-intercepts are $(-2/3, 0)$ and $(-1, 0)$. The graph appears in Figure 2–8.

75 POLYNOMIAL AND RATIONAL FUNCTIONS

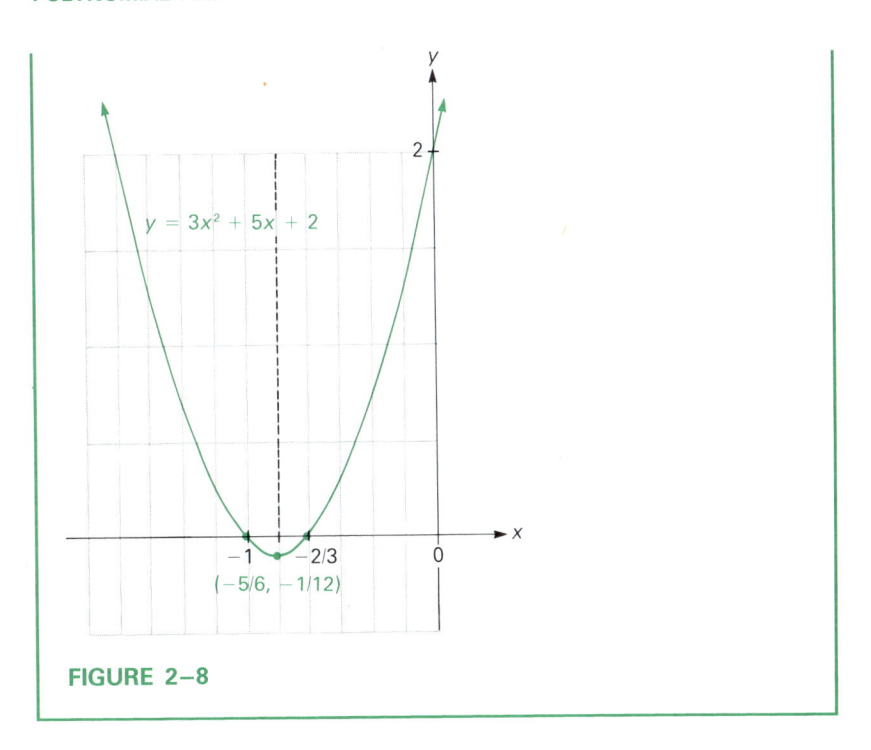

FIGURE 2–8

[Graph showing parabola $y = 3x^2 + 5x + 2$ with x-intercepts at -1 and $-2/3$, and vertex at $(-5/6, -1/12)$]

Quadratic Formula

When finding the x-intercepts of a parabola, we must often solve a quadratic equation of the form

$$ax^2 + bx + c \neq 0$$

where a, b, and c are constant real numbers and $a \neq 0$. Sometimes, such an equation may be solved by factoring, as was the case in Example 2–2. A more general method is given by the **quadratic formula.**

Quadratic Formula

$$x = \frac{-b \pm \sqrt{b^2 - 4ac}}{2a}$$

We will review the use of the quadratic formula by calculating the x-intercepts of the parabola of Example 2–2. Substituting $a = 3$, $b = 5$, and $c = 2$ into the quadratic formula

$$x = \frac{-b \pm \sqrt{b^2 - 4ac}}{2a}$$

gives us

$$x = \frac{-5 \pm \sqrt{5^2 - 4(3)(2)}}{2(3)}$$

$$= \frac{-5 \pm \sqrt{25 - 24}}{6} = \frac{-5 \pm \sqrt{1}}{6}$$

$$= \frac{-5 \pm 1}{6} \quad \begin{matrix} \frac{-5+1}{6} = -\frac{2}{3} \\ \frac{-5-1}{6} = -1 \end{matrix}$$

Thus, the x-intercepts are $(-2/3, 0)$ and $(-1, 0)$.

The expression $b^2 - 4ac$, which appears under the square root sign in the quadratic formula, determines the character of the solutions. Hence, it is called the **discriminant**. Specifically,

1. If $b^2 - 4ac > 0$, there are two real solutions (two x-intercepts).
2. If $b^2 - 4ac = 0$, there is one real solution (one x-intercept).
3. If $b^2 - 4ac < 0$, there are no real solutions (no x-intercepts).

EXERCISES

1. Graph each of the following:
   (A) $y = 5x^2$
   (B) $y = -5x^2$
   (C) $y = \frac{1}{2}x^2$
   (D) $y = -\frac{1}{2}x^2$

2. Graph the following on the same set of axes: $y = x^2$, $y = 6x^2$, $y = (1/4)x^2$.

3. Graph the following on the same set of axes: $y = -x^2$, $y = -4x^2$, $y = -(1/3)x^2$.

4. Graph the following on the same set of axes: $f(x) = x^2$, $f(x) = 3x^2$, $f(x) = 7x^2$.

5. Graph each of the following:
   (A) $y = x^2 + 7$
   (B) $f(x) = 3x^2 + 5$
   (C) $y = x^2 - 9$
   (D) $y = -x^2 + 9$
   (E) $f(x) = 2x^2 - 8$
   (F) $y = -4x^2 + 36$
   (G) $y = -5x^2 + 1$
   (H) $f(x) = -3x^2 + 12$
   (I) $f(x) = 5x^2 + 1$
   (J) $y = 2x^2 - 6$
   (K) $y = -3x^2 + 6$
   (L) $f(x) = \frac{1}{2}x^2 - 4$

6. The Good Morning Cereal Company sells cereal. The equation

   $$p(x) = -\frac{1}{200}x^2 + 18 \quad (x \geq 0)$$

   expresses the price per case, p, charged for an order consisting of x cases of cereal.
   (A) Sketch $p(x) = -(1/200)x^2 + 18$.
   (B) If a customer places an order for 10 cases of cereal, what will be the price per case?

77 POLYNOMIAL AND RATIONAL FUNCTIONS

    (C) If a customer places an order for 20 cases of cereal, what will be the price per case?

7. A company's annual profits, P (in millions of dollars), are related to the time, t (in years), it has been in business by the function

$$P(t) = \frac{1}{5}t^2 + 20 \quad (t \geq 0)$$

    (A) Sketch $P(t) = (1/5)t^2 + 20$.
    (B) Find this company's annual profits for the fifth year.

8. A textile company produces amounts x and y of two different jeans using the same production process. The amounts x and y are related by

$$y = -\frac{1}{10}x^2 + 30 \quad (0 \leq x \leq 17)$$

The graph of this equation is called a **product transformation curve**.
    (A) Sketch $y = -(1/10)x^2 + 30$.
    (B) If the company wants to produce amount x such that it is twice as much as amount y ($x = 2y$), what amounts x and y should be produced?

9. Graph each of the following:
    (A) $y = (x - 3)^2$ (B) $y = (x + 3)^2$
    (C) $y = 4(x - 1)^2$ (D) $y = 4(x + 1)^2$
    (E) $f(x) = -2(x + 5)^2$ (F) $f(x) = -2(x - 3)^2$
    (G) $f(x) = -(x + 2)^2$ (H) $f(x) = -(x - 1)^2$

10. Graph each of the following:
    (A) $y = 2(x - 1)^2 + 3$ (B) $y = -(x - 4)^2 - 1$
    (C) $f(x) = 3(x - 2)^2 - 1$ (D) $f(x) = -3(x - 2)^2 + 4$
    (E) $y = -3(x + 2)^2 - 5$ (F) $y = 4(x + 3)^2 - 2$
    (G) $f(x) = -(x + 1)^2 + 2$ (H) $f(x) = (x + 5)^2 - 3$
    (I) $f(x) = 5(x - 1)^2 - 2$ (J) $f(x) = 5(x - 4)^2 + 3$

11. Graph each of the following:
    (A) $f(x) = (x + 3)^2 - 1$ (B) $f(x) = (x - 2)^2 + 4$
    (C) $f(x) = -(x + 2)^2 + 1$ (D) $f(x) = -(x - 3)^2 + 2$
    (E) $y = 4(x - 1)^2 + 2$ (F) $y = 2(x + 3)^2 + 1$
    (G) $y = 3(x - 2)^2 + 1$ (H) $y = 3(x - 2)^2 - 1$
    (I) $f(x) = -2(x - 4)^2 + 3$ (J) $y = -2(x + 4)^2 - 3$

12. Graph each of the following:
    (A) $y = 2x^2 + 8x$ (B) $f(x) = -3x^2 + 7x$
    (C) $f(x) = x^2 - 5x$ (D) $y = -x^2 + 8x$
    (E) $f(x) = 5x^2 - 3x$ (F) $f(x) = 7x^2 - 2x$
    (G) $y = x^2 + 9x$ (H) $y = -x^2 + 7x$
    (I) $y = -2x^2 + 6x$ (J) $y = 3x^2 - 2x$
    (K) $f(x) = x^2 + x$ (L) $y = x^2 - x$

13. Graph each of the following:
    (A) $y = x^2 - 4x - 5$ (B) $y = x^2 + 7x + 6$
    (C) $f(x) = 2x^2 + x - 3$ (D) $f(x) = -x^2 + 7x - 12$
    (E) $f(x) = -5x^2 + 6x + 4$ (F) $y = x^2 - 8x + 16$

(G) $y = -x^2 + 8x - 16$
(H) $y = -2x^2 - x + 3$
(I) $y = x^2 + x + 3$
(J) $f(x) = x^2 + x + 1$
(K) $f(x) = x^2 - 10x + 26$
(L) $y = 2x^2 - 3x + 6$
(M) $y = 4x^2 + 5x + 6$
(N) $f(x) = x^2 - 6x + 7$
(O) $f(x) = 2x^2 + 4x + 1$
(P) $y = -x^2 + 5x + 2$
(Q) $f(x) = x^2 - 10x + 25$
(R) $f(x) = x^2 + 10x + 25$
(S) $y = x^2 - 6x + 9$
(T) $f(x) = x^2 + 6x + 9$

14. Graph each of the following:
(A) $y = (x - 2)(x + 3)$
(B) $f(x) = (x + 1)(x - 5)$
(C) $y = 4(x - 2)(x + 3)$
(D) $y = -2(x + 1)(x - 5)$
(E) $f(x) = (x - 6)^2$
(F) $y = -(x - 6)^2$
(G) $y = -5(x + 1)^2$
(H) $f(x) = (x + 6)(x - 6)$
(I) $y = (x - 2)(x + 2)$
(J) $f(x) = (x + 6)(x + 6)$
(K) $y = -3(x - 2)(x + 2)$
(L) $f(x) = 4(x + 1)(x - 1)$

2–3

APPLICATIONS OF QUADRATIC FUNCTIONS

Recall that if a parabola opens down, the vertex is its highest or maximum point. Thus, the y-coordinate of the vertex is the **maximum value** of the quadratic function (see Figure 2–9). If a parabola opens up, the vertex is its lowest or minimum point. Hence, the y-coordinate of the vertex is the **minimum value** of the quadratic function (see Figure 2–10). These concepts of maximum value and minimum value will be used in the applications of this section.

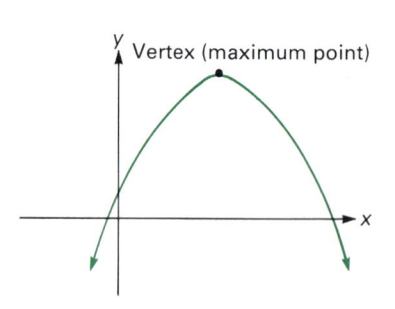

FIGURE 2–9

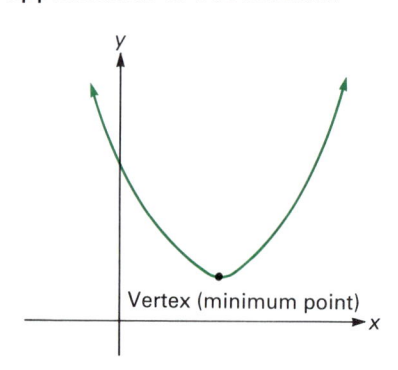

FIGURE 2–10

EXAMPLE 2–3 A manufacturer of children's watches finds that the demand, x, for watches is related to price per watch, p, by the equation

$$p = 10 - x$$

(A) Determine the equation that expresses total sales revenue, R, as a function of x.

POLYNOMIAL AND RATIONAL FUNCTIONS

(B) What is the maximum sales revenue? Also, how many watches must be sold in order to achieve the maximum sales revenue?

Solutions

(A) Here, we use

sales revenue = (price per watch)(number of watches sold)

$$R = px$$
$$= (10 - x)(x)$$
$$= 10x - x^2$$

Thus, the sales revenue equation is

$$R(x) = -x^2 + 10x$$

Its graph appears in Figure 2–11. Since x and R are nonnegative, the parabola is drawn as a solid line in the first quadrant.

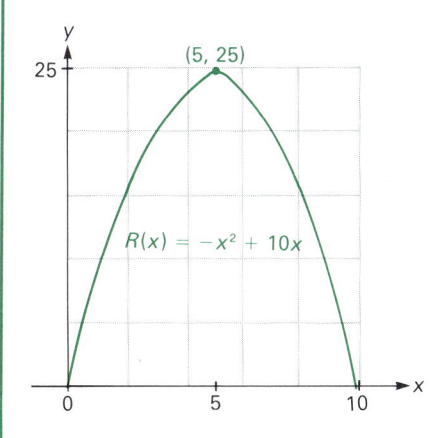

FIGURE 2–11

(B) Studying Figure 2–11, note that the y-coordinate of the vertex is the maximum value of the revenue function. Thus, the maximum sales revenue is $25. The corresponding x-coordinate, $x = 5$, indicates that 5 watches must be sold in order to achieve the maximum sales revenue of $25.

EXAMPLE 2–4 Suppose the manufacturer of Example 2–3 has a fixed cost of $14 and a variable cost of $1 per watch.
(A) Determine the cost equation.

EXAMPLE 2–4 (continued)

(B) Sketch both cost and sales revenue functions on the same set of axes.
(C) Determine the break-even points.

Solutions
(A) If $C(x)$ = total cost of producing x watches, then

$$C(x) = x + 14$$

(B) See Figure 2–12.

FIGURE 2–12

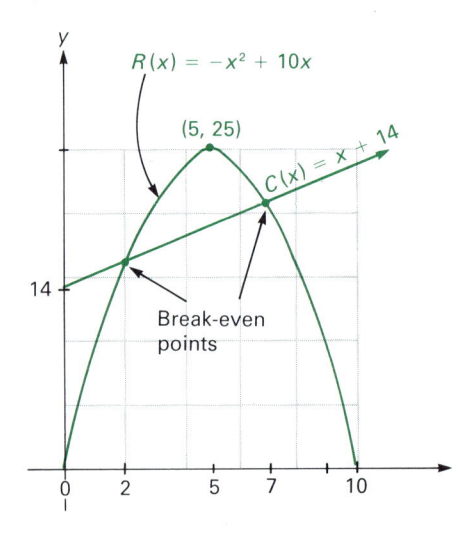

(C) In Section 1–6, we learned that the intersection of the graphs of a revenue function and a cost function is a break-even point. In Figure 2–12, we observe two such break-even points. To find them, we equate $R(x)$ and $C(x)$. Hence,

$$R(x) = C(x)$$
$$-x^2 + 10x = x + 14$$
$$0 = x^2 - 9x + 14$$
$$0 = (x - 2)(x - 7)$$

$x - 2 = 0 \qquad x - 7 = 0$
$\quad x = 2 \qquad\qquad x = 7$

x-coordinates of break-even points

Thus, the break-even points are at $x = 2$ and $x = 7$ (see

POLYNOMIAL AND RATIONAL FUNCTIONS

Figure 2–13). Studying Figure 2–13, note that the total sales revenue is greater than the total cost [i.e., $R(x) > C(x)$] for $2 < x < 7$. Thus, as long as the firm produces and sells between 2 and 7 watches, there will be a positive profit.

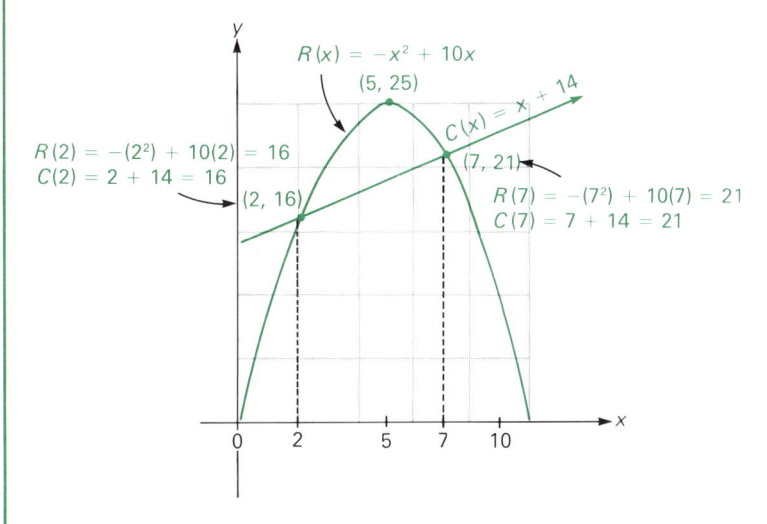

$R(x) = -x^2 + 10x$
(5, 25)
$C(x) = x + 14$
(7, 21)
$R(2) = -(2^2) + 10(2) = 16$
$C(2) = 2 + 14 = 16$
(2, 16)
$R(7) = -(7^2) + 10(7) = 21$
$C(7) = 7 + 14 = 21$

FIGURE 2–13

EXAMPLE 2–5 Assume the revenue and cost functions of Examples 2–3 and 2–4.
(A) Determine the profit equation.
(B) What is the maximum profit? How many watches must be produced and sold in order to achieve the maximum profit?

Solutions
(A) We have

$$\text{profit} = \text{revenue} - \text{cost}$$
$$P(x) = R(x) - C(x)$$
$$= (-x^2 + 10x) - (x + 14)$$
$$= -x^2 + 10x - x - 14$$
$$= -x^2 + 9x - 14$$

The graph of this equation appears in Figure 2–14. Since x and P are nonnegative, the parabola is drawn as a solid line in the first quadrant.

EXAMPLE 2-5 (continued)

FIGURE 2-14

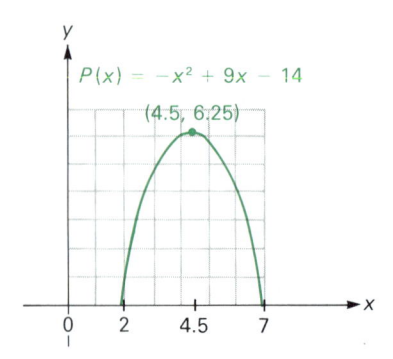

Graph showing P(x) = −x² + 9x − 14, a downward parabola with vertex at (4.5, 6.25) and x-intercepts at 2 and 7.

(B) Studying Figure 2-14, note that the y-coordinate of the vertex is the maximum value of the profit function. Thus, the maximum profit is $6.25. The corresponding x-coordinate, $x = 4.5$, indicates that 4.5 watches must be sold in order to attain the maximum profit of $6.25.

Observe in Figure 2-14 that the x-intercepts of the profit function are also the x-coordinates of the break-even points of Figure 2-13. Thus, $P(x) = 0$ or $R(x) = C(x)$ at $x = 2$ and $x = 7$.

EXAMPLE 2-6 The manager of Fruity's Frogurt Stand finds that the daily demand, q, for cinnamon frogurt is related to the price per cup, p, by the equation

$$q = 100 - 200p$$

(A) Determine the equation that expresses total sales revenue, R, as a function of p.
(B) What is the maximum sales revenue? What price per cup yields the maximum sales revenue?
(C) How many cups will be sold at a price of $0.25 per cup?

Solutions
(A) Here, we use

sales revenue = (price per cup)(number of cups sold)
$$R = pq$$
$$= p(100 - 200p)$$
$$= 100p - 200p^2$$

The graph of this equation appears in Figure 2-15. Since p and R are nonnegative, the parabola is drawn as a solid line in the first quadrant.

POLYNOMIAL AND RATIONAL FUNCTIONS

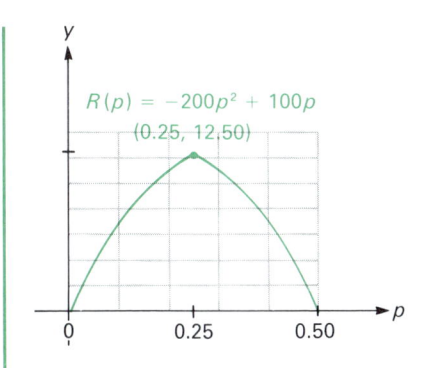

$$R(p) = -200p^2 + 100p$$
$$(0.25, 12.50)$$

FIGURE 2-15

(B) Studying Figure 2-15, note that the y-coordinate of the vertex is the maximum sales revenue. Thus, the maximum sales revenue is $12.50. The corresponding p-coordinate, $p = 0.25$, indicates that a price of $0.25 per cup will yield the maximum sales revenue.

(C) Since

$$q = 100 - 200p$$

then

$$q = 100 - 200(0.25) = 50$$

Thus, 50 cups will be sold at $0.25 per cup.

EXAMPLE 2-7 The supply, $S(p)$, of watches is given by the equation

$$S(p) = 2p^2 - 8p + 12 \qquad (p \geq 2)$$

and the demand, $D(p)$, is given by the equation

$$D(p) = p^2 - 18p + 68 \qquad (0 < p \leq 5)$$

where p is the price per watch.
(A) Find the equilibrium price.
(B) Calculate the supply and the demand at the equilibrium price.

Solutions
(A) The equilibrium price is such that supply = demand. Hence,

$$S(p) = D(p)$$
$$2p^2 - 8p + 12 = p^2 - 18p + 68$$

EXAMPLE 2–7 (continued)

Thus,
$$p^2 + 10p - 56 = 0$$
$$(p + 14)(p - 4) = 0$$
$$p + 14 = 0 \qquad p - 4 = 0$$
$$p = -14 \qquad p = 4$$

↗ Disregard this negative solution since price cannot be negative.

↑ This positive solution is the equilibrium price.

Therefore, at a price of $4 per watch, supply will equal demand, as shown in part (B).

(B) Supply and demand at the equilibrium price are as follows:
$$S(4) = 2(4^2) - 8(4) + 12$$
$$= 12$$
$$D(4) = 4^2 - 18(4) + 68$$
$$= 12$$

Therefore, at a price of $4 per watch, 12 watches will be both demanded and supplied. The supply and demand functions are graphed in Figure 2–16.

FIGURE 2–16

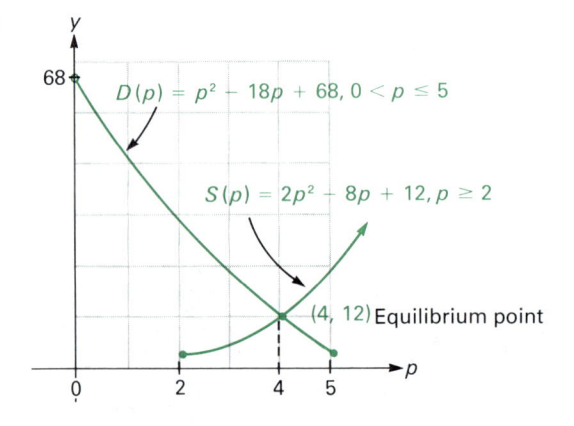

[Graph showing $D(p) = p^2 - 18p + 68$, $0 < p \le 5$ and $S(p) = 2p^2 - 8p + 12$, $p \ge 2$, intersecting at equilibrium point (4, 12).]

EXAMPLE 2–8 A park manager has 400 feet of fence which will be used to enclose a rectangular area, A, alongside a river. No fence is needed along the river (see Figure 2–17).

POLYNOMIAL AND RATIONAL FUNCTIONS

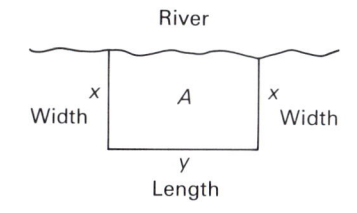

```
                    River
           ~~~~~~~~~~~~~~~~~~~~
         x |                | x
       Width|       A       | Width
           |                |
           ~~~~~~~~~~~~~~~~~~
                    y
                  Length
```

FIGURE 2–17

(A) Find the equation that expresses the enclosed area, A, exclusively in terms of x.
(B) If the park manager wishes to maximize the enclosed area, what must be its dimensions?

Solutions

(A) For a rectangle,

$$\text{area} = (\text{length})(\text{width})$$
$$A = yx$$

Note here that A is expressed in terms of the two variables x and y. That is, it is not expressed exclusively in terms of x. We now need a second equation that relates y with x. This equation can be solved for y in terms of x and the result substituted for y into our main equation, $A = yx$.

This second equation relating y and x is found by using the fact that the total length of the fence is 400 feet. Observing Figure 2–17, note that

$$\text{length of fence} = x + x + y$$
$$= 2x + y$$
$$400 = 2x + y$$

We now have our second equation relating y and x. Solving for y, we obtain

$$y = 400 - 2x$$

Substituting $400 - 2x$ for y into the equation $A = yx$ yields

$$A = (400 - 2x)(x)$$

Thus, A is now expressed exclusively in terms of x as

$$A = -2x^2 + 400x$$

The graph of this quadratic function appears in Fig-

EXAMPLE 2–8 (continued)

ure 2–18. Since x and A are nonnegative, the parabola is drawn as a solid line in the first quadrant.

FIGURE 2–18

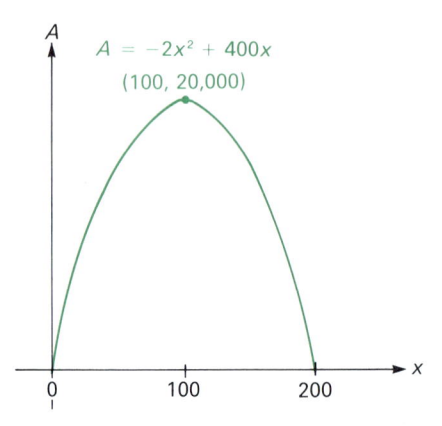

$A = -2x^2 + 400x$
(100, 20,000)

(B) Referring to Figure 2–18, the vertex, (100, 20,000), is the maximum point of the parabola. Thus, when x = 100 feet, the enclosed area will be maximized at 20,000 square feet. Recall that the length y of the enclosed area is given by the equation

$$y = 400 - 2x$$

Therefore, when x = 100,

$$y = 400 - 2(100) = 200 \text{ feet}$$

Hence, the dimensions of the maximum enclosed area are 100 feet by 200 feet.

EXAMPLE 2–9 In the face of risk and uncertainty, a decision maker's attitude toward gaining or losing various amounts of money represents that person's **utility** for money. Often, decision makers define a function that assigns utility values to monetary values. Such a function is called a **utility function**. The utility values indicate the subjective true worth of various amounts of money to a decision maker in terms of an abstract unit of measure called a **utile**.

As an example, the equation

$$U(x) = -0.3x^2 + 3x + 0.5 \qquad (-1 \leq x \leq 5)$$

87 POLYNOMIAL AND RATIONAL FUNCTIONS

defines a utility function that assigns utility values of $U(x)$ to various investment payoffs x (in millions of dollars).

(A) Calculate the utility value for an investment payoff of $2 million.

(B) What investment payoff maximizes utility, and what is the maximum utility value?

Solutions

(A) The utility value here is calculated as

$$U(2) = -0.3(2^2) + 3(2) + 0.5 = 5.3$$

Thus, an investment payoff of $2 million is assigned a utility value of 5.3 utiles. The point is indicated in Figure 2–19, which shows the graph of $U(x)$.

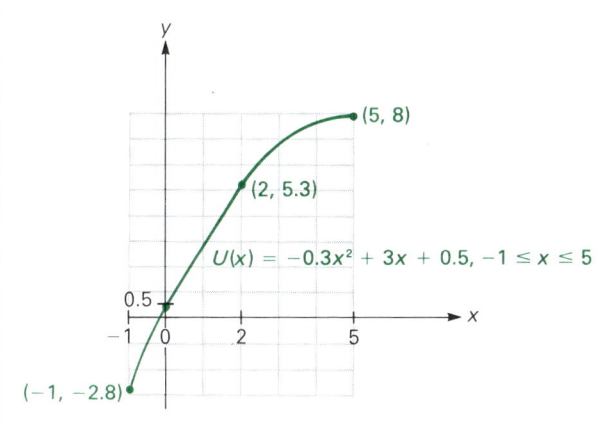

FIGURE 2–19

(B) Observing Figure 2–19, note that the vertex, (5, 8), is the maximum point on the parabola. Thus, an investment payoff of $5 million yields the maximum utility value of 8 utiles.

EXAMPLE 2–10 (This problem appeared on a past Uniform CPA Examination.) MacKenzie Park sells its trivets for $0.25 per unit and during 1969 has net sales of $500,000 and a net income of $35,000. Production capacity is limited to 15,000 trivets per day and trivets are produced 300 days each year. The variable cost is $0.10 per trivet.

The company does not maintain an inspection system but has an agreement to reimburse the wholesaler $0.50 for each defective unit the wholesaler finds. The wholesaler

EXAMPLE 2–10 (continued)

uses a method of inspection that detects all defective units. The number of defective units in lots of 300 units is equal to the daily unit production rate divided by 200. Let x = daily production in units.

(A) Determine the algebraic expression that represents the number of defective units per day.
(B) Determine the function that expresses the total daily contribution to profit, including the reimbursement to the wholesaler for defective units.
(C) What is the maximum daily profit? How many units are produced daily in order to yield the maximum daily profit?

Solutions

(A) The number of lots of size 300 is $x/300$. The number of defective units in each of the $x/300$ lots is $x/200$. Thus, the number of defective units per day is

$$\frac{x}{200} \cdot \frac{x}{300} = \frac{x^2}{60,000}$$

(B) First, we have

profit = sales revenue − cost − reimbursement

$$P(x) = 0.25x - 0.10x - 0.50\left(\frac{x^2}{60,000}\right)$$

$$= 0.15x - \frac{x^2}{120,000}$$

Thus, the profit function is defined by

$$P(x) = -\frac{x^2}{120,000} + 0.15x \quad (0 \leq x \leq 15,000)$$

Its graph appears in Figure 2–20. Since $0 \leq x \leq 15,000$, the parabola is drawn as a solid line over this interval.

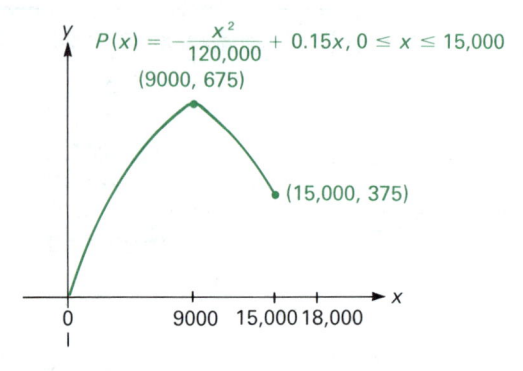

FIGURE 2–20

$P(x) = -\frac{x^2}{120,000} + 0.15x, \; 0 \leq x \leq 15,000$

(9000, 675)

(15,000, 375)

0   9000   15,000 18,000

POLYNOMIAL AND RATIONAL FUNCTIONS

> (C) As seen in Figure 2-20, the vertex, (9000, 675), is the maximum point on the parabola. Thus, when 9000 units are produced daily, the daily profit will be maximized at $675.

EXERCISES

1. A manufacturer of tricycles finds that the demand, x, for tricycles is related to the price per tricycle, p, by the equation

$$p = -2x + 40$$

   (A) Determine the equation that expresses total sales revenue, R, as a function of x.
   (B) Graph the revenue function of part (A).
   (C) Determine the maximum sales revenue.
   (D) How many tricycles must be sold in order to maximize sales revenue?

2. Assume that the firm in Exercise 1 has fixed costs of $72 and a variable cost of $10 per tricycle.
   (A) Find the cost equation.
   (B) Graph both cost and sales revenue functions on the same set of axes.
   (C) Calculate and interpret the break-even points.
   (D) Find the profit equation.
   (E) Graph the profit equation on a separate set of axes.
   (F) Find the maximum profit.
   (G) How many units must be produced and sold in order to maximize profit?
   (H) If the firm produces and sells 8 units, what is the corresponding profit?

3. The Great Taste Cookie Company has determined the demand function for its cookies to be defined by the linear equation

$$q = -3p + 48$$

   where p = selling price per box of cookies and q = number of boxes of cookies demanded.
   (A) Determine the sales revenue equation, $R(p)$.
   (B) Graph the sales revenue function.
   (C) Find the maximum sales revenue.
   (D) At how much should a package of cookies be priced in order to maximize sales revenue?

4. The GBW Waffle Company has defined the demand function for its package of waffles by the linear equation

$$q = -3p + 24$$

   where p = selling price per package of waffles demanded.
   (A) Determine the sales revenue equation, $R(p)$.
   (B) Graph the sales revenue function.
   (C) Find the maximum sales revenue.

(D) What price should be given to a package of waffles in order to maximize sales revenue?

5. The Haskins Company manufactures ornamental bells. The equation

$$C(x) = x^2 - 100x + 2900$$

relates total daily production cost, C, with daily production, x, of bells.
   (A) Graph this cost function.
   (B) How many bells should the company produce daily in order to minimize daily production cost?
   (C) What is the minimum daily production cost?
   (D) At a production level of $x = 60$ bells, one additional bell will cost how much to produce?

6. The supply, $S(p)$, of jewelry rings is given by the equation

$$S(p) = 2p^2 + 12p + 20 \qquad (0 \leq p \leq 5)$$

where p = price per ring. The demand, $D(p)$, is given by the equation

$$D(p) = p^2 - 10p + 28 \qquad (0 \leq p \leq 5)$$

   (A) Graph both supply and demand functions on the same set of axes.
   (B) Find the equilibrium point.
   (C) State the equilibrium price and the equilibrium demand.

7. The supply, Y_s, of scarfs is given by the equation

$$Y_s = 1.5x^2 - 12x + 26.5 \qquad (x \geq 4)$$

where x = price per scarf. The demand is

$$Y_d = 0.5x^2 - 10x + 61.5 \qquad (0 \leq x \leq 10)$$

   (A) Graph both supply and demand equations on the same set of axes.
   (B) Find the equilibrium point.
   (C) State the equilibrium price and the equilibrium demand.

8. A farmer has 800 feet of fence which will be used to enclose a rectangular area, A, alongside a river. No fence is needed along the river.

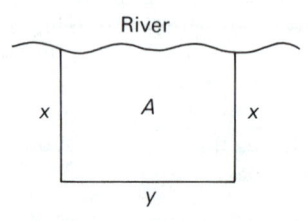

   (A) Find the equation that expresses the enclosed area, A, exclusively in terms of x.
   (B) Graph the equation of part (A).

91 POLYNOMIAL AND RATIONAL FUNCTIONS

(C) For what values of x and y will A be maximum? What is the maximum value of A?

9. A park manager has 2000 feet of fence with which to enclose a rectangular area, A.
   (A) Find the equation that expresses the enclosed area, A, exclusively in terms of x, the length of one of the sides.
   (B) Graph the equation of part (A).
   (C) Find the dimensions of the rectangle that yield the maximum area. What is the maximum area?

10. Show that the maximum area enclosed by a rectangle will always result in a square. (*Hint:* Follow the procedure of Exercise 9 except let L represent the length of the fence.)

11. A ball is projected vertically into the air. The function defined by
    $$S(t) = -16t^2 + 192t$$
    gives the height of the ball (in feet) above 0 at time t (in seconds).
    (A) Graph this function.
    (B) When does the ball reach maximum height? What is the maximum height?
    (C) When is the ball at zero height?

12. The Container Corporation manufactures wooden barrels. Each barrel costs $20 to produce and sells for $27. The quality control manager has observed that in each lot of 500 barrels, there are x/20 defectives, where x = monthly production volume. Each defective barrel costs the company an additional $5.
    (A) Find the equation that relates profit, P, with monthly production volume, x.
    (B) Graph the equation of part (A).
    (C) For what monthly production volume, x, is profit, P, maximum? What is the maximum profit?

13. If an apple grower harvests his crop now, he will pick, on the average, 120 pounds per tree. He will get $0.48 per pound for his apples. From past experience, he knows that for each additional week he waits, the yield per tree will increase by about 10 pounds, while the price will decrease by about $0.03 per pound.
    (A) Express total revenue, R, in terms of x, the number of weeks he waits.
    (B) Graph the equation of part (A).
    (C) How many weeks should the grower wait in order to maximize sales revenue? What is the maximum sales revenue?

14. An apartment complex contains 100 apartments. If the rent is $300 per month, all apartments will be rented. However, for each additional $50 increase in the monthly rent, 5 additional apartments will become vacant.
    (A) Find the equation relating total monthly rental income, R, with the number, x, of $50 increases in monthly rent.
    (B) Graph the equation of part (A).

(C) At what monthly rent is the total monthly rental income maximized? What is the maximum total monthly rental income?

15. A corporation executive's utility for money is given by the utility function defined by

$$U(x) = -0.4x^2 + 5.6x - 9.6 \quad (0 \leq x \leq 7)$$

where x is given in millions of dollars.

(A) Calculate the utility value for an investment payoff of $5 million.
(B) A utility value of 0 utiles corresponds to an investment payoff of how many million dollars?
(C) What investment payoff maximizes utility? What is the maximum utility value?

2-4 POLYNOMIAL FUNCTIONS

In Chapter 1, we discussed linear functions. Recall that a linear function is defined by an equation of the form

$$f(x) = mx + b$$

Since the highest powered term, mx, is of the first degree, linear functions are sometimes called **first-degree polynomial functions**. The graphs of first-degree polynomial functions are straight lines, as we learned in Chapter 1.

In Sections 2-2 and 2-3, we discussed quadratic functions. Recall that a quadratic function is defined by an equation of the form

$$f(x) = ax^2 + bx + c \quad (a \neq 0)$$

Since the highest powered term, ax^2, is of the second degree, quadratic functions are sometimes called **second-degree polynomial functions**. The graphs of second-degree polynomial functions are parabolas (see Figure 2-21), as we learned in Sections 2-2 and 2-3. Note that a second-degree quadratic function (a parabola) has one point, the vertex, where its graph changes direction (see Figure 2-21). Also, a second-degree polynomial function has at most two x-intercepts.

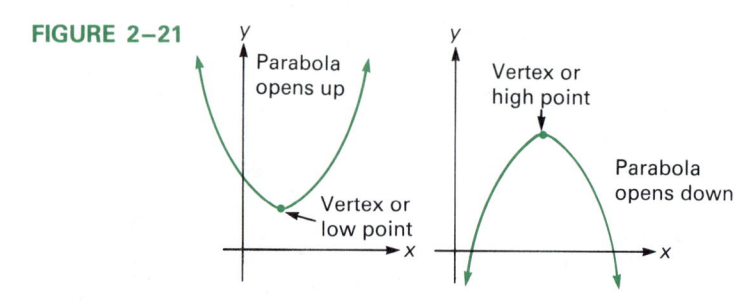

FIGURE 2-21

POLYNOMIAL AND RATIONAL FUNCTIONS

An equation such as

$$f(x) = x^3 - 5x^2 + 3x + 9$$

defines a **third-degree polynomial function** since the highest powered term, x^3, is of the third degree. The general form of the equation of a third-degree polynomial function is

$$f(x) = a_3x^3 + a_2x^2 + a_1x + a_0$$

where a_3, a_2, a_1, and a_0 are constant real numbers and $a_3 \neq 0$. Figure 2–22 illustrates possible shapes of graphs of third-degree polynomial functions. The points labeled A, B, C, and D in Figure 2–22 are places where the graph of $f(x)$ changes direction. A third-degree polynomial function has at most two points where its graph changes direction, and at most three x-intercepts.

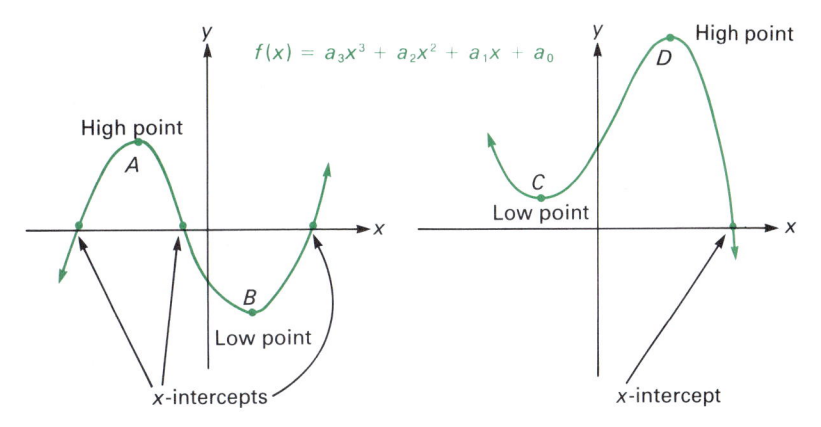

FIGURE 2–22

In general, an equation of the form

$$f(x) = a_nx^n + a_{n-1}x^{n-1} + \ldots + a_2x^2 + a_1x + a_0$$

where n is a positive integer, a_n, a_{n-1}, ..., a_2, a_1, a_0 are constant real numbers, and $a_n \neq 0$, defines the **nth-degree polynomial function**. An nth-degree polynomial function has at most n − 1 points where its graph changes direction, and at most n x-intercepts. In this section, we will develop and illustrate a procedure for sketching the graphs of factorable polynomial equations of at least the third degree.

We now consider the function defined by

$$f(x) = 2x^3 + 2x^2 - 10x + 6$$

This equation may be written in factored form as

$$f(x) = 2(x + 3)(x - 1)^2$$

Note that the y-intercept is $f(0) = 6$. To determine the x-intercepts, we set $f(x) = 0$. Hence,

$$0 = 2(x + 3)(x - 1)^2$$

Setting each factor equal to 0 and solving for x yields

$$x + 3 = 0 \qquad (x - 1)^2 = 0$$
$$x = -3 \qquad x = 1$$

Thus, the x-intercepts are $(-3, 0)$ and $(1, 0)$.

The x-intercepts divide the x-axis into various subintervals. Important information about the graph of $f(x)$ is obtained by analyzing the sign of $f(x)$ on these subintervals. Figure 2–23 shows the sign of each factor of $f(x)$ and the sign of $f(x)$ for values of x. Studying this figure, note that the sign of $x + 3$ is negative for $x < -3$ and positive for $x > -3$. Of course, $x + 3 = 0$ at $x = -3$. Observe that the sign of $(x - 1)^2$ is positive at all values of x except $x = 1$, where it is 0. Multiplying the sign of $x + 3$ by the sign of $(x - 1)^2$, we obtain the sign of $(x + 3)(x - 1)^2$. The sign of $(x + 3)(x - 1)^2$ is the same as the sign of $f(x) = 2(x + 3)(x - 1)^2$ since the positive constant multiplier 2 does not change the sign of the original product, $(x + 3)(x - 1)^2$. Thus, $f(x)$ is nonnegative for $x > -3$ and negative for $x < -3$.

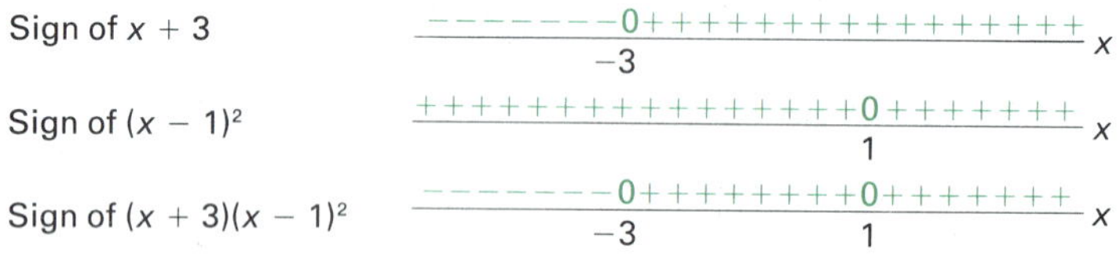

Sign of $x + 3$

Sign of $(x - 1)^2$

Sign of $(x + 3)(x - 1)^2$

FIGURE 2–23

Notice that $f(x)$ changes sign at the x-intercept of $x = -3$. Thus, the graph of $f(x)$ *crosses* the x-axis at $x = -3$. However, since $f(x)$ does not change sign at the x-intercept of $x = 1$, the graph of $f(x)$ *is tangent to,* and does not cross, the x-axis at $x = 1$ (see Figure 2–24). Note that $f(x)$ does not change sign at $x = 1$ because the corresponding factor $(x - 1)^2$ has an *even exponent.* Similarly, $f(x)$ does change sign at $x = -3$ because the factor $x + 3$ has an *odd exponent.* In general, we state the following rule to determine whether the graph of a function either crosses or is tangent to the x-axis at an x-intercept.

95 POLYNOMIAL AND RATIONAL FUNCTIONS

$(x + 3)^1 = 0$
Odd exponent:
Function *crosses* here

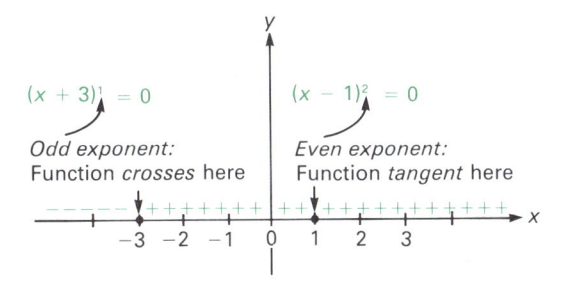

$(x - 1)^2 = 0$
Even exponent:
Function *tangent* here

FIGURE 2-24

> If the factor yielding the x-intercept has an *odd exponent*, then the graph of the function *crosses* the x-axis at that x-intercept. If the factor yielding the x-intercept has an *even exponent*, then the graph of the function *is tangent to*, and does not cross, the x-axis at that x-intercept.

Lastly, we determine the behavior of $f(x)$ as x increases without limit and as x becomes more and more negative without limit. We will show that the highest powered term determines the behavior of a polynomial function at such values of x. We begin by factoring out $2x^3$, the highest powered term of

$$f(x) = 2x^3 + 2x^2 - 10x + 6$$

and obtain

$$f(x) = 2x^3\left(1 + \frac{1}{x} - \frac{5}{x^2} + \frac{3}{x^3}\right)$$

As x takes on larger and larger positive values, the terms $1/x$, $-5/x^2$, and $3/x^3$ all approach 0. Thus, the highest powered term, $2x^3$, determines the behavior of $f(x)$ as x increases without limit. Since the highest powered term, $2x^3$, takes on larger and larger positive values as x takes on larger and larger positive values, $f(x)$ gets larger and larger as x gets larger and larger. This implies that for positive values of x far away from the origin, the graph of $f(x)$ is located in the upper right-hand corner of the first quadrant, as illustrated in Figure 2-25.

As x takes on negative values of larger and larger magnitude, the highest powered term, $2x^3$, takes on negative values of larger and larger magnitude. Thus, $f(x)$ becomes more and more negative without limit as x becomes more and more negative without

CHAPTER TWO

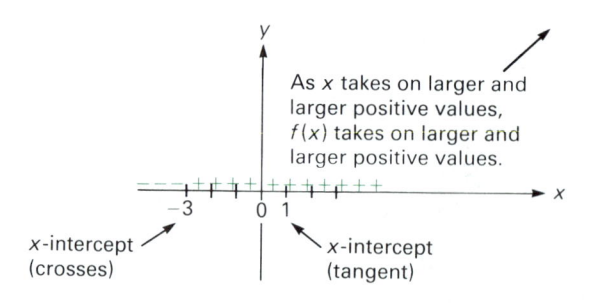

FIGURE 2-25

(As x takes on larger and larger positive values, f(x) takes on larger and larger positive values. x-intercept (crosses) at -3; x-intercept (tangent) at 1.)

limit. This implies that for negative values of x far away from the origin, the graph of $f(x)$ is located in the lower left-hand corner of the third quadrant, as is shown in Figure 2–26.

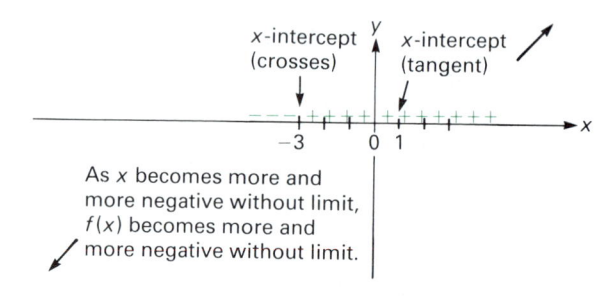

FIGURE 2-26

(x-intercept (crosses) at -3; x-intercept (tangent) at 1. As x becomes more and more negative without limit, f(x) becomes more and more negative without limit.)

Combining the information gained in the previous paragraphs, we sketch the graph of $f(x)$ in Figure 2–27. In Figure 2–27, we have labeled as A and B the points where the graph of $f(x)$ changes direction. The table of x and $f(x)$ values in the figure is used to approximate the location of point A.

$f(x) = 2x^3 + 2x^2 - 10x + 6$

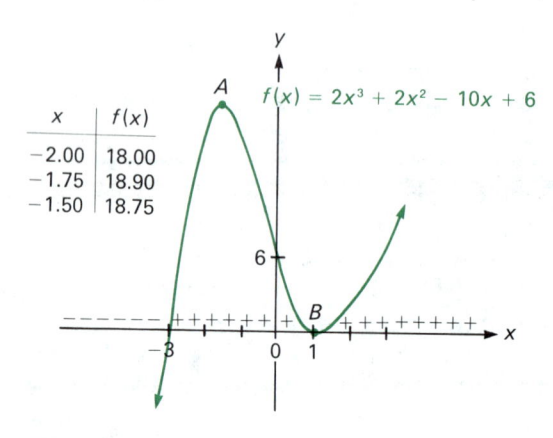

x	f(x)
-2.00	18.00
-1.75	18.90
-1.50	18.75

FIGURE 2-27

POLYNOMIAL AND RATIONAL FUNCTIONS

EXAMPLE 2–11
Graph $f(x) = x^4 + 4x^3 - 3x^2 - 18x = x(x - 2)(x + 3)^2$.
Solution The y-intercept is $f(0) = 0$. To determine the x-intercepts, we set $f(x) = 0$. Hence,

$$0 = x(x - 2)(x + 3)^2$$

Thus, the x-intercepts are $(0, 0)$, $(2, 0)$, and $(-3, 0)$. Since the factors yielding the x-intercepts $(0, 0)$ and $(2, 0)$ have odd exponents, the graph of $f(x)$ crosses the x-axis at these x-intercepts. Since the factor yielding the x-intercept $(-3, 0)$ has an even exponent, the graph of $f(x)$ is tangent to the x-axis at $(-3, 0)$. Figure 2–28 shows the sign of each factor of $f(x)$ and the sign of $f(x)$ for values of x.

Sign of x   ``----------------0++++++++++++++`` x
                                              0

Sign of x − 2   ``-------------------0+++++++++`` x
                                                 2

Sign of $(x + 3)^2$ ``+++++0++++++++++++++++++++++++`` x
                                −3

Sign of $f(x)$ ``+++++0+++++++0----0+++++++++`` x
                                −3   0   2

FIGURE 2–28

Then, we determine the behavior of $f(x)$ as x increases without limit. Note that the highest powered term, x^4, takes on larger and larger positive values as x takes on larger and larger positive values. This implies that for positive values of

x	y
−1.4	12.19
−1.2	12.44
−1.0	12.00

High point A

$y = f(x) = x^4 + 4x^3 - 3x^2 - 18x$

y-intercept

x-intercept (tangent)   x-intercept (crosses)

B Low point

x-intercept (crosses)

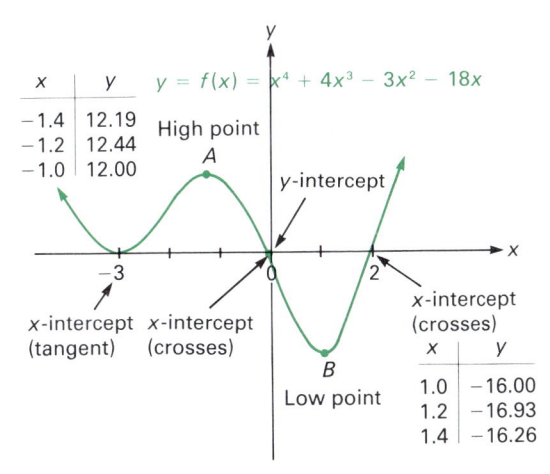

x	y
1.0	−16.00
1.2	−16.93
1.4	−16.26

FIGURE 2–29

EXAMPLE 2–11 (continued)

x far away from the origin, the graph of $f(x)$ is located in the upper right-hand corner of the first quadrant.

Also, as x takes on negative values of larger and larger magnitude, the highest powered term, x^4, takes on larger and larger positive values. This implies that for negative values of x far away from the origin, the graph of $f(x)$ is located in the upper left-hand corner of the second quadrant.

Combining all of the preceding information, we sketch the graph of $f(x)$ in Figure 2–29 (page 97). Observe that we have labeled as A and B the points where the graph of $f(x)$ changes direction. The tables of x and y values of Figure 2–29 are used to approximate the location of points A and B.

EXERCISES

1. Graph each of the following:
   - (A) $f(x) = x^3 - 7x + 6 = (x - 1)(x + 3)(x - 2)$
   - (B) $y = x^4 + x^3 - 21x^2 - x + 20 = (x - 1)(x - 4)(x + 1)(x + 5)$
   - (C) $f(x) = x^3 - 7x^2 + 11x - 5 = (x - 1)^2(x - 5)$
   - (D) $f(x) = x^3 - 4x^2 - 3x + 18 = (x - 3)^2(x + 2)$
   - (E) $y = 3x^4 - 3x^3 - 9x^2 + 15x - 6 = 3(x - 1)^3(x + 2)$
   - (F) $y = 5x^6 + 10x^5 - 15x^4$

2. Graph each of the following:
   - (A) $f(x) = x^3 + 3x^2 - 9x + 5 = (x - 1)^2(x + 5)$
   - (B) $y = x^4 + 2x^3 - 15x^2$
   - (C) $f(x) = -x^3 + 36x$
   - (D) $f(x) = x^4 - 2x^3 + 15x^2$
   - (E) $f(x) = x^3 - 19x - 30 = (x + 2)(x - 5)(x + 3)$
   - (F) $f(x) = x^3 - 4x^2 - 3x + 18 = (x + 2)(x - 3)^2$
   - (G) $f(x) = x^3 + x^2 - 5x + 3 = (x - 1)^2(x + 3)$
   - (H) $f(x) = x^5 + x^4 - 5x^3 + 3x^2$
   - (I) $y = x^3 + 3x^2 - 16x + 12 = (x - 1)(x + 6)(x - 2)$
   - (J) $f(x) = x^4 + 6x^3 - 7x^2 - 60x = x(x - 3)(x + 4)(x + 5)$

3. Each of the following is a polynomial equation $f(x) = a_n x^n + a_{n-1} x^{n-1} + \ldots + a_2 x^2 + a_1 x + a_0$ expressed in factored form. Graph each one.
   - (A) $f(x) = \frac{1}{2}(x - 2)^3(x + 5)$
   - (B) $y = -\frac{1}{4}(x - 3)^2(x + 1)$
   - (C) $f(x) = -3x^5(x - 2)^3(x + 4)$
   - (D) $f(x) = x^2(x - 2)(x + 5)$

4. Each of the following defines a polynomial function expressed in factored form. Graph each one.

99 POLYNOMIAL AND RATIONAL FUNCTIONS

(A) $f(x) = 2x^3(x - 5)(x + 6)^2$
(B) $f(x) = x^2(x - 5)^4(x - 2)$
(C) $y = (x - 2)(x + 3)^4(x - 1)$
(D) $y = (x - 1)^2(x + 3)(x + 1)^2$

2-5

APPLICATIONS OF POLYNOMIAL FUNCTIONS

In this section, various applications of polynomial functions will be presented.

EXAMPLE 2–12 *(Quality Control and Profits)* The Sharp Company manufactures table knives. Each knife costs $12 to produce and sells for $16. The quality control manager has determined from past data that the equation

$$d = \frac{x^2}{20{,}000{,}000}$$

relates the fraction of defectives produced, *d*, with daily production volume, *x*. Each defective knife costs the company an additional $20.
(A) Determine the equation that expresses daily profit, *P*, as a function of daily volume, *x*.
(B) Referring to Figure 2–30, for which values of *x* is $P(x)$ nonnegative?
(C) Referring to Figure 2–30, for which value of *x* within the interval $0 \leq x \leq 2000$ will the daily profit be maximal?

Solutions
(A) The number of defective knives produced in a day is found by multiplying the fraction defective by the daily production volume, or *d* times *x*, to obtain

$$\frac{x^2}{20{,}000{,}000} \cdot x = \frac{x^3}{20{,}000{,}000}$$

Since each defective knife costs an additional $20, then $x^3/20{,}000{,}000$ defective knives will cost an additional

$$20\left(\frac{x^3}{20{,}000{,}000}\right) = \frac{x^3}{1{,}000{,}000}$$

A unit profit of $16 − $12 = $4 is made on each knife.

EXAMPLE 2-12 (continued)

Thus, if x knives are produced and sold, the company makes a daily profit of

$$P(x) = 4x - \begin{pmatrix} \text{additional cost of the} \\ \text{defective knives} \end{pmatrix}$$

Hence,

$$P(x) = 4x - \frac{x^3}{1,000,000}$$

The graph of this function appears in Figure 2-30. Note that since x and P are nonnegative, the function is drawn as a solid line in the first quadrant.

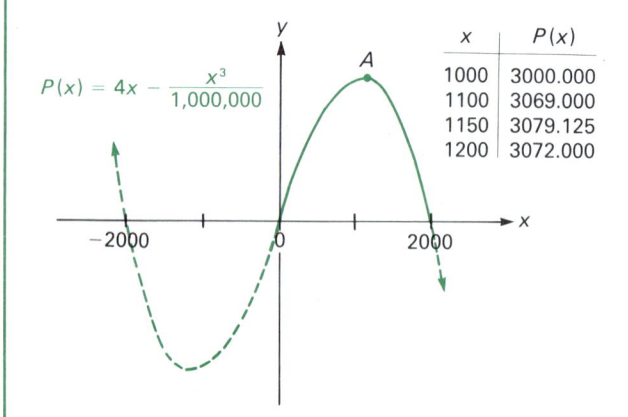

x	$P(x)$
1000	3000.000
1100	3069.000
1150	3079.125
1200	3072.000

FIGURE 2-30

(B) If $0 \leq x \leq 2000$, then $P(x) \geq 0$. Thus, the company's daily profit will be nonnegative as long as daily production does not exceed 2000 knives.

(C) The daily profit will be maximal at the x-coordinate of point A. Using the table of x and $P(x)$ values of Figure 2-30, we see that the x-coordinate of point A is approximately 1150.

EXAMPLE 2-13 (Volume)
A company manufactures open boxes by beginning with a square piece of tin 16 inches on each side, cutting equal squares from the corners of this piece of tin, and folding up the flaps to form sides (see Figure 2-31).

101 POLYNOMIAL AND RATIONAL FUNCTIONS

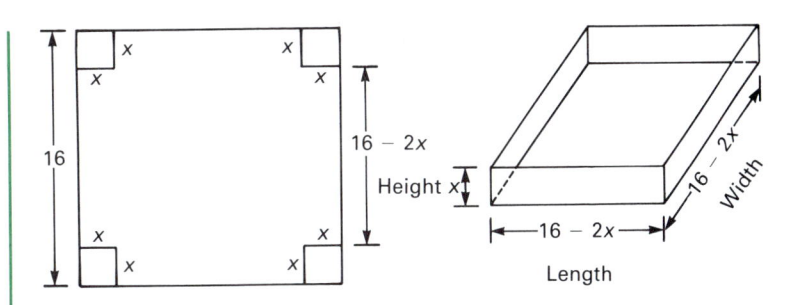

FIGURE 2–31

(A) Express the volume, V, of the box as a function of x, the length of each side of the cut-out square.
(B) Referring to Figure 2–32, for what values of x is V(x) nonnegative? Does this make sense in terms of Figure 2–31?
(C) Referring to Figure 2–32, for which positive value of x within the interval $0 \leq x \leq 8$ will the volume of the box be minimal?
(D) Referring to Figure 2–32, for which value of x within the interval $0 \leq x \leq 8$ will the volume of the box be maximal?

FIGURE 2–32

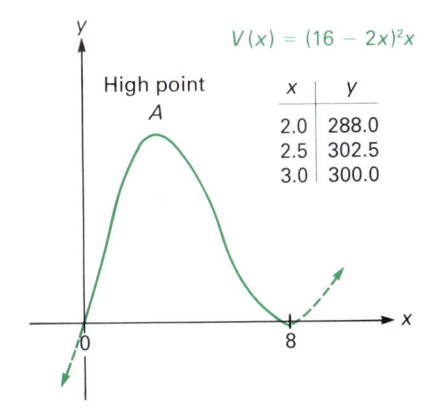

$V(x) = (16 - 2x)^2 x$

x	y
2.0	288.0
2.5	302.5
3.0	300.0

Solutions

(A) The volume of a box is the product of its length, width, and height (see Figure 2–31). Thus,

$$V(x) = (\text{length})(\text{width})(\text{height})$$
$$= (16 - 2x)(16 - 2x)(x)$$
$$= (16 - 2x)^2 x$$

The graph of this function appears in Figure 2–32.

102 CHAPTER TWO

EXAMPLE 2–13 (continued)

(B) If $x \geq 0$, then $V(x) \geq 0$. However, referring to Figure 2–31, note that if $x = 8$, then all the tin would be cut away. And if $x > 8$, then more tin would be cut away than exists—an impossibility. Hence, the function "makes sense" only for $0 \leq x \leq 8$. Thus, that portion of the graph is drawn with a solid line in Figure 2–32.
(C) It will be minimal at $x = 8$, the x-coordinate of the point (8, 0).
(D) It will be maximal at the x-coordinate of point A. Observing the table of x and $V(x)$ values of Figure 2–32, we see that the x-coordinate of point A is approximately 2.5.

EXAMPLE 2–14 *(Unit Pricing and Revenue)* A company manufactures jewelry beads. The function

$$p(x) = -\frac{1}{100} x^2 + 16$$

relates the selling price per case, p, with the number of cases sold, x.

(A) Determine the sales revenue, R, as a function of the number of cases sold, x.
(B) Referring to Figure 2–33, for which value of x will sales revenue, $R(x)$, be maximal?

FIGURE 2–33

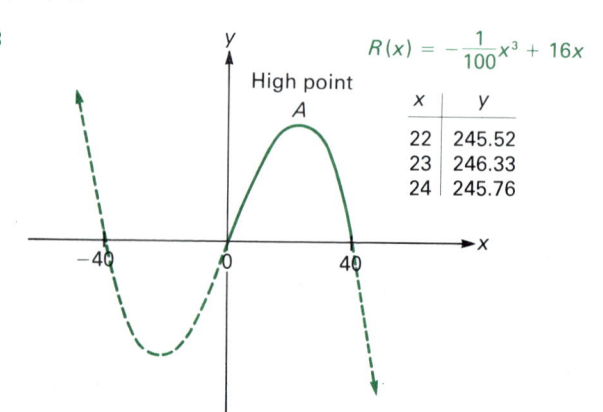

$R(x) = -\frac{1}{100}x^3 + 16x$

High point A

x	y
22	245.52
23	246.33
24	245.76

Solutions
(A) Since

$$\text{revenue} = \begin{pmatrix}\text{number of}\\\text{cases sold}\end{pmatrix}\begin{pmatrix}\text{price per}\\\text{case}\end{pmatrix}$$

then, we have

$$R = x \cdot p(x)$$
$$= x\left(-\frac{1}{100}x^2 + 16\right)$$

Hence,

$$R(x) = -\frac{1}{100}x^3 + 16x$$

The graph of this function appears in Figure 2–33.

(B) Sales revenue will be maximal at the x-coordinate of point A. The table of x and $R(x)$ values of Figure 2–33 shows that this x-coordinate is approximately 23.

EXAMPLE 2–15 *(Batch Process and Unit Profits)* The Super Oar Company manufactures plastic paddles using the batch process (i.e., no production takes place until enough orders are received to produce a batch). The size of a batch is variable and may range from 0 to 30 paddles. The company statistician has found unit profit, p, and batch size, x, to be related by the polynomial equation

$$p(x) = -0.0001x^4 + 0.005x^3 - 0.07x^2 + 0.3x \quad (0 \le x \le 30)$$

When factored, this becomes

$$p(x) = -0.0001x(x-10)^2(x-30)$$

The graph of this function is shown in Figure 2–34.

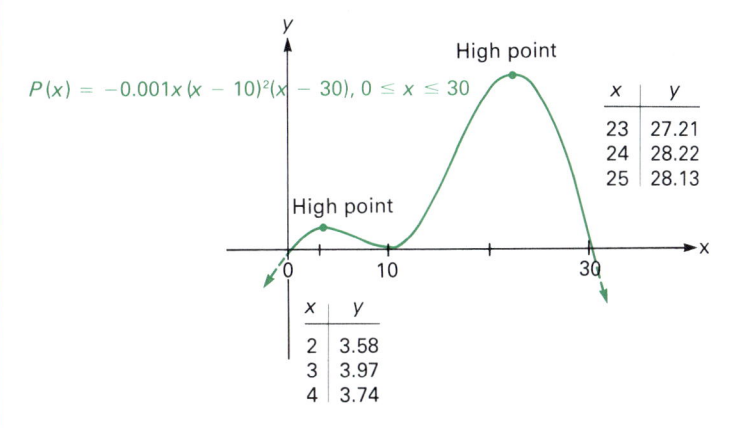

$P(x) = -0.001x(x-10)^2(x-30),\ 0 \le x \le 30$

x	y
23	27.21
24	28.22
25	28.13

x	y
2	3.58
3	3.97
4	3.74

FIGURE 2–34

EXAMPLE 2–15 (continued)

(A) For which batch sizes will the company make zero unit profit?
(B) Will Super Oar Company make a positive unit profit if batch size, x, falls within the interval $0 < x < 10$?
(C) Will Super Oar Company make a positive unit profit if batch size, x, falls within the interval $10 < x < 30$?
(D) Will the company make a positive unit profit if batch size, x, falls outside the interval $0 \leq x \leq 30$?
(E) Calculate the profit per paddle at a batch size of $x = 20$.

Solutions

(A) Zero unit profit occurs at batch sizes corresponding to the x-intercepts: $x = 0$, $x = 10$, and $x = 30$.
(B) Since the graph of $p(x)$ lies above the x-axis over the interval $0 < x < 10$, Super Oar Company makes a positive unit profit for those corresponding batch sizes.
(C) Since the graph of $p(x)$ lies above the x-axis over the interval $10 < x < 30$, Super Oar Company makes a positive unit profit for those corresponding batch sizes.
(D) Since the graph of $p(x)$ lies below the x-axis for values of x outside the interval $0 \leq x \leq 30$, Super Oar Company makes a negative profit (i.e., loses money) for those corresponding batch sizes. Thus, we are only interested in the graph of $p(x)$ over the interval $0 \leq x \leq 30$. That portion of the graph is drawn as a solid line in Figure 2–34.
(E) Since

$$p(x) = -0.0001x(x - 10)^2(x - 30)$$

then

$$p(20) = -0.0001(20)(20 - 10)^2(20 - 30)$$
$$= \$2$$

Thus, at a batch size of 20 paddles, the company makes a profit of $2 per paddle.

EXERCISES

1. The Time Company manufactures watches. Each watch costs $30 to produce and sells for $38. From past observations, it has been determined that the equation

$$d = \frac{x^2}{10,000,000}$$

relates the fraction of defectives produced, d, with daily production volume, x. Each defective watch costs the company an additional $5.
- (A) Determine the equation that expresses daily profit, P, as a function of daily production volume, x.
- (B) Graph the profit function, $P(x)$.
- (C) For which values of x is $P(x)$ nonnegative?
- (D) Graphically indicate the location of the value of x, within the interval $0 \leq x \leq 4000$, at which daily profit is maximized.

2. A company manufactures electric switches. The equation

$$p(x) = -\frac{1}{400}x^2 + 90 \qquad (0 \leq x \leq 185)$$

represents the selling price per case, p, as a function of the number of cases sold, x.
- (A) Determine the sales revenue, R, as a function of x.
- (B) Graph the revenue function, $R(x)$.
- (C) For which values of x is sales revenue nonnegative?
- (D) Graphically indicate the location of the value of x at which sales revenue is maximal.

3. The factored polynomial equation

$$p(x) = \left(\frac{1}{10,000}x^2\right)(x - 20)(x - 30)^2 \qquad (0 \leq x \leq 40)$$

relates unit profit, p, with batch size, x, for Supergo Corporation.
- (A) Sketch the graph of this function.
- (B) For which batch sizes will Supergo Corporation make zero unit profit?
- (C) Will Supergo make a unit profit if batch size, x, falls within the interval $0 \leq x \leq 20$?
- (D) Will Supergo make a unit profit if batch size, x, falls within the interval $20 < x < 30$?
- (E) Will Supergo make a unit profit if batch size, x, falls within the interval $30 < x \leq 40$?
- (F) Calculate the unit profit at each of the following batch sizes: $x = 25$, $x = 35$, $x = 40$.

4. A company manufactures open boxes from rectangular pieces of tin of dimension 10 inches by 20 inches. The process involves cutting equal squares from the corners of each piece of tin and folding up the flaps to form sides.
- (A) Express the volume, V, of the box as a function of x, the length of each side of the cut-out square.
- (B) Graph the function $V(x)$ of part (A).
- (C) For which values of x is volume, V, nonnegative? Does this make sense?
- (D) Graphically indicate the location of the value of x at which volume will be maximal.

5. A particle starts at point 0 and moves along a horizontal line as illustrated here:

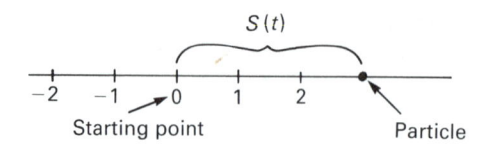

```
           S(t)
    ⌢⎯⎯⎯⎯⎯⎯⎯⎯⌢
─┼────┼────┼────┼────┼●────────
 −2  −1   0    1    2
      Starting point      Particle
```

The equation

$$S(t) = t^3 - 16t^2$$

expresses the distance $S(t)$ between the particle and its starting point, 0, after t seconds have elapsed.

(A) Graph $S(t)$.
(B) Where is the particle in relation to its starting point after 5 seconds have elapsed?
(C) Where is the particle after 20 seconds have elapsed?
(D) At what points in time is the particle at the starting point?

2-6

RATIONAL FUNCTIONS

In this section, we will learn to graph equations like

$$y = f(x) = \frac{6x + 12}{2x - 1}$$

Such an equation defines a **rational function** since it is a quotient of two polynomial functions.

A graph of $y = f(x) = (6x + 12)/(2x - 1)$ may be obtained by finding some ordered pairs (x, y) satisfying the equation $y = (6x + 12)/(2x - 1)$ and plotting their corresponding points on the rectangular coordinate system. Table 2–1 shows values of y, or $f(x)$, for various values of x. The corresponding ordered pairs (x, y) are plotted in Figure 2–35. Studying the graph of Figure 2–35, note that this rational function is undefined at $x = 1/2$. As x takes on values closer and closer to 1/2, $f(x)$ gets larger and larger in magnitude. The graph of $f(x)$ gets closer and closer to the vertical line $x = 1/2$. Such a vertical line is called a **vertical asymptote**. Note that *vertical asymptotes occur at values of x for which the denominator of a rational function = 0 and the numerator ≠ 0.*

POLYNOMIAL AND RATIONAL FUNCTIONS

TABLE 2-1

x	f(x)
-50	2.85
-10	2.29
-3	0.86
-2	0
0	-12
1/3	-42
1	18
3	6
50	3.15

$y = f(x) = \dfrac{6x + 12}{2x - 1}$

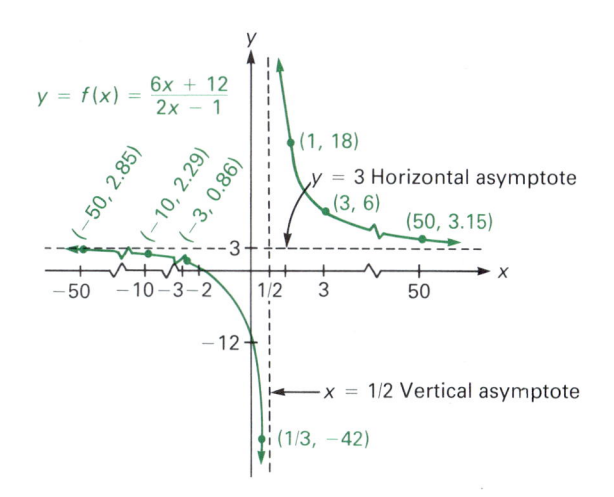

(1, 18)
(-50, 2.85)
(-10, 2.29)
(-3, 0.86)
$y = 3$ Horizontal asymptote
(3, 6)
(50, 3.15)
(1/3, -42)
$x = 1/2$ Vertical asymptote

FIGURE 2-35

Further study of the graph of Figure 2-35 shows us that as x takes on larger and larger positive values, $f(x)$ gets closer and closer to 3. Thus, for positive values of x far away from the origin, the graph of $f(x)$ approaches the horizontal line $y = 3$. As x takes on negative values of larger and larger magnitude, $f(x)$ again gets closer and closer to 3. Thus, for negative values of x far away from the origin, the graph of $f(x)$ also approaches the horizontal line $y = 3$. Such a horizontal line is called a **horizontal asymptote**. Given the equation of a rational function, the horizontal asymptote (if it exists) is found by determining the behavior of the rational function at values of x far away from the origin. We will see that the behavior of a rational function at values of x far away from the origin is determined by the quotient

$$\frac{\text{highest powered term of numerator}}{\text{highest powered term of denominator}}$$

We begin by factoring out the highest powered term of the numerator and also the highest powered term of the denominator of our rational function

$$y = f(x) = \frac{6x + 12}{2x - 1}$$

This operation gives us

$$y = f(x) = \frac{6x \left(1 + \dfrac{2}{x}\right)}{2x \left(1 - \dfrac{1}{2x}\right)}$$

As x takes on larger and larger positive values, the terms $2/x$ and $1/2x$ approach 0. Thus, the quotient

$$\frac{\text{highest powered term of numerator}}{\text{highest powered term of denominator}} = \frac{6x}{2x} = 3$$

determines the behavior of $f(x)$ as x takes on larger and larger positive values. Also, as x takes on negative values of larger and larger magnitude, the quotient of the highest powered terms of both numerator and denominator determines the behavior of the graph of $f(x)$ at negative values of x far away from the origin. For this example, the quotient is the constant 3, indicating that the horizontal asymptote is $y = 3$.

Although we sketched the rational function $y = f(x) = (6x + 12)/(2x - 1)$ by choosing arbitrary x-values and plotting their corresponding points, there is a more methodical procedure, outlined by the following steps:

Step 1 Find the y-intercept.
Step 2 Find the x-intercept(s).
Step 3 Find the vertical asymptote(s).
Step 4 Determine the behavior of the function as x takes on values farther and farther away from the origin.

We now illustrate this procedure by sketching the graph of the rational function defined by

$$y = f(x) = \frac{2(x-1)^3(x+4)}{(x-3)^2(x+1)}$$

Step 1 Find the y-intercept. Here, we use

$$y = f(0) = \frac{2(0-1)^3(0+4)}{(0-3)^2(0+1)} = \frac{2(-1)^3(4)}{(-3)^2(1)} = \frac{-8}{9}$$

Thus, the y-intercept is $(0, -8/9)$.

Step 2 Find the x-intercept(s). Set $y = 0$. Hence,

$$0 = \frac{2(x-1)^3(x+4)}{(x-3)^2(x+1)}$$

$$0 = 2(x-1)^3(x+4)$$

$(x-1)^3 = 0 \qquad\qquad (x+4)^1 = 0$

$x = 1 \qquad\qquad\qquad x = -4$

odd exponent: graph *crosses* x-axis here

odd exponent: graph *crosses* x-axis here

109 POLYNOMIAL AND RATIONAL FUNCTIONS

Step 3 Find the vertical asymptote(s). Vertical asymptotes exist at those values of x for which the denominator of a rational function equals 0. Thus, we set the denominator equal to 0. This gives us

$$0 = (x - 3)^2(x + 1)$$

$$(x - 3)^2 = 0 \qquad x + 1 = 0$$
$$x = 3 \qquad\qquad x = -1$$

vertical asymptotes

Before leaving step 3, we must determine whether the function approaches the *same end* of the vertical asymptote from both sides, or whether it approaches *different ends* of the vertical asymptote from both sides (see Figure 2–36). The behavior of a rational function at a vertical asymptote is determined by noting whether the exponent of the factor from which the vertical asymptote was calculated is odd or even. The following rule is applied.

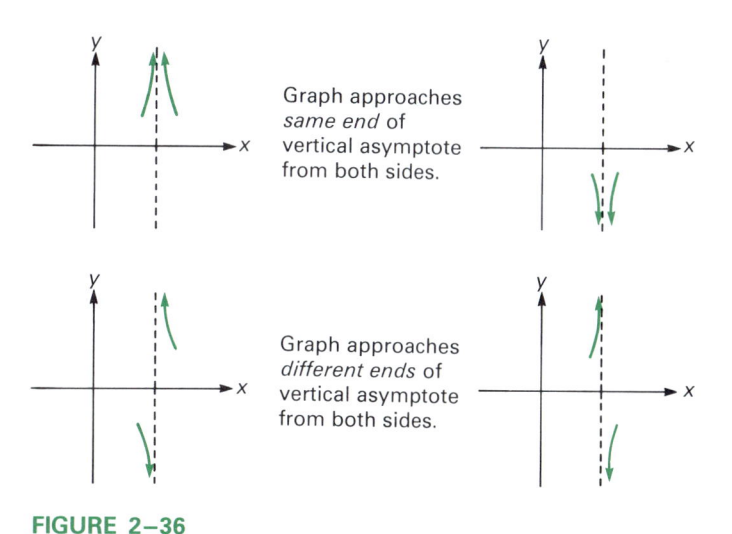

Graph approaches *same end* of vertical asymptote from both sides.

Graph approaches *different ends* of vertical asymptote from both sides.

FIGURE 2–36

If the exponent of the factor from which the vertical asymptote was calculated is *even*, then the graph of the rational function will approach the *same end* of the vertical asymptote from both sides, as illustrated in Figure 2–37.

FIGURE 2-37

Even exponent: Graph approaches *same end* of vertical asymptote from both sides

If the exponent of the factor from which the vertical asymptote was calculated is *odd*, then the graph of the rational function will approach *different ends* of the vertical asymptote from both sides, as illustrated in Figure 2-38.

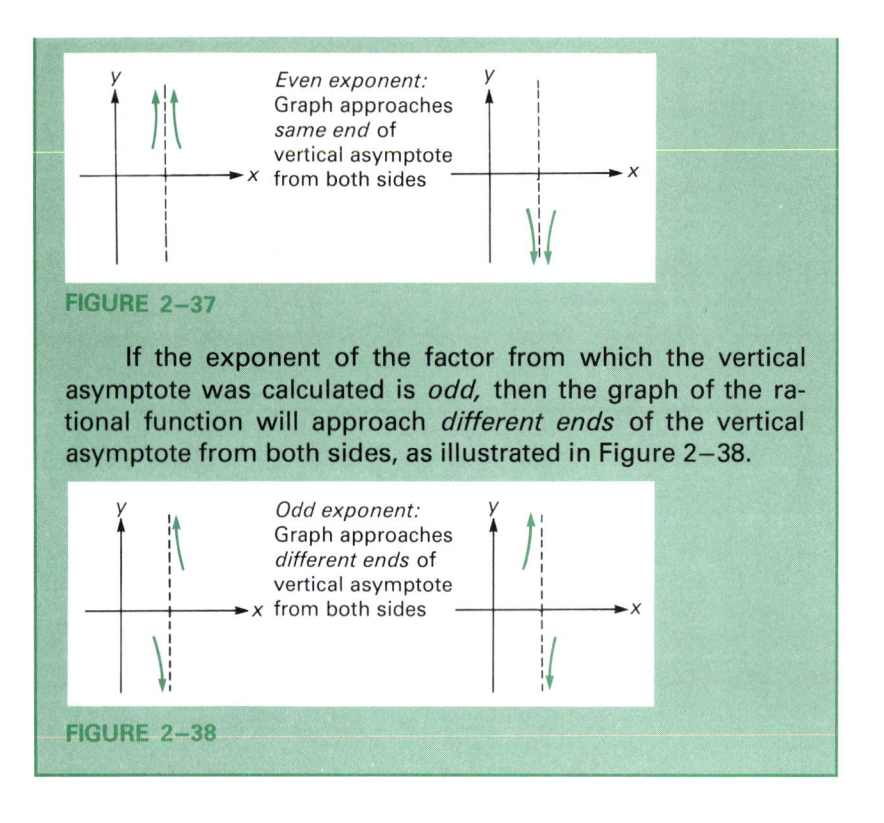

Odd exponent: Graph approaches *different ends* of vertical asymptote from both sides

FIGURE 2-38

Returning to our example, note that the vertical asymptotes were calculated as follows:

$$(x-3)^2 = 0 \qquad (x+1)^1 = 0$$
$$x = 3 \qquad\qquad x = -1$$

even exponent: graph approaches *same end* of vertical asymptote $x = 3$ from both sides

odd exponent: graph approaches *different ends* of vertical asymptote $x = -1$ from both sides

One can see why this rule works by noting that for the rational function defined by

$$f(x) = \frac{2(x-1)^3(x+4)}{(x-3)^2(x+1)}$$

the values of $f(x)$ do not change sign as x takes on values from either side of 3 since the exponent of the factor $(x-3)^2$ is 2, an

111 POLYNOMIAL AND RATIONAL FUNCTIONS

even number. Hence, $f(x)$ approaches the vertical asymptote $x = 3$ from the same end. However, for values of x smaller than -1, the factor $x + 1$ is negative, whereas for values of x larger than -1, the factor $x + 1$ is positive, thereby causing a sign change in $f(x)$. Hence, $f(x)$ approaches the vertical asymptote $x = -1$ from different ends.

Step 4 Determine the behavior of the function as x takes on values farther and farther away from the origin. The quotient of the highest powered terms of both numerator and denominator determines the behavior of $f(x)$ at values of x far away from the origin. In our example, this quotient is

$$\frac{2x^4}{x^3} = 2x$$

As x takes on larger and larger positive values, the quotient $2x$ takes on larger and larger positive values. Thus, for positive values of x far away from the origin, the graph of $f(x)$ is located in the upper right-hand corner of the first quadrant. As x takes on negative values of larger and larger magnitude, the quotient $2x$ becomes more and more negative. Thus, for negative values of x, the graph of $f(x)$ is located in the lower left-hand corner of the third quadrant.

The information obtained in the preceding four steps is illustrated in Figure 2–39. The graph of $f(x)$ appears in Figure 2–40.

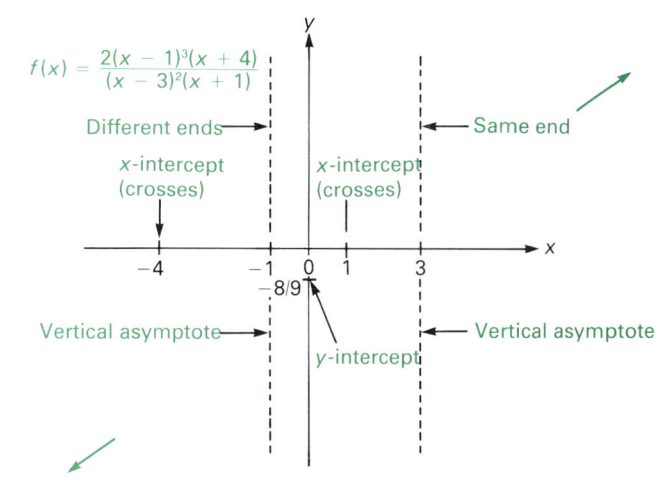

FIGURE 2–39

112 CHAPTER TWO

$$y = f(x) = \frac{2(x-1)^3(x+4)}{(x-3)^2(x+1)}$$

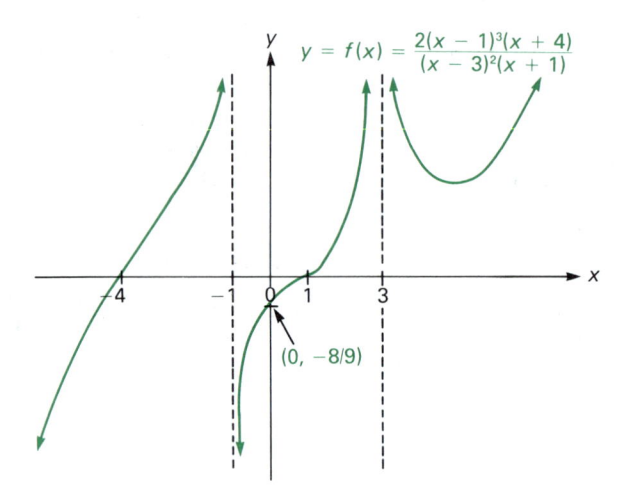

(0, −8/9)

FIGURE 2–40

EXAMPLE 2–16 The equation

$$y = f(x) = \frac{20x}{104 - x} \quad (0 \leq x \leq 100)$$

expresses the cost, y (in thousands of dollars), of removing $x\%$ of a certain pollutant from the atmosphere of a large city. The graph of such a function is called a **cost–benefit curve.**
(A) Sketch the cost–benefit curve.
(B) Find the cost of removing 90% of the pollutant.
(C) Find the cost of removing 95% of the pollutant.
(D) Find the cost of removing 100% of the pollutant.
Solutions
(A) *Step 1* Find the y-intercept. If $x = 0$, then

$$y = \frac{20(0)}{104 - 0} = \frac{0}{104} = 0$$

Thus, the y-intercept is (0, 0).
Step 2 Find the x-intercept(s). Set $y = 0$ and solve for x. This leads to

$$0 = \frac{20x}{104 - x}$$

$$0 = 20x$$

$$x = 0 \qquad \text{\textit{odd exponent:}}$$
$$ \qquad \text{\textit{graph crosses x-axis}}$$
$$ \qquad \text{\textit{here}}$$

x-intercept

113 POLYNOMIAL AND RATIONAL FUNCTIONS

Step 3 Find the vertical asymptote(s). Set the denominator equal to 0. Hence,

$$104 - x = 0$$
$$x = 104$$

and the line $x = 104$ is a vertical asymptote. Since the exponent of the factor $104 - x$ is odd, the graph of $f(x)$ approaches different ends of the vertical asymptote from both sides.

Step 4 Determine the behavior of $f(x)$ as x takes on values farther and farther away from the origin. The quotient of the highest powered terms of both numerator and denominator is

$$\frac{20x}{-x} = -20$$

Since this quotient is the constant -20, then $y = -20$ is a horizontal asymptote.

The information of steps 1 through 4 is summarized in Figure 2–41.

$y = f(x) = \dfrac{20x}{104 - x}$

Different ends

x-intercept (crosses)

Vertical asymptote

104

$y = -20$

Horizontal asymptote

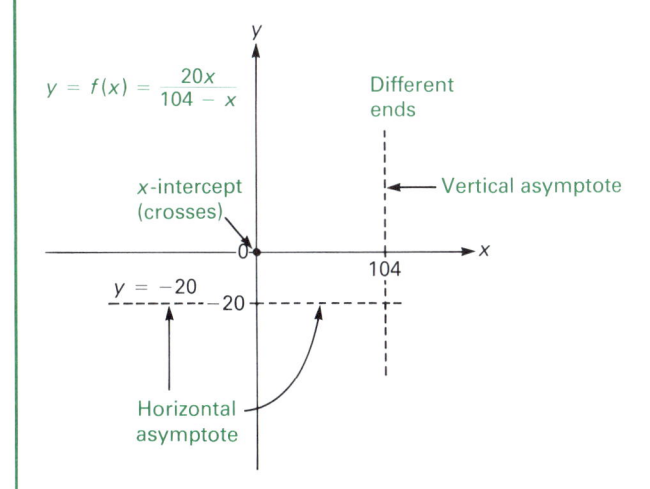

FIGURE 2–41

Since this function has a horizontal asymptote, we must determine whether the graph of $f(x)$ approaches the horizontal asymptote from *above* or *below* as x takes on larger and larger positive values, and also as x takes on negative values of larger magnitude. This is

EXAMPLE 2-16 (continued)

accomplished by evaluating the function at relatively large values of x.

Specifically, to get some idea of whether the function approaches the horizontal asymptote from above or below as x takes on larger and larger positive values, we will evaluate $f(x)$ at $x = 1000$. Here,

$$y = f(1000) = \frac{20(1000)}{104 - 1000} = \frac{20{,}000}{-896} = -22.32$$

Since $-22.32 < -20$, then for large positive values of x, the graph of $f(x)$ appears to approach the horizontal asymptote $y = -20$ from below (see Figure 2-42).

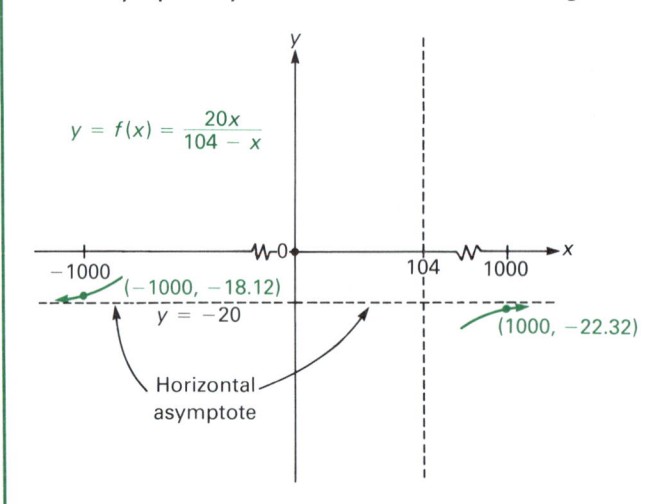

FIGURE 2-42

Similarly, to get some idea of whether the graph approaches the horizontal asymptote from above or below as x takes on negative values of larger and larger magnitude, we evaluate $f(x)$ at $x = -1000$. In this case,

$$y = f(-1000) = \frac{20(-1000)}{104 - (-1000)} = \frac{-20{,}000}{1104} = -18.12$$

Since $-18.12 > -20$, then as x takes on negative values far away from the origin, the graph of $f(x)$ appears to approach the horizontal asymptote $y = -20$ from above (see Figure 2-42).

The completed graph is sketched in Figure 2-43. Note that the graph of the function is drawn as a solid line for values of x within the interval $0 \leq x \leq 100$.

115 POLYNOMIAL AND RATIONAL FUNCTIONS

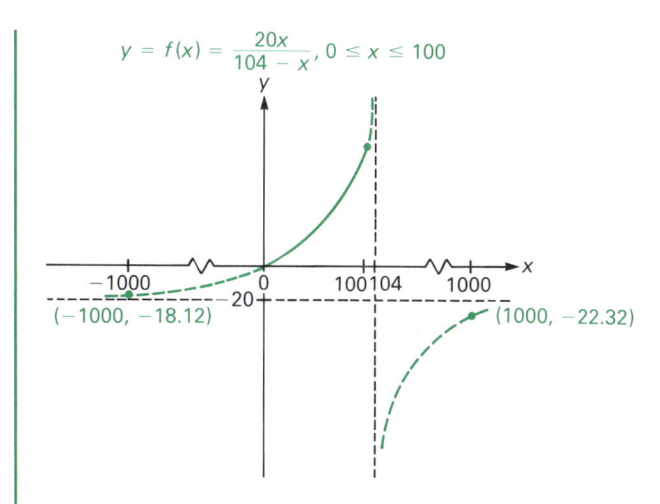

$$y = f(x) = \frac{20x}{104 - x}, \; 0 \leq x \leq 100$$

FIGURE 2–43

(B) The cost of removing 90% of the pollutant is

$$y = f(90) = \frac{20(90)}{104 - 90} = \frac{1800}{14} \approx \$129 \text{ thousand}*$$

(C) The cost of removing 95% of the pollutant is

$$y = f(95) = \frac{20(95)}{104 - 95} = \frac{1900}{9} \approx \$211 \text{ thousand}$$

(D) The cost of removing 100% of the pollutant is

$$y = f(100) = \frac{20(100)}{104 - 100} = \frac{2000}{4} = \$500 \text{ thousand}$$

Note that it gets increasingly expensive to remove the last small portion of the pollutant.

*The symbol $\approx$ means "approximately equal to."

EXERCISES

1. Sketch each of the following:

   (A) $y = \dfrac{(x - 3)^2(x + 8)}{(x - 1)^2}$

   (B) $f(x) = \dfrac{(x + 5)(x - 1)^4}{x(x - 2)}$

   (C) $f(x) = \dfrac{(x - 1)(x + 3)^2}{(x + 1)^3(x + 4)}$

   (D) $y = \dfrac{(x - 1)(x + 3)^2}{x^4(x + 4)}$

   (E) $y = \dfrac{1}{x - 4}$

   (F) $f(x) = \dfrac{5}{(x + 3)^2}$

   (G) $y = \dfrac{3x + 12}{x - 2}$

   (H) $f(x) = \dfrac{2x - 10}{x + 3}$

   (I) $f(x) = \dfrac{2x - 16}{x + 1}$

   (J) $f(x) = \dfrac{5x - 30}{x^2 - 81}$

(K) $y = \dfrac{x^2 - 9x}{3x - 15}$ (L) $f(x) = \dfrac{x^3 - 8x^2}{x^2 - 36}$

2. The demand, D (in pounds), for peanuts is related to the price per pound, p (in cents), by the equation

$$D(p) = \dfrac{500}{p - 40} \quad (p > 40)$$

(A) Sketch this demand function.
(B) Find the demand at a price of 42 cents per pound.

3. The Strong Steel Company produces two types of steel beams: I shaped and T shaped. The rational function defined by

$$y = \dfrac{420(x - 200)}{x - 210} \quad (0 \leq x \leq 200)$$

relates the number x of I-shaped beams and the number y of T-shaped beams produced during a month. Such a function's graph is called a **production possibility curve.**
(A) Sketch this production possibility curve.
(B) If no I-shaped beams are produced, how many T-shaped are produced?
(C) If no T-shaped beams are produced, how many I-shaped beams are produced?
(D) If 70 I-shaped beams are produced, how many T-shaped beams are produced?

4. Keep-Cool Corporation manufactures two types of air conditioners: Model SM401 and Model LG535. The production possibility curve

$$y = \dfrac{500(x - 300)}{x - 320} \quad (0 \leq x \leq 300)$$

expresses the relationship between the number x of Model SM401 and the number y of Model LG535 produced during a month.
(A) Sketch this production possibility curve.
(B) If Keep-Cool Corporation produces no Model SM401s, how many Model LG535s can it produce?
(C) If Keep-Cool Corporation produces no Model LG535s, how many Model SM401s can it produce?
(D) If Keep-Cool Corporation produces 220 Model SM401s, how many Model LG535s can it produce?

5. A company produces tires for trucks. If $C =$ total production cost, then the equation

$$C(x) = 50x + 30{,}000 \quad (x \geq 0)$$

relates the total production cost, C, with x, the number of tires produced. If

$$\overline{C} = \dfrac{\text{total production cost}}{\text{number of units produced}}$$

117 POLYNOMIAL AND RATIONAL FUNCTIONS

then $\overline{C}$ represents the average cost per tire. The equation

$$\overline{C}(x) = \frac{50x + 30{,}000}{x} \quad (x > 0)$$

relates the average cost per tire, $\overline{C}$, with x, the number of tires produced. The function $\overline{C}(x)$ is called an **average cost function**.
(A) Sketch $\overline{C}(x)$.
(B) Find the average cost per tire if 10 tires are produced.
(C) Find the average cost per tire if 1000 tires are produced.
(D) As the number of tires produced, x, becomes very large, the average cost per tire approaches what value?

6. The total production cost, C, and the number of units produced, x, are related by the cost equation

$$C(x) = 100x + 50{,}000 \quad (x \geq 0)$$

(A) Find the equation defining the average cost function, $\overline{C}(x)$.
(B) Sketch the graph of $\overline{C}(x)$.
(C) Find the average cost per unit if 20 units are produced.
(D) Find the average cost per unit if 2000 units are produced.
(E) As the number of units produced, x, becomes very large, the average cost per unit approaches what value?

7. The total production cost, C, and the number of units produced, x, are related by the cost equation

$$C(x) = -\frac{1}{2}x^2 + 10x + \frac{125}{2} \quad (0 \leq x \leq 10)$$

(A) Find the equation defining the average cost function, $\overline{C}(x)$.
(B) Sketch the graph of $\overline{C}(x)$.
(C) Find the average cost per unit if 2 units are produced.
(D) Find the average cost per unit if 5 units are produced.
(E) Find the average cost per unit if 10 units are produced.

8. The rational function defined by

$$y = \frac{40x}{110 - x} \quad (0 \leq x \leq 100)$$

expresses the relationship between the cost y (in millions of dollars) of removing $x\%$ of a pollutant from the atmosphere. This is another example of a cost–benefit curve.
(A) Sketch the graph of this function.
(B) Find the cost of removing the following amounts of pollution: 10%, 20%, 50%, 80%, 100%.

9. The rational function defined by

$$y = \frac{20x}{100 - x} \quad (0 \leq x < 100)$$

expresses the relationship between the cost y (in thousands of dollars) of removing $x\%$ of a pollutant.

(A) Sketch this cost–benefit curve.
(B) Find the cost of removing the following amounts of pollution: 10%, 20%, 50%, 80%, 90%.
(C) According to this function, is it possible to remove 100% of the pollutant?

10. Graph the function defined by

$$y = f(x) = \frac{(x-3)(x+3)}{x^2+1}$$

This rational function does not have a vertical asymptote. Why not? Compute $f(-2)$, $f(-1)$, $f(-1/2)$, $f(0)$, $f(1/2)$, $f(1)$, and $f(2)$ to determine the behavior of the graph of $f(x)$ over the interval $-3 < x < 3$.

11. Graph the function defined by

$$y = f(x) = \frac{x^2-16}{x^2+2}$$

EXPONENTIAL AND LOGARITHMIC FUNCTIONS

3

$y = 3e^x$

CHAPTER THREE

In the previous chapters, we studied linear, polynomial, and rational functions. In this chapter, we will discuss functions used to describe growth and decay of various quantities. Such functions are called **exponential functions**. We will also consider **logarithms** and **logarithmic functions**.

3–1

EXPONENTIAL FUNCTIONS

Consider two quantities, y and x, related by the equation

$$y = 2^x$$

Since the variable x is an exponent, such an equation defines an **exponential function**. Its graph and a table of x and y values are illustrated in Figure 3–1 and Table 3–1, respectively. Observing the graph of Figure 3–1, note that (0, 1) is the y-intercept. Since 2^x approaches 0 as x takes on negative values of larger and larger magnitude, the graph of $y = 2^x$ approaches the x-axis as x takes on negative values far away from the origin. Thus, the x-axis is a horizontal asymptote.

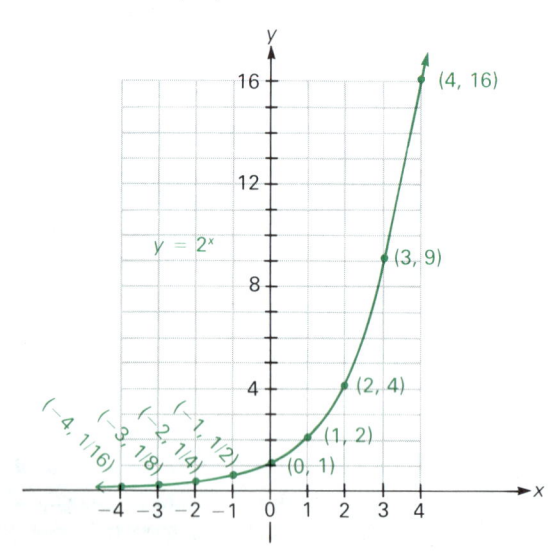

FIGURE 3–1

Note that Table 3–1 shows values of 2^x for rational values of x. It is possible to evaluate 2^x for irrational values of x by approximating each irrational value of x by rational values. For example, $2^{\sqrt{3}}$ may be approximated to varying degrees of accuracy by considering the sequence of rational powers

$$2^1, 2^{1.7}, 2^{1.73}, 2^{1.732}, \ldots$$

121 EXPONENTIAL AND LOGARITHMIC FUNCTIONS

TABLE 3–1

x	$y = 2^x$
−4	$y = 2^{-4} = (1/2)^4 = 1/16$
−3	$y = 2^{-3} = (1/2)^3 = 1/8$
−2	$y = 2^{-2} = (1/2)^2 = 1/4$
−1	$y = 2^{-1} = (1/2)^1 = 1/2$
0	$y = 2^0 = 1$
1	$y = 2^1 = 2$
2	$y = 2^2 = 4$
3	$y = 2^3 = 8$
4	$y = 2^4 = 16$

Recall that $\sqrt{3} \approx 1.732$. Thus, as the exponents get closer and closer to $\sqrt{3}$, the corresponding powers of 2 get closer and closer to $2^{\sqrt{3}}$. Therefore, the exponential function defined by $y = 2^x$ has the real numbers as its domain, and its graph is the continuous curve illustrated in Figure 3–1.

We now consider the function defined by

$$y = 5 \cdot 2^x$$

Since the variable x appears as an exponent, this equation also defines an exponential function. Its graph and a table of x and y values are illustrated in Figure 3–2 and Table 3–2, respectively. Looking at Figure 3–2, note that (0, 5) is the y-intercept and that the x-axis is a horizontal asymptote.

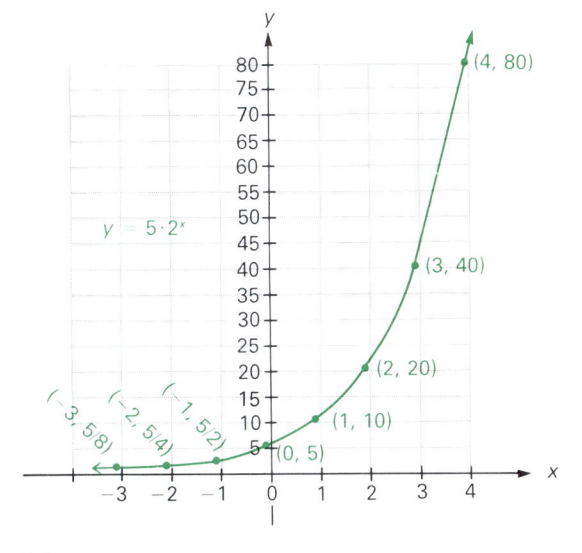

TABLE 3–2

x	$y = 5 \cdot 2^x$
−3	$y = 5 \cdot 2^{-3} = 5 \cdot (1/2)^3 = 5/8$
−2	$y = 5 \cdot 2^{-2} = 5 \cdot (1/2)^2 = 5/4$
−1	$y = 5 \cdot 2^{-1} = 5 \cdot (1/2)^1 = 5/2$
0	$y = 5 \cdot 2^0 = 5 \cdot 1 = 5$
1	$y = 5 \cdot 2^1 = 5 \cdot 2 = 10$
2	$y = 5 \cdot 2^2 = 5 \cdot 4 = 20$
3	$y = 5 \cdot 2^3 = 5 \cdot 8 = 40$
4	$y = 5 \cdot 2^4 = 5 \cdot 16 = 80$

FIGURE 3–2

If we compare the graphs of $y = 2^x$ and $y = 5 \cdot 2^x$ by sketching both functions on the same axis system, we obtain Figure 3-3. Note that the x-axis is the horizontal asymptote for both functions. Also, (0, 1) is the y-intercept of $y = 2^x$, whereas (0, 5) is the y-intercept of $y = 5 \cdot 2^x$.

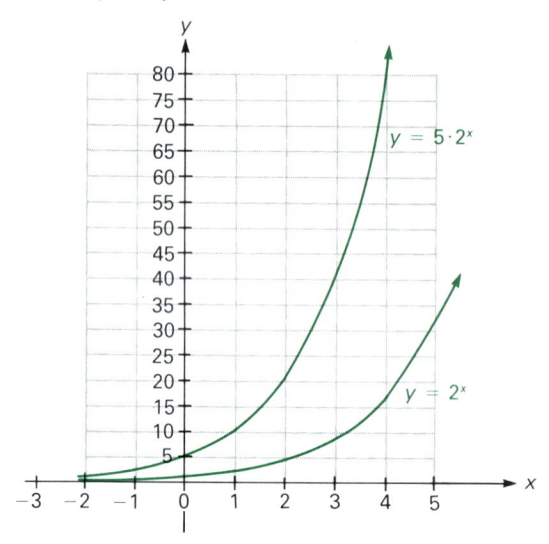

FIGURE 3-3

In general, the equation

$$y = ab^x$$

defines an exponential function. If $b > 1$ and $a > 0$, the graph of Figure 3-4 is typical of exponential functions of this form. If $x = 0$, then $y = ab^0 = a \cdot 1 = a$. Hence, (0, a) is the y-intercept. Thus, the graph of any exponential function defined by an equation of the form $y = ab^x$, where $b > 1$ and $a > 0$, will resemble that of Figure 3-4. Note specifically that the graphs of $y = 2^x$ and $5 \cdot 2^x$ resemble that of Figure 3-4 (compare the graphs of Figure 3-3 with that of Figure 3-4).

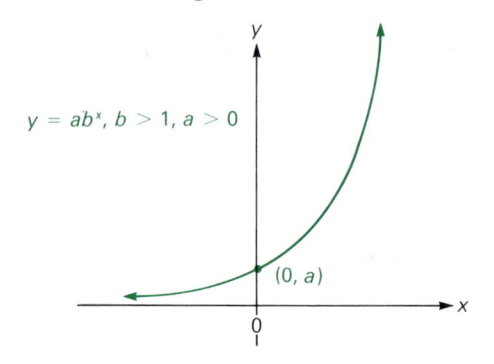

FIGURE 3-4

EXPONENTIAL AND LOGARITHMIC FUNCTIONS

Frequently, we will encounter exponential functions defined by equations of the form

$$y = ae^x$$

where e represents the irrational number whose decimal expansion begins

$$e = 2.7182818 \ldots$$

One way of obtaining e is to evaluate numbers of the form $(1 + 1/x)^x$ for arbitrarily large positive values of x. As x gets larger and larger, $(1 + 1/x)^x$ gets closer and closer to e. Observe that for a few integral values of x, we have

$$\left(1 + \frac{1}{100}\right)^{100} \approx 2.7048138$$

$$\left(1 + \frac{1}{1000}\right)^{1000} \approx 2.7169239$$

$$\left(1 + \frac{1}{10{,}000}\right)^{10{,}000} \approx 2.7181459$$

Comparing the equation $y = ae^x$ with the form $y = ab^x$, note that $b = e$ and $e > 1$. Hence, the graph of $y = ae^x$ resembles that of $y = ab^x$, illustrated in Figure 3–4. Table 3 of the Appendix at the end of the text gives values of e^x for various values of x.

EXAMPLE 3–1 Sketch $f(x) = e^x$.
Solution Since this equation is of the form $y = ab^x$ with $a = 1$ and $b = e$, its graph, given in Figure 3–5, resembles that of Figure 3–4.

FIGURE 3–5

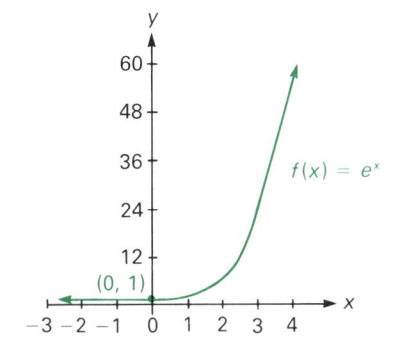

Up to this point, we have graphed equations of the form $y = ab^x$, $b > 1$, where the coefficient a is a *positive* number. All y-

values of such a function are positive, and the graph is located above the x-axis (see Figures 3–4 and 3–5). We now consider exponential functions defined by equations of the form $y = ab^x$, $b > 1$, where the coefficient a is a *negative* number. For such an exponential function, all y-values are negative. Thus, its graph appears below the x-axis. Figure 3–6 compares the graph of $y = ab^x$, $b > 1$, $a > 0$, to that of $y = ab^x$, $b > 1$, $a < 0$. After studying the graphs of Figure 3–6, we conclude that to graph an equation of the form $y = ab^x$, $b > 1$, $a < 0$, we begin with the graph of $y = ab^x$, $b > 1$, $a > 0$, and *draw it upside down*.

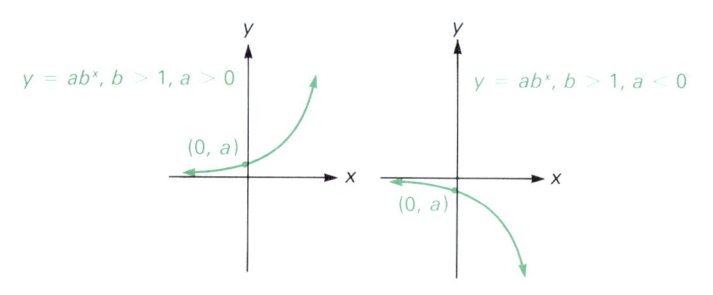

FIGURE 3–6

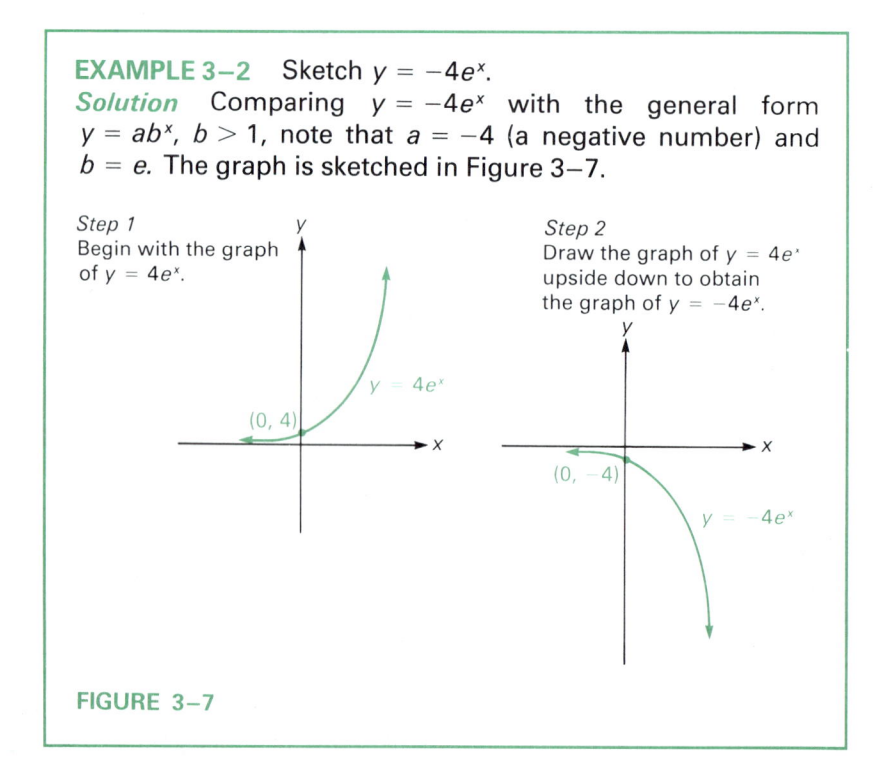

EXAMPLE 3–2 Sketch $y = -4e^x$.

Solution Comparing $y = -4e^x$ with the general form $y = ab^x$, $b > 1$, note that $a = -4$ (a negative number) and $b = e$. The graph is sketched in Figure 3–7.

Step 1
Begin with the graph of $y = 4e^x$.

Step 2
Draw the graph of $y = 4e^x$ upside down to obtain the graph of $y = -4e^x$.

FIGURE 3–7

EXPONENTIAL AND LOGARITHMIC FUNCTIONS

Exponential Functions Defined by $y = ab^x + c$, $b > 1$

Sometimes we encounter an exponential function defined by an equation such as

$$y = 3e^x + 2$$

Since the constant 2 is added to $y = 3e^x$, the y-values of $y = 3e^x + 2$ are *2 greater* than those of $y = 3e^x$. Hence, $y = 3e^x + 2$ may be sketched by beginning with the graph of $y = 3e^x$ and *lifting it vertically by 2 units* (see Figure 3–8). Note that the horizontal line $y = 2$ is the horizontal asymptote for $y = 3e^x + 2$.

Step 1
Begin with the graph of $y = 3e^x$.

Step 2
Lift the graph of $y = 3e^x$ vertically by 2 units to obtain the graph of $y = 3e^x + 2$.

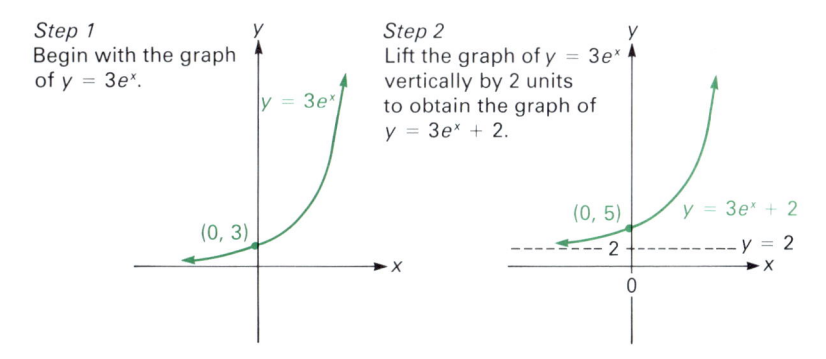

FIGURE 3–8

Thus, to graph an exponential function defined by an equation of the form $y = ab^x + c$, $b > 1$, we begin with the graph of $y = ab^x$, $b > 1$, and either lift it (if $c > 0$) or lower it (if $c < 0$) vertically by $|c|$ units.

EXAMPLE 3–3 Sketch $f(x) = -2e^x - 1$.

Solution Compare this equation with the general form $y = ab^x + c$, where $a = -2$ (a negative number), $b = e$, and $c = -1$. The graph is sketched in Figure 3–9.

Step 1
Begin with the graph of $y = 2e^x$.

Step 2
Draw the graph of $y = 2e^x$ upside down to obtain the graph of $y = -2e^x$.

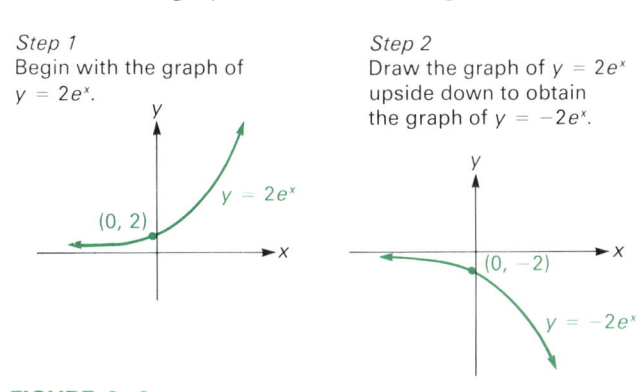

FIGURE 3–9

EXAMPLE 3–3 (continued)

FIGURE 3–9 (continued) *Step 3* Lower the graph of $y = -2e^x$ vertically by 1 unit to obtain the graph of $y = -2e^x - 1$.

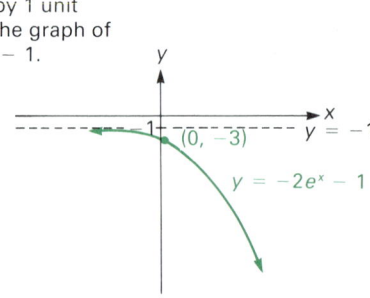

Exponential Functions Defined by $y = ab^{-x}$, $b > 1$

The Long Haul Moving Company has recently purchased a new truck. The truck depreciates in such a way that it loses one-half its value each year. If the truck cost $5000, then after 1 year, the truck's value is

$$5 \cdot \frac{1}{2} = 5 \cdot 2^{-1} \text{ thousand dollars}$$

after 2 years, the truck's value is

$$(5 \cdot 2^{-1})\left(\frac{1}{2}\right) = 5 \cdot 2^{-2} \text{ thousand dollars}$$

after 3 years, the truck's value is

$$(5 \cdot 2^{-2})\left(\frac{1}{2}\right) = 5 \cdot 2^{-3} \text{ thousand dollars}$$

after 4 years, the truck's value is

$$(5 \cdot 2^{-3})\left(\frac{1}{2}\right) = 5 \cdot 2^{-4} \text{ thousand dollars}$$

and after x years, the truck's value is

$$5 \cdot 2^{-x} \text{ thousand dollars}$$

Thus, if y represents the truck's value (in thousands of dollars) at the end of the xth year, y and x are related by the equation

$$y = 5 \cdot 2^{-x}$$

Since the variable x appears as an exponent, this equation defines an exponential function. Note that the exponent, $-x$, contains a negative sign. The graph of $y = 5 \cdot 2^{-x}$ and a table of x- and y-values are illustrated in Figure 3–10 and Table 3–3, respectively.

127 EXPONENTIAL AND LOGARITHMIC FUNCTIONS

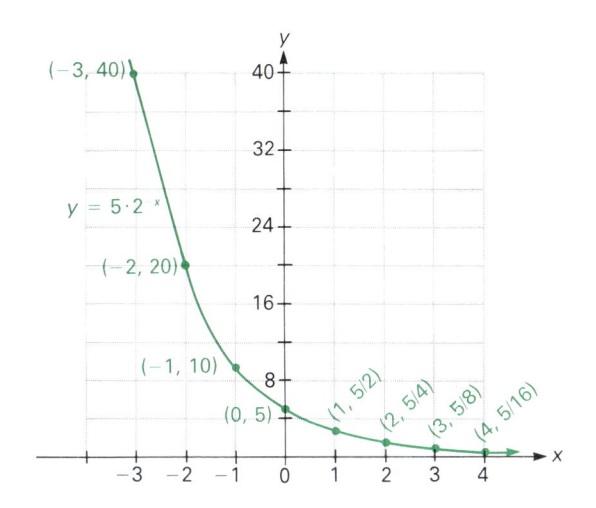

TABLE 3-3

x	$y = 5 \cdot 2^{-x}$
-3	$y = 5 \cdot 2^{-(-3)} = 5 \cdot 2^3 = 40$
-2	$y = 5 \cdot 2^{-(-2)} = 5 \cdot 2^2 = 20$
-1	$y = 5 \cdot 2^{-(-1)} = 5 \cdot 2^1 = 10$
0	$y = 5 \cdot 2^0 = 5 \cdot 1 = 5$
1	$y = 5 \cdot 2^{-1} = 5 \cdot 1/2 = 5/2$
2	$y = 5 \cdot 2^{-2} = 5 \cdot 1/4 = 5/4$
3	$y = 5 \cdot 2^{-3} = 5 \cdot 1/8 = 5/8$
4	$y = 5 \cdot 2^{-4} = 5 \cdot 1/16 = 5/16$

FIGURE 3-10

Studying Figure 3-10, we see that (0, 5) is the y-intercept. Since $5 \cdot 2^{-x}$ approaches 0 as x takes on larger and larger positive values, the graph of $y = 5 \cdot 2^{-x}$ approaches the x-axis for positive values of x far away from the origin. Thus, the x-axis is a horizontal asymptote. Although only that portion of the function for which $x \geq 0$ applies to the truck's value, the function has been sketched for all real values of x.

The equation $y = 5 \cdot 2^{-x}$ defines an exponential function of the general form

$$y = ab^{-x} \quad (b > 1, a > 0)$$

The graph of Figure 3-11 is typical of exponential functions of this form. If $x = 0$, then $y = ab^0 = a \cdot 1 = a$. Hence, (0, a) is the y-intercept. Since $b > 1$, then ab^{-x} approaches 0 as x gets larger

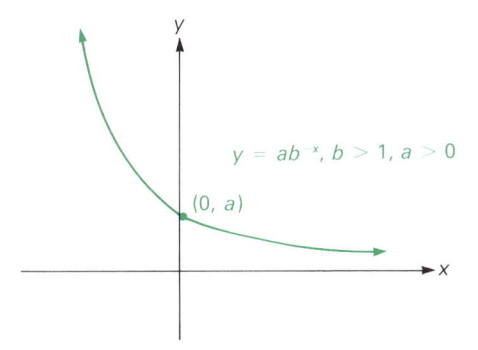

FIGURE 3-11

128 CHAPTER THREE

and larger. Thus, the x-axis is a horizontal asymptote. If the coefficient a were negative, the graph would be drawn upside down. Also, the addition of a constant, c, to the function $y = ab^{-x}$ would either lift (if $c > 0$) or lower (if $c < 0$) the graph vertically by $|c|$ units. Therefore, for an exponential function defined by an equation of the form $y = ab^{-x} + c$, the horizontal line $y = c$ is the horizontal asymptote.

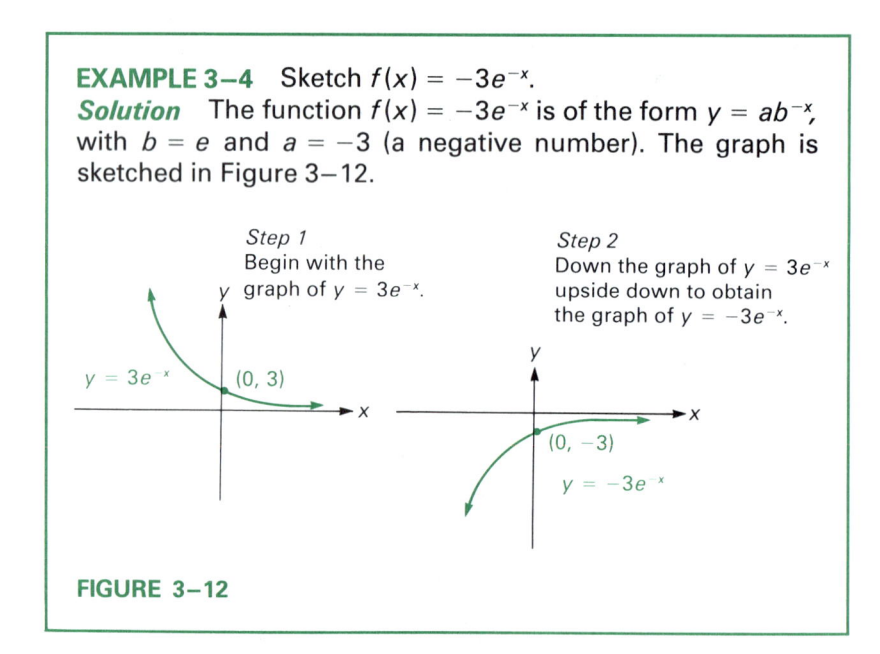

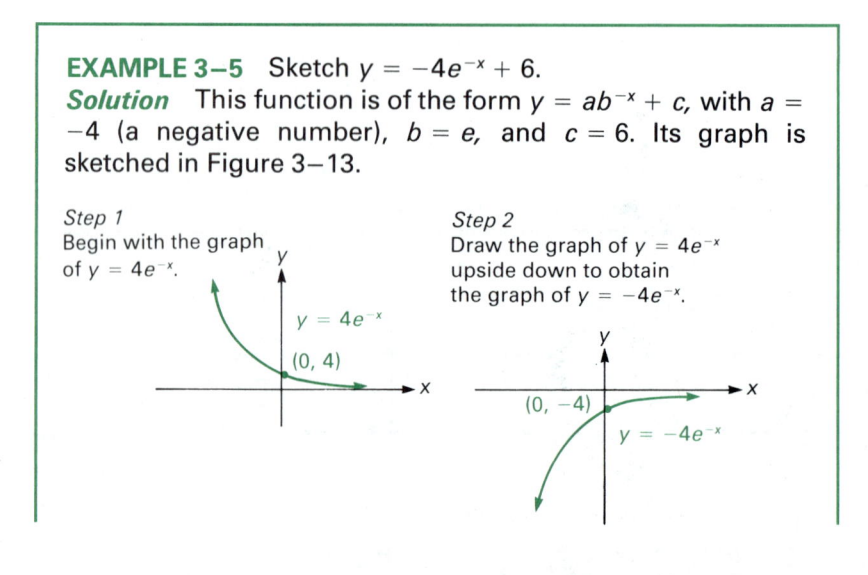

EXAMPLE 3-4 Sketch $f(x) = -3e^{-x}$.
Solution The function $f(x) = -3e^{-x}$ is of the form $y = ab^{-x}$, with $b = e$ and $a = -3$ (a negative number). The graph is sketched in Figure 3-12.

Step 1
Begin with the graph of $y = 3e^{-x}$.

Step 2
Down the graph of $y = 3e^{-x}$ upside down to obtain the graph of $y = -3e^{-x}$.

FIGURE 3-12

EXAMPLE 3-5 Sketch $y = -4e^{-x} + 6$.
Solution This function is of the form $y = ab^{-x} + c$, with $a = -4$ (a negative number), $b = e$, and $c = 6$. Its graph is sketched in Figure 3-13.

Step 1
Begin with the graph of $y = 4e^{-x}$.

Step 2
Draw the graph of $y = 4e^{-x}$ upside down to obtain the graph of $y = -4e^{-x}$.

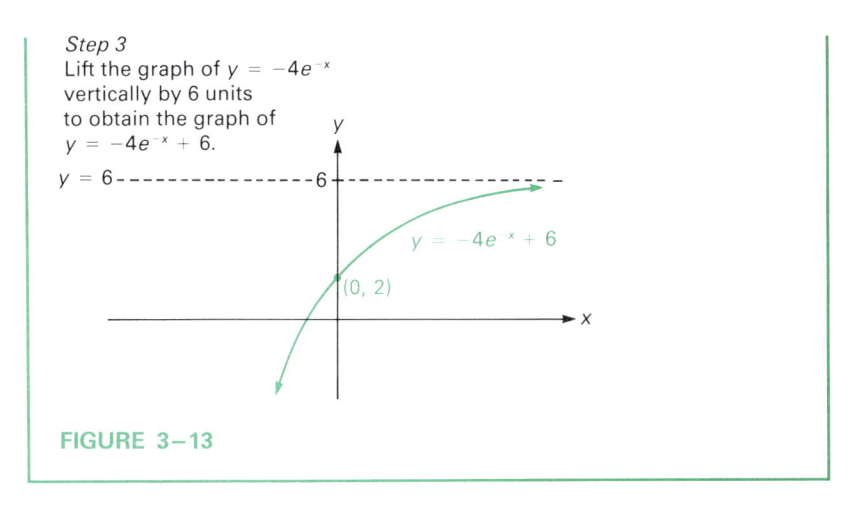

EXPONENTIAL AND LOGARITHMIC FUNCTIONS

Step 3
Lift the graph of $y = -4e^{-x}$ vertically by 6 units to obtain the graph of $y = -4e^{-x} + 6$.

FIGURE 3–13

All of the equations defining exponential functions of this section have appeared in either of the following two forms:

$$y = ab^x + c \qquad (b > 1)$$

or

$$y = ab^{-x} + c \qquad (b > 1)$$

Note that for each form, $b > 1$. We now consider the cases for which $b \leq 1$. If $b = 1$, then $b^x = 1^x = 1$ for all finite x-values. Thus, the equation $y = ab^x + c$ becomes the horizontal line $y = a + c$. The same happens to the form $y = ab^{-x} + c$. We have little interest in such exponential functions.

If $0 < b < 1$, then b^x may be written as b_1^{-x}, with $b_1 = 1/b$ and $b_1 > 1$. Also, b^{-x} becomes b_1^x, with $b_1 = 1/b$ and $b_1 > 1$. Thus, the equation $y = ab^x + c$, with $0 < b < 1$, may be written as $y = ab_1^{-x} + c$, with $b_1 > 1$, since $b_1 = 1/b$. Also, the equation $y = ab^{-x} + c$, with $0 < b < 1$, may be rewritten as $y = ab_1^x + c$, with $b_1 > 1$, since $b_1 = 1/b$.

If $b \leq 0$, we have little interest in equations of the form $y = ab^x + c$ or $y = ab^{-x} + c$ since the function is not defined at many x-values. Specifically, if $b = 0$ and $x = -1$, then $y = ab^x + c$ is undefined; if $b = -4$ and $x = 1/2$, then $y = ab^x + c$ is undefined.

Exponential Functions with Terms e^{kx} and e^{-kx}

If $b > 0$, there is a value k such that $b = e^k$. In other words, any positive number b may be expressed as e^k.* Thus, since

$$b^x = (e^k)^x = e^{kx}$$

---

*We will learn how to find k in Section 3–3.

equations of the forms

$$y = ab^x + c \quad (b > 1)$$

and

$$y = ab^{-x} + c \quad (b > 1)$$

may be restated as

$$y = ae^{kx} + c \quad (k > 0)$$

and

$$y = ae^{-kx} + c \quad (k > 0)$$

respectively. Therefore, the procedures of this section also apply to sketching exponential functions expressed in the latter forms that contain the terms e^{kx} and e^{-kx}.

EXAMPLE 3-6 Sketch $y = -3e^{-0.05x} + 7$.

Solution The graph is sketched in Figure 3-14.

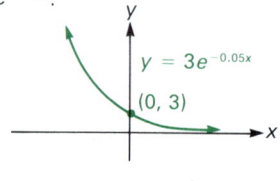

Step 1
Begin with the graph of $y = 3e^{-0.05x}$.

Step 2
Draw the graph of $y = 3e^{-0.05x}$ upside down to obtain the graph of $y = -3e^{-0.05x}$.

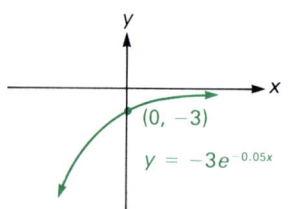

Step 3
Lift the graph of $y = -3e^{-0.05x}$ vertically by 7 units to obtain the graph of $y = -3e^{-0.05x} + 7$.
$y = 7$

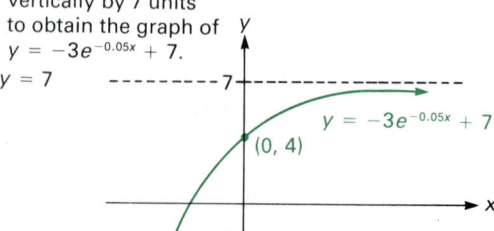

FIGURE 3-14

131 EXPONENTIAL AND LOGARITHMIC FUNCTIONS

EXERCISES

1. Sketch the following. Identify the y-intercept and the horizontal asymptote in each case.
   - (A) $f(x) = 5^x$
   - (B) $y = 3^x$
   - (C) $y = 4^x$
   - (D) $f(x) = 4 \cdot 5^x$
   - (E) $y = 2 \cdot 3^x$
   - (F) $y = 3 \cdot 4^x$
   - (G) $f(x) = -4 \cdot 5^x$
   - (H) $f(x) = -2 \cdot 3^x$
   - (I) $y = -3 \cdot 4^x$
   - (J) $y = 3e^x$
   - (K) $f(x) = 7e^x$
   - (L) $y = 10e^x$
   - (M) $y = -3e^x$
   - (N) $y = -7e^x$
   - (O) $y = -10e^x$

2. Sketch the following. Identify the y-intercept and the horizontal asymptote in each case.
   - (A) $y = 2 \cdot 3^x + 5$
   - (B) $f(x) = -3 \cdot 4^x + 8$
   - (C) $f(x) = 4e^x + 1$
   - (D) $y = 4e^x - 1$
   - (E) $y = -2e^x + 8$
   - (F) $f(x) = -2e^x - 5$

3. Sketch the following. Identify the y-intercept and the horizontal asymptote in each case.
   - (A) $y = 5^{-x}$
   - (B) $f(x) = 3^{-x}$
   - (C) $y = 4^{-x}$
   - (D) $y = 4 \cdot 5^{-x}$
   - (E) $f(x) = 2 \cdot 3^{-x}$
   - (F) $y = 3 \cdot 4^{-x}$
   - (G) $y = -4 \cdot 5^{-x}$
   - (H) $f(x) = -2 \cdot 3^{-x}$
   - (I) $y = -3 \cdot 4^{-x}$
   - (J) $y = 3e^{-x}$
   - (K) $y = 7e^{-x}$
   - (L) $y = 10e^{-x}$
   - (M) $y = -3e^{-x}$
   - (N) $f(x) = -7e^{-x}$
   - (O) $y = -10e^{-x}$

4. Sketch the following. Identify the y-intercept and the horizontal asymptote in each case.
   - (A) $y = 2 \cdot 3^{-x} + 5$
   - (B) $f(x) = -3 \cdot 4^{-x} + 8$
   - (C) $f(x) = 4e^{-x} + 1$
   - (D) $y = 4e^{-x} - 1$
   - (E) $y = -2e^{-x} + 30$
   - (F) $f(x) = -2e^{-x} - 5$
   - (G) $y = 30e^{-x} + 30$
   - (H) $f(x) = 50 - 50e^{-x}$
   - (I) $y = 10(1 - e^{-x})$
   - (J) $f(x) = -5(1 - e^{-x})$

5. Sketch the following. Identify the y-intercept and the horizontal asymptote in each case.
   - (A) $y = \left(\dfrac{1}{2}\right)^x$
   - (B) $y = \left(\dfrac{2}{3}\right)^x$
   - (C) $y = \left(\dfrac{1}{2}\right)^{-x}$
   - (D) $f(x) = \left(\dfrac{2}{3}\right)^{-x}$
   - (E) $y = 5 \cdot \left(\dfrac{1}{5}\right)^x + 1$
   - (F) $f(x) = -4 \cdot \left(\dfrac{1}{6}\right)^{-x} + 2$

6. Sketch the following. Identify the y-intercept and the horizontal asymptote in each case.
   - (A) $y = 3e^{2x} + 1$
   - (B) $f(x) = 4e^{-3x} + 1$
   - (C) $f(x) = -2e^{0.05x} + 1$
   - (D) $y = 3e^{-0.02x} - 5$
   - (E) $y = 10e^{0.07x}$
   - (F) $y = -5e^{0.03x}$
   - (G) $f(x) = 7e^{-0.10x}$
   - (H) $y = -3e^{-0.06x}$
   - (I) $y = 10(1 - e^{-0.20x})$
   - (J) $y = -2e^{-0.3x} - 6$
   - (K) $y = -8e^{-0.40x}$
   - (L) $y = 70e^{-0.50x} + 10$

7. A certain type of bacteria triples its numbers each day. Initially, there were 500,000 such bacteria present. Let y represent the number of bacteria present and t represent the number of days elapsed.
   - (A) Find the equation that relates y with t.

(B) Sketch the graph of the equation of part (A).
(C) How many bacteria are present after four days?

8. A certain car depreciates in such a way that it loses two-thirds of its value each year. The car initially cost $4000. Let y represent the car's value (in dollars) at the end of the xth year.
   (A) Find the equation that relates y and x.
   (B) Sketch the graph of the equation of part (A).
   (C) Find the car's value at the end of the fourth year.

3-2

APPLICATIONS OF EXPONENTIAL FUNCTIONS

The exposition of this section will illustrate many of the practical applications of exponential functions. Applications to compound interest will be discussed in Chapter 4.

Earnings per Share

The annual **earnings per share** (frequently abbreviated **EPS** and expressed in dollars) of a corporation is calculated by dividing the corporation's annual earnings by the number of shares of its stock outstanding. That is,

$$\text{EPS} = \frac{\text{annual earnings}}{\text{number of shares of stock outstanding}}$$

Anyone interested in evaluating the performance of a corporation usually looks at its EPS. Specifically, a potential investor usually tries to find a relationship between the annual EPS and time (in years). Such relationships are often expressed by exponential functions.

Consider the exponential function defined by

$$y = \frac{1}{4} \cdot 3^t \quad (t \geq 0)$$

where y represents the annual EPS and t represents time measured in years. Thus, for the year represented by $t = 0$, the EPS is

$$y = \frac{1}{4} \cdot 3^0 = \frac{1}{4} \cdot 1 = \$0.25$$

And for the year represented by $t = 4$, the EPS is

$$y = \frac{1}{4} \cdot 3^4 = \frac{1}{4} \cdot 81 = \$20.25$$

133 EXPONENTIAL AND LOGARITHMIC FUNCTIONS

The graph of $y = (1/4) \cdot 3^t$ is illustrated in Figure 3–15.

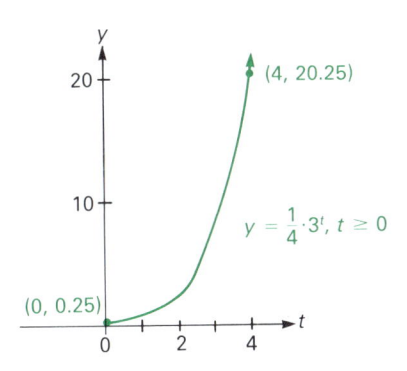

FIGURE 3–15

Employee Growth

Managers are often concerned with determining the number of employees of a company at a given time. Frequently, the number of employees is related to a company's annual sales (in millions of dollars) by an exponential function. Hence, we say that the number of employees is increasing exponentially with respect to annual sales.

Suppose the number of employees, $P(x)$, of a certain company is related to annual sales, x (in millions of dollars), by the equation

$$P(x) = 100{,}000 e^{0.06x} \quad (x \geq 0)$$

Here, then

$$P(0) = 100{,}000 e^{0.06(0)}$$
$$= 100{,}000$$

Thus, at $x = 0$, the company's number of employees is 100,000. Similarly,

$$P(3) = 100{,}000 e^{0.06(3)}$$
$$= 100{,}000 e^{0.18}$$
$$= 100{,}000(1.197217)$$
$$\approx 119{,}722$$

Therefore, when annual sales are $3 million, the number of employees is approximately 119,722. The graph of $P(x) = 100{,}000 e^{0.06x}$ is illustrated in Figure 3–16.

FIGURE 3–16

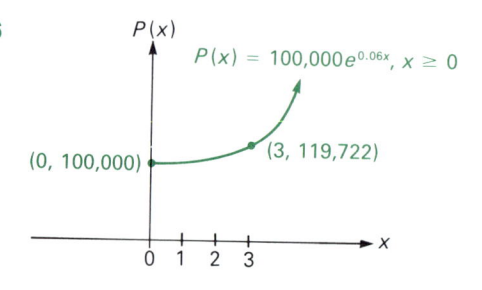

$P(x) = 100{,}000e^{0.06x}$, $x \geq 0$

(0, 100,000) (3, 119,722)

Radioactive Decay

A radioactive substance loses its mass (or, if you prefer, weight) as time passes. Specifically, its mass, y, is related to the time elapsed, t, by an equation of the form

$$y = ae^{-kt} \quad (k > 0)$$

Note that at $t = 0$, the initial mass is

$$y = ae^{-k(0)}$$
$$= a \cdot 1$$
$$= a$$

The letter k represents a constant associated with the specific radioactive substance being considered. If, for a particular radioactive substance, the initial mass is $a = 1000$ grams and $k = 0.70$, then the equation

$$y = 1000e^{-0.70t}$$

relates its mass, y (in grams), with the time elapsed, t (in minutes). Thus, at $t = 1$ (after 1 minute has elapsed), the mass is

$$y = 1000e^{-0.70(1)}$$
$$= 1000e^{-0.70}$$
$$= 1000(0.496585)$$
$$\approx 496.59 \text{ grams}$$

The graph of $y = 1000e^{-0.70t}$ is illustrated in Figure 3–17.

FIGURE 3–17

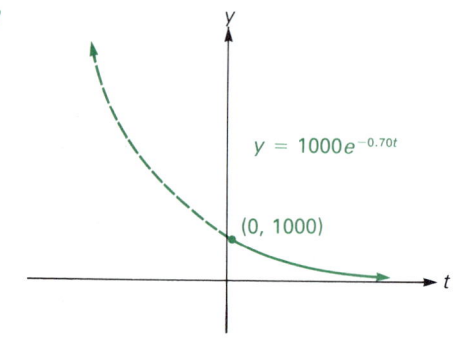

$y = 1000e^{-0.70t}$

(0, 1000)

135 EXPONENTIAL AND LOGARITHMIC FUNCTIONS

Learning Curve Psychologists have found that when a person learns a new task, learning is rapid at first. Then, as time passes, learning tends to taper off. Once the task is mastered, the person's level of performance approaches an upper limit. The function that relates a learner's performance with the time elapsed is called a **learning curve**. Learning curves are often expressed by exponential functions.

Consider the learning curve expressed by the exponential function

$$y = 40 - 40e^{-0.2x}$$

Here, the performance, y, is the number of items produced by the worker during the xth day following the training period. Thus, if $x = 0$, then

$$y = 40 - 40e^{-0.2(0)}$$
$$= 40 - 40(1)$$
$$= 0$$

Hence, during the training period, the worker produces 0 units. And if $x = 3$, then

$$y = 40 - 40e^{-0.2(3)}$$
$$= 40 - 40e^{-0.60}$$
$$= 40 - 40(0.548812)$$
$$= 40 - 21.95248$$
$$\approx 18.05$$

Therefore, during the third day following the training period, the worker produces approximately 18 units. The graph of $y = 40 - 40e^{-0.2x}$ is sketched in Figure 3–18.

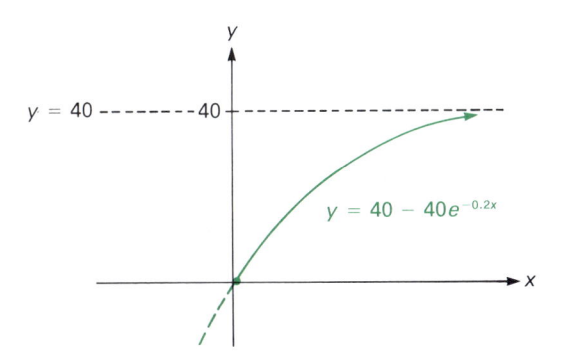

FIGURE 3–18

Observing the learning curve of Figure 3–18, note that the worker's daily production approaches 40 units as x gets larger and larger. Thus, according to this learning curve, the worker's daily production will never exceed 40 units.

Newton's Law of Cooling

Newton's law of cooling expresses the relationship between the temperature of a cooling object and the time elapsed since cooling first began. According to Newton's law of cooling, if

y = temperature of a cooling object after t units of time

c = temperature of the medium surrounding the cooling object

then the exponential function

$$y = ae^{-kt} + c$$

relates temperature y with time t. The letters a and k represent constants associated with the cooling object. A sketch of the general function $y = ae^{-kt} + c$ appears in Figure 3–19. The constants a and k are nonnegative. Studying Figure 3–19, we note that the temperature y of the cooling object approaches the temperature c of the surrounding medium as t gets larger and larger. Thus, the temperature of a cooling object will not fall below the temperature of the surrounding medium.

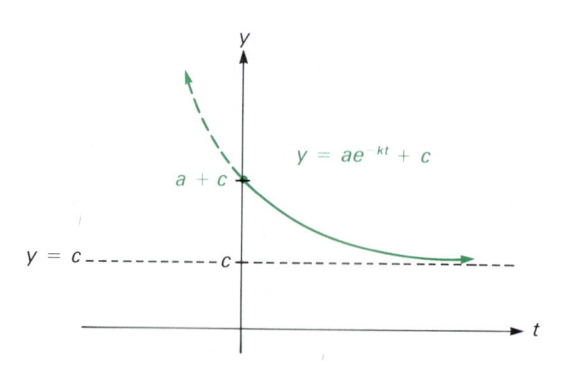

FIGURE 3–19

As a specific example, consider a cup of tea heated to 210°F. The room temperature is 70°F. Here, the exponential function

$$y = 140e^{-0.01t} + 70$$

relates the temperature y of the tea with the time elapsed, t (in minutes). Hence, after 10 minutes have elapsed ($t = 10$), we find that

137 EXPONENTIAL AND LOGARITHMIC FUNCTIONS

$$y = 140e^{-0.01(10)} + 70$$
$$= 140e^{-0.1} + 70$$
$$= 140(0.904837) + 70$$
$$= 126.67718 + 70$$
$$\approx 196.7$$

Thus, the object at this point in time has cooled to a temperature of approximately 197°F. The graph of $y = 140e^{-0.01t} + 70$ is sketched in Figure 3–20.

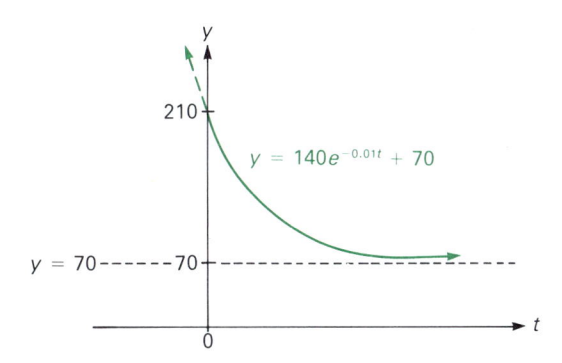

FIGURE 3–20

EXERCISES

1. The annual sales, y (in millions of dollars), of a particular company are related to time, t, by the equation

   $$y = 3 \cdot 2^t$$

   with $t = 0$ corresponding to the year 19x0, $t = 1$ corresponding to the year 19x1, etc.
   (A) Sketch $y = 3 \cdot 2^t$.
   (B) Find the annual sales for years 19x0, 19x1, and 19x2.

2. The annual earnings per share of AKD Corporation are related to time by the equation

   $$y(t) = e^{0.1t}$$

   where $y(t)$ represents the earnings per share for year t. Note that $t = 0$ corresponds to the year 19x0, $t = 1$ corresponds to the year 19x1, etc.
   (A) Sketch $y(t) = e^{0.1t}$.
   (B) Find the earnings per share for years 19x0, 19x1, 19x2, and 19x3.

3. The annual maintenance cost, y, of a machine is related to the number of years it is run, t, by the equation

   $$y = 1000e^{0.05t} \qquad (t \geq 0)$$

(A) Sketch $y = 1000e^{0.05t}$, $t \geq 0$.
(B) Find the annual maintenance cost after the machine has run for 2 years.

4. The population, $P(t)$, of a certain city is related to time, t (in years), by the exponential function

$$P(t) = 10,000e^{-0.03t}$$

Note that $t = 0$ corresponds to the year 19x0, $t = 1$ corresponds to the year 19x1, etc.
(A) Sketch $P(t) = 10,000e^{-0.03t}$.
(B) Find this city's population for 19x0, 19x1, 19x2, and 19x3.

5. A certain radioactive substance decays in accordance with the equation

$$y(t) = 2000e^{-0.60t}$$

where $y(t)$ represents the mass (in grams) at time t (in hours).
(A) Sketch $y(t) = 2000e^{-0.60t}$.
(B) Calculate the initial mass of the substance.
(C) Calculate the mass of the substance after 1/2 hour has elapsed.
(D) Calculate the mass of the substance after 3 hours have elapsed.

6. The graph of the exponential function defined by

$$N(x) = 50 - 50e^{-0.3x}$$

is a learning curve, where $N(x)$ represents the number of items produced by an assembly line worker during the xth day after the training period.
(A) Sketch $N(x) = 50 - 50e^{-0.3x}$.
(B) How many items are produced during the fifth day following the training period?
(C) This worker's daily production will never exceed how many units?

7. The temperature, y, of a heated cup of coffee is related to the time elapsed, t (in minutes), by the equation

$$y = 150e^{-0.02t} + 65$$

(A) Sketch $y = 150e^{-0.02t} + 65$.
(B) Calculate the temperature of the coffee before cooling began.
(C) Calculate the temperature of the coffee after 5 minutes have elapsed.
(D) What is the room temperature?
(E) The temperature of the coffee will not decline below what value?

8. A coroner determines that the temperature, T, of a murder victim's body is related to the time elapsed, t (in hours), since death by the equation

$$T = 38.6e^{-0.05t} + 60$$

(A) Sketch $T = 38.6e^{-0.05t} + 60$.
(B) What is the room temperature?

EXPONENTIAL AND LOGARITHMIC FUNCTIONS

(C) Calculate the dead body's temperature after 2 hours have elapsed.

9. The function defined by

$$y = 3.6e^{0.02x}$$

approximates the relationship between sales, y (in billions of dollars), and advertising expenditure, x (in millions of dollars).
(A) Sketch $y = 3.6e^{0.02x}$.
(B) Calculate the expected sales for an advertising expenditure of $5 million.

3-3

LOGARITHMS If three numbers, L, b, and N (with $b > 0$, $b \neq 1$, and $N > 0$), are related in such a way that

$$N = b^L$$

then the exponent, L, is defined as "the logarithm of N to the base b." This definition is written in shorthand notation as

$$L = \log_b N$$

Thus, the statement $N = b^L$ has the same meaning as the statement $L = \log_b N$. The statement $N = b^L$ is written in **exponential notation**. The statement $L = \log_b N$ is written in **logarithmic notation**. Specifically, the statement $9 = 3^2$ may be expressed in logarithmic notation as $2 = \log_3 9$, read "2 equals the log of 9 to the base 3."

EXAMPLE 3–7 Rewrite the statement $5^3 = 125$ in logarithmic notation.
Solution $3 = \log_5 125$, read "3 equals the log of 125 to the base 5."

EXAMPLE 3–8 Rewrite the statement $10^4 = 10,000$ in logarithmic notation.
Solution $4 = \log_{10} 10,000$, read "4 equals the log of 10,000 to the base 10."

Studying the preceding examples, we note that a **logarithm** is an *exponent* of a number called its **base**. Thus, the statement

$$2 = \log_{10} 100$$

has the same meaning as the statement

$$10^2 = 100$$

Note that 2, the logarithm, is the exponent of 10, the base.

EXAMPLE 3–9 Find y if $y = \log_2 8$.
Solution Since $y = \log_2 8$ means $2^y = 8$, then $y = 3$. Thus, 3 is the log of 8 to the base 2.

EXAMPLE 3–10 Find $\log_{10} 100$.
Solution Let $y = \log_{10} 100$. Translated into exponential notation, the statement becomes $10^y = 100$. Hence, $y = 2$. Thus, 2 is the log of 100 to the base 10.

Different Bases We again emphasize that a logarithm is an exponent of a number called its base (study Examples 3–7 through 3–10). Two bases are most commonly used: **base 10** and **base e**. Base-10 logarithms are called **common logarithms**. Base-e logarithms are called **natural logarithms,** or **Napierian logarithms**. The following notation is used to distinguish between common logarithms and natural logarithms. Specifically, the common logarithm of x, $\log_{10} x$, is abbreviated log x; the natural logarithm of x, $\log_e x$, is abbreviated ln x. Thus, the statement $y = \log x$ means $10^y = x$; the statement $y = \ln x$ means $e^y = x$.

EXAMPLE 3–11 Find log 10.
Solution Let $L = \log 10$. Rewriting the statement in exponential notation, we have

$$10^L = 10$$

Hence, $L = 1$. Thus, the common logarithm of 10 is 1.

EXAMPLE 3–12 Find ln e.
Solution Let $L = \ln e$. Rewriting this statement in exponential notation gives us

$$e^L = e$$

Hence, $L = 1$. Thus, the natural logarithm of e is 1.

141 EXPONENTIAL AND LOGARITHMIC FUNCTIONS

> **EXAMPLE 3–13** In Section 3–1, we stated that for any positive number b, there is a value k such that $b = e^k$. Find k.
> **Solution** Rewriting the statement $b = e^k$ in logarithmic form, we have
>
> $$k = \ln b$$
>
> Thus, $b = e^k = e^{\ln b}$

Logarithmic Functions

Consider two variables, x and y, related by the equation

$$y = \log x \quad (x > 0)$$

Since this is an equation with two variables, it may be graphed by computing a table of x and y values. Such a table is easily obtained by rewriting the equation $y = \log x$ in exponential form as $10^y = x$ and then choosing arbitrary y-values (see Table 3–4). The graph is sketched in Figure 3–21.

TABLE 3–4

$y = \log x$, or $x = 10^y$	Ordered Pairs (x, y)
If $y = -3$, then $x = 10^{-3} = 1/1000$	$(1/1000, -3)$
If $y = -2$, then $x = 10^{-2} = 1/100$	$(1/100, -2)$
If $y = -1$, then $x = 10^{-1} = 1/10$	$(1/10, -1)$
If $y = 0$, then $x = 10^0 = 1$	$(1, 0)$
If $y = 1$, then $x = 10^1 = 10$	$(10, 1)$
If $y = 2$, then $x = 10^2 = 100$	$(100, 2)$
If $y = 3$, then $x = 10^3 = 1000$	$(1000, 3)$

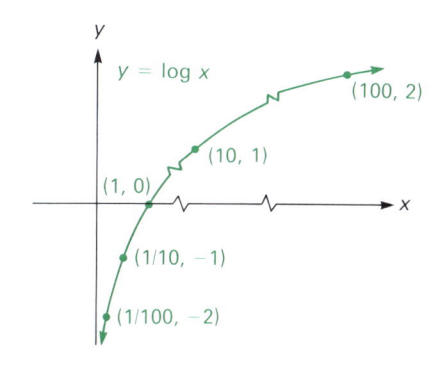

FIGURE 3–21

The equation $y = \log x$ defines a logarithmic function. The graph of $y = \log x$, illustrated in Figure 3–21, is typical of the graph of a logarithmic function with a *base greater than 1*. Studying Figure 3–21, we see that $(1, 0)$ is the x-intercept and that the y-axis is a vertical asymptote. Also, note that $y = \log x$ is undefined for $x \leq 0$. A few trial values of x will indicate why. If $x = 0$, then $y = \log 0$ means $10^y = 0$. But no value of y exists such that $10^y = 0$. Thus, $\log 0$ is undefined. However, as y takes on negative values of larger magnitude, 10^y approaches 0 and the y-axis is a vertical asymptote. If x is a negative number, such as $x = -3$, then $y = \log -3$ means $10^y = -3$. Again, no value of y exists such that $10^y = -3$.

EXAMPLE 3–14 Sketch the graph of $y = \ln x$.
Solution Since this logarithmic function has base e (a number greater than 1), its graph will resemble that of $y = \log x$. See Figure 3–22.

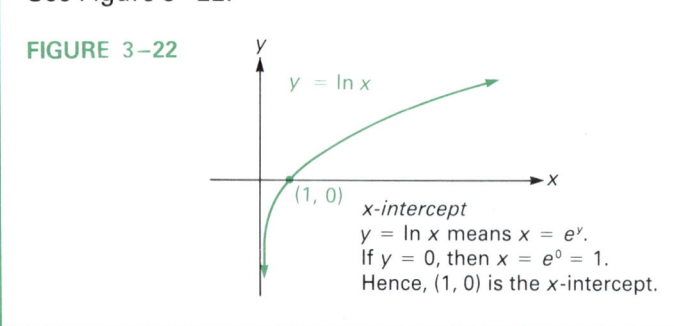

FIGURE 3–22

$y = \ln x$

(1, 0) *x-intercept*
$y = \ln x$ means $x = e^y$.
If $y = 0$, then $x = e^0 = 1$.
Hence, (1, 0) is the x-intercept.

Properties of Logarithms

Logarithms obey certain rules called **properties of logarithms**. These properties simplify much of our work with logarithms. They also enable us to perform certain arithmetic operations using logarithms. We now state the three properties of logarithms.

Properties of Logarithms

Let x, y, and b be any positive real numbers with $b \neq 1$. Also, let p be any real number.

Property 1 $\log_b xy = \log_b x + \log_b y$

Property 2 $\log_b \dfrac{x}{y} = \log_b x - \log_b y$

Property 3 $\log_b x^p = p \log_b x$

Since it is important to understand the meaning of these properties of logarithms, we will consider each in turn.

Property 1 $\log_b xy = \log_b x + \log_b y$

This property states that *the logarithm of a product of two numbers is equal to the sum of their logarithms.* The following are a few numerical illustrations of this property:

$\log_2 (8 \cdot 32) = \log_2 8 + \log_2 32$
$\log (5.23 \times 100) = \log 5.23 + \log 100$
$\ln (3 \cdot 7) = \ln 3 + \ln 7$

EXPONENTIAL AND LOGARITHMIC FUNCTIONS

Property 2 $\quad \log_b \dfrac{x}{y} = \log_b x - \log_b y$

This property states that *the logarithm of a quotient of two numbers is equal to the difference of their logarithms.* The following are a few numerical illustrations of this property:

$$\log_2 \dfrac{64}{16} = \log_2 64 - \log_2 16$$

$$\log \dfrac{5.63}{100} = \log 5.63 - \log 100$$

$$\ln \dfrac{40}{3} = \ln 40 - \ln 3$$

Property 3 $\quad \log_b x^p = p \log_b x$

This property states that *the logarithm of the pth power of a number is equal to p times the logarithm of the number.* The following are a few numerical illustrations of this property:

$$\log_2 4^3 = 3 \log_2 4$$
$$\ln 8^2 = 2 \ln 8$$
$$\log 3^{1/2} = \dfrac{1}{2}(\log 3)$$

The proofs of the three properties of logarithms appear in the appendix at the end of this section. The interested student who wishes to find out why these properties hold should study this material.

EXAMPLE 3–15 Given that log 3 = 0.4771 and log 5 = 0.6990, find each of the following using the properties of logarithms:
(A) log 15 (B) log 625
(C) log $\sqrt{5}$ (D) log 0.6

Solutions
(A) log 15 = log (5 · 3)
 = log 5 + log 3   Property 1
 = 0.6990 + 0.4771
 = 1.1761
(B) log 625 = log 5^4
 = 4 log 5   Property 3
 = 4(0.6990)
 = 2.7960

EXAMPLE 3–15 (continued)

(C) $\log \sqrt{5} = \log 5^{1/2}$
$= \frac{1}{2}(\log 5)$ Property 3
$= \frac{1}{2}(0.6990)$
$= 0.3495$

(D) $\log 0.6 = \log \frac{3}{5}$
$= \log 3 - \log 5$ Property 2
$= 0.4771 - 0.6990$
$= -0.2219$

EXAMPLE 3–16 In Example 3–11, we learned than log 10 = 1. With this information, and given that log 7.23 = 0.8591, use the properties of logarithms to calculate each of the following:

(A) log 72.3   (B) log 723
(C) log 723,000   (D) log 0.00723

Solutions

(A) $\log 72.3 = \log (7.23 \times 10)$
$= \log 7.23 + \log 10$
$= 0.8591 + 1$
$= 1.8591$

(B) $\log 723 = \log (7.23 \times 10^2)$
$= \log 7.23 + \log 10^2$
$= \log 7.23 + 2 \log 10$
$= 0.8591 + 2(1)$
$= 2.8591$

(C) $\log 723,000 = \log (7.23 \times 10^5)$
$= \log 7.23 + \log 10^5$
$= \log 7.23 + 5 \log 10$
$= 0.8591 + 5(1)$
$= 5.8591$

(D) $\log 0.00723 = \log (7.23 \times 10^{-3})$
$= \log 7.23 + \log 10^{-3}$
$= \log 7.23 - 3 \log 10$
$= 0.8591 - 3(1)$
$= -2.1409$

145 EXPONENTIAL AND LOGARITHMIC FUNCTIONS

The appendix to this text contains common logarithm tables to find common logarithms of numbers and natural logarithm tables to find natural logarithms of numbers. However, if a student has a calculator with "log" and "ln" buttons, then the common log of a number may be easily found by using the "log" button and the natural log of a number may be determined by using the "ln" button.

Antilogarithms: Finding the Number Given Its Logarithm

Consider the problem of finding the value x such that

$$\log x = 3.2625$$

Note that we wish to find that value of x having a common logarithm of 3.2625. Such a problem can be solved by using the "y^x" button of a calculator. Rewriting the equation $\log x = 3.2625$ in exponential form, we have

$$10^{3.2625} = x$$

Using a calculator with a "y^x" button, we find that

$$x = 10^{3.2625} = 1830.2061$$

The value $x = 1830.2061$ is called the **antilogarithm** of 3.2625.

EXAMPLE 3–17 Find x if $\log x = 1.9568$.
Solution Rewriting the equation $\log x = 1.9568$ in exponential form, we have

$$10^{1.9568} = x$$

Using the "y^x" button on the calculator, we determine that

$$x = 10^{1.9568} = 90.5316$$

EXAMPLE 3–18 Find x if $\ln x = 1.9457$.
Solution Rewriting the equation $\ln x = 1.9457$ in exponential form gives us

$$e^{1.9457} = x$$

Using a calculator with an "e^x" button, we find that

$$x = e^{1.9457} = 6.9985$$

APPENDIX   PROOFS OF PROPERTIES OF LOGARITHMS

Note that the properties of logarithms are simply translations of laws of exponents.

Property 1

Let $L_1 = \log_b x$ and $L_2 = \log_b y$. Rewriting these statements in exponential form, we have

$$x = b^{L_1} \qquad y = b^{L_2}$$

Thus,

$$xy = b^{L_1} \cdot b^{L_2}$$
$$= b^{L_1 + L_2}$$

Rewriting the statement $xy = b^{L_1 + L_2}$ in logarithmic form, we obtain

$$\log_b xy = L_1 + L_2$$

Since $L_1 = \log_b x$ and $L_2 = \log_b y$, the preceding statement may be rewritten as

$$\log_b xy = \log_b x + \log_b y$$

Property 2

Let $L_1 = \log_b x$ and $L_2 = \log_b y$. Rewriting these statements in exponential form results in

$$x = b^{L_1} \qquad y = b^{L_2}$$

Thus,

$$\frac{x}{y} = \frac{b^{L_1}}{b^{L_2}}$$
$$= b^{L_1 - L_2}$$

Rewriting the statement $x/y = b^{L_1 - L_2}$ in logarithmic form, we have

$$\log_b \frac{x}{y} = L_1 - L_2$$

Since $L_1 = \log_b x$ and $L_2 = \log_b y$, the preceding statement may be rewritten as

$$\log_b \frac{x}{y} = \log_b x - \log_b y$$

EXPONENTIAL AND LOGARITHMIC FUNCTIONS

Property 3

Let $L = \log_b x$. Rewriting this statement in exponential form, we have

$$x = b^L$$

Raising both sides to the pth power, we obtain

$$x^p = (b^L)^p$$
$$= b^{Lp}$$
$$= b^{pL}$$

Rewriting the statement $x^p = b^{pL}$ in logarithmic form gives us

$$\log_b x^p = pL$$

Since $L = \log_b x$, this statement may be rewritten as

$$\log_b x^p = p \log_b x$$

EXERCISES

1. Rewrite each of the following statements in logarithmic notation:
   - (A) $5^2 = 25$
   - (B) $4^2 = 16$
   - (C) $2^6 = 64$
   - (D) $10^5 = 100{,}000$
   - (E) $10^{-2} = 0.01$
   - (F) $10^1 = 10$
   - (G) $t^w = S$
   - (H) $4^3 = 64$
   - (I) $b^{x+y} = N$

2. Find y for each of the following:
   - (A) $y = \log_3 9$
   - (B) $y = \log_9 81$
   - (C) $y = \log_2 8$
   - (D) $y = \log_3 1$
   - (E) $y = \log_2 16$
   - (F) $y = \log_7 7$
   - (G) $y = \log_{10} 1$
   - (H) $y = \log_{10} 10$
   - (I) $y = \log_{10} 100$
   - (J) $y = \log_{10} 1000$
   - (K) $y = \log_{10} 10{,}000$
   - (L) $y = \log_{10} 100{,}000$
   - (M) $y = \ln 1$
   - (N) $y = \ln e^2$

3. Find each of the following logarithms:
   - (A) $\log_3 81$
   - (B) $\log_2 32$
   - (C) $\log_4 16$
   - (D) $\log_3 1$
   - (E) $\log_5 1$
   - (F) $\log_8 1$
   - (G) $\log 1$
   - (H) $\log 10$
   - (I) $\log 100$
   - (J) $\log 1000$
   - (K) $\log 10{,}000$
   - (L) $\log 100{,}000$

4. Sketch the graph of $y = \log_2 x$. What is the x-intercept?

5. Sketch the graph of $y = \log_3 x$. What is the x-intercept?

6. Given that $\log 2 = 0.3010$ and $\log 7 = 0.8451$, find each of the following using the properties of logarithms:
   - (A) $\log 14$
   - (B) $\log 3.5$
   - (C) $\log \dfrac{2}{7}$
   - (D) $\log 49$
   - (E) $\log 98$
   - (F) $\log 56$
   - (G) $\log \sqrt{2}$
   - (H) $\log \sqrt[3]{2}$
   - (I) $\log \sqrt{7}$
   - (J) $\log \sqrt{14}$
   - (K) $\log \sqrt[5]{7}$
   - (L) $\log \sqrt{98}$

148 CHAPTER THREE

7. Given that log 3.71 = 0.5694, find each of the following using the properties of logarithms. (Remember that log 10 = 1.)
   (A) log 37.1 (B) log 371 (C) log 3710
   (D) log 37,100 (E) log 371,000 (F) log 0.371
   (G) log 0.0371 (H) log 0.00371 (I) log 0.000371

8. Given that ln 3 = 1.098612 and ln 2 = 0.693147, find each of the following using the properties of logarithms:
   (A) ln 6 (B) ln 1.5 (C) ln $\frac{2}{3}$
   (D) ln 81 (E) ln 8 (F) ln 12
   (G) ln $\sqrt{3}$ (H) ln $\sqrt[3]{2}$ (I) ln 0.75

9. Using a calculator with a "log" button, find each of the following:
   (A) log 4.76 (B) log 8.73 (C) log 92.1
   (D) log 4760 (E) log 0.0673 (F) log 0.80

10. Using a calculator with an "ln" button, find each of the following:
    (A) ln 2.8 (B) ln 10 (C) ln 25
    (D) ln 0.15 (E) ln 0.60 (F) ln 80

11. Find x for each of the following:
    (A) log x = 0.7738 (B) log x = 0.9047
    (C) log x = 2.7738 (D) log x = 3.9047
    (E) log x = 1.4698 (F) log x = 5.4099

12. Find x for each of the following:
    (A) ln x = 0.6419 (B) log x = 1.3610
    (C) ln x = 3.6889 (D) ln x = 1.7047

13. Express each of the following numbers as e^k:
    (A) 7 (B) 3.4 (C) 14 (D) 5.5 (E) 750

14. A company's sales, $S(x)$, are related to advertising expenditure, x (in thousands of dollars), by the equation

    $$S(x) = 100{,}000 + 8000 \log(x + 2)$$

    (A) Calculate the sales associated with each of the following advertising expenditures: $1000, $3000, $10,000, $20,000.
    (B) Calculate $S(21) - S(20)$ and interpret the answer.
    (C) If the company increases its advertising expenditure from $30,000 to $31,000, find the corresponding increase in sales.

15. A company's sales, $S(x)$, are related to advertising expenditures, x (in thousands of dollars), by the equation

    $$S(x) = 200{,}000 + 10{,}000 \ln x$$

    (A) Calculate the sales associated with each of the following advertising expenditures: $2000, $5000, $10,000, $30,000.
    (B) Calculate $S(13) - S(12)$ and interpret the answer.
    (C) If the company increases its advertising expenditure from $19,000 to $20,000, find the corresponding increase in sales.

16. A company's sales revenue, $R(x)$, is related to the number of units sold, x, by the equation

    $$R(x) = 10 \log(20x + 1)$$

149 EXPONENTIAL AND LOGARITHMIC FUNCTIONS

(A) Calculate and interpret each of the following: $R(1)$, $R(2)$, $R(5)$, $R(10)$.
(B) Calculate $R(11) - R(10)$ and interpret the answer.
(C) If the company increases its sales from 20 units to 21 units, calculate the increase in sales revenue.

17. A company's total production cost, $C(x)$, is related to the number of units produced, x, by the equation

$$C(x) = 9000 + 10 \ln(x + 1)$$

(A) Calculate and interpret each of the following: $C(0)$, $C(1)$, $C(5)$, $C(10)$.
(B) Calculate $C(19) - C(18)$ and interpret the answer.
(C) If the company increases its production from 12 units to 13 units, calculate the increase in total production cost.

3–4
APPLICATIONS OF LOGARITHMS

Solving Exponential Equations Using Logarithms

Sometimes, we have to solve exponential equations such as

$$e^{x+2} = 100$$

for x. The usual procedure is to rewrite the equation in logarithmic form. Thus, our exponential statement becomes

$$x + 2 = \log_e 100$$
$$= \ln 100$$

Solving for x, we obtain

$$x = \ln 100 - 2$$
$$= 4.605170 - 2$$
$$\approx 2.61$$

EXAMPLE 3–19 Solve $3e^{0.2t+1} = 120$ for t.
Solution Dividing both sides by 3, we have

$$e^{0.2t+1} = 40$$

Rewriting the result in logarithmic notation gives us

$$0.2t + 1 = \log_e 40$$
$$= \ln 40$$

EXAMPLE 3–19 (continued)

Solving for t yields

$$0.2t = \ln 40 - 1$$
$$t = \frac{\ln 40 - 1}{0.2}$$
$$= \frac{3.688879 - 1}{0.2}$$
$$\approx 13.44$$

EXAMPLE 3–20 Solve $10^{3x+2} = 80$ for x.
Solution Rewriting the exponential equation in logarithmic notation results in

$$3x + 2 = \log 80$$

Solving for x, we have

$$3x = \log 80 - 2$$
$$x = \frac{\log 80 - 2}{3}$$
$$= \frac{1.9031 - 2}{3}$$
$$= -0.0323$$

Note that the preceding exponential equations contained either base e or base 10. Thus, when rewritten in logarithmic form, the logarithms were either natural logarithms or common logarithms, respectively, and solution values were calculated by using either natural logarithm or common logarithm tables.

However, it is possible to encounter an exponential equation with a base different from either e or 10. Consider, for example, solving the following exponential equation for x:

$$5^x = 40$$

Rewriting the equation in logarithmic form yields

$$x = \log_5 40$$

Finding the solution for x would be easy if we had base-5 logarithm tables. However, only base-e (natural logarithm) and base-10 (common logarithm) tables are available. For an exponential equation containing a base other than e or 10, we must take either the

EXPONENTIAL AND LOGARITHMIC FUNCTIONS

natural or common logarithm of each side. In the present example, $5^x = 40$, we take the common logarithm of both sides and obtain

$$\log 5^x = \log 40$$

Using the third property of logarithms, we have

$$x \log 5 = \log 40$$

Solving for x yields

$$x = \frac{\log 40}{\log 5}$$
$$= \frac{1.6021}{0.6990}$$
$$\approx 2.2920$$

EXAMPLE 3–21 Solve $7^{3t+2} = 960$ for t.

Solution Taking the natural logarithm of both sides, we have

$$\ln 7^{3t+2} = \ln 960$$

Using the third property of logarithms yields

$$(3t + 2)\ln 7 = \ln 960$$

Solving for t, we get

$$3t + 2 = \frac{\ln 960}{\ln 7}$$

$$3t = \frac{\ln 960}{\ln 7} - 2$$

$$t = \frac{\frac{\ln 960}{\ln 7} - 2}{3}$$

$$= \frac{\frac{6.8669}{1.9459} - 2}{3}$$

$$\approx 0.5096$$

EXAMPLE 3–22 The demand, x (in millions), for a product is related to its unit price, p (in dollars), by the equation

$$p = e^{5-0.2x}$$

EXAMPLE 3-22 (continued)

(A) Express the demand, x, in terms of unit price, p.
(B) Calculate the demand associated with a unit price of $6.

Solution

(A) Rewriting the equation in logarithmic form, we have

$$5 - 0.2x = \ln p$$

Solving for x yields

$$x = \frac{\ln p - 5}{-0.2}$$

(B) Since $p = \$6$, then

$$\begin{aligned} x &= \frac{\ln p - 5}{-0.2} \\ &= \frac{\ln 6 - 5}{-0.2} \\ &= \frac{1.791759 - 5}{-0.2} \\ &\approx 16.04 \text{ million units} \end{aligned}$$

Exponential Functions and Semilogarithmic Graph Paper

Consider an axis system where the vertical axis is graduated in terms of the common logarithms of real numbers and the horizontal axis is a real number line. The graph paper of Figure 3-23 illustrates such an axis system. It is called **semilogarithmic graph paper**. Again, we note that the scale of the vertical axis is measured in terms of the common logarithms of real numbers. Thus, when we plot a point (x, y) on semilogarithmic graph paper, we are really plotting the point $(x, \log y)$.

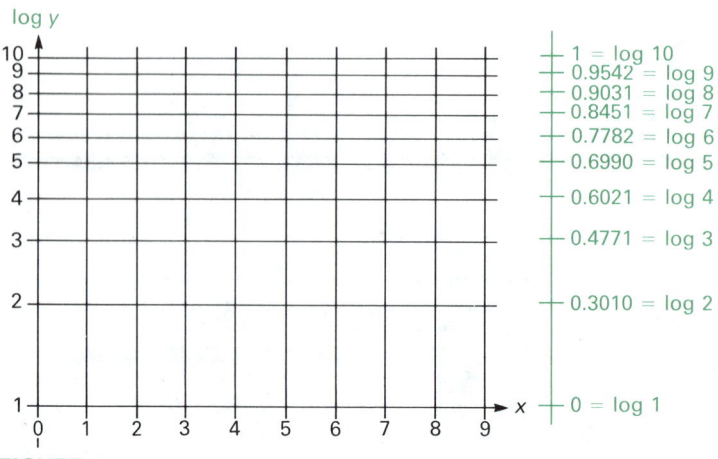

FIGURE 3-23

EXPONENTIAL AND LOGARITHMIC FUNCTIONS

If the points of an exponential function of the form

$$y = ab^x$$

are plotted on semilogarithmic graph paper, they fall in a straight line. Figure 3-24 illustrates such a plotting of points of the exponential function $y = 0.5(2^x)$.

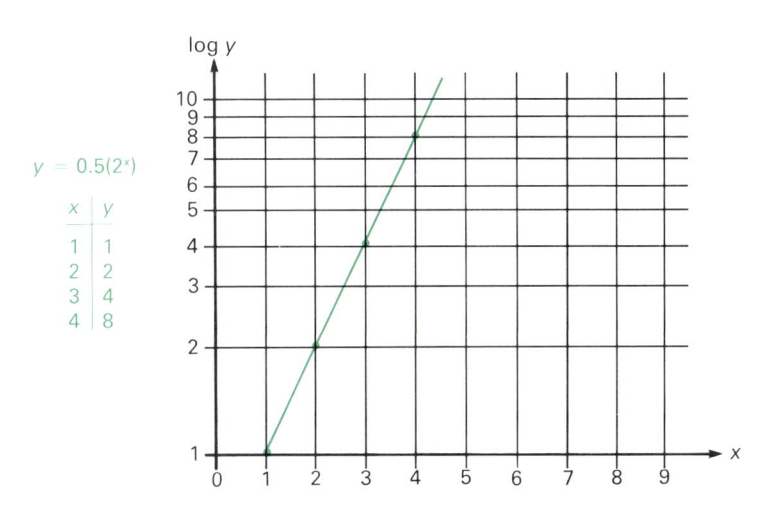

$y = 0.5(2^x)$

x	y
1	1
2	2
3	4
4	8

FIGURE 3-24

This can be explained algebraically by restating the exponential function $y = ab^x$ in logarithmic notation. Taking the common logarithm of both sides, the equation $y = ab^x$ becomes

$$\log y = \log ab^x$$
$$= \log a + \log b^x$$
$$= \log a + x \log b$$

Thus, the exponential function $y = ab^x$ may be restated in logarithmic form as $\log y = \log a + x \log b$. Observe that the logarithmic form expresses a *linear relationship* between the variables $\log y$ and x. Thus, when we plot (on semilogarithmic graph paper) a point (x, y) of the exponential function $y = ab^x$, we are really plotting the point $(x, \log y)$. Since the variables x and $\log y$ are linearly related by the logarithmic equation $\log y = \log a + x \log b$, the result is a straight line.

Semilogarithmic graph paper is quite useful if one is trying to determine whether a set of data points (x, y) is approximated by an exponential function. A quick test is to plot the data on semilogarithmic graph paper. The closer the points fall in a straight line, the more likely the data can be approximated by an exponential function (see Exercise 7 at the end of this section).

154 CHAPTER THREE

EXERCISES

1. Solve each of the following for x:
   - (A) $e^{x+5} = 73$
   - (B) $e^{2x+4} = 15$
   - (C) $4e^{3x-1} = 80$
   - (D) $7e^{-2x+6} = 42$
   - (E) $10^{5x+1} = 36$
   - (F) $10^{4x-3} = 800$

2. Solve each of the following for x:
   - (A) $7^x = 81$
   - (B) $6^{x+2} = 34$
   - (C) $2^{3x-1} = 15$
   - (D) $e^{3x-5} = 712$
   - (E) $e^{x-4} = 83$
   - (F) $4^{3x-2} = 47$

3. The demand, x (in billions), for a product is related to its unit price, p (in dollars), by the equation

   $$p = e^{8-0.3x}$$

   - (A) Express demand, x, in terms of unit price, p.
   - (B) Find the demand associated with a unit price of $9.

4. The supply, y (in millions), for a product is related to its unit price, p (in dollars), by the supply equation

   $$p = e^{1+0.3y}$$

   - (A) Solve for y in terms of p.
   - (B) Calculate the supply corresponding to a unit price of $8.

5. Calculate the y-coordinates of the following points of the exponential function $y = (1/4)2^x$: $(1, \_)$, $(2, \_)$, $(3, \_)$, $(4, \_)$, $(5, \_)$. Plot these points on semilogarithmic graph paper. The points fall in the path of what type of line?

6. Rewrite the exponential function $y = 5 \cdot 3^x$ in logarithmic form.

7. A set of data points relating some quantity of interest with time is called a **time series**. Determine whether the time series data of Table 3–5 may be approximated by an exponential function of the form $y = ab^x$.

TABLE 3–5

| Annual Earnings per Share ABC Corporation ||
Year	EPS
1980	0.90
1981	1.90
1982	5.00
1983	8.50
1984	15.80

EXPONENTIAL AND LOGARITHMIC FUNCTIONS

8. The population, P, of a certain town is related to time, t (in years), by the exponential function

$$P = 1000e^{0.05t}$$

(A) How long will it take for the population to double itself?
(B) How long will it take for the population to triple itself?

CASE A STOCK PRICE FORECASTING—MICROCHIPS, INC.

For a given stock, the ratio

$$\frac{\text{price per share}}{\text{earnings per share}}$$

is called the **price earnings ratio** (denoted **P/E ratio**). The P/E ratio indicates the amount of money a potential buyer (i.e., the market) is willing to pay for a dollar of the firm's earnings. Thus, part of the process of forecasting the price of a stock involves multiplying the firm's P/E ratio by its projected earnings per share.

A security analyst is forecasting the price of a high-technology growth stock called Microchips, Inc. The following data express the earnings per share history of Microchips, Inc.:

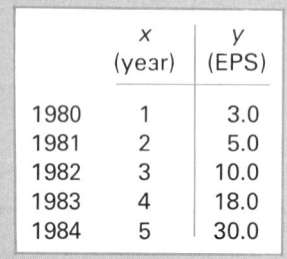

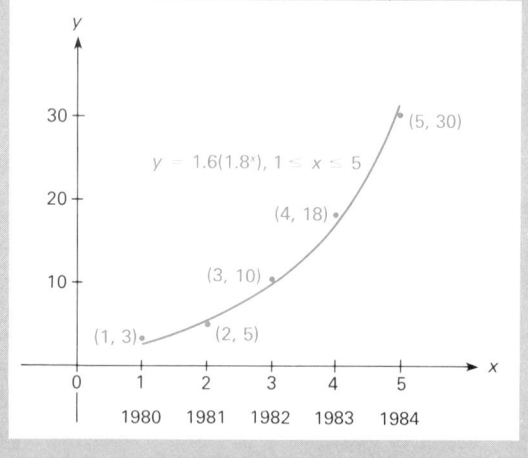

	x (year)	y (EPS)
1980	1	3.0
1981	2	5.0
1982	3	10.0
1983	4	18.0
1984	5	30.0

FIGURE 1

The data points (x, y) are plotted in Figure 1. Using a curve-fitting technique, the analyst obtains

$$y = 1.6(1.8^x) \quad (1 \leq x \leq 5)$$

as the equation of the best-fitting curve to the set of data points (x, y) of Figure 1. Note that this equation is valid for $1 \leq x \leq 5$. However, the analyst makes the assumption that the relationship between EPS and time [as defined by the equation $y = 1.6(1.8^x)$] will continue for the next 3 years and uses the equation

$$y = 1.6(1.8^x)$$

to forecast the earnings per share for 1985 by substituting $x = 6$, to obtain

$$y = 1.6(1.8^6)$$
$$= 54.42$$

Thus, Microchips, Inc.'s projected EPS for 1985 is $54.42. Multiplying this result by Microchip's P/E ratio of 7 yields

$$7(54.42) = \$380.94$$

as the forecasted price per share for 1985. Since Microchips, Inc. is currently selling for $285 per share and has successfully met a checklist of the analyst's investment criteria, the analyst issues a "buy" recommendation.

EXERCISES

1. Forecast the 1986 EPS and price per share for Microchips, Inc.
2. Forecast the 1987 EPS and price per share for Microchips, Inc.
3. Often, amid speculation of EPS increases, the market places a higher value on a dollar of a firm's earnings by increasing its P/E ratio. Suppose the P/E ratio of Microchips, Inc. increases to 8. Find its projected prices per share for 1985, 1986, and 1987.

MATHEMATICS OF FINANCE

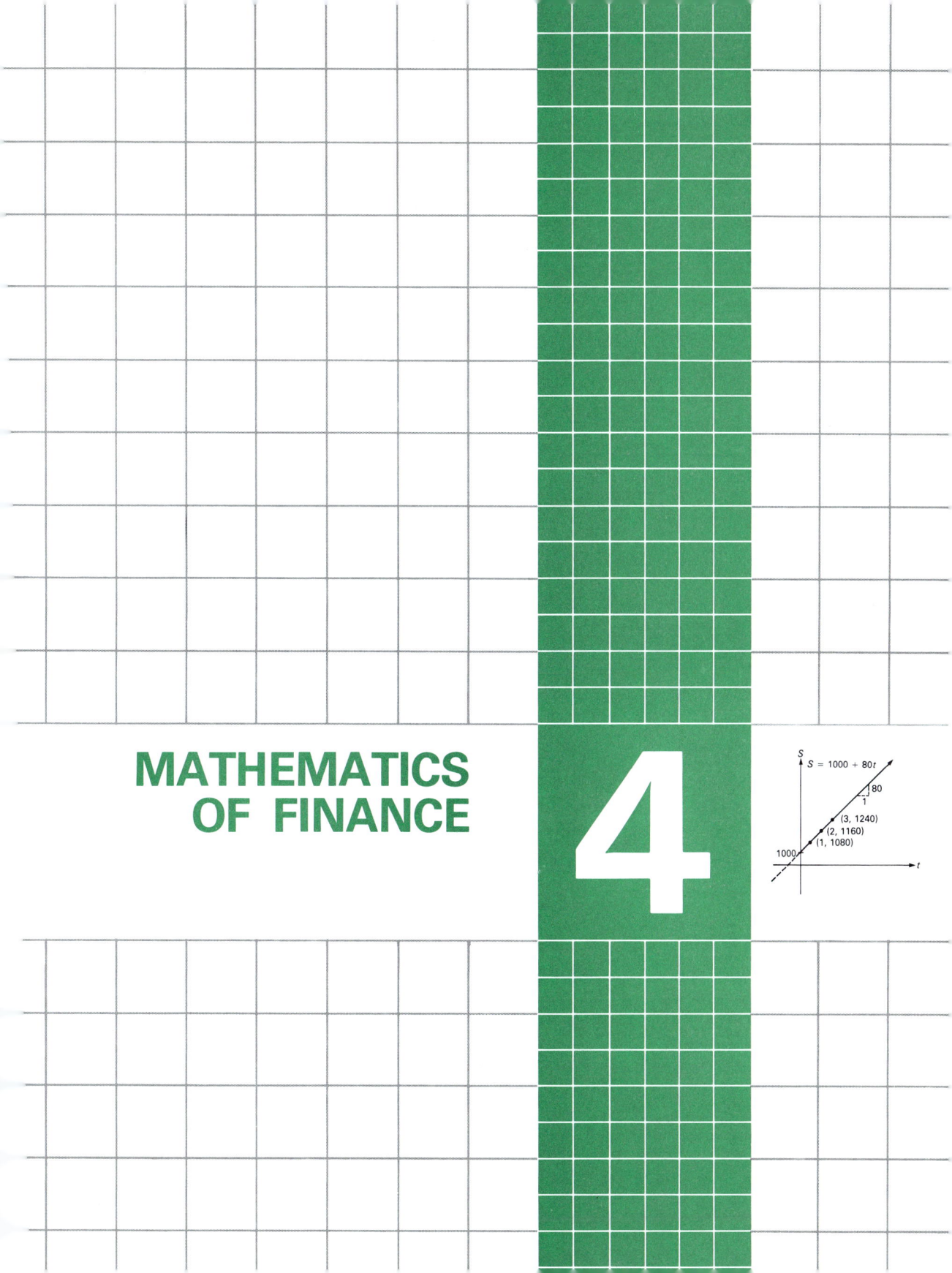

4

$S = 1000 + 80t$

(3, 1240)
(2, 1160)
(1, 1080)
1000

4–1

SIMPLE INTEREST

Interest is the price paid for the use of money. The amount of money lent (or borrowed or invested) is called the **principal**. After a given time period, the borrower must repay the principal plus interest. The interest is computed as a percentage of the principal. This percentage is called the **interest rate.** An interest rate is stated as a specified percentage per unit of time. In this section, the unit of time will be a year. Thus, an interest rate of 5% means 5% per year. Specifically, if Henry Orwell borrows $1000 for 2 years at an interest rate of 5%, then the principal is $1000, the interest rate is 5% per year, and the time period is 2 years. The following notation will be used:

P = principal (original amount lent, borrowed, or invested)
r = interest rate per year
t = time (duration of loan or investment in years)
I = amount of simple interest in dollars
S = total amount (principal + interest) due at the end of the time period

Simple interest is computed by the following formula.

$$I = Prt$$

Thus, the interest on Henry Orwell's 2-year, $1000 loan (at 5% per year) is

$$I = 1000(0.05)(2) = \$100$$

The total amount, S, that Henry must repay is calculated by the formula

$$S = P + I$$

Hence, at the end of 2 years, Henry must repay

$$S = \$1000 + \$100 = \$1100$$

The bank views this loan as an investment. Its principal of $1000 is worth $1100 at the end of 2 years. Thus, the total amount (in this case, $1100) is often called the **future value,** whereas the prin-

MATHEMATICS OF FINANCE

cipal (in this case, $1000) is called the **present value** (see the time diagram of Figure 4–1).

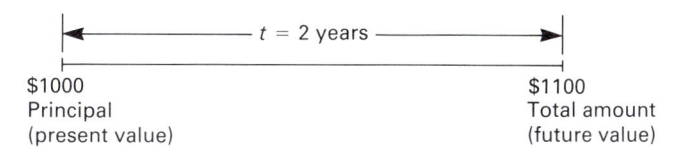

```
|←――――――― t = 2 years ―――――――→|
$1000                              $1100
Principal                          Total amount
(present value)                    (future value)
```

FIGURE 4–1

EXAMPLE 4–1 An investor lent $10,000 to a business associate for 6 months at an interest rate of 8% per year.
(A) Calculate the simple interest.
(B) Calculate the total amount (future value).
Solutions Here, $P = \$10,000$, $r = 0.08$, and $t = 1/2$ year.
(A) $I = Prt$

$$= 10{,}000(0.08)\left(\frac{1}{2}\right)$$

$$= \$400$$

(B) $S = P + I$

$$= \$10{,}000 + \$400$$

$$= \$10{,}400$$

Thus, the investor's $10,000 is worth $10,400 at the end of 6 months (see the time diagram of Figure 4–2).

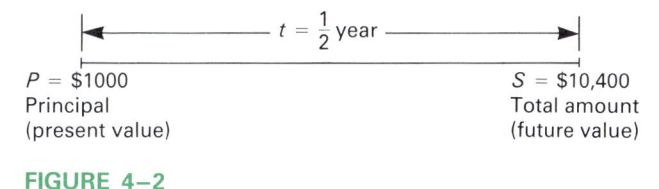

```
|←―――――― t = ½ year ――――――→|
P = $1000                          S = $10,400
Principal                          Total amount
(present value)                    (future value)
```

FIGURE 4–2

Value of Money at Simple Interest

If a sum of money, P, is invested at simple interest, its value increases by the same amount each year. This can be shown by beginning with the formula for total amount

$$S = P + I$$

and replacing I with the equivalent expression Prt. The result is

$$S = P + Prt$$

Given values for P and r, this formula expresses a linear relationship between total amount, S, and time, t. The slope is Pr. Specifically, if a person invests $1000 at 8% per year, after 1 year,

$$S = 1000 + 1000(0.08)(1)$$
$$= \$1080$$

after 2 years,

$$S = 1000 + 1000(0.08)(2)$$
$$= \$1160$$

after 3 years,

$$S = 1000 + 1000(0.08)(3)$$
$$= \$1240$$

and after t years,

$$S = 1000 + 1000(0.08)t$$
$$= 1000 + 80t$$

Note that the slope, 80, is the yearly increase in S. The linear function defined by $S = 1000 + 80t$ is sketched in Figure 4–3.

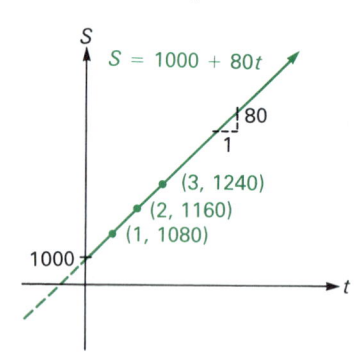

FIGURE 4–3

Calculating the Present Value

Sometimes, it is necessary to calculate the present value (principal) given the future value (total amount). Thus, we now derive a formula for present value, P. We begin with

$$S = P + I$$

and replace I with the equivalent expression Prt. The result is

$$S = P + Prt$$

MATHEMATICS OF FINANCE

Factoring out P, we obtain

$$S = P(1 + rt)$$

Solving for P yields the equation

$$P = \frac{S}{1 + rt}$$

EXAMPLE 4-2 What amount of money should be invested now at 6% per year to yield a future value of $8000 seven months from now?
Solution Since $S = \$8000$, $r = 0.06$, and $t = 7/12$ year, then

$$P = \frac{S}{1 + rt}$$

$$= \frac{8000}{1 + 0.06\left(\frac{7}{12}\right)}$$

$$= \frac{8000}{1 + 0.035}$$

$$= \$7729.47 \quad \text{(See Figure 4-4.)}$$

$$\longleftarrow t = \frac{7}{12} \text{ year} \longrightarrow$$

$P = \$7729.47$ $S = \$8000$
Present value   Future value

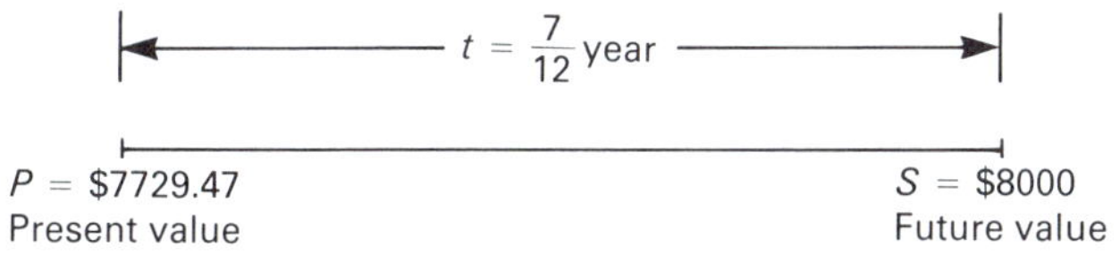

FIGURE 4-4

Sometimes, a loan is transacted by the borrower giving the lender a signed paper promising to pay a specified amount by a given date. Such an instrument is called a **note**. The future value of the loan is called the **future value,** or the **maturity value,** of the note. To illustrate, suppose Henry Orwell transacted his 2-year, $1000 loan by giving the bank a note. Then, $1100, the future value of the loan, is also the future or maturity value of the note. The bank is called the **holder** of the note.

The holder of a note may sell the note prior to maturity. The buyer usually pays an amount less than the maturity value. When the borrower repays the loan (with interest), the money (maturity

164 CHAPTER FOUR

value of the note) is automatically transferred to the buyer (now the present holder) of the note.

EXAMPLE 4–3 Willis Harcase wishes to buy a note that has a maturity value of $3000. The note is due 3 years from today. If Willis wants to earn 9% per year on his invested money, how much should he pay for the note?

Solution The time diagram of Figure 4–5 illustrates this problem. Willis must calculate the present value of the note.

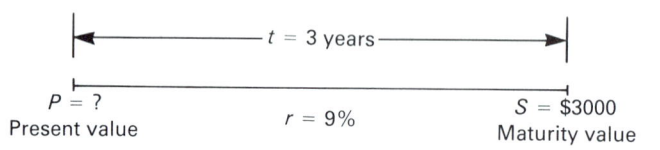

$P = ?$ Present value
$t = 3$ years
$r = 9\%$
$S = \$3000$ Maturity value

FIGURE 4–5

Here,

$$P = \frac{S}{1 + rt}$$

$$= \frac{3000}{1 + 0.09(3)}$$

$$= \frac{3000}{1 + 0.27}$$

$$= \$2362.20$$

Thus, if Willis pays $2362.20 for the note today, he will earn 9% per year on his money when the note is repaid in 3 years.

EXAMPLE 4–4 Melanie borrows a sum of money from Harry at 10% per year. She gives Harry a 5-year note with a maturity value of $15,000. How much did Melanie borrow from Harry?

Solution See the time diagram of Figure 4–6. Calculating the present value, we have

$$P = \frac{S}{1 + rt}$$

$$= \frac{15{,}000}{1 + 0.10(5)}$$

$$= \$10{,}000$$

MATHEMATICS OF FINANCE

Therefore, Melanie borrowed $10,000 from Harry.

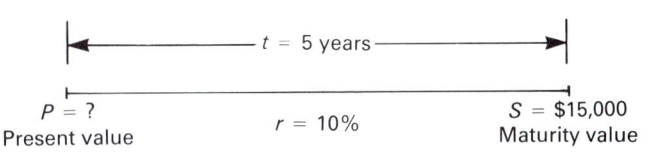

$P = ?$ Present value
$r = 10\%$
$t = 5$ years
$S = \$15,000$ Maturity value

FIGURE 4–6

EXAMPLE 4–5 Referring to Example 4–4, suppose Harry sold the note to Jacob three years before maturity. What amount did Jacob pay for the note if his invested money is earning 11% per year?

Solution The time diagram of Figure 4–7 illustrates this problem. We must calculate the present value of $15,000 three years before maturity. The simple interest rate is 11%.

$P = \$10,000$
Present value
5 years before maturity

$P = ?$
Present value
3 years before maturity

$r = 11\%$

$S = \$15,000$
Maturity value

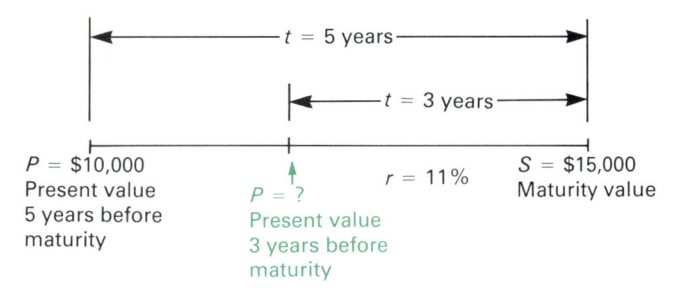

$t = 5$ years
$t = 3$ years

FIGURE 4–7

Hence,

$$P = \frac{S}{1 + rt}$$

$$= \frac{15{,}000}{1 + 0.11(3)}$$

$$= \$11{,}278.20$$

Thus, Jacob paid $11,278.20 for the note. His profit is $15,000.00 − $11,278.20 = $3721.80. Harry's profit is $11,278.20 − $10,000.00 = $1278.20.

Simple Discount Note

In simple interest problems, the simple interest rate, r, is applied to the principal, P. Sometimes, a loan is transacted by a **discount note**. In such a case, the cost of borrowing is called the **discount, D**. The discount is computed as a percentage of the maturity value,

S. This percentage is called the **discount rate, d.** Thus if t is the length of time (in years) that the money is borrowed, then

$$\text{discount} = (\text{maturity value})(\text{discount rate})(\text{length of time})$$

or

$$D = Sdt$$

Specifically, consider the following example. Sam Schultz borrows a sum of money for 9 months by giving his bank a discount note for $1000. The discount rate is 8%. Thus, we have $S = \$1000$, $d = 0.08$, and $t = 9/12 = 3/4$ year. Hence, the discount is

$$D = Sdt$$
$$= 1000(0.08)\left(\frac{3}{4}\right)$$
$$= \$60$$

Although the note is written in terms of the maturity value of $1000, Sam only receives

$$\$1000 - \$60 = \$940$$

This amount is called the **proceeds, B,** of the note. In general, the proceeds of a discount note are calculated by the equation

$$\text{proceeds} = \text{maturity value} - \text{discount}$$

or

$$B = S - D$$

This situation is illustrated by the time diagram of Figure 4–8.

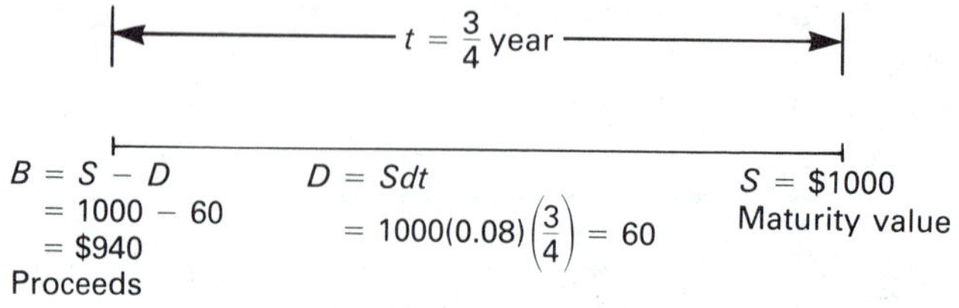

$$\xleftarrow{\qquad t = \tfrac{3}{4} \text{ year} \qquad}$$

$B = S - D$ $D = Sdt$ $S = \$1000$
$= 1000 - 60$ $= 1000(0.08)\left(\frac{3}{4}\right) = 60$ Maturity value
$= \$940$
Proceeds

FIGURE 4–8

MATHEMATICS OF FINANCE

Since a discount note is written in terms of its maturity value, but the borrower receives the lesser amount, called the proceeds, then the following problem arises when a discount note is written:

What should the maturity value, S, of a discount note be in order to yield a given amount of proceeds, B, at a specified discount rate, d?

To answer this question, we determine a formula for S in the following manner. We start with the expression

$$B = S - D$$

Since $D = Sdt$, substitution into this equation yields

$$B = S - Sdt$$
$$= S(1 - dt)$$

Solving for S, we have

$$S = \frac{B}{1 - dt}$$

Thus, if Sam Schultz wishes to receive proceeds of $950, the note must be written with maturity value

$$S = \frac{950}{1 - 0.08\left(\frac{3}{4}\right)}$$
$$= \frac{950}{1 - 0.06}$$
$$= \$1010.64$$

If a discount note is analyzed from a simple interest point of view, the simple interest rate is usually larger than the discount rate. For example, if we consider Sam Schultz's first discount rate for $1000 with proceeds of $940, we realize that Sam paid $60 interest to borrow $940 for 3/4 year. If we begin with the formula for simple interest

$$I = Prt$$

and solve for the simple interest rate, r, we have

$$r = \frac{I}{Pt}$$

$$= \frac{60}{940\left(\frac{3}{4}\right)}$$

$$= \frac{60}{705}$$

$$= 0.0851 = 8.51\%$$

Thus, a discount rate of 8% for 3/4 year is equivalent to a simple interest rate of 8.51% for that same time period.

EXERCISES

1. Find the simple interest and total amount of each of the following loans:
   (A) $1000 for 3 years at 7%
   (B) $10,000 for 2 years at 6%
   (C) $5000 for 6 months at 8%
   (D) $8000 for 3 months at 12%
   (E) $2000 for 4 months at 9%
   (F) $9000 for 1 year at 10%

2. A man invests $10,000 at 9% per year.
   (A) Determine the equation that expresses total amount, S, as a function of time, t (in years).
   (B) Graph the equation of part (A).

3. What amount of money should be invested now at 8% per year to yield a future value of $10,000 nine months from now?

4. A note with a maturity value of $1000 matures in 5 months. If one wishes to earn 9% per year, how much should be paid for the note now?

5. Sam Smith plans to buy a note with a maturity value of $10,000. The note is due 5 years from now. If Sam wishes to earn 10% per year on his invested money, how much should he pay for the note?

6. Helen borrows a sum of money from Tom at 9% per year by giving Tom a 6-month note with a maturity value of $9000.
   (A) How much money did Helen borrow from Tom?
   (B) Two years before maturity, Tom sells the note to Susan. What amount does Susan pay for the note if she is earning 10% per year?

7. Ellen Rydell borrows a sum of money for 5 years by giving her bank a discount note for $6000. The discount rate is 9%.
   (A) Find the discount.
   (B) Find the proceeds.
   (C) Find the equivalent simple interest rate.

8. A woman borrows a sum of money for 6 months by giving her bank a discount note for $8000. The discount rate is 10%.

(A) Find the discount.
(B) Find the proceeds.
(C) How much money did the woman borrow?
(D) Find the equivalent simple interest rate.

9. Glenn Nash borrows $5000 for 4 years by giving his bank a discount note. The discount rate is 7%.
   (A) Find the maturity value of the note.
   (B) Find the equivalent simple interest rate.

10. A man borrows $5000 from his bank for 9 months by using a discount note. The discount rate is 12%.
    (A) Find the maturity value of the note.
    (B) Find the equivalent simple interest rate.

11. Find the maturity value of an interest note for $600 issued for 4 months at an interest rate of 6% per year.

12. Find the present value of a 3-month interest note with a maturity value of $6000. Assume that the interest rate is 8% per year.

4-2

COMPOUND INTEREST

In Section 4-1, we learned that the value of a sum of money, P, invested at simple interest increases by the same amount each year. This is due to the fact that the interest rate, r, is applied to the original principal, P. In this section, we will discuss a method of computing interest in which the value of a sum of money, P, increases by a larger amount each year. Here, the interest rate, r, will be applied to the original principal plus interest rather than to just the original principal. Such a method of computing interest is called **compounding**. The result is **compound interest**.

Compound interest is usually computed periodically throughout the year. If compound interest is computed every month (12 times a year), it is said to be **compounded monthly**. Each month is called a **conversion period**, or **interest period**. If compound interest is computed every 3 months (4 times a year), it is said to be **compounded quarterly**. Thus, each 3-month time interval between successive compoundings is a conversion or interest period. If compound interest is computed every 6 months (2 times a year), it is said to be **compounded semiannually**. Hence, each 6-month time interval between successive compoundings is a conversion or interest period. Interest may also be compounded annually, weekly, daily, and continuously.

We now compute the total amount of $1000 invested for 2 years at 8% compounded semiannually. Note that $P = \$1000$, and $r = 0.08$. Since the interest is compounded semiannually, there will be two compoundings per year (see Figure 4-9).

170 CHAPTER FOUR

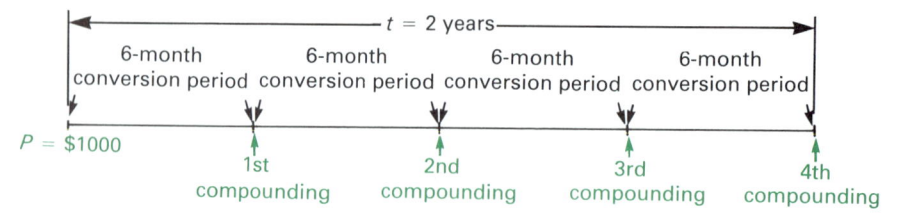

```
          |←―――――――――― t = 2 years ――――――――――→|
   6-month        6-month         6-month         6-month
conversion period conversion period conversion period conversion period
      ↓               ↓               ↓               ↓
P = $1000 ↑           ↑               ↑               ↑
         1st          2nd             3rd             4th
      compounding  compounding    compounding     compounding
```

FIGURE 4–9

Thus, there will be four compoundings and four conversion periods during the 2-year time interval. We now show each compounding.

1. **First Compounding** (at End of First Conversion Period)—The interest is calculated and added to the original principal. Hence,

$$I = Prt$$
$$= 1000(0.08)\left(\frac{1}{2}\right)$$
$$= 1000(0.04)$$
$$= \$40$$
$$S = P + I$$
$$= \$1000 + \$40$$
$$= \$1040$$

Thus, the original investment is now worth $1040 (see Figure 4–10).

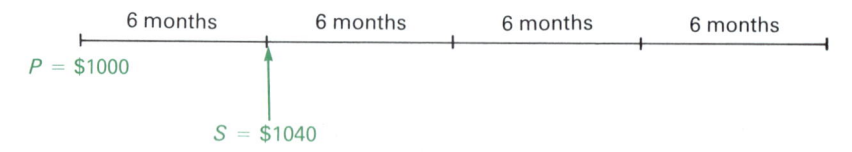

```
       6 months        6 months        6 months        6 months
      |――――――――|――――――――|――――――――|――――――――|
P = $1000      ↑
            S = $1040
```

FIGURE 4–10

2. **Second Compounding** (at End of Second Conversion Period)—The total amount from the previous compounding becomes the *new principal.* Interest is calculated on the new principal. The total amount is calculated by adding the interest to the new principal. Hence, we have

$$I = \text{new principal} \cdot r \cdot t$$
$$= 1040(0.08)\left(\frac{1}{2}\right)$$
$$= 1040(0.04)$$
$$= \$41.60$$

MATHEMATICS OF FINANCE

$$S = \text{new principal} + I$$
$$= \$1040.00 + \$41.60$$
$$= \$1081.60$$

Thus, the original investment is now worth $1081.60 (see Figure 4–11).

```
     6 months      6 months      6 months      6 months
|---------------|---------------|---------------|---------------|
P = $1000       ↑              ↑
              S = $1040
                          S = $1081.60
```

FIGURE 4–11

3. **Third Compounding** (at End of Third Conversion Period)—The process is repeated, and we obtain

$$I = \text{new principal} \cdot r \cdot t$$
$$= 1081.60(0.08)\left(\frac{1}{2}\right)$$
$$= 1081.60(0.04)$$
$$= \$43.26$$

$$S = \text{new principal} + I$$
$$= \$1081.60 + \$43.26$$
$$= \$1124.86$$

Hence, the original investment is now worth $1124.86 (see Figure 4–12).

```
     6 months      6 months      6 months      6 months
|---------------|---------------|---------------|---------------|
P = $1000       ↑              ↑              ↑
              S = $1040
                          S = $1081.60
                                        S = $1124.86
```

FIGURE 4–12

4. **Fourth Compounding** (at End of Fourth Conversion Period)—The process is repeated, giving us

$$I = \text{new principal} \cdot r \cdot t$$
$$= 1124.86(0.08)\left(\frac{1}{2}\right)$$
$$= 1124.86(0.04)$$
$$= \$44.99$$

$$S = \text{new principal} + I$$
$$= \$1124.86 + \$44.99$$
$$= \$1169.85$$

Thus, the original investment is now worth $1169.85 (see Figure 4–13). This total is often called the **compound amount**.

```
     6 months      6 months      6 months      6 months
|---------------|---------------|---------------|---------------|
P = $1000       ↑               ↑               ↑               ↑
             S = $1040
                            S = $1081.60
                                            S = $1124.86
                                                            S = $1169.85
```

FIGURE 4–13

Observe that, for each compounding in our example, the new principal is always multiplied by $(0.08)(1/2) = 0.04$. This value is called the **interest rate per conversion period**. In general, the interest rate per conversion period will be indicated by the symbol i. Hence, for the previous example, we write

$$i = 0.08\left(\frac{1}{2}\right) = 0.04$$

Thus, 8% compounded semiannually is equivalent to 4% per conversion period. Note that the interest rate per conversion period may be calculated by dividing the quoted interest rate by the number of conversion periods per year. Thus, if

$r = $ quoted interest rate (or **nominal rate**)

$m = $ number of conversion periods per year

then

$$i = \frac{r}{m}$$

Observe, also, that the number of conversion periods, n, may be calculated by multiplying the number of conversion periods per year, m, by the number of years, t. Hence,

$$n = mt$$

MATHEMATICS OF FINANCE

Thus, for the illustrated example,

$$m = 2 \text{ conversion periods per year}$$
$$t = 2 \text{ years}$$
$$n = 2(2) = 4 \text{ conversion periods}$$

EXAMPLE 4–6 A sum of money is invested for 4 years at 6% compounded monthly. Calculate i and n.

Solution Since $r = 6\%$, $m = 12$ conversion periods per year, and $t = 4$ years, then

$$i = \frac{r}{m}$$
$$= \frac{6\%}{12} = \frac{1}{2}\% \text{ per month}$$
$$n = mt$$
$$= 12(4) = 48 \text{ conversion periods}$$

We will now derive a general formula for the compound amount, S. If

$$P = \text{original principal}$$
$$i = \text{interest rate per conversion period}$$
$$n = \text{total number of conversion periods}$$
$$S = \text{compound amount (or total amount or maturity value or accumulated value)}$$

then, at the end of the first conversion period,

$$I = Pi$$
$$S = P + I$$
$$= P + Pi = P(1 + i)$$

At the end of the second conversion period,

$$\text{new principal} = P(1 + i)$$
$$I = \text{new principal} \cdot i$$
$$= P(1 + i)i$$
$$S = \text{new principal} + I$$
$$= P(1 + i) + P(1 + i)i$$
$$= P(1 + i)(1 + i) = P(1 + i)^2$$

At the end of the third conversion period,

$$\text{new principal} = P(1 + i)^2$$
$$I = \text{new principal} \cdot i$$
$$= P(1 + i)^2 i$$
$$S = \text{new principal} + I$$
$$= P(1 + i)^2 + P(1 + i)^2 i$$
$$= P(1 + i)^2(1 + i) = P(1 + i)^3$$

And at the end of the nth conversion period,

$$S = P(1 + i)^n$$

The time diagram of Figure 4–14 summarizes the preceding calculations.

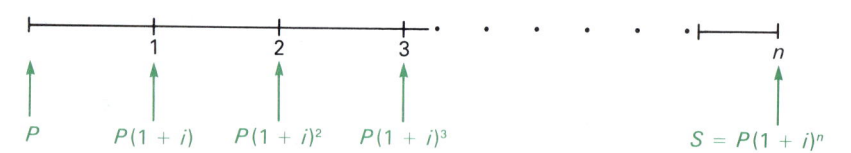

FIGURE 4–14

Thus, if we invest P dollars for n periods at an interest rate per conversion period of i, the compound amount, S, is given by

$$S = P(1 + i)^n$$

Table 4 in the Appendix at the end of the text gives values of $(1 + i)^n$ for various rates i and number of periods n. Using the formula $S = P(1 + i)^n$ to compute the compound amount for the introductory problem of this section, we have

$$S = 1000(1 + 0.04)^4$$

We now use Table 4 of the Appendix to determine the value of $(1 + 0.04)^4$. Looking at the top of the table, we locate the "$i = 4\%$" column. Then, moving four periods down in that column, we find $(1 + 0.04)^4 = 1.169859$. Hence,

$$S = 1000(1.169859)$$
$$= \$1169.86$$

EXAMPLE 4–7 Find the compound amount of $10,000 invested for 5 years at 8% compounded quarterly.
Solution Here, $P = \$10{,}000$, $r = 8\%$, $m = 4$, and $t = 5$ years. Thus,

$$i = \frac{r}{m} = \frac{8\%}{4} = 2\% \text{ per conversion period}$$

$$n = mt = 4(5) = 20 \text{ conversion periods}$$

$$\begin{aligned} S &= P(1 + i)^n \\ &= 10{,}000(1 + 0.02)^{20} \quad \text{From Appendix Table 4,} \\ &= 10{,}000(1.485947) \quad\ \ (1 + 0.02)^{20} = 1.485947. \\ &= \$14{,}859.47 \quad \text{(See Figure 4–15.)} \end{aligned}$$

|←——————— $t = 5$ years ———————→|

$P = \$10{,}000$ $S = \$14{,}859.47$

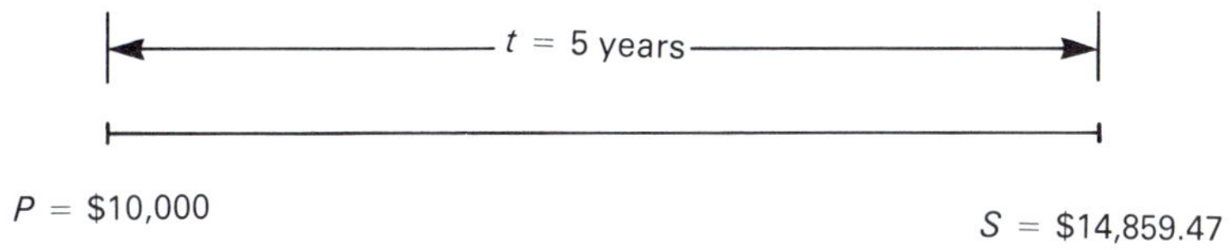

FIGURE 4–15

Value of Money at Compound Interest

If a sum of money is invested at compound interest, its value increases exponentially with time. This can be shown by beginning with the formula for compound amount

$$S = P(1 + i)^n$$

and replacing n with the equivalent expression mt. The result is

$$S = P(1 + i)^{mt}$$

where m is an integer. Applying a law of exponents, the equation becomes

$$S = P[(1 + i)^m]^t$$

Thus, S is a function of t. Given values for P, i, and m, the function is an exponential equation of the form $y = ab^x$, $b > 1$. It is sketched in Figure 4–16.

FIGURE 4–16

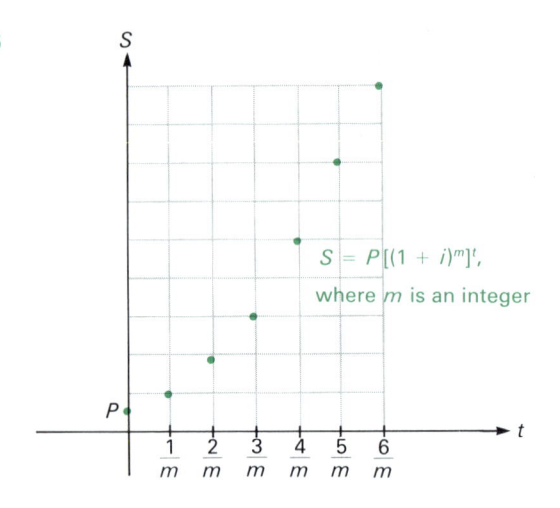

$S = P[(1 + i)^m]^t$, where m is an integer

Present Value at Compound Interest

Sometimes, it is necessary to calculate the present value, P, given the compound amount. A formula for P may be derived by beginning with the equation for compound amount

$$S = P(1 + i)^n$$

and solving for P. The result is

$$P = S(1 + i)^{-n}$$

This formula is used to find the present value of a given compound amount. Appendix Table 5 gives values of $(1 + i)^{-n}$ for various values of i and n.

EXAMPLE 4–8 What sum of money should be invested for 4 years at 9% compounded monthly in order to provide a compound amount of $8000?
Solution Here, $S = \$8000$, $r = 9\%$, $m = 12$, and $t = 4$ years. Thus,

$$i = \frac{r}{m} = \frac{9\%}{12} = \frac{3}{4}\% \text{ per month}$$

$n = mt = 12(4) = 48$ conversion periods

$P = S(1 + i)^{-n}$

$ = 8000(1 + 0.0075)^{-48}$ From Appendix Table 5,

$ = 8000(0.698614)$ $(1 + 0.0075)^{-48} = 0.698614$.

$ = \$5588.91$ (See Figure 4–17.)

MATHEMATICS OF FINANCE

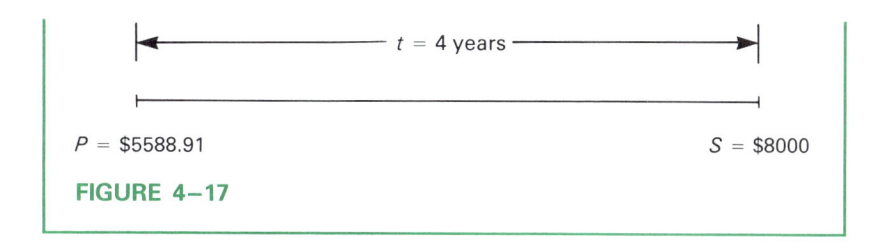

```
|←——————— t = 4 years ———————→|
|—————————————————————————————|
P = $5588.91                                    S = $8000
```

FIGURE 4–17

Effective Annual Interest Rate

Consider the following question:

What percentage compounded annually is equivalent to 8% compounded quarterly?

The answer is called the **effective annual interest rate**. This interest rate may be computed by the ratio

$$\frac{\text{interest earned in 1 year}}{\text{principal at the beginning of the year}}$$

If a sum of P dollars is compounded m times a year, then at the end of the year, it will accumulate into

$$P(1 + i)^m$$

The interest earned in 1 year is

$$P(1 + i)^m - P$$

Thus, the effective annual interest rate is

$$\frac{P(1 + i)^m - P}{P} = (1 + i)^m - 1$$

We therefore have the following definition.

---

effective annual interest rate $= (1 + i)^m - 1$

where $m =$ number of conversion periods for 1 year and $i =$ interest rate per conversion period.

Specifically, if money is invested at 8% compounded quarterly, then $m = 4$, $r = 8\%$, and $i = r/m = 8\%/4 = 2\%$, then we calculate

$$(1 + i)^m - 1 = (1 + 0.02)^4 - 1$$
$$= 1.082432 - 1$$
$$= 8.24\%$$

Thus, 8% compounded quarterly is equivalent to 8.24% compounded annually.

"Double, Triple, Quadruple... Your Money"

Consider the following question:

How long does it take a principal to double itself at 8% compounded quarterly?

The answer is found by beginning with the formula for compound amount

$$S = P(1 + i)^n$$

and substituting $2P$ for S. Hence,

$$2P = P(1 + i)^n$$

Dividing by P, we obtain

$$2 = (1 + i)^n$$

We must now solve the equation for n. Taking the common logarithm of both sides, we get

$$\log 2 = \log(1 + i)^n$$
$$= n \log(1 + i)$$

Solving for n yields

$$n = \frac{\log 2}{\log(1 + i)}$$

Since $i = 8\%/4 = 2\% = 0.02$, then

$$n = \frac{\log 2}{\log 1.02}$$
$$= \frac{0.3010}{0.0086}$$
$$= 35 \text{ conversion periods}$$

Thus, the principal, P, will double itself after $n = 35$ conversion periods or $35/4 = 8\frac{3}{4}$ years (since there are 4 conversion periods per year).

EXERCISES

1. Find the compound amount and interest for each of the following situations:
   - (A) $P = \$1000$, $r = 0.08$, $m = 4$, $n = 40$
   - (B) $P = \$5000$, $r = 0.08$, $m = 2$, $n = 30$
   - (C) $P = \$8000$, $r = 0.06$, $m = 1$, $n = 10$
   - (D) $P = \$3000$, $r = 0.12$, $m = 3$, $n = 36$

2. Find the compound amount and interest for each of the following investments:
   (A) $5000 invested at 4% compounded quarterly for 6 years
   (B) $10,000 invested at 8% compounded semiannually for 15 years
   (C) $3000 invested at 12% compounded monthly for 4 years
   (D) $20,000 invested at 6% compounded semiannually for 11 years

3. A man borrows $10,000 from a bank. The bank charges interest at the rate of 8% compounded quarterly. Ten years later, the man repays the loan in a lump-sum payment.
   (A) Find the lump-sum payment.
   (B) Find the interest.

4. During a 10-year period, the population of a city increased at a rate of 4% a year (i.e., 4% compounded annually). If the initial population was 500,000, what was the population 10 years later? What was the increase?

5. The day a girl was born, her father deposited $500 into a bank account paying 5% compounded annually. How much will be in this account on the girl's twentieth birthday?

6. Find the present value of each of the following:
   (A) $8000 due in 4 years with money worth 8% compounded quarterly
   (B) $4000 due in 10 years with money worth 5% compounded annually
   (C) $20,000 due in 20 years with money worth 8% compounded semiannually
   (D) $10,000 due in 3 years with money worth 12% compounded monthly

7. How much money should be invested for 6 years at 8% compounded semiannually in order to provide a compound amount of $10,000?

8. An investment contract with a maturity value of $1000 matures in 5 years. If one wishes to earn interest at 6% compounded semiannually, how much should be paid for the investment contract now?

9. Hank Jackson plans to buy a note with a maturity value of $9000. The note is due 6 years from now. If Hank wishes to earn interest at 12% compounded monthly, how much should he pay for the note?

10. A woman deposits $10,000 into a bank account. During the first 4 years, the account earns interest at 5% compounded annually. During the last 6 years, the account earns interest at 6% compounded semiannually. Find the total amount.

11. Find the effective annual interest rate corresponding to each of the following:
    (A) 6% compounded semiannually
    (B) 4% compounded quarterly
    (C) 12% compounded monthly
    (D) 8% compounded semiannually

12. How long will it take money to double itself at
    (A) 6% compounded semiannually?
    (B) 8% compounded semiannually?

(C) 8% compounded annually?
(D) 12% compounded monthly?

13. How long will it take money to triple itself at
    (A) 8% compounded quarterly?
    (B) 6% compounded semiannually?

14. How long will it take money to quadruple itself at
    (A) 8% compounded quarterly?
    (B) 6% compounded semiannually?

15. Jean Scott deposits $5000 into a bank account earning interest at the rate of 8% compounded semiannually. Three years later, Jean deposits an additional $6000. Also, at this time, the bank's interest rate is increased to 10% compounded semiannually. How much is in the account 10 years after the initial $5000 deposit was made? Assume that no withdrawals have been made.

16. Harry Mazzuri deposited $900 in a savings account that paid 6% compounded semiannually. Seven years later, the bank changed its interest rate to 10% compounded quarterly. How much will Harry's account be worth 10 years after the original deposit of $900?

17. Lucy Maceratta deposited $600 in a savings account. Six years later, she deposited an additional $300. How much will be in Lucy's account 11 years after the original deposit of $600 if the interest rate is 8% compounded quarterly?

18. A 3-year note has a maturity value of $10,000. The interest rate is 12% compounded quarterly.
    (A) How much money is needed to pay the debt now?
    (B) How much money is needed to pay the debt 2 years from now?

19. How much money should be deposited now in order to accumulate into $8000 in 10 years at an interest rate of 12% compounded quarterly?

4–3

CONTINUOUS COMPOUNDING

So far in this chapter, we have worked examples where interest has been compounded annually, semiannually, quarterly, and monthly. Interest may also be compounded weekly, daily, and hourly. When interest is compounded daily, most financial institutions have used a 360-day year. However, with the increasing use of computers and electronic calculators, many are using a 365-day year. In this text, we will employ a 365-day year. Hence, if interest is compounded daily, then $i = r/365$.

EXAMPLE 4–9 *(Calculator Exercise)* A man invests $18,000 for 3 years at 10% compounded daily. Find the compound amount.

MATHEMATICS OF FINANCE

Solution Here, we use
$$S = P(1 + i)^n$$
where $P = \$18,000$, $i = 0.10/365$, and $n = 365(3) = 1095$. Hence,
$$S = 18,000\left(1 + \frac{0.10}{365}\right)^{1095}$$

Let us first evaluate $(1 + 0.10/365)^{1095}$. Using an electronic calculator, we determine that $(1 + 0.10/365)^{1095} = 1.3498025$. Thus, the compound amount is
$$S = 18,000(1.3498025)$$
$$= \$24,296.45$$

EXAMPLE 4–10 *(Calculator Exercise)* Find the effective annual interest rate corresponding to 12% compounded daily.
Solution We must evaluate
$$\left(1 + \frac{0.12}{365}\right)^{365} - 1$$

Using our electronic calculator, we determine that $(1 + 0.12/365)^{365} = 1.127475$. Hence, the effective annual interest rate is
$$\left(1 + \frac{0.12}{365}\right)^{365} - 1 = 1.127475 - 1$$
$$= 0.127475$$

Thus, 12% compounded daily is equivalent to 12.75% compounded annually.

The previous two examples have involved daily compounding of interest. We may go further and compound every minute, every second, every half-second, etc. As the number of compoundings per year, m, increases without bound, the interest is said to be **compounded continuously.**

To determine the formula for the future value of an amount P that is compounded continuously, we begin with
$$S = P\left(1 + \frac{r}{m}\right)^{mt}$$

where r = nominal interest rate, m = number of conversion periods per year, and t = number of years. As m increases without bound, the preceding formula for S is rewritten as

$$S = P\left[\left(1 + \frac{1}{\frac{m}{r}}\right)^{m/r}\right]^{rt}$$

Hence, as m increases, then m/r increases. Eventually, the factor

$$\left(1 + \frac{1}{\frac{m}{r}}\right)^{m/r}$$

approaches $e = 2.71828\ldots$. Thus, when interest is compounded continuously at a nominal rate, r, the future value, S, is given by

$$S = Pe^{rt}$$

EXAMPLE 4–11 A financial analyst invests $10,000 at 7% compounded continuously for 10 years. Find the compound amount.
Solution Here, $P = \$10,000$, $r = 0.07$, and $t = 10$ years. Thus,

$$\begin{aligned}S &= Pe^{rt} \\ &= 10{,}000e^{0.07(10)} \\ &= 10{,}000(2.013753) \\ &= \$20{,}137.53\end{aligned}$$

From Appendix Table 3, $e^{0.70} = 2.013753$.

EXAMPLE 4–12 Find the effective annual interest rate corresponding to 10% compounded continuously.
Solution We seek the interest earned by investing $1 at 10% compounded continuously for 1 year. This is given by

$$\begin{aligned}e^{0.10(1)} - 1 &= 1.105171 - 1 \\ &= 0.105171 \\ &\approx 10.52\%\end{aligned}$$

Thus, 10% compounded continuously is equivalent to approximately 10.52% compounded annually.

MATHEMATICS OF FINANCE

EXAMPLE 4–13 How long does it take money to double itself at 10% compounded continuously?
Solution Beginning with the compound amount formula for S, we have

$$S = Pe^{0.10t}$$

where t is the number of years. Since P is to double, we replace S with $2P$ to obtain

$$2P = Pe^{0.10t}$$

Dividing both sides by P gives us

$$2 = e^{0.10t}$$

Restating this result in logarithmic form, we get

$$0.10t = \ln 2$$

Solving for t yields

$$\begin{aligned} t &= \frac{\ln 2}{0.10} \\ &= \frac{0.693147}{0.10} \\ &= 6.93147 \end{aligned}$$

Thus, it takes approximately 6.9 years for money to double itself at 10% compounded continuously.

Present Value at Continuous Compounding

The formula for present value at continuous compounding is found by beginning with the formula for compound amount

$$S = Pe^{rt}$$

and solving for P. Thus,

$$P = Se^{-rt}$$

EXAMPLE 4–14 What sum of money should be invested for 5 years at 6% compounded continuously in order to provide a compound amount of $9000?

EXAMPLE 4-14 (continued)

Solution In this case, we have $S = \$9000$, $r = 0.06$, and $t = 5$ years. Therefore,

$$P = Se^{-rt}$$
$$= 9000e^{-0.06(5)}$$
$$= 9000(0.740818)$$
$$= \$6667.36$$

From Appendix Table 3, $e^{-0.30} = 0.740818$.

EXERCISES

1. Find the compound amount of each of the following:
   (A) $1000 invested for 3 years at 6% compounded continuously
   (B) $5000 invested for 8 years at 5% compounded continuously
   (C) $10,000 invested for 10 years at 8% compounded continuously
   (D) $6000 invested for 5 years at 6% compounded continuously

2. Find the present value of each of the following:
   (A) $6000 due in 3 years at 6% compounded continuously
   (B) $10,000 due in 7 years at 10% compounded continuously
   (C) $8000 due in 10 years at 8% compounded continuously
   (D) $4000 due in 2 years at 7% compounded continuously

3. Find the effective annual interest rate corresponding to each of the following:
   (A) 6% compounded continuously
   (B) 7% compounded continuously
   (C) 8% compounded continuously
   (D) 9% compounded continuously

4. How long does it take money to double itself at
   (A) 6% compounded continuously?
   (B) 7% compounded continuously?
   (*Hint:* Use the natural logarithm tables.)

5. How long does it take money to triple itself at
   (A) 6% compounded continuously?
   (B) 7% compounded continuously?
   (*Hint:* Use the natural logarithm tables.)

6. (*Calculator Exercise*) A financial advisor invests $15,000 for 2 years at 12% compounded daily. Find the total amount.

7. (*Calculator Exercise*) Find the compound amount factor $(1 + i)^n$ for an interest rate of 18% compounded daily and a time period of 4 years.

4–4

GEOMETRIC SERIES AND ANNUITIES

A **geometric series** is an expression of the form

$$a + ar + ar^2 + \ldots + ar^{n-1}$$

Geometric Series

Each term is a constant multiple, r, of the preceding term. If S_n denotes the sum of the first n terms of a geometric series, then

$$S_n = \underset{\text{1st term}}{a} + \underset{\text{2nd term}}{ar} + \underset{\text{3rd term}}{ar^2} + \ldots + \underset{n\text{th term}}{ar^{n-1}}$$

An alternate formula for evaluating S_n is derived as follows. Take the equation

$$S_n = a + ar + ar^2 + \ldots + ar^{n-1}$$

and multiply both sides by r to obtain

$$rS_n = ar + ar^2 + \ldots + ar^{n-1} + ar^n$$

Now, consider both equations:

$$S_n = a + ar + ar^2 + \ldots + ar^{n-1}$$
$$rS_n = ar + ar^2 + \ldots + ar^{n-1} + ar^n$$

Note that

$$S_n - rS_n = a - ar^n$$

Factoring both sides of this last equation gives us

$$S_n(1 - r) = a(1 - r^n)$$

Hence,

$$S_n = \frac{a(1 - r^n)}{1 - r}$$

Thus, for the geometric series

$$\underset{\text{1st term}}{9} + \underset{\text{2nd term}}{9 \cdot 2} + \underset{\text{3rd term}}{9 \cdot 2^2} + \underset{\text{4th term}}{9 \cdot 2^3} + \underset{\text{5th term}}{9 \cdot 2^4} + \underset{\text{6th term}}{9 \cdot 2^5}$$

$a = 9$, $r = 2$, and $n = 6$. Then,

$$S_6 = \frac{a(1 - r^n)}{1 - r}$$

$$= \frac{9(1 - 2^6)}{1 - 2}$$

$$= \frac{9(1 - 64)}{1 - 2}$$

$$= \frac{9(-63)}{-1}$$

$$= 567$$

Note that if the series were added term by term, we would obtain the same result.

EXAMPLE 4–15 Find the sum of the seven terms of the geometric series

$$5 + 5 \cdot 3 + 5 \cdot 3^2 + 5 \cdot 3^3 + 5 \cdot 3^4 + 5 \cdot 3^5 + 5 \cdot 3^6$$

Solution Comparing this geometric series with the general form

$$a + ar + ar^2 + \ldots + ar^{n-1}$$

we have $a = 5$, $r = 3$, and $n = 7$. Hence,

$$S_7 = \frac{a(1 - r^n)}{1 - r}$$

$$= \frac{5(1 - 3^7)}{1 - 3}$$

$$= \frac{5(1 - 2187)}{1 - 3}$$

$$= \frac{5(-2186)}{-2}$$

$$= 5465$$

Annuities An **annuity** is a series of equal payments made at equal intervals of time. Many everyday business transactions are annuities. Mortgage payments, premium payments on insurance plans, rental payments on a lease, and installment purchases are a few examples. In general, any series of equal payments made at equal intervals of time is an annuity. Each payment is called the **periodic**

payment, or **periodic rent.** Periodic payment will be denoted by R. The total time during which these payments are made is called the **term** of the annuity.

Specifically, if a person deposits $100 at the end of each 6-month period for 2 years, then the periodic payment, R, is $100, the payment period is 6 months, and the term is 2 years (see Figure 4–18). Note that each payment is made *at the end* of the payment period. Such an annuity is called an **ordinary annuity.** If each payment were made *at the beginning* of the payment period, the annuity would be called an **annuity due.** In this section, we will begin with ordinary annuities. Annuities due will be covered later in the section.

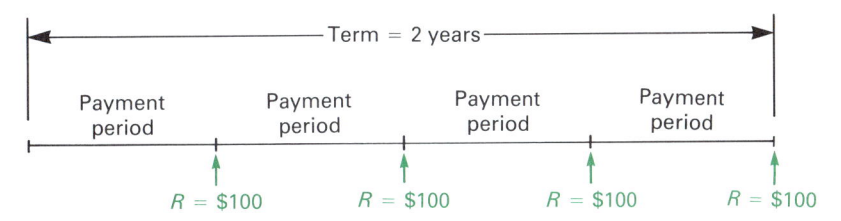

```
|←─────────── Term = 2 years ───────────→|
| Payment  | Payment  | Payment  | Payment  |
| period   | period   | period   | period   |
|----------|----------|----------|----------|
           ↑          ↑          ↑          ↑
        R = $100   R = $100   R = $100   R = $100
```

FIGURE 4–18

Each payment R of an annuity earns compound interest. Here, and in Sections 4–5 through 4–8, we will discuss only those annuities for which the payment period and conversion period coincide. Returning to our example, since each $100 payment is deposited semiannually, the interest will be compounded semiannually. We will use an interest rate of 6% compounded semiannually. Hence, $i = 3\%$ per conversion period.

The sum of all periodic payments R plus their interest is called the **total amount** of the annuity. It will be denoted by S. To calculate the total amount, S, of an annuity, we must realize that each periodic payment R is made at a different point in time. Hence, each payment R earns compound interest for the duration of its term in the annuity. This is illustrated in Figure 4–19 for our previous annuity of $100 deposited semiannually for 2 years at an interest rate of 6% compounded semiannually. Studying Figure 4–19, note that the total amount, S, of the annuity is equal to the sum of the compound amounts of the payments of $100. Hence,

$$S = 100 + 100(1 + 0.03)^1 + 100(1 + 0.03)^2 + 100(1 + 0.03)^3$$

Observe that the expression for S is a geometric series with $a = 100$, $r = 1 + 0.03 = 1.03$, and $n = 4$. We will not evaluate S at this time, however.

188 CHAPTER FOUR

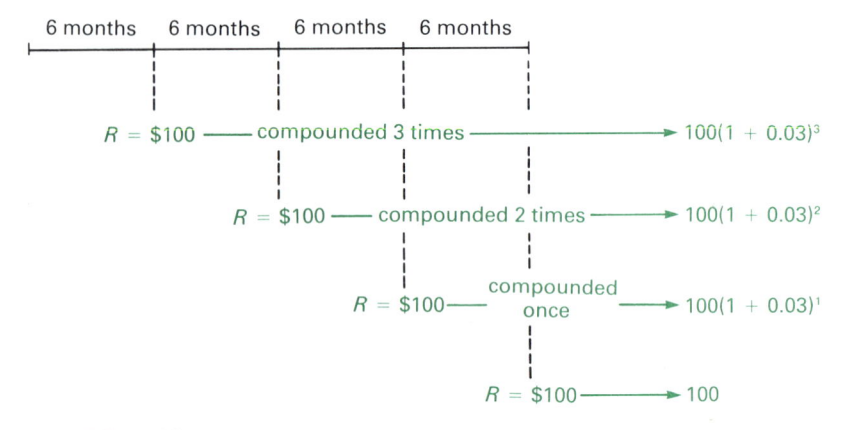

```
6 months   6 months   6 months   6 months
```

R = $100 ——— compounded 3 times ——————→ $100(1 + 0.03)^3$

R = $100 ——— compounded 2 times ——————→ $100(1 + 0.03)^2$

R = $100 ——— compounded once ——————→ $100(1 + 0.03)^1$

R = $100 ——————————————→ 100

FIGURE 4-19

Instead, we will derive a general formula for the total amount, S, of an ordinary annuity. The time diagram of Figure 4-20 illustrates an ordinary annuity of n payments of R dollars each.

```
|←——————— n payment periods ———————→|
  1    2    3   . . . . . .   n-2  n-1   n
```

R ——— compounded n − 1 times ——————→ $R(1 + i)^{n-1}$

R ——— compounded n − 2 times ——————→ $R(1 + i)^{n-2}$

R ——————————————→ $R(1 + i)^2$

R ——————————————→ $R(1 + i)$

R ——→ R

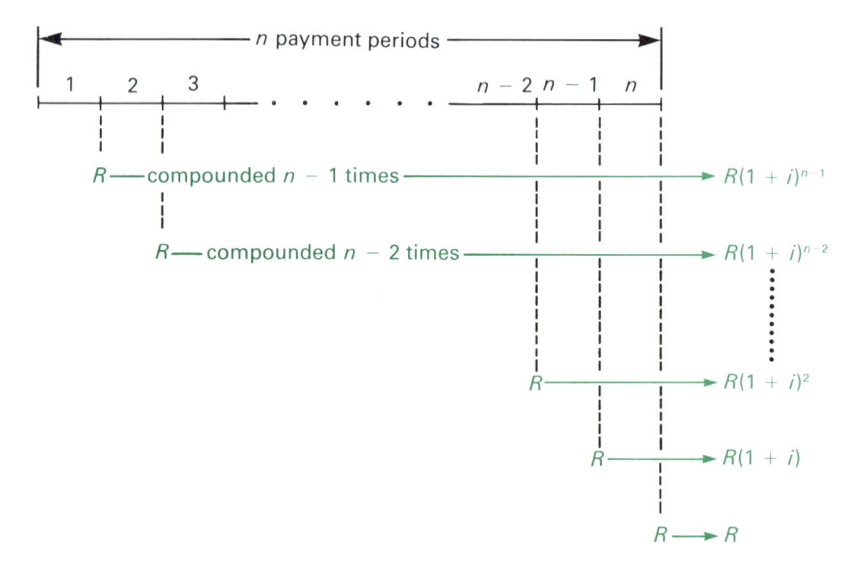

FIGURE 4-20

Thus, there are n payment periods (or conversion periods). As usual, $i =$ interest rate per conversion period. Studying Figure 4-20, we note that the total amount, S, of the annuity is equal to the sum of the compound amounts of the payments R. Therefore,

$$S = R + R(1 + i) + R(1 + i)^2 + \ldots + R(1 + i)^{n-2} + R(1 + i)^{n-1}$$

MATHEMATICS OF FINANCE

Observe that this expression is a geometric series of the form

$$a + ar + ar^2 + \ldots + ar^{n-1}$$

with $a = R$ and $r = 1 + i$. Recall that the sum of the n terms of this latter geometric series is given by the formula

$$\frac{a(1 - r^n)}{1 - r}$$

Substituting R for a and $1 + i$ for r yields

$$\frac{R[1 - (1 + i)^n]}{1 - (1 + i)}$$

which simplifies to

$$R\left[\frac{(1 + i)^n - 1}{i}\right]$$

Thus, the formula for the total amount, S, of an ordinary annuity of n payments of R dollars each is

$$S = R\left[\frac{(1 + i)^n - 1}{i}\right]$$

where i = interest rate per conversion period.

Appendix Table 6 lists the tabulations of the quantity $[(1 + i)^n - 1]/i$ for various values of i and n. For brevity, this quantity is usually denoted by the symbol $s_{\overline{n}|i}$, read "s angle n at i." Thus, the preceding formula for the total amount, S, of an annuity is usually written

$$S = R \cdot s_{\overline{n}|i}$$

Returning to the previous example, where $R = \$100$, $i = 0.03$, and $n = 4$, the total amount is calculated as

$$S = R \cdot s_{\overline{4}|0.03}$$
$$= 100(4.183627)$$
$$= \$418.36$$

From Appendix Table 6, $s_{\overline{4}|0.03} = 4.183627$.

EXAMPLE 4–16 Mr. Haskins deposits $200 at the end of each quarter into a pension fund earning interest at 8% compounded quarterly.

EXAMPLE 4–16 (continued)

(A) Find the total amount at the end of 10 years.
(B) How much interest was earned?

Solutions Here, we have $R = \$200$, $i = 8\%/4 = 2\% = 0.02$, and $n = 4(10) = 40$.

(A) The calculation is

$$S = R \cdot s_{\overline{n}|i}$$
$$= 200 \cdot s_{\overline{40}|0.02} \qquad \text{From Appendix Table 6,}$$
$$= 200(60.401983) \qquad s_{\overline{40}|0.02} = 60.401983.$$
$$= \$12,080.40$$

(B) Mr. Haskins actually deposited 40 payments of $200 each, or 40($200) = $8000. Thus, the interest earned is

$$\$12,080.40 - \$8000.00 = \$4080.40$$

Annuity Due

As was stated earlier, if each payment, R, of an annuity is made *at the beginning of* the payment period, the annuity is called an *annuity due*. The time diagram of Figure 4–21 compares an ordinary annuity and an annuity due.

Ordinary annuity

Annuity due

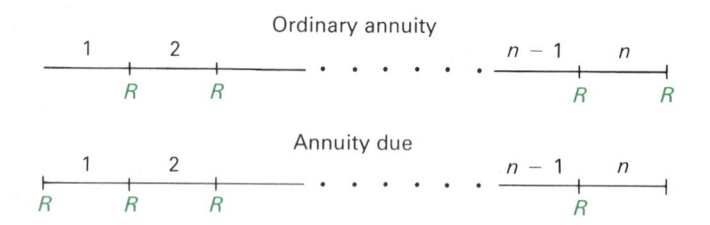

FIGURE 4–21

Note that both annuities have n payments of R dollars each. However, since each payment of an annuity due is made at the beginning of the period, then each payment earns interest for one additional period. We now derive the formula for the total amount of an annuity due (see Figure 4–22).

Studying Figure 4–22, note that

$$S = R(1 + i) + R(1 + i)^2 + \ldots + R(1 + i)^{n-1} + R(1 + i)^n$$
$$= \underbrace{[R + R(1 + i) + R(1 + i)^2 + \ldots + R(1 + i)^n]}_{\text{geometric series}} - R$$

MATHEMATICS OF FINANCE

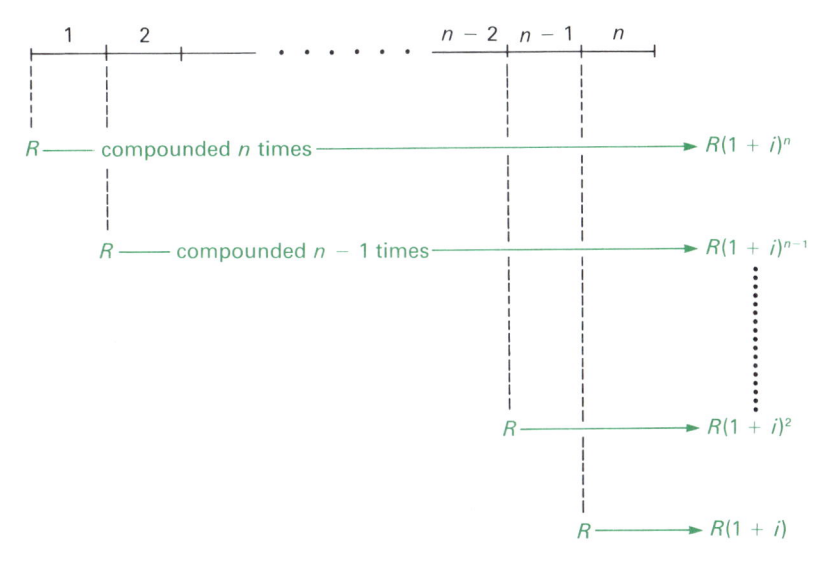

```
    1   2              n-2  n-1   n
    ├───┼──── · · · ───┼────┼────┤
    R ──── compounded n times ───────────────→ R(1 + i)ⁿ

        R ──── compounded n − 1 times ───────→ R(1 + i)ⁿ⁻¹

                                              ⋮

                              R ─────────────→ R(1 + i)²

                                   R ─────────→ R(1 + i)
```

FIGURE 4–22

Applying the formula for the sum of a geometric series, we have

$$S = R\left[\frac{1 - (1 + i)^{n+1}}{1 - (1 + i)}\right] - R$$

$$= R\left[\frac{(1 + i)^{n+1} - 1}{i}\right] - R$$

$$= R \cdot s_{\overline{n+1}|i} - R$$

$$= R(s_{\overline{n+1}|i} - 1)$$

Thus, the formula for the total amount, S, of an annuity due is

$$S = R(s_{\overline{n+1}|i} - 1)$$

EXAMPLE 4–17 Ms. Jones deposits $100 at the beginning of each quarter into a bank account earning interest at 8% compounded quarterly. Find the total amount at the end of 9 years.

Solution Here, $R = \$100$, $i = 8\%/4 = 2\% = 0.02$, and $n = 4(9) = 36$. Therefore,

EXAMPLE 4–17 (continued)

$$S = R(s_{\overline{n+1}|i} - 1)$$
$$= 100(s_{\overline{37}|0.02} - 1) \qquad \text{From Appendix Table 6,}$$
$$= 100(54.034255 - 1) \qquad s_{\overline{37}|0.02} = 54.034255.$$
$$= 100(53.034255)$$
$$= \$5303.43$$

EXERCISES

1. For each of the following geometric series, find the sum of the indicated terms in two ways:
   (A) $2 + 2 \cdot 5 + 2 \cdot 5^2 + 2 \cdot 5^3$
   (B) $3 + 3 \cdot 2 + 3 \cdot 2^2 + 3 \cdot 2^3 + 3 \cdot 2^4 + 3 \cdot 2^5$
   (C) $7 + 7 \cdot 4 + 7 \cdot 4^2 + 7 \cdot 4^3 + 7 \cdot 4^4 + 7 \cdot 4^5 + 7 \cdot 4^6$

2. Find the total amount of each of the following ordinary annuities:
   (A) $100 each quarter for 5 years at 7% compounded quarterly
   (B) $1000 semiannually for 20 years at 6% compounded semiannually
   (C) $500 monthly for 4 years at 12% compounded monthly
   (D) $5000 annually for 20 years at 5% compounded annually

3. A woman deposits $100 at the end of each quarter into a fund earning interest at 4% compounded quarterly.
   (A) Find the total amount at the end of 12 years.
   (B) How much interest was earned?

4. A man deposits $1500 at the end of each year into a pension fund earning interest at 5% compounded annually.
   (A) Find the total amount at the end of 20 years.
   (B) How much interest was earned?

5. Sally Smith deposits $1000 at the end of each six-month period for 15 years into a fund earning interest at 6% compounded semiannually.
   (A) Find the total amount at the end of 15 years.
   (B) If Sally leaves the total amount in the fund for 5 more years without making any additional deposits, how much is her fund worth if it earns interest at 6% compounded semiannually?

6. Repeat Exercises 2 through 5 under the assumption that each annuity is an annuity due.

7. For 4 years, a man deposits $100 into a retirement account at the beginning of each month. If the interest rate is 12% compounded monthly, how much is in the account after 4 years?

8. A woman deposits $1000 at the beginning of each quarter for 5 years. Each deposit earns interest at 12% compounded quarterly.
   (A) Find the total amount at the end of 5 years.
   (B) If the woman leaves the total amount in the fund for 4 more years without making any additional deposits, how much is her fund worth if the interest rate is 12% compounded quarterly?

9. Jane deposits $100 at the end of each month into a bank account earning interest at 12% compounded monthly. How much is in the

193 MATHEMATICS OF FINANCE

account at the end of 4 years? Assume that no withdrawals have been made.

10. Referring to Exercise 9, suppose that Jane, after 4 years, deposits $200 at the end of each month for the next 3 years. If the $200 deposits earn interest at 12% compounded monthly and the accumulated amount of the $100 deposits also earns interest at 12% compounded monthly, how much does Jane have in her account at the end of 7 years (i.e., 7 years after she first opened the account)?

11. A man deposits $500 at the end of each six-month period into a bank account for 4 years. At that time, his deposits are changed to $600 semiannually for the next 5 years. If all deposits earn interest at 12% compounded semiannually, how much is in the account at maturity?

12. Repeat Exercise 11 under the assumption that each payment is made at the beginning of the 6-month period.

13. *(Calculator Exercise)* Using the formula

$$s_{\overline{n}|i} = \frac{(1+i)^n - 1}{i}$$

find the following:

(A) $s_{\overline{100}|0.015}$
(B) $s_{\overline{360}|0.024}$
(C) $s_{\overline{n}|i}$ for an interest rate of 20% compounded monthly for 10 years

4–5

PRESENT VALUE OF AN ANNUITY

The **present value of an annuity** is the sum of the present values of all the periodic payments R (see Figure 4–23).

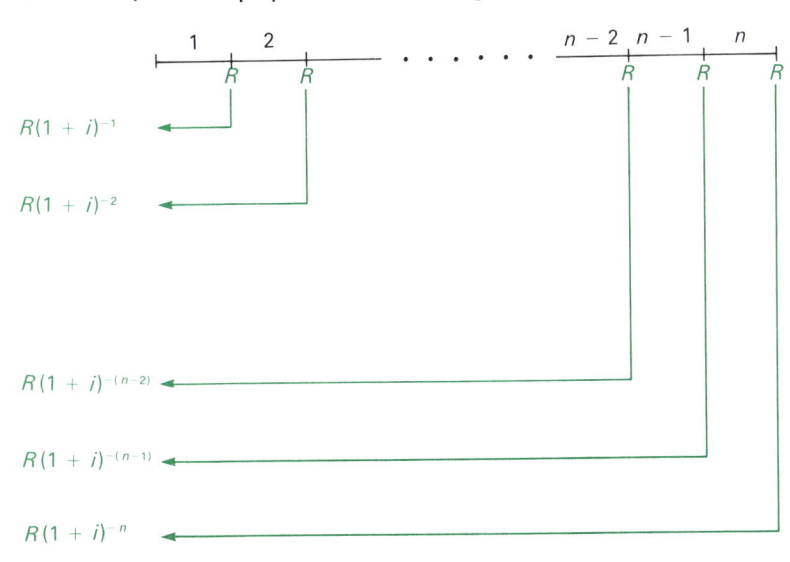

FIGURE 4–23

Observing Figure 4–23, the present value, A, of an annuity is

$$A = R(1+i)^{-n} + R(1+i)^{-(n-1)} + R(1+i)^{-(n-2)} + \ldots \\ + R(1+i)^{-2} + R(1+i)^{-1}$$

Thus, an annuity of R dollars per period for n periods is worth A dollars now. In other words, a lump-sum investment of A dollars now will provide payments of R dollars per period for the next n periods.

The preceding formula for A may be simplified. Multiplying the right-hand side by $(1+i)^n/(1+i)^n$ yields

$$A = \frac{R + R(1+i) + R(1+i)^2 + \ldots + R(1+i)^{n-2} + R(1+i)^{n-1}}{(1+i)^n}$$

Note that the numerator is a geometric series. Applying the formula for its sum, we obtain

$$A = \frac{R\left[\dfrac{(1+i)^n - 1}{i}\right]}{(1+i)^n}$$

$$= R\left[\frac{1 - (1+i)^{-n}}{i}\right]$$

Thus, the formula for the present value, A, of an annuity of n payments of R dollars each is

$$A = R\left[\frac{1 - (1+i)^{-n}}{i}\right]$$

where i = interest rate per conversion period.

Appendix Table 8 lists the tabulations of the quantity $[1 - (1+i)^{-n}]/i$ for various values of i and n. For brevity, this quantity will be denoted by the symbol $a_{\overline{n}|i}$, read "a angle n at i." Thus, the preceding formula for the present value, A, of an annuity may be written

$$A = R \cdot a_{\overline{n}|i}$$

EXAMPLE 4–18 A father wishes to provide for quarterly payments of $400 each for the next 4 years. The payments will be made at the end of each quarter to his daughter who

MATHEMATICS OF FINANCE

will be attending college. How much should the father invest at 8% compounded quarterly?

Solution The answer is the present value, A, of the annuity. Note that $R = \$400$, $i = 8\%/4 = 2\% = 0.02$, and $n = 4(4) = 16$. Thus,

$$A = R \cdot a_{\overline{n}|i}$$
$$= 400 \cdot a_{\overline{16}|0.02} \quad \text{From Appendix Table 8,}$$
$$= 400(13.577709) \quad a_{\overline{16}|0.02} = 13.577709.$$
$$= \$5431.08$$

Thus, the father's investment of $5431.08 will provide for 16 payments of $400, or 16($400) = $6400, over the next 4 years.

EXAMPLE 4–19 A woman buys a car by agreeing to pay $200 at the end of each month for the next 4 years. This includes interest at 12% compounded monthly. Suppose that instead of financing the car, the woman decides to pay cash now. How much should she pay?

Solution The answer is the present value, A, of the annuity. Note that $R = \$200$, $i = 12\%/12 = 1\% = 0.01$, and $n = 12(4) = 48$. Thus,

$$A = R \cdot a_{\overline{n}|i}$$
$$= 200 \cdot a_{\overline{48}|0.01} \quad \text{From Appendix Table 8,}$$
$$= 200(37.973959) \quad a_{\overline{48}|0.01} = 37.973959.$$
$$= \$7594.79$$

We should also understand that the present value of an annuity, A, if invested for the duration of the annuity, will yield the same total amount, S, as the annuity. In other words, if Mr. Johnson deposits n payments of R dollars each and Mr. Thomas deposits a lump sum of A dollars now, then after n conversion periods, both will have the same total amount (assuming that the interest rate, i, is the same for both). The time diagrams of Figure 4–24 illustrate this example.

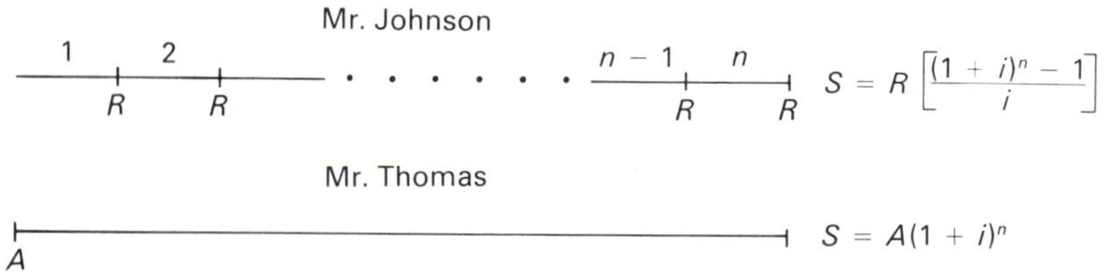

CHAPTER FOUR

Mr. Johnson

```
 1   2           n-1   n
 +   +   . . . . +     +     S = R [((1 + i)^n - 1)/i]
 R   R           R     R
```

Mr. Thomas

```
+-----------------------------+     S = A(1 + i)^n
A
```

FIGURE 4–24

Observe that if we equate the formulas for the total amount, S, we have

$$A(1 + i)^n = R\left[\frac{(1 + i)^n - 1}{i}\right]$$

Solving for A yields

$$A = R\left[\frac{1 - (1 + i)^{-n}}{i}\right]$$

Note that this is the formula for the present value of an annuity.

EXAMPLE 4–20 Henry deposits $100 at the end of each month for 3 years into a savings account. If Cindy wishes to have the same amount in her account at the end of 3 years, what single sum of money should she deposit now? Assume that both accounts earn interest at 6% compounded monthly.
Solution The answer is the present value, A, of the annuity. Note that $R = \$100$, $i = 6\%/12 = (1/2)\% = 0.005$, and $n = 12(3) = 36$. Thus,

$$A = R \cdot a_{\overline{n}|i}$$
$$= 100 \cdot a_{\overline{36}|0.005}$$
$$= 100(32.871016)$$
$$= \$3287.10$$

EXAMPLE 4–21 *(Leases)* A corporation can either lease a machine with a 5-year useful life for $2000 per year or buy it for a lump sum of $8000.
(A) If money is worth 7% compounded annually, which alternative is preferable?
(B) If money is worth 8% compounded annually, which alternative is preferable?

Solutions

(A) The leasing alternative involves an annuity with $R = \$200$, $n = 5$, and $i = 0.07$. The present value of this annuity is

$$A = R \cdot a_{\overline{n}|i}$$
$$= 2000 \cdot a_{\overline{5}|0.07}$$
$$= 2000(4.100197)$$
$$= \$8200.39$$

Since this amount exceeds the purchase price, purchasing the machine is cheaper.

(B) In this case,

$$A = R \cdot a_{\overline{n}|i}$$
$$= 2000 \cdot a_{\overline{5}|0.08}$$
$$= 2000(3.992710)$$
$$= \$7985.42$$

Since this amount is less than the purchase price, leasing the machine is cheaper.

EXAMPLE 4–22 *(Capital Expenditure Analysis)* A corporation wants to modernize its equipment in order to reduce labor costs. It has a chance of purchasing two machines for a certain assembly operation. Machine A costs $7000, will save $1600 annually, and has a useful life of 6 years. Machine B costs $9000, will save $1900 annually, and has a useful life of 7 years. If money is worth 8% compounded annually, which machine should the company buy?

Solution Machine A saves $1600 annually for 6 years. The present value of this annuity at 8% compounded annually is

$$A = R \cdot a_{\overline{n}|i}$$
$$= 1600 \cdot a_{\overline{6}|0.08}$$
$$= 1600(4.622880)$$
$$= \$7396.61$$

Thus, the annual savings of $1600 for 6 years are equivalent to a lump-sum savings of $7396.61 now. Since Machine A costs $7000, the net saving is

$$\$7396.61 - \$7000 = \$396.61$$

> **EXAMPLE 4–22** (continued)
>
> Machine B saves $1900 annually for 7 years. The present value of this annuity at 8% compounded annually is
>
> $$A = R \cdot a_{\overline{n}|i}$$
> $$= 1900 \cdot a_{\overline{7}|0.08}$$
> $$= 1900(5.206370)$$
> $$= \$9892.10$$
>
> Thus, the annual savings of $1900 for 7 years are equivalent to a lump-sum savings of $9892.10 now. Since Machine B costs $9000, the net saving is
>
> $$\$9892.10 - \$9000 = \$892.10$$
>
> Since Machine B has the larger net saving, this is the one the corporation should buy.

Present Value of an Annuity Due The formula for the present value of an annuity due is derived by observing the time diagram of Figure 4–25 and equating the total amounts. Thus, we have

$$A(1+i)^n = R\left[\frac{(1+i)^{n+1} - 1}{i} - 1\right]$$

$$= R\left[\frac{(1+i)^{n+1} - 1 - i}{i}\right]$$

$$= R\left[\frac{(1+i)^{n+1} - (1+i)}{i}\right]$$

Dividing both sides by $(1+i)^n$ yields

$$A = R\left[\frac{(1+i) - (1+i)^{-(n-1)}}{i}\right]$$

$$= R\left[1 + \frac{1 - (1+i)^{-(n-1)}}{i}\right]$$

Since

$$\frac{1 - (1+i)^{-(n-1)}}{i} = a_{\overline{n-1}|i}$$

then the present value of an annuity due is found by the formula

MATHEMATICS OF FINANCE

$$A = R(1 + a_{\overline{n-1}|i})$$

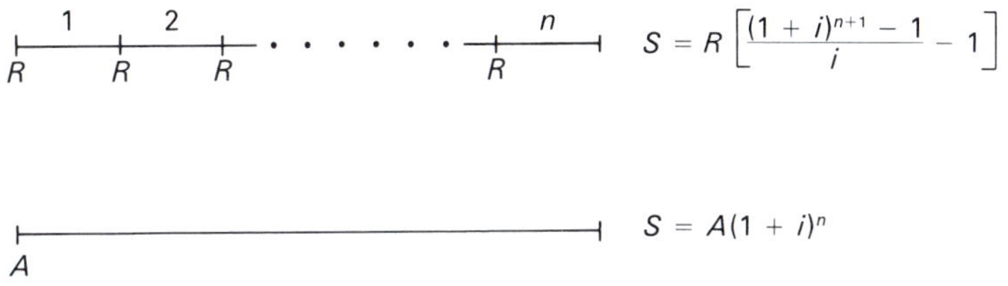

$$S = R\left[\frac{(1+i)^{n+1} - 1}{i} - 1\right]$$

$$S = A(1 + i)^n$$

FIGURE 4–25

EXAMPLE 4–23 Find the present value of an annuity due of $1000 quarterly for 5 years at 8% compounded quarterly.
Solution Here, we have $R = \$1000$, $i = 8\%/4 = 2\% = 0.02$, and $n = 4(5) = 20$. Thus,

$$A = R(1 + a_{\overline{n-1}|i})$$
$$= 1000(1 + a_{\overline{19}|0.02})$$
$$= 1000(1 + 15.678462)$$
$$= 1000(16.678462)$$
$$= \$16,678.46$$

EXERCISES

1. Find the present value of each of the following ordinary annuities:
   (A) $3000 quarterly for 6 years at 8% compounded quarterly
   (B) $10,000 annually for 4 years at 5% compounded annually
   (C) $2000 semiannually for 10 years at 6% compounded semiannually
   (D) $6000 monthly for 3 years at 12% compounded monthly

2. Fara Fields wishes to provide for a semiannual payment of $2000 at the end of each six-month period. The payments will be made to her daughter who will be attending college and medical school. How much money should Fara invest now at 10% compounded semiannually?

3. Mr. Johnson buys a car by agreeing to pay $500 at the end of each quarter for the next 5 years. This includes interest at 12% compounded quarterly. If instead of financing the car Mr. Johnson decides to pay cash now, how much should he pay?

4. Holly deposits $400 at the end of each quarter into a savings account. If Sara wishes to have the same amount in her account at the end of 5 years, what single sum of money should she deposit now? Assume that both accounts earn interest at 8% compounded quarterly.

5. A company is planning a project that will generate a cash inflow of $10,000 a year for 8 years. If the company wants a rate of return on its invested capital of at least 10% compounded annually, what should be the maximum amount invested in this project now?

6. Repeat Exercises 1 through 4 under the assumption that each annuity is an annuity due.

7. An investment contract promises to pay its holder $500 at the beginning of each month for 3 years. If one wishes to earn 12% compounded monthly, how much should be paid for this contract now?

8. A certain investment contract promises to pay $1000 at the end of each quarter for the next 7 years. If we wish to buy this investment contract and earn 12% compounded quarterly, how much should we pay for it now?

9. Repeat Exercise 8 under the assumption that the payments are made at the beginning of each quarter.

10. Mr. Smith wishes to provide for payments of $400 at the end of each quarter to his son for the next 5 years. What sum should be deposited now at 8% compounded quarterly to attain Mr. Smith's objective?

11. Repeat Exercise 10 under the assumption that the payments are made at the beginning of each quarter.

4–6

SINKING FUNDS AND AMORTIZATION

Often, a person decides to accumulate a sum of money by making periodic deposits into a fund. At the end of a specified time period, the deposits plus the interest earned equal the desired accumulated amount. Such a fund is called a **sinking fund**.

Sinking Funds

As an example, consider a contractor foreseeing the need for a new truck 4 years from now. The price of the truck is forecasted to be $20,000. The contractor wishes to accumulate this amount by setting aside semiannual payments of R dollars each for 4 years. Each payment of this sinking fund earns interest at 10% compounded semiannually. The contractor must determine the semiannual payment R. This situation is illustrated by the time diagram of Figure 4–26.

MATHEMATICS OF FINANCE

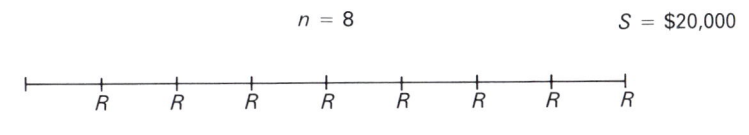

$$n = 8 \qquad S = \$20{,}000$$

```
├───┼───┼───┼───┼───┼───┼───┼───┤
    R   R   R   R   R   R   R   R
```

FIGURE 4–26

Since the semiannual payments constitute an annuity with a total amount of $20,000, then

$$S = R \cdot s_{\overline{n}|i}$$
$$20{,}000 = R \cdot s_{\overline{8}|0.05}$$

Solving for R yields

$$R = \frac{20{,}000}{s_{\overline{8}|0.05}}$$
$$= \frac{20{,}000}{9.549109}$$
$$= \$2094.44$$

Thus, the series of semiannual payments $R = \$2094.44$ plus interest will accumulate to $S = \$20{,}000$. Note that the contractor will make eight payments of $2094.44 each, or 8($2094.44) = $16,755.52. Therefore, the interest earned is

$$\$20{,}000 - \$16{,}755.52 = \$3244.48$$

These results are summarized in Table 4–1.

TABLE 4–1 Sinking Fund Schedule

Payment Number	Payment	Interest	Total
1	$2,094.44	$ 0	$ 2,094.44
2	2,094.44	104.72	4,293.60
3	2,094.44	214.68	6,602.72
4	2,094.44	330.14	9,027.30
5	2,094.44	451.37	11,573.11
6	2,094.44	578.66	14,246.21
7	2,094.44	712.31	17,052.96
8	2,094.44	852.65	20,000.05
		$3,244.53	

Studying the sinking fund schedule of Table 4–1, note that the interest for each period is determined by multiplying $i = 0.05$ times the previous period's total. Observe that the total interest is $3244.53. The $0.05 discrepancy from the previous calculation is due to round-off error.

EXAMPLE 4–24 A business executive wishes to set aside semiannual payments to purchase machinery 2 years from now. The machinery's estimated cost is $5000. Each payment earns interest at 12% compounded semiannually.
(A) Find the semiannual payment.
(B) Find the total interest earned.
(C) Prepare a sinking fund schedule similar to that of Table 4–1.

Solutions
(A) Here, $S = \$5000$, $i = 12\%/2 = 6\% = 0.06$, and $n = 2(2) = 4$. We must determine R. Since

$$S = R \cdot s_{\overline{n}|i}$$

then

$$5000 = R \cdot s_{\overline{4}|0.06}$$

Solving for R yields

$$R = \frac{5000}{s_{\overline{4}|0.06}}$$

$$= \frac{5000}{4.374616}$$

$$= \$1142.96$$

(B) The business executive will make four payments of $1142.96 each, or 4($1142.96) = $4571.84. Thus, the interest earned is

$$\$5000.00 - \$4571.84 = \$428.16$$

(C) The sinking fund schedule is shown in Table 4–2. Observing this sinking fund schedule, note that the interest for each period is determined by multiplying $i = 0.06$ times the previous period's total. Here, the total interest is $428.17. The $0.01 discrepancy from the previous calculation is due to round-off error.

MATHEMATICS OF FINANCE

TABLE 4–2 Sinking Fund Schedule

Payment Number	Payment	Interest	Total
1	$1142.96	$ 0	$1142.96
2	1142.96	68.58	2354.50
3	1142.96	141.27	3638.73
4	1142.96	218.32	5000.01
		$428.17	

Amortization

Often, a loan is repaid by a series of equal payments made at equal intervals of time—an annuity. The *amount of the loan is the present value of the annuity.* A portion of each payment is applied against the principal, and the remainder is applied against the interest. When a loan is repaid by an annuity, it is said to be **amortized**.

Consider a person borrowing $7000 to buy a car. The loan plus interest is to be repaid in equal quarterly installments made at the end of each quarter during a 2-year interval. The interest rate is 16% compounded quarterly. We must determine the quarterly payment R. This situation is illustrated by the time diagram of Figure 4–27.

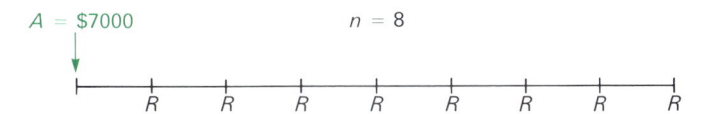

FIGURE 4–27

Since the quarterly payments constitute an annuity with a present value of $7000, then

$$A = R \cdot a_{\overline{n}|i}$$
$$7000 = R \cdot a_{\overline{8}|0.04}$$

Solving for R yields

$$R = \frac{7000}{a_{\overline{8}|0.04}}$$
$$= \frac{7000}{6.732745}$$
$$= \$1039.69$$

204 CHAPTER FOUR

Thus, the borrower will make eight payments of $1039.69 each, or 8($1039.69) = $8317.52, to repay the $7000 loan. Thus, the interest is

$$\$8317.52 - \$7000.00 = \$1317.52$$

These results are summarized by the amortization schedule of Table 4–3.

TABLE 4–3 Amortization Schedule

Payment Number	Payment R	Interest	Principal Reduction	Equity	Balance
1	$1039.69	$ 280.00	$759.69	$ 759.69	$6240.31
2	1039.69	249.61	790.08	1549.77	5450.23
3	1039.69	218.01	821.68	2371.45	4628.55
4	1039.69	185.14	854.55	3226.00	3774.00
5	1039.69	150.96	888.73	4114.73	2885.27
6	1039.69	115.41	924.28	5039.01	1960.99
7	1039.69	78.44	961.25	6000.26	999.74
8	1039.69	39.99	999.70	6999.96	0.04
		$1317.56			

Studying the amortization schedule of Table 4–3, note that the interest for each period is determined by multiplying $i = 0.04$ times the previous period's balance. The amount of **principal reduction** for a period is the difference between the payment R and the interest for that period. The **equity** column is the cumulation of the principal reductions. The **balance** column may be determined by either of two methods:

1. As the difference between the amount of the loan and the equity
2. As the difference between the previous period's balance and the principal reduction for the given period

EXAMPLE 4–25 *(Mortgage)* Mr. and Mrs. Morro have purchased a home for $100,000. They have put $40,000 down and will obtain a 20-year mortgage for $60,000 at an interest rate of 12% compounded monthly. If $a_{\overline{240}|0.01} = 90.819416$, then

(A) Find the monthly mortgage payment.
(B) How much will the Morros pay out for the loan after 20 years, and what will the total interest be?
(C) What is the balance after 15 years? In other words, if after 15 years the Morros wish to pay off the loan, what amount must they pay at that time?
(D) What is their equity after 15 years?

Solutions

(A) We must determine R. Since $i = 12\%/12 = 1\% = 0.01$, $n = 12(20) = 240$, and the amount of the mortgage is the present value of the annuity, then

$$A = R \cdot a_{\overline{n}|i}$$
$$60{,}000 = R \cdot a_{\overline{240}|0.01}$$

Solving for R, we obtain

$$R = \frac{60{,}000}{a_{\overline{240}|0.01}}$$
$$= \frac{60{,}000}{90.819416}$$
$$= \$660.65$$

(B) The Morros will make 240 payments of $660.65 each, or 240($660.65) = $158,556. Thus, the total interest is

$$\$158{,}556 - \$60{,}000 = \$98{,}556$$

(C) The balance of the mortgage after 15 years is the present value of the remaining 60 payments. Hence,

$$A = R \cdot a_{\overline{60}|0.01}$$
$$= 660.65(44.955038)$$
$$= \$29{,}699.55$$

(D) The equity after 15 years is the difference between the amount of the mortgage and the balance, or

$$\$60{,}000.00 - \$29{,}699.55 = \$30{,}300.45$$

This is, of course, in addition to the down payment of $40,000. Thus, the Morro's total equity in the $100,000 house is

$$\$40{,}000.00 + \$30{,}300.45 = \$70{,}300.45$$

EXERCISES

1. A grocer anticipates a need for a new freezer 5 years from now. The price is expected to be $10,000. If the grocer wishes to accumulate this amount by setting aside quarterly payments earning interest at 8% compounded quarterly for the next 5 years, how much should be set aside at the end of each quarter?

2. A person wishes to accumulate $20,000 in 20 years by setting aside annual payments earning interest at 5% compounded annually. How much should be set aside at the end of each year?

3. Six years from now, Gerry Grumble must pay Brian Broker $5000. Gerry wishes to set aside, at the end of each six-month period, a payment earning interest at 6% compounded semiannually. How much should each payment be in order to retire the debt in 6 years?

4. John Gaylor borrows $10,000. This loan will be repaid by a monthly installment at the end of each month over the next 4 years. If the interest rate is 12% compounded monthly, what is John's monthly payment?

5. Tom Thrift buys a house with a purchase price of $80,000. He makes a down payment of $20,000 and finances the remainder with a mortgage requiring a quarterly payment at the end of each quarter for the next 12 years. If the interest rate is 8% compounded quarterly, what is Tom's quarterly payment?

6. A debt of $20,000 is to be amortized over 10 years by a payment at the end of each six-month period. If the interest rate is 6% compounded semiannually, how much is each payment?

7. A company sets aside a payment at the end of each six-month period to provide for the replacement of equipment 3 years from now. Each payment earns interest at 10% compounded semiannually, and the equipment's projected cost is $20,000.
   (A) Find the semiannual payment.
   (B) Find the total interest earned.
   (C) Prepare a sinking fund schedule similar to that of Table 4–1.

8. Joan Grimes has purchased a car for $10,000. She has made a down payment of $4000 and will finance the balance by making a payment at the end of each quarter for 2 years. The interest rate is 12% compounded quarterly.
   (A) Find the quarterly payment.
   (B) How much will Joan pay out for the loan after 2 years, and what will the total interest be?
   (C) What is the balance after $1\frac{1}{4}$ years?
   (D) What is the equity after $1\frac{1}{4}$ years?
   (E) Prepare an amortization schedule similar to that of Table 4–3.

9. Mr. and Mrs. Spencer have purchased a home for $120,000. They have put $30,000 down and will obtain a 20-year mortgage for $90,000 at an interest rate of 18% compounded monthly. If $a_{\overline{240}|0.015} = 64.79573$, then
   (A) Find the payment due at the end of each month.
   (B) How much will the Spencer's pay out for the loan after 20 years, and what will be the total interest?

(C) What is the balance after 14 years?
(D) What is the equity after 14 years?

10. *(Calculator Exercise)* Using the formula

$$a_{\overline{n}|i} = \frac{1 - (1 + i)^{-n}}{i}$$

compute each of the following:
(A) $a_{\overline{360}|0.015}$
(B) $a_{\overline{360}|0.01}$
(C) $a_{\overline{n}|i}$ for an interest rate of 15% compounded monthly for 20 years

4–7

EQUATIONS OF VALUE

Before going on, let us summarize some basic concepts from previous sections of this chapter. For any financial transaction, the value of an amount of money changes with time as a result of the application of interest. Thus, to *accumulate* or *bring forward* a single payment R for n periods at an interest rate i per period, we multiply R by $(1 + i)^n$, as illustrated in Figure 4–28. To *bring back* a single payment R for n periods at an interest rate of i per period, we multiply R by $(1 + i)^{-n}$, as shown in Figure 4–29. To accumulate or bring forward an annuity of n payments of R dollars each, we multiply R by $s_{\overline{n}|i}$, where $s_{\overline{n}|i} = [(1 + i)^n - 1]/i$, as pictured in Figure 4–30. To bring back an annuity of n payments of R dollars each, we multiply R by $a_{\overline{n}|i}$, where $a_{\overline{n}|i} = [1 - (1 + i)^{-n}]/i$, as illustrated in Figure 4–31.

Bringing forward a single payment R

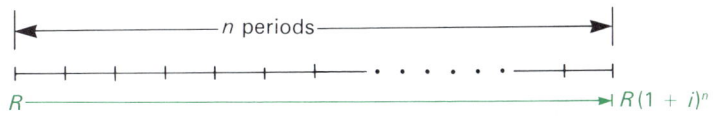

FIGURE 4–28

Bringing back a single payment R

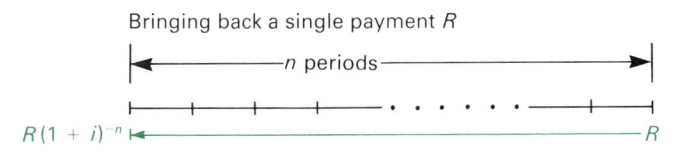

FIGURE 4–29

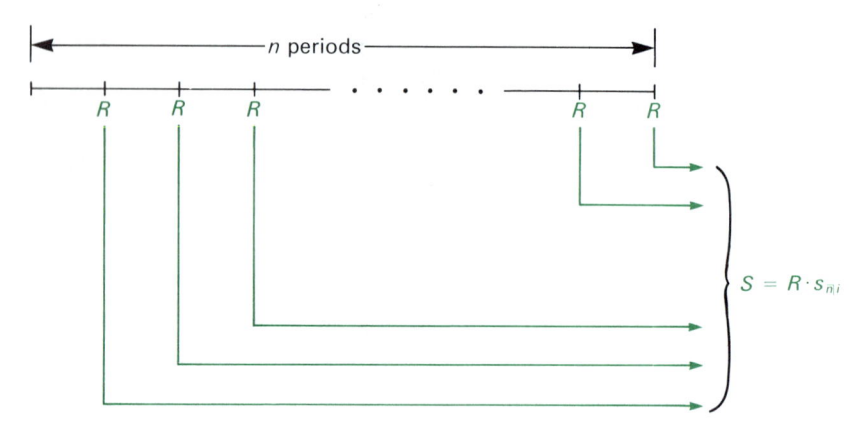

FIGURE 4–30

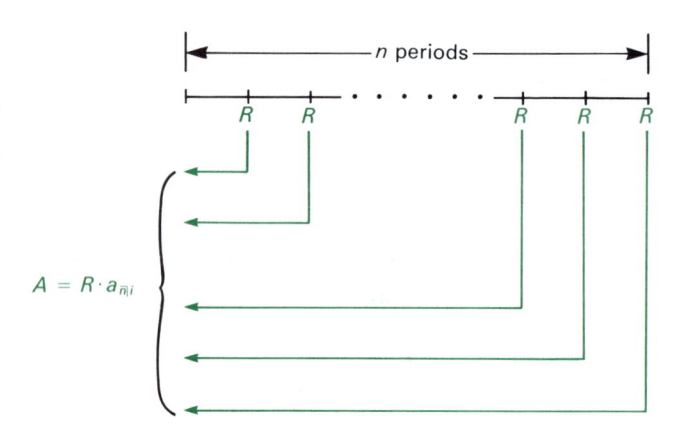

$$S = R \cdot s_{\overline{n}|i}$$

FIGURE 4–31

$$A = R \cdot a_{\overline{n}|i}$$

We now consider the following problem:

A business person has a debt of $4000 due in 3 years. He wants to repay this debt by making a $3000 payment 2 years from now and a last payment 5 years from now. If the interest rate is 12% compounded annually, what must be the amount of the last payment?

This situation is illustrated in the time diagram of Figure 4–32. Note that the last payment is denoted by *x*. Thus, we want to determine *x* so that the value of the two payments on the lower time line is equivalent to the value of the single payment on the upper time line if the interest rate is 12% compounded annually.

MATHEMATICS OF FINANCE

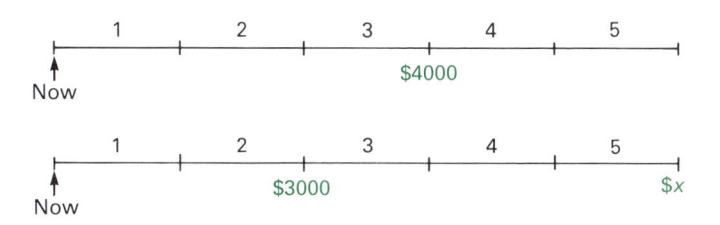

FIGURE 4-32

Since the value of any amount of money changes with time as a result of the application of interest, we must choose a point on both time lines at which we will equate the values of the payments of both time lines. Such a point is called a **comparison point** and may be chosen arbitrarily.

If, for the situation of Figure 4-32, we choose the comparison point to be "Now," we must:

1. Bring back the $4000 payment on the upper time line three periods
2. Bring back the $3000 payment on the lower time line two periods and the unknown payment five periods (see Figure 4-33)

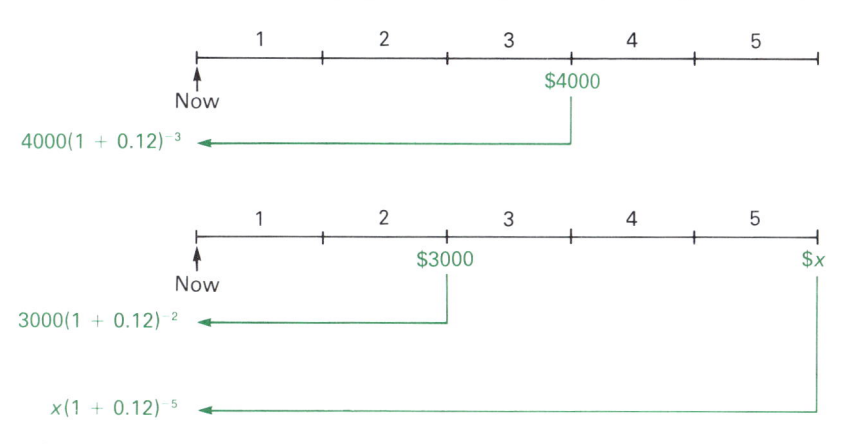

FIGURE 4-33

Equating the values of both time lines at the comparison point "Now," we obtain the equation

$$4000(1 + 0.12)^{-3} = 3000(1 + 0.12)^{-2} + x(1 + 0.12)^{-5}$$

This equation is called an **equation of value.** Solving for x, we multiply both sides by $(1 + 0.12)^5$ to obtain

$$4000(1 + 0.12)^2 = 3000(1 + 0.12)^3 + x$$

Hence,

$$x = 4000(1 + 0.12)^2 - 3000(1 + 0.12)^3$$

Using Appendix Table 4 to obtain the needed powers of $(1 + 0.12)$, we have

$$x = 4000(1.254400) - 3000(1.404928)$$
$$= \$802.82$$

Thus, the last payment is $802.82.

As mentioned before, the comparison point may be chosen arbitrarily. If the comparison point for the preceding problem is chosen to be the end of the fifth year, then each payment must be brought forward to the end of the fifth year. The resulting equation of value is

$$4000(1 + 0.12)^2 = 3000(1 + 0.12)^3 + x$$

Note that this equation of value is equivalent to the first one. If we multiply both sides of the first equation of value by $(1 + 0.12)^5$ (which we did in order to solve it), we obtain the latter one. Again, the last payment, x, is $802.82.

EXAMPLE 4–26 A business person wishes to borrow $10,000 today and $6000 three years from today. She wishes to repay both loans with equal annual payments at the end of each year for the next four years. If the interest rate is 15% compounded annually, what is the annual payment?

Solution This problem is illustrated by the time lines of Figure 4–34. The annual payment is denoted by x. If the beginning of the first year is chosen as the comparison point, then the equation of value is

$$10{,}000 + 6000(1 + 0.15)^{-3} = x \cdot a_{\overline{4}|0.15}$$

Solving for x, we have

$$x = \frac{10{,}000 + 6000(1 + 0.15)^{-3}}{a_{\overline{4}|0.15}}$$

$$= \$4884.48$$

Thus, the annual payment is $4884.48.

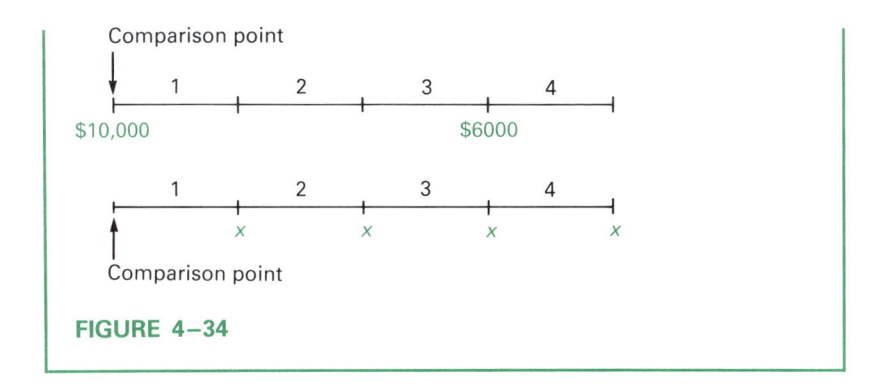

MATHEMATICS OF FINANCE

```
Comparison point
    ↓
    |---1---+---2---+---3---+---4---|
$10,000              $6000

    |---1---+---2---+---3---+---4---|
    ↑       x       x       x       x
Comparison point
```

FIGURE 4–34

Deferred Annuities A **deferred annuity** is an annuity whose payments begin later than at the end of the first period.

EXAMPLE 4–27 An investment contract promises to pay $1000 at the end of each year beginning with the end of the fifth year and ending with the end of the twelfth year. If an investor wishes to earn 10% compounded annually, how much should he pay for this contract now?

Solution The time lines of Figure 4–35 illustrate this problem. Since the first payment occurs later than the end of the first period, this is a deferred annuity. The investor must determine the present value of this deferred annuity. This quantity is denoted by *x*. If the end of the fourth year is chosen as the comparison point, the present value, *x*, must be brought forward four periods. This resulting value equals the present value of the annuity at the comparison point.

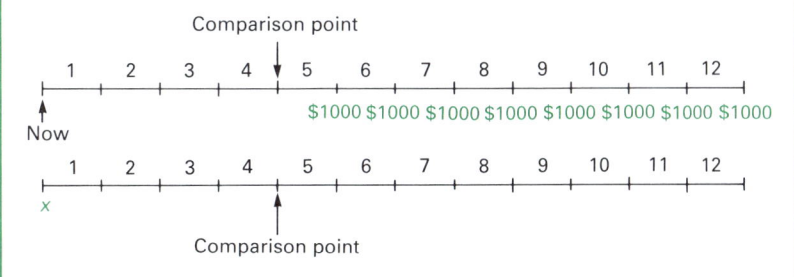

```
                    Comparison point
                          ↓
  1   2   3   4   5   6   7   8   9  10  11  12
  |---+---+---+---+---+---+---+---+---+---+---|
  ↑              $1000 $1000 $1000 $1000 $1000 $1000 $1000 $1000
 Now

  1   2   3   4   5   6   7   8   9  10  11  12
  |---+---+---+---+---+---+---+---+---+---+---|
  x              ↑
          Comparison point
```

FIGURE 4–35

Hence, the equation of value is

$$x(1 + 0.10)^4 = 1000 \cdot a_{\overline{8}|0.10}$$

EXAMPLE 4–27 (continued)

Solving for x, we have

$$x = \frac{1000 \cdot a_{\overline{8}|0.10}}{(1 + 0.10)^4}$$
$$= \$3643.83$$

Thus, the investor should pay $3643.83 for this contract if he wishes to earn 10% compounded annually on his investment.

Variable Annuities

Some financial institutions offer graduated payment loans. The repayment of such a loan involves an annuity whose later payments are larger than earlier ones. Such an annuity is called a **variable annuity**.

EXAMPLE 4–28
Mrs. Logan finances the purchase of a new car with a cash price of $10,000 by a variable annuity. During the first 3 years, she will make payments of a certain amount at the end of each year. During the last 4 years, her annual payment will be twice as large. If the interest rate is 12% compounded annually, what are the annual payments?

Solution The time lines of Figure 4–36 illustrate this problem.

```
              Comparison point
                    ↓
    1     2     3     4     5     6     7
    +-----+-----+-----+-----+-----+-----+
       x     x     x    2x    2x    2x    2x

          1     2     3     4     5     6     7
    +-----+-----+-----+-----+-----+-----+-----+
 $10,000                  ↑
                    Comparison point
```

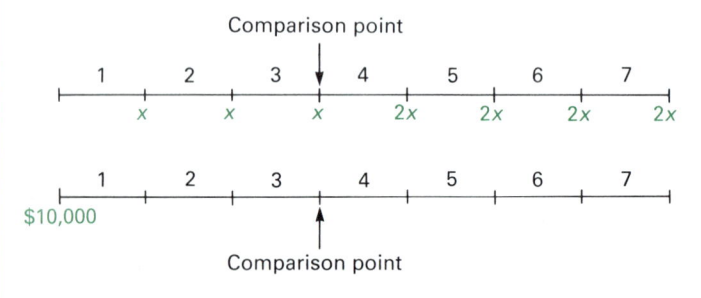

FIGURE 4–36

Observe that x denotes the first three annual payments, and $2x$ denotes the remaining payments. If the comparison point is chosen to be the end of the third year, the equation of value is

$$x \cdot s_{\overline{3}|0.12} + 2x \cdot a_{\overline{4}|0.12} = 10{,}000(1 + 0.12)^3$$

Using either tables or a calculator to determine $s_{\overline{3}|0.12}$, $a_{\overline{4}|0.12}$, and $(1 + 0.12)^3$, this equation becomes

$$x(3.374400) + 2x(3.037349) = 10{,}000(1.404928)$$

Solving for x, we have

$$3.3744x + 6.074698x = 14049.28$$
$$x = \$1486.84$$

Thus, the first three annual payments are $1486.84, and the remaining annual payments are 2($1486.84) = $2973.68.

EXERCISES

1. Mr. Evans has a debt of $8000 due in four years. He wants to repay this debt by making a $2000 payment one year from now, a $1000 payment three years from now, and a last payment six years from now. If the interest rate is 10% compounded annually, what must be the amount of the last payment?

2. A business person's debt is payable as follows: $2000 one year from now and $5000 five years from now. The business person wants to repay the debt as follows: a $1000 payment now, a $2000 payment two years from now, a $1000 payment three years from now, and the last payment four years from now. If the interest rate is 12% compounded annually, find the amount of the last payment.

3. A woman wishes to borrow $5000 now and $4000 two years from now. She wishes to repay both loans with equal annual payments at the end of each year for the next five years. If the interest rate is 10% compounded annually, find the annual payment.

4. If money is worth 12% compounded annually, what single payment made 2 years from now can replace the following two payments: $3000 due 1 year from now and $5000 due 4 years from now.

5. Find the present value of an annuity of $2000 at the end of each year with the first payment occurring 4 years from now and the last occurring 10 years from now if the interest rate is 8% compounded annually.

6. An investment contract promises to pay $5000 at the end of each year beginning with the end of the third year and ending with the end of the ninth year. If the investor wishes to earn 10% compounded annually on his money, how much should he pay for this contract now?

7. A company is considering purchasing new equipment which will result in an increased cash flow of $10,000 per year beginning 4 years from now and ending 9 years from now. If the company wants a rate of return of 12% compounded annually on its investment, how much should be spent on the new equipment?

214 CHAPTER FOUR

8. Mary Smith signed a mortgage for $10,000. The mortgage is to be repaid with equal monthly payments for the first 2 years, and equal monthly payments twice as large for the next 3 years. If the interest rate is 12% compounded monthly, find the monthly payments. Assume each payment is made at the end of the month.

9. Find the present value of a variable annuity consisting of $500 at the end of each year for the first 4 years and $800 at the end of each year for the next 5 years. The interest rate is 10% compounded annually.

10. A man buys a car with a cash price of $12,000 by financing it over a 5-year period. At the end of each of the first 2 years, he will make equal payments. At the end of each of the next 3 years, his equal payments will be $1\frac{1}{2}$ times as large. If the interest rate is 12% compounded annually, find his annual payments.

4–8

DEFERRED ANNUITIES

In this section, we will formally present formulas for the present value of a deferred annuity. As stated in the preceding section, in the case of a deferred annuity, the first payment occurs later than at the end of the first conversion period. If a person finances the purchase of some item by agreeing to pay 20 quarterly payments of $60 each, *with the first payment due 2 years from now,* then this series of payments constitutes a deferred annuity. The length of time from the present to the beginning of the first payment time interval is called the **period of deferment**. If d = number of periods of deferment, then as illustrated in Figure 4–37, $d = 7$ and the annuity consists of $n = 20$ payments, each payment made at the end of a quarter.

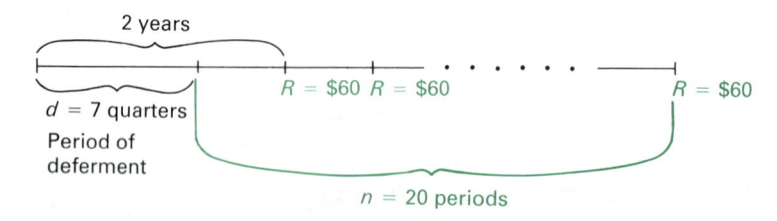

FIGURE 4–37

The total amount of a deferred annuity is its future value at the end of n periods and is determined by the usual formula for the total amount of an annuity,

$$S = R \cdot s_{\overline{n}|i}$$

For the annuity of Figure 4–37, if the interest rate is 12% compounded quarterly, then $i = 0.03$ and

MATHEMATICS OF FINANCE

$$S = 60 \cdot s_{\overline{20}|0.03}$$
$$= 60(26.870374)$$
$$= \$1612.22$$

Present Value of a Deferred Annuity

In general, the present value, A, of a deferred annuity of n payments with d periods of deferment (see Figure 4–38) at an interest rate i per conversion period may be determined by the formula

$$A = R \cdot a_{\overline{n}|i}(1 + i)^{-d}$$

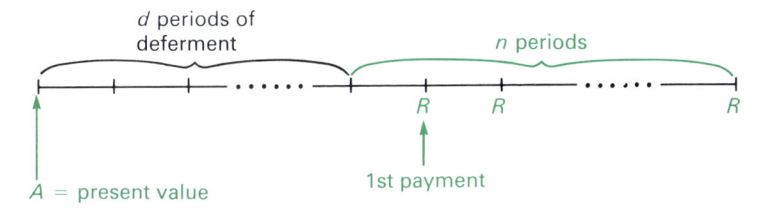

FIGURE 4–38

Observing Figure 4–39, note that the term $R \cdot a_{\overline{n}|i}$ gives the present value of the annuity at the beginning of the first payment period. Multiplying $R \cdot a_{\overline{n}|i}$ by $(1 + i)^{-d}$ brings this result back d periods to yield the present value of the deferred annuity. Thus, for the deferred annuity of Figure 4–37, the present value is

$$A = R \cdot a_{\overline{n}|i}(1 + i)^{-d}$$
$$= 60 \cdot a_{\overline{20}|0.03}(1 + 0.03)^{-7}$$
$$= 60(14.877475)(0.813092)$$
$$= \$725.81$$

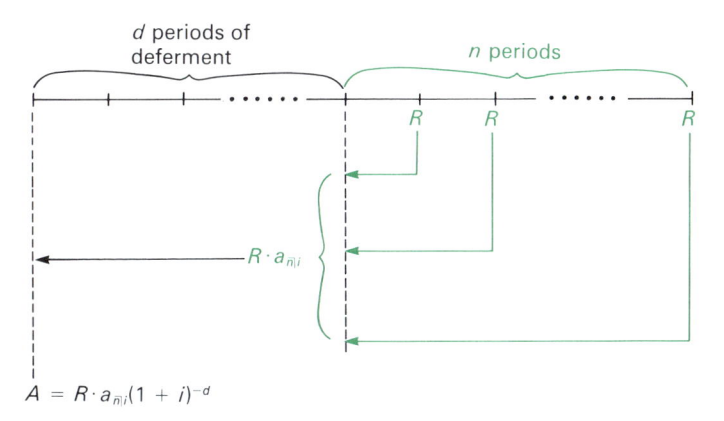

FIGURE 4–39

EXAMPLE 4–29 Sara Smith agrees to repay a loan by making $200 payments at the end of each month for 3 years. The first payment is due at the end of 6 months. If the interest rate is 12% compounded monthly, then find the amount of the loan.

Solution The amount of the loan is the present value of this deferred annuity. Observing Figure 4–40, note that $d = 5$, $n = 12(3) = 36$, and $R = \$200$. Also, $i = 0.12/12 = 0.01$. Hence, the present value is

$$A = R \cdot a_{\overline{n}|i}(1 + i)^{-d}$$
$$= 200 \cdot a_{\overline{36}|0.01}(1 + 0.01)^{-5}$$
$$= 200(30.107505)(0.951466)$$
$$= \$5729.25$$

FIGURE 4–40

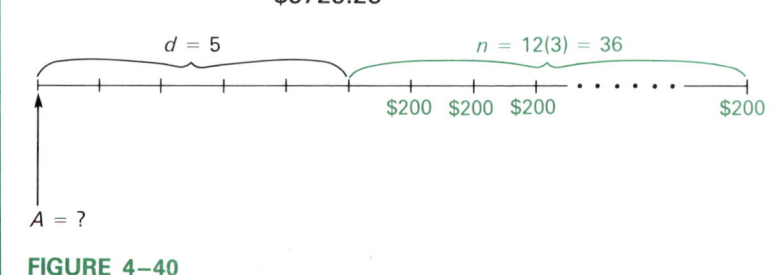

Another Method The present value, A, of a deferred annuity of n periods with d periods of deferment (as illustrated in Figure 4–41) at an interest rate i per conversion period may also be found by first finding the present value of the annuity consisting of $(d + n)$ payments and then subtracting from this amount the present value of the annuity consisting of d payments (see Figure 4–41). Thus, the formula for the present value, A, is

$$A = R \cdot a_{\overline{d+n}|i} - R \cdot a_{\overline{d}|i}$$

or

$$A = R(a_{\overline{d+n}|i} - a_{\overline{d}|i})$$

Many people prefer this method for finding the present value of a deferred annuity over the previous one because it involves the use of only one set of tables.

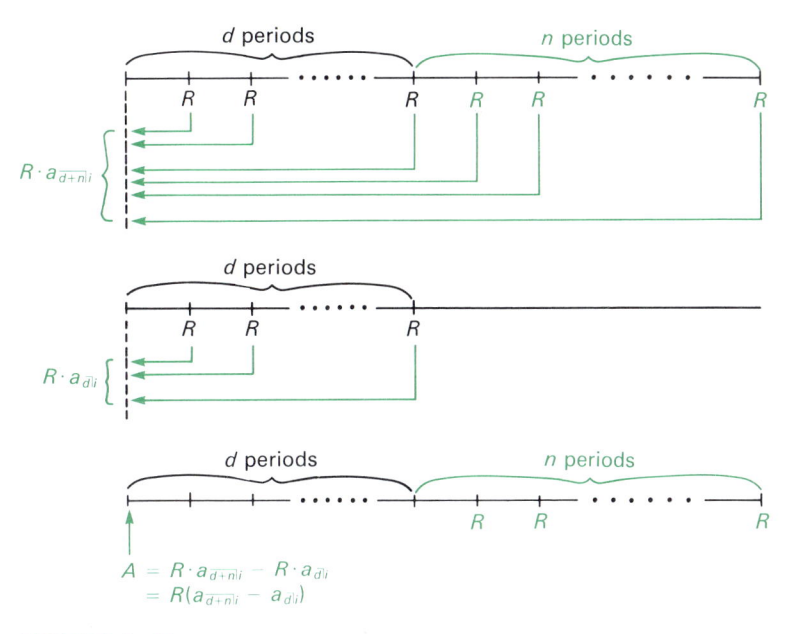

217 MATHEMATICS OF FINANCE

$$A = R \cdot a_{\overline{d+n}|i} - R \cdot a_{\overline{d}|i}$$
$$= R(a_{\overline{d+n}|i} - a_{\overline{d}|i})$$

FIGURE 4–41

EXAMPLE 4–30 A man wishes to set up a fund that will provide for quarterly payments of $500 each at the end of each quarter for 5 years. If the first payment is to be made at the end of 9 months and the interest rate is 16% compounded quarterly, how much should the man deposit into the fund now?

Solution We seek the present value of a deferred annuity with $R = \$500$, $n = 4(5) = 20$, $d = 2$, and $i = 0.16/4 = 0.04$ (see Figure 4–42). Thus, we have

$$A = R(a_{\overline{d+n}|i} - a_{\overline{d}|i})$$
$$= 500(a_{\overline{22}|0.04} - a_{\overline{2}|0.04})$$
$$= 500(14.451115 - 1.886095)$$
$$= 500(12.565020)$$
$$= \$6282.51$$

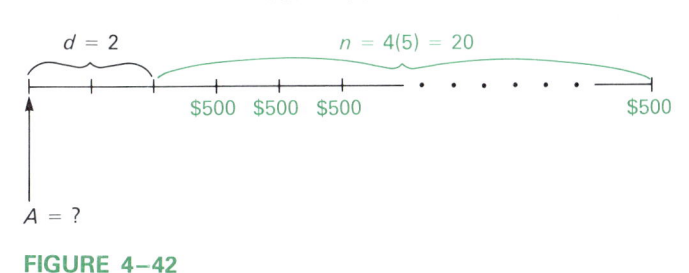

FIGURE 4–42

Determining the Periodic Payment, R

Another problem often encountered in the business world is finding the periodic payment, R, of a deferred annuity. Example 4–31 illustrates such a problem and its solution.

EXAMPLE 4–31 A loan of $10,000 is to be repaid by 20 equal quarterly payments at the end of each quarter. Find the size of each payment if the first payment is due 2 years from now and the interest rate is 20% compounded quarterly.

Solution The time line of Figure 4–43 illustrates this deferred annuity.

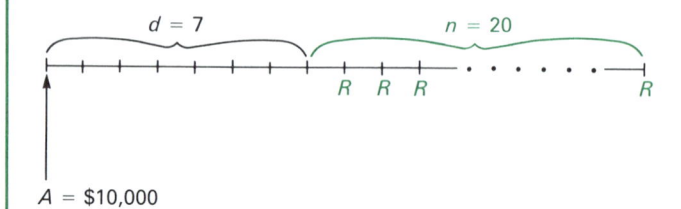

```
       d = 7                    n = 20
  |+++++++|+++.......+|
               R R R              R
  ↑
A = $10,000
```

FIGURE 4–43

Note that $d = 7$, $n = 20$, and the amount of the loan, $10,000, is the present value of this annuity. Hence, $A = \$10{,}000$. We may use either of the two present-value formulas to determine R. Using the first present-value formula of this section, we have

$$A = R \cdot a_{\overline{n}|i}(1 + i)^{-d}$$
$$10{,}000 = R \cdot a_{\overline{20}|0.05}(1 + 0.05)^{-7}$$

Solving this for R, we obtain

$$R = \frac{10{,}000}{a_{\overline{20}|0.05}(1 + 0.05)^{-7}}$$
$$= \frac{10{,}000}{(12.462210)(0.710681)}$$
$$= \$1129.09$$

We will also solve this problem using the second present-value formula of this section, as follows:

$$A = R(a_{\overline{d+n}|i} - a_{\overline{d}|i})$$
$$10{,}000 = R(a_{\overline{27}|0.05} - a_{\overline{7}|0.05})$$
$$10{,}000 = R(14.643034 - 5.786373)$$
$$10{,}000 = R(8.856661)$$

Solving for R, we obtain

MATHEMATICS OF FINANCE

$$R = \frac{10{,}000}{8.856661}$$
$$= \$1129.09$$

EXERCISES

1. A loan is repaid by paying $500 at the end of each quarter for 10 years. The first payment is due 1 year from now. If the interest rate is 16% compounded quarterly, what is the amount of the loan?

2. If a person repays a loan with monthly payments of $400 at the end of each month for 3 years, what is the amount of the loan if the first payment is due in 6 months and the interest rate is 24% compounded monthly?

3. Ms. James is considering buying an investment contract that promises to pay $1000 at the end of each 6 months for 10 years. The first payment is due in 18 months. If Ms. James wishes to earn 16% compounded semiannually on her investment, how much should she pay for this investment contract?

4. Harry Morgan wishes to set up a fund to provide for $2000 payments at the end of each quarter for 4 years. The payments will be made to Harry's college-age daughter who will enter college 1 year from now. Thus, the first payment will be made 1 year from now. If Harry earns 16% compounded quarterly on his money, how much should be deposited into this fund now?

5. A loan of $50,000 will be repaid by quarterly payments made at the end of each quarter for 5 years. If the interest rate is 20% compounded quarterly and the first payment is due $1\frac{1}{2}$ years from now, what is the quarterly payment?

6. A mortgage of $70,000 will be repaid by monthly payments made at the end of each month for 8 years. If the interest rate is 18% compounded monthly and the first payment is due in 1 year, find the monthly payment.

4–9

COMPLEX ANNUITIES

Up to this point, we have been considering annuities in which the payment period coincides with the conversion (interest) period. Such an annuity is called a **simple annuity.** If we have an annuity where the payment period does not coincide with the conversion (interest) period, we have a **complex annuity,** or **general annuity.** In this section, we will discuss formulas for the present and future values of complex annuities.

For notation, we will let

n = total number of payments of an annuity

c = number of conversion (interest) periods in one payment period

220 CHAPTER FOUR

Thus,

$$nc = \text{total number of conversion (interest) periods for a complex annuity}$$

To illustrate these terms, we consider in Figure 4–44 an annuity of three semiannual payments R at an interest rate that is compounded monthly. Note that $n = 3$ payments, $c = 6$ conversion (interest) periods per payment period, and $nc = 3(6) = 18$ conversion (interest) periods for the entire complex annuity.

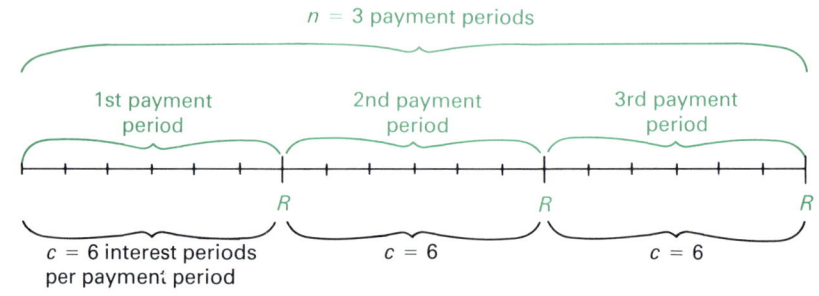

$n = 3$ payment periods

1st payment period | 2nd payment period | 3rd payment period

$c = 6$ interest periods per payment period $c = 6$ $c = 6$

$nc = 3(6) = 18$ interest periods

FIGURE 4–44

EXAMPLE 4–32 Draw a time diagram and determine n, c, and nc for a complex annuity of 24 monthly payments R at an interest rate that is compounded semiannually.

Solution Studying the time diagram of Figure 4–45, note that we have six payment periods in one conversion period.

$n = 24$ payments

1st payment period, 2nd payment period, ..., 6th payment period

6-month conversion period

$c = \dfrac{1}{6}$ interest period per payment period

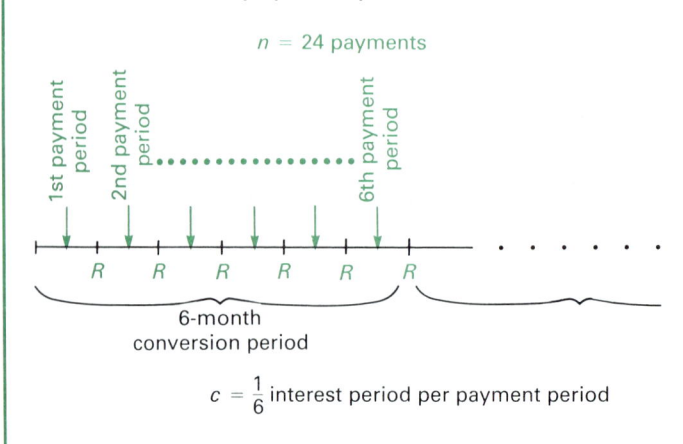

FIGURE 4–45

MATHEMATICS OF FINANCE

> Since c is the number of conversion periods in one payment period, then c takes on the fractional value 1/6. Hence, $c = 1/6$ and $nc = 24(1/6) = 4$ conversion periods for this complex annuity.

In general, if there are m payments in a given conversion (interest) period, then

$$c = \frac{1}{m}$$

Total Amount and Present Value of a Complex Annuity

We now derive formulas for the total amount and present value of a complex annuity. Let us consider an annuity of seven payments of $1000 apiece made at the end of each quarter. The interest rate is 12% compounded monthly. Studying the time diagram of Figure 4–46, note that $n = 7$ payments, $c = 3$ interest periods per payment period, and $nc = 7(3) = 21$ interest periods for this annuity.

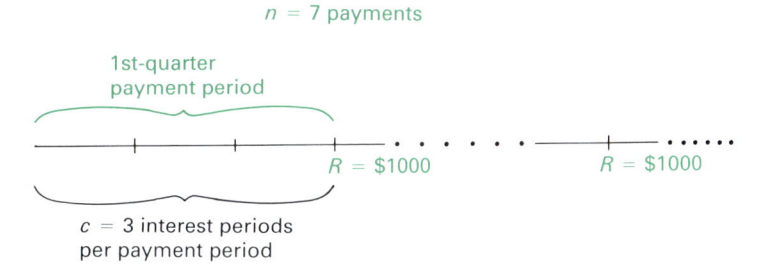

FIGURE 4–46

Our goal is to convert this complex annuity into an equivalent simple annuity where the interest period coincides with the payment period. If we focus on an individual payment period of the annuity of Figure 4–46, we may convert the original payment R into an equivalent series of $c = 3$ payments payable at the end of each month. Thus, each payment R is the total amount of an annuity of $c = 3$ payments of E dollars each, as illustrated in Figure 4–47.

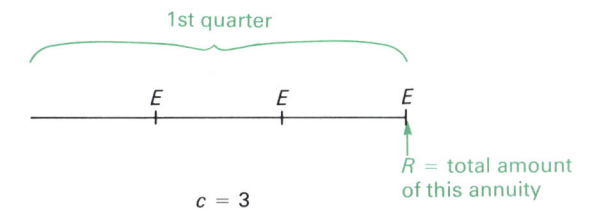

FIGURE 4–47

222　CHAPTER FOUR

Using the formula for the total amount of a simple annuity

$$S = R \cdot s_{\overline{n}|i}$$

we replace S with R, R with E, and n with c to obtain

$$R = E \cdot s_{\overline{c}|i}$$

Solving for E yields

$$E = \frac{R}{s_{\overline{c}|i}}$$

If we do this for each payment period, we will convert the original complex annuity into a simple annuity consisting of $nc = 7(3) = 21$ monthly payments of E dollars each at the end of each month earning interest at $i = 12\%/12 = 1\%$ per month (see Figure 4–48).

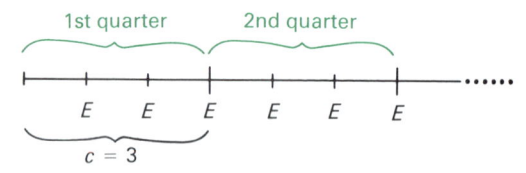

FIGURE 4–48

Thus, the total amount of this new annuity is

$$S = E \cdot s_{\overline{nc}|i}$$

Substituting $R/s_{\overline{c}|i}$ for E into this equation, we obtain

$$S = \left(\frac{R}{s_{\overline{c}|i}}\right) s_{\overline{nc}|i}$$

as the formula for the total amount of a complex annuity of n payments of R dollars each with c interest periods per payment period and an interest rate of i per interest or conversion period.

In a similar manner, we derive the formula for the present value, A, of a complex annuity. The present value of the annuity of nc payments of E dollars each of Figure 4–48 is

$$A = E \cdot a_{\overline{nc}|i}$$

MATHEMATICS OF FINANCE

Substituting $R/s_{\overline{c}|i}$ for E into this equation yields as the formula for the present value of a complex annuity of n payments of R dollars each with c interest periods per payment period and interest rate i per interest or conversion period.

$$A = \left(\frac{R}{s_{\overline{c}|i}}\right) a_{\overline{nc}|i}$$

EXAMPLE 4–33 Find the total amount and present value of an annuity of $500 payable at the end of each year for 8 years if the interest rate is 18% compounded monthly.

Solution Here, $R = \$500$, $c = 12$ interest or conversion periods per payment period, $n = 8$ payments, $nc = 8(12) = 96$ interest or conversion periods, and $i = 18\%/12 = 1\frac{1}{2}\%$ per conversion period. Hence, the total amount of this complex annuity is

$$S = \left(\frac{R}{s_{\overline{c}|i}}\right) s_{\overline{nc}|i}$$

$$= \left(\frac{500}{s_{\overline{12}|1\frac{1}{2}\%}}\right) s_{\overline{96}|1\frac{1}{2}\%}$$

$$= \left(\frac{500}{13.041211}\right) 211.720235$$

$$= \$8117.35$$

The present value of this complex annuity is

$$A = \left(\frac{R}{s_{\overline{c}|i}}\right) a_{\overline{nc}|i}$$

$$= \left(\frac{500}{s_{\overline{12}|1\frac{1}{2}\%}}\right) a_{\overline{96}|1\frac{1}{2}\%}$$

$$= \left(\frac{500}{13.041211}\right) 50.701675$$

$$= \$1943.90$$

EXAMPLE 4–34 Find the total amount and present value of an annuity of $700 payable at the end of each quarter for 5 years if the interest rate is 18% compounded semiannually.

EXAMPLE 4–34 (continued)

Solution Here, $R = \$700$, $c = 1/2$ interest period per payment period, $n = 20$ payments, $nc = 20(1/2) = 10$ interest or conversion periods, and $i = 18\%/2 = 9\%$ per conversion period. Hence, the total amount of this complex annuity is

$$S = \left(\frac{R}{s_{\overline{c}|i}}\right) s_{\overline{nc}|i}$$

$$= \left(\frac{700}{s_{\overline{1/2}|9\%}}\right) s_{\overline{10}|9\%}$$

Note that we must find $s_{\overline{c}|i}$ for a fractional value of c. Appendix Table 7 gives values of $s_{\overline{c}|i}$ for fractional values of c. Using this table, we find that $s_{\overline{1/2}|9\%} = 0.489229$. Thus, the preceding formula for S becomes

$$S = \left(\frac{700}{0.489229}\right) 15.192930$$

$$= \$21{,}738.39$$

The present value of this complex annuity is

$$A = \left(\frac{R}{s_{\overline{c}|i}}\right) a_{\overline{nc}|i}$$

$$= \left(\frac{700}{s_{\overline{1/2}|9\%}}\right) a_{\overline{10}|9\%}$$

$$= \left(\frac{700}{0.489229}\right) 6.417658$$

$$= \$9182.53$$

EXAMPLE 4–35 A \$40,000 loan is to be repaid with 10 semiannual payments payable at the end of each 6-month period. If the interest rate is 12% compounded monthly, find the semiannual payment.

Solution Here, $c = 6$ conversion periods per payment period, $n = 10$ payments, $nc = 10(6) = 60$ conversion periods, $i = 12\%/12 = 1\%$ per conversion period, and the amount of the loan, \$40,000, is the present value of this complex annuity. We must determine R. The equation

$$A = \left(\frac{R}{s_{\overline{c}|i}}\right) a_{\overline{nc}|i}$$

relates the present value, A, of a complex annuity with the periodic payment, R. Substituting $40,000 for A into the preceding equation gives us

$$40{,}000 = \left(\frac{R}{s_{\overline{6}|1\%}}\right) a_{\overline{60}|1\%}$$

Solving for R, we obtain

$$R = \frac{40{,}000 \cdot s_{\overline{6}|1\%}}{a_{\overline{60}|1\%}}$$

$$= \frac{40{,}000(6.152015)}{44.955038}$$

$$= \$5473.93$$

EXAMPLE 4–36 A loan is to be repaid with four annual payments of $4000 each. The first payment is due 3 years from now. If the interest rate is 10% compounded semiannually, find the amount of the loan.

Solution We seek the present value of this complex deferred annuity. Note that $R = \$4000$, $n = 4$, $c = 2$ interest periods per payment period, $nc = 4(2) = 8$ conversion or interest periods, and $i = 10\%/2 = 5\%$ per conversion period.

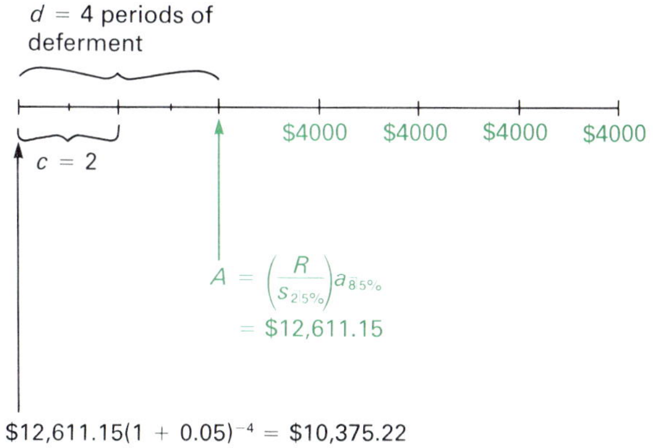

$d = 4$ periods of deferment

$c = 2$

$4000 \quad \$4000 \quad \$4000 \quad \$4000$

$$A = \left(\frac{R}{s_{\overline{2}|5\%}}\right) a_{\overline{8}|5\%}$$

$$= \$12{,}611.15$$

$\$12{,}611.15(1 + 0.05)^{-4} = \$10{,}375.22$

FIGURE 4–49

EXAMPLE 4–36 (continued)

Studying the time diagram of Figure 4–49 (page 225), we see that the formula

$$A = \left(\frac{R}{s_{\overline{c}|i}}\right) a_{\overline{nc}|i}$$

$$= \left(\frac{4000}{s_{\overline{2}|5\%}}\right) a_{\overline{8}|5\%}$$

$$= \left(\frac{4000}{2.050000}\right) 6.463213$$

$$= \$12{,}611.15$$

gives the present value of the complex annuity at the end of the period of deferment. Multiplying this result by the factor $(1 + i)^{-d}$ yields the present value at the beginning of the period of deferment. Thus, the present value of this deferred annuity is

$$12{,}611.15(1 + 0.05)^{-4} = 12{,}611.15(0.822702)$$
$$= \$10{,}375.22$$

Therefore, the amount of the loan is $10,375.22.

EXERCISES

1. Determine n, c, and nc for each of the following complex annuities:
   (A) 21 monthly payments at 6% compounded quarterly
   (B) Quarterly payments over 5 years at 12% compounded monthly
   (C) Semiannual payments over 8 years at 16% compounded quarterly
   (D) Annual payments over 10 years at 18% compounded monthly

2. Find the total amount and present value of an annuity of $1000 payable at the end of each quarter for 5 years at an interest rate of 24% compounded monthly.

3. Find the total amount and present value of an annuity of $3500 payable at the end of each 6 months for 10 years. The interest rate is 16% compounded quarterly.

4. Find the total amount and present value of an annuity of $5600 payable at the end of each month for 6 years. The interest rate is 20% compounded quarterly.

5. Find the total amount and present value of an annuity of $1300 payable at the end of each quarter for 10 years. The interest rate is 15% compounded annually.

MATHEMATICS OF FINANCE

6. A loan is to be repaid with quarterly payments of $650 at the end of each quarter for 6 years. If the interest rate is 18% compounded monthly, find the amount of the loan.

7. A loan is repaid by monthly payments of $250 at the end of each month for 5 years. If the interest rate is 18% compounded semiannually, find the amount of the loan.

8. A borrower will repay a loan with 21 monthly payments of $1000 each. The first payment is due 10 months from now. If the interest rate is 12% compounded quarterly, find the amount of the loan.

9. A loan will be repaid by six annual payments of $4500 each. The first payment is due 2 years from now. If the interest rate is 20% compounded quarterly, find the amount of the loan.

10. A loan of $12,500 is to be repaid by semiannual payments at the end of each 6 months for 7 years. If the interest rate is 18% compounded monthly, find the semiannual payment.

11. A loan of $16,000 is to be repaid with 36 monthly payments, the first one due 1 month from now. If the interest rate is 20% compounded quarterly, find the monthly payment.

12. A loan of $8600 will be repaid by annual payments at the end of each year for 6 years. If the interest rate is 16% compounded quarterly, find the annual payment.

13. Mr. Evans wishes to accumulate $9000 by making quarterly payments into a fund at the end of each quarter for 6 years. If the interest rate is 18% compounded monthly, find the quarterly payment.

14. A woman wishes to accumulate $10,000 by making monthly payments into a sinking fund at the end of each month for 5 years. If the interest rate is 20% compounded semiannually, find the monthly payment.

4–10

COMPLEX ANNUITIES DUE

If each payment R of a complex annuity is payable at the *beginning* of the payment period, then we have a **complex annuity due**. We will now derive formulas for the total amount and present value of a complex annuity due.

Consider an annuity of 10 semiannual payments of $3000 each payable at the beginning of each payment period. The interest rate is 16% compounded quarterly. Studying the time diagram of Figure 4–50, note that $n = 10$ payments, $c = 2$ interest periods per payment period, $nc = 10(2) = 20$ interest periods, and $i = 16\%/4 = 4\%$ per interest period.

CHAPTER FOUR

$n = 10$ payment periods

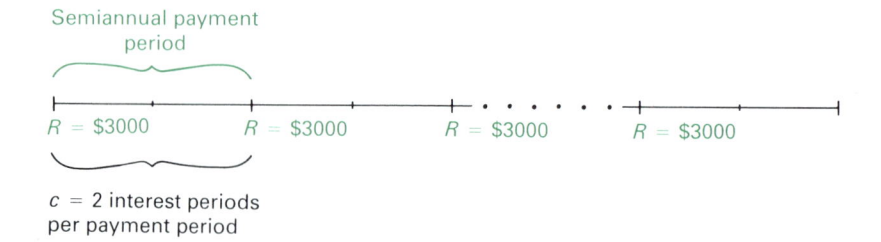

Semiannual payment period

$R = \$3000 \quad R = \$3000 \quad R = \$3000 \quad R = \$3000$

$c = 2$ interest periods per payment period

$i = 4\%$ per quarter

FIGURE 4–50

We may convert this complex annuity due into an equivalent simple annuity where the interest period and payment period coincide by focusing on an individual payment period (see Figure 4–51) and recognizing that the periodic payment, R, is the present value of an annuity of $c = 2$ quarterly payments of E dollars each, as illustrated in Figure 4–51.

Semiannual payment period

$E \qquad E$

$c = 2$

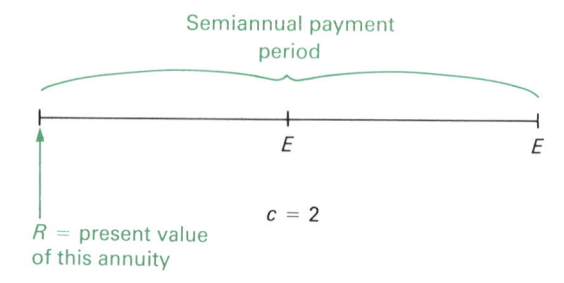

$R = $ present value of this annuity

FIGURE 4–51

Using the formula for the present value of a simple annuity

$$A = R \cdot a_{\overline{n}|i}$$

we replace A with R, R with E, and n with c to obtain

$$R = E \cdot a_{\overline{c}|i}$$

Solving for E, we get

$$E = \frac{R}{a_{\overline{c}|i}}$$

If we do this for each payment period, we will convert the original complex annuity due into a simple annuity consisting of

MATHEMATICS OF FINANCE

$nc = 10(2) = 20$ quarterly payments of E dollars each payable at the end of each quarter and earning interest at $i = 4\%$ per quarter. Thus, the total amount of this new annuity is

$$S = E \cdot s_{\overline{nc}|i}$$

Substituting $R/a_{\overline{c}|i}$ for E into this equation, we obtain

$$S = \left(\frac{R}{a_{\overline{c}|i}}\right) s_{\overline{nc}|i}$$

as the formula for the total amount of a complex annuity due of n payments of R dollars each with c interest periods per payment period and interest rate i per interest period. In a similar manner, the formula for the present value, A, is derived to be

$$A = \left(\frac{R}{a_{\overline{c}|i}}\right) a_{\overline{nc}|i}$$

Thus, for the complex annuity due of Figure 4–50, the total amount is

$$S = \left(\frac{R}{a_{\overline{c}|i}}\right) s_{\overline{nc}|i}$$

$$= \left(\frac{3000}{a_{\overline{2}|4\%}}\right) s_{\overline{20}|4\%}$$

$$= \left(\frac{3000}{1.886095}\right) 29.778079$$

$$= \$47{,}364.65$$

and the present value is

$$A = \left(\frac{R}{a_{\overline{c}|i}}\right) a_{\overline{nc}|i}$$

$$= \left(\frac{3000}{a_{\overline{2}|4\%}}\right) a_{\overline{20}|4\%}$$

$$= \left(\frac{3000}{1.886095}\right) 13.590326$$

$$= \$21{,}616.61$$

EXAMPLE 4–37 Find the total amount and present value of an annuity of $800 payable at the beginning of each quarter for 6 years if the interest rate is 24% compounded monthly.
Solution Note that $R = \$800$, $c = 3$ conversion periods per payment period, $n = 24$ payments, $nc = 24(3) = 72$ conversion periods, and $i = 24\%/12 = 2\%$ per conversion period. Thus, the total amount of this complex annuity due is

$$S = \left(\frac{R}{a_{\overline{c}|i}}\right) s_{\overline{nc}|i}$$

$$= \left(\frac{800}{a_{\overline{3}|2\%}}\right) s_{\overline{72}|2\%}$$

$$= \left(\frac{800}{2.883883}\right) 158.057019$$

$$= \$43,845.61$$

and present value is

$$A = \left(\frac{R}{a_{\overline{c}|i}}\right) a_{\overline{nc}|i}$$

$$= \left(\frac{800}{a_{\overline{3}|2\%}}\right) a_{\overline{72}|2\%}$$

$$= \left(\frac{800}{2.883883}\right) 37.984063$$

$$= \$10,536.92$$

EXAMPLE 4–38 Find the total amount and present value of an annuity of $900 payable at the beginning of each month for 2 years if the interest rate is 20% compounded quarterly.
Solution Here, $R = \$900$, $c = 1/3$ interest period per payment period, $n = 24$ payments, $nc = 24(1/3) = 8$ interest periods, and $i = 20\%/4 = 5\%$ per interest or conversion period. Thus, the total amount of this complex annuity due is

$$S = \left(\frac{R}{a_{\overline{c}|i}}\right) s_{\overline{nc}|i}$$

$$= \left(\frac{900}{a_{\overline{1/3}|5\%}}\right) s_{\overline{8}|5\%}$$

Note that we must find $a_{\overline{c}|i}$ for a fractional value of c. Appendix Table 9 gives values of $a_{\overline{c}|i}$ for fractional values of c. Using this table, we find that $a_{\overline{1/3}|5\%} = 0.322637$. Thus, the preceding formula for S becomes

$$S = \left(\frac{900}{0.322637}\right) 9.549109$$
$$= \$26{,}637.36$$

The present value of this complex annuity due is

$$A = \left(\frac{R}{a_{\overline{c}|i}}\right) a_{\overline{nc}|i}$$

$$= \left(\frac{900}{a_{\overline{1/3}|5\%}}\right) a_{\overline{8}|5\%}$$

$$= \left(\frac{900}{0.322637}\right) 6.463213$$

$$= \$18{,}029.22$$

EXAMPLE 4–39 A loan of \$35,000 is to be repaid by annual payments payable at the beginning of each year for 8 years. If the interest rate is 24% compounded monthly, find the annual payment.

Solution Here, $c = 12$ conversion periods per payment period, $n = 8$ payments, $nc = 8(12) = 96$ conversion periods, $i = 24\%/12 = 2\%$ per conversion period, and the amount of the loan, \$35,000, is the present value of this complex annuity due. We must determine the annual payment, R. The equation

$$A = \left(\frac{R}{a_{\overline{c}|i}}\right) a_{\overline{nc}|i}$$

relates the present value, A, of a complex annuity due with the periodic payment, R. Substituting \$35,000 for A into this equation yields

$$35{,}000 = \left(\frac{R}{a_{\overline{12}|2\%}}\right) a_{\overline{96}|2\%}$$

EXAMPLE 4–39 (continued)

Solving for R, we obtain

$$R = \frac{35{,}000 \cdot a_{\overline{12}|2\%}}{a_{\overline{96}|2\%}}$$

$$= \frac{35{,}000(10.575341)}{42.529434}$$

$$= \$8703.08$$

EXERCISES

1. Find the total amount and present value of an annuity of $1200 payable at the beginning of each half-year if the interest rate is 18% compounded monthly. The payments continue for 8 years.
2. Find the total amount and present value of an annuity due of $150 monthly for 4 years if the interest rate is 20% compounded semiannually.
3. Find the total amount and present value of an annuity due of $890 quarterly for 5 years if the interest rate is 12% compounded monthly.
4. Find the total amount and present value of an annuity due of $500 semiannually for 10 years if the interest rate is 15% compounded annually.
5. A loan of $8500 is to be repaid by semiannual payments at the beginning of each 6-month time period over a period of 8 years. If the interest rate is 18% compounded monthly, find the semiannual payment.
6. A loan of $4600 is to be repaid by monthly payments at the beginning of each month for 5 years. If the interest rate is 20% compounded quarterly, find the monthly payment.
7. A person wishes to accumulate $10,000 during a 5-year period by making monthly deposits into a fund earning interest at 16% compounded quarterly. Find the monthly deposit if payments are made at the beginning of each month.
8. A person wishes to accumulate $8000 over a 4-year period by making annual deposits into a fund earning interest at 20% compounded quarterly. Find the annual deposit if payments are made at the beginning of each year.

CHAPTER EXERCISES

1. The maturity value of a 9-month discount note is $10,000. The note is discounted at 8% by a bank.
   (A) Find the discount.
   (B) Find the proceeds.
2. Mary Jones wishes to borrow $8000 for 6 months by using a discount note with a discount rate of 8%.
   (A) Find the maturity value of the discount note.
   (B) Mary's note is sold to a third party 2 months before maturity

at a discount rate of 6%. How much does the third party pay for the note? How much did the original holder earn on the note?

3. A person invests $1000 at 9% per year for 6 years.
   (A) Find the simple interest earned.
   (B) Find the future value.

4. A person earns $20 on a principal of $500 during a 3-month time interval. Find the rate of interest.

5. A person invests $10,000 at 8% per year for 10 years.
   (A) Find the simple interest earned.
   (B) Find the future value.

6. Carl Johnson deposits $10,000 into a bank that pays 8% compounded semiannually. What amount will he have at the end of 11 years?

7. How much money does R. T. White need now if he can invest the money at 6% compounded quarterly for 7 years and receive $6000 at the end of the period?

8. A 5-year note has a maturity value of $10,000. If we wish to purchase this note now and earn 12% compounded quarterly on our investment, how much should be paid for the note now (5 years before maturity)?

9. A person invests $9000 at 8% compounded quarterly for 9 years. Find the maturity value.

10. How much money should be deposited now at 12% compounded monthly in order to accumulate into $80,000 in 8 years?

11. Miss Smith deposits $600 at the end of each quarter for 11 years. Each deposit earns interest at 12% compounded quarterly. Find the total amount in this account at the end of 11 years.

12. Mr. Jones wants to accumulate the same total amount as Miss Smith in Exercise 11. However, he wants to deposit one lump-sum payment now. Assuming that the interest rate and time period are the same as in Miss Smith's case, how much should Mr. Jones deposit now?

13. A man deposits $100 in a bank at the end of each month for 3 years and 9 months. If the money earns interest at 12% compounded monthly, how much money does he have in his account at the end of the period?

14. A woman wishes to withdraw $400 at the end of every 6 months for 10 years. If the money earns interest at 6% compounded semiannually, how much must she deposit now?

15. A certain investment contract promises to pay $1000 at the end of each quarter for the next 7 years. If we wish to buy this investment contract and earn 12% compounded quarterly on our money, how much should we pay for it now?

16. A grocer anticipates an expenditure of $10,000 for a new freezer 5 years from now. How much should he deposit at the end of each month into a sinking fund earning interest at 6% compounded monthly?

17. A woman buys a piece of real estate selling for $60,000 by putting $40,000 down and financing the balance with a 20-year mortgage having an interest rate of 16% compounded quarterly. Find the payment due at the end of each quarter.

18. A certain investment contract promises to pay $500 at the end of each quarter for the next two years and a lump sum of $6000 three years thereafter. How much is this investment contract worth now at 12% compounded quarterly?

19. Joan Hill deposits $100 at the end of each quarter for 12 years into a savings account paying interest at 8% compounded quarterly. How much money will be in her account at the end of 12 years? How much interest is earned?

20. John owes Harry $20,000 10 years from now. Instead of waiting 10 years, John wishes to pay an equal sum at the end of each quarter for the next 10 years. If the interest is compounded quarterly at 8%, find the quarterly payment

21. A man has a 12-year mortgage of $20,000 at an interest rate of 8% compounded quarterly. If a payment is made at the end of each quarter, find the payment.

22. A company is considering purchasing new equipment that will result in an increased cash flow of $10,000 per year for the next 5 years. If the goal is a rate of return of 8% compounded annually on the investment, how much should be spent now on the new equipment?

23. A person borrows $7000 to buy a car. The loan is to be paid off by a payment at the end of each month over a 4-year period. The interest rate is 24% compounded monthly.
 (A) Find the borrower's monthly payment.
 (B) At the end of 4 years, how much did the borrower actually pay for the car?
 (C) Find the amount of interest paid.

24. Mrs. Elipe purchased equipment that required a $1000 down payment, and she agreed to pay $100 at the end of each month for the next 5 years. If money is worth 12% compounded semiannually, what would have been the price had Mrs. Elipe decided to pay cash for the entire purchase?

25. In anticipation of an expenditure of $30,000 six years from now, the H. J. Ellis Co. establishes a sinking fund into which a payment is made at the end of each month. Assuming that the interest rate is 12% compounded monthly, find the monthly payment.

26. A woman purchased a house for $70,000. She made a down payment of $20,000 and financed the balance with a 20-year mortgage calling for equal semiannual payments. If the interest rate is 8% compounded semiannually and the first payment is due in six months, find the semiannual payment.

27. A man desires to have a $12,000 fund at the end of 10 years. If his savings can be invested at 8% compounded quarterly, how much must he invest at the end of each quarter for 10 years?

235 MATHEMATICS OF FINANCE

28. A grocer anticipates an expenditure of $10,000 for a new freezer 5 years from now. How much should she deposit at the end of each month into a sinking fund earning interest at 12% compounded monthly?

29. A man deposits $100 in a bank at the end of each month for 3 years and 9 months. If the money earns interest at 12% compounded monthly, how much money does he have in his account at the end of the period?

30. A woman wishes to withdraw $400 at the end of every 6 months for 10 years. If the money earns interest at 6% compounded semiannually, how much must she deposit now?

31. A businessperson wants to accumulate $1 million in 10 years from now by depositing a payment at the end of each year into an account earning interest at the rate of 8% compounded annually. Find the annual payment.

32. Mr. Johnson deposits $100 at the beginning of each month into an account earning interest at 24% compounded monthly. How much is in the account at the end of 4 years?

33. Susan Sullivan wants to accumulate $40,000 ten years from now by depositing a payment at the end of each year into an account earning interest at the rate of 10% compounded annually. Find the annual payment.

34. A loan is to be repaid by making $400 payments at the end of each quarter for 5 years. The first payment is due 1 year from now. If the interest rate is 12% compounded quarterly, find the amount of the loan.

35. Find the total amount and present value of an annuity of $600 payable at the end of each year for 5 years if the interest rate is 8% compounded quarterly.

36. Find the total amount and present value of an annuity of $800 payable at the beginning of each year for 3 years if the interest rate is 12% compounded monthly.

37. Find the total amount and present value of an annuity of $1000 payable at the end of each month for 4 years if the interest rate is 12% compounded semiannually.

38. Find the total amount and present value of an annuity of $1000 payable at the beginning of each month for 3 years if the interest rate is 10% compounded annually.

39. A loan of $35,000 is to be repaid by quarterly payments at the beginning of each quarter for 5 years. The first payment is due 1 year from now. If the interest rate is 18% compounded monthly, find the quarterly payment.

CASE B CAPITAL INVESTMENT DECISION—WRITE GRAPHICS, INC.

Write Graphics, Inc. is planning to invest $350,000 in new computer equipment which is expected to reduce labor costs by $100,000 per year (after taxes) for the next 5 years. The management of Write Graphics, Inc. wishes to determine if the investment is cost-effective (assuming money is currently worth 10% compounded annually).

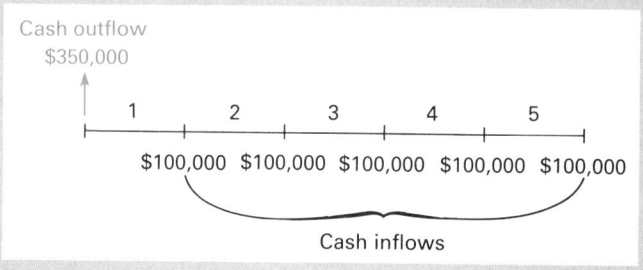

```
Cash outflow
$350,000
     ↑
     |____1____2____3____4____5
          $100,000 $100,000 $100,000 $100,000 $100,000
                        Cash inflows
```

FIGURE 1

The time diagram of Figure 1 illustrates the cash flows involved in this investment decision. To determine whether the investment of $350,000 is cost-effective at 10% compounded annually, management must determine the **net present value** (denoted by **NPV**) of the cash flows. The net present value is defined by

$$\text{NPV} = \text{present value of cash inflows} - \text{present value of cash outflows}$$

The present value of cash outflows is the initial investment of $350,000. The present value of cash inflows at 10% compounded annually is determined by the formula for the present value of an annuity

$$A = R \cdot a_{\overline{5}|0.10}$$
$$= 100{,}000(3.790787)$$
$$= \$379{,}078.70$$

Thus, the net present value of this investment is

$$\text{NPV} = \frac{\text{present value of}}{\text{cash inflows}} - \frac{\text{present value of}}{\text{cash outflows}}$$

$$= \$379{,}078.70 - \$350{,}000.00$$

$$= \$29{,}078.70$$

The net present value is positive because the present value of the cash inflows exceeds the present value of cash outflow. Thus, the investment is cost-effective at an interest rate of 10% compounded annually. This means that the investment of $350,000 is earning a rate of return greater than the quoted interest rate of 10% compounded annually.

If the present value of the cash inflows had been less than the present value of the cash outflow, the net present value would have been negative and the investment would not have been cost-effective at the quoted interest rate of 10% compounded annually. In other words, the investment of $350,000 would be earning a rate of return less than the quoted interest rate of 10% compounded annually.

If the present value of the cash inflows had equaled the present value of the cash outflow, the net present value would have been 0 and the investment would have been cost-effective at the quoted interest rate of 10% compounded annually. In other words, the investment of $350,000 would be earning a rate of return of 10% compounded annually.

EXERCISES

1. If the preceding investment resulted in cash inflows of $70,000, $80,000, $100,000, $110,000, and $130,000 at the end of the first, second, third, fourth, and fifth years, respectively, calculate the net present value if money is worth 10% compounded annually. Is the investment earning a rate of return of at least 10% compounded annually?

2. Assume that the initial investment is changed to an amount different from $350,000 and that the resulting cash inflows are $80,000, $100,000, $110,000, $120,000, and $140,000 at the end of the first, second, third, fourth, and fifth years, respectively. If the net present value is −$1000, find the amount invested assuming money is worth 12% compounded annually.

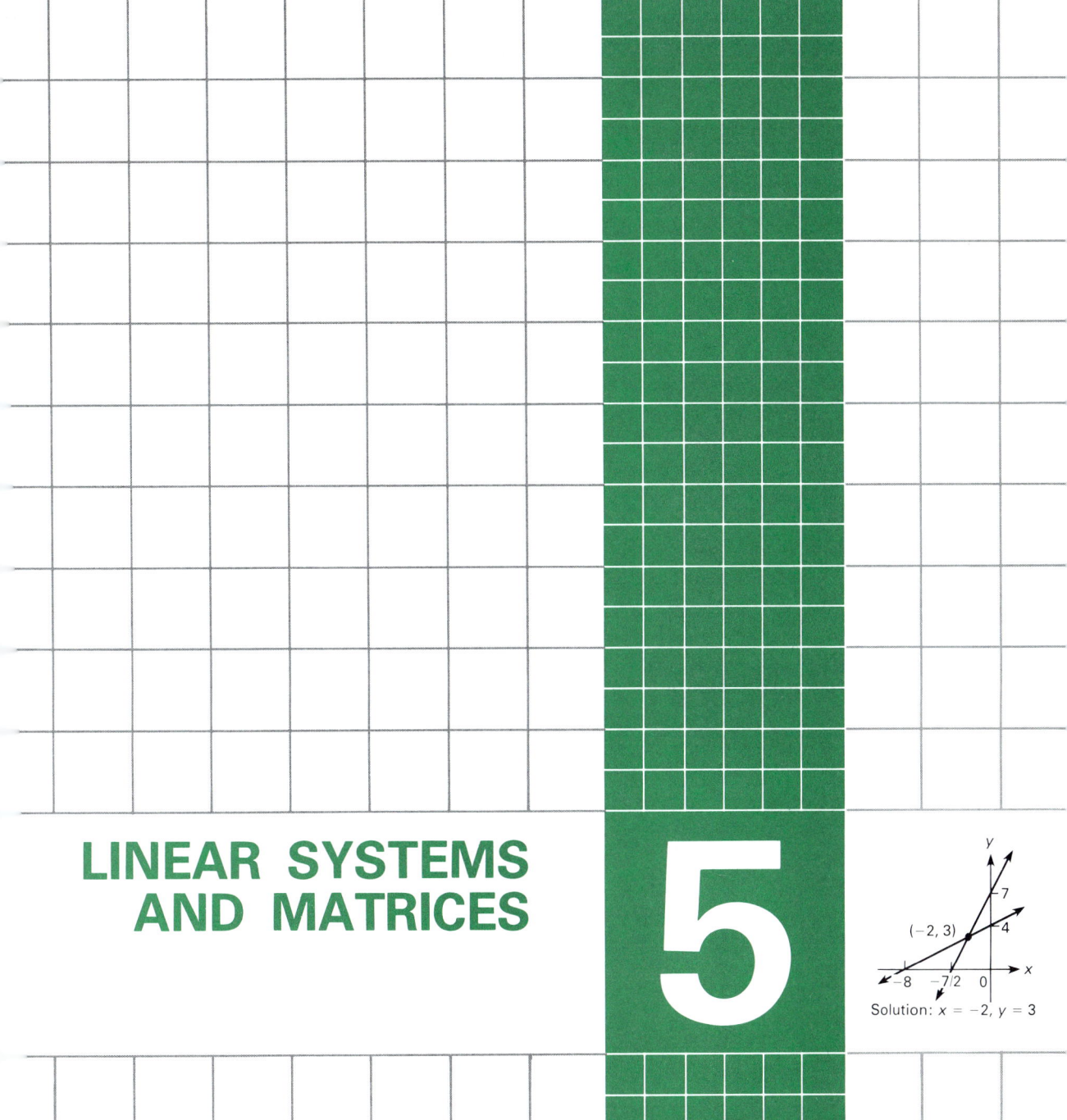

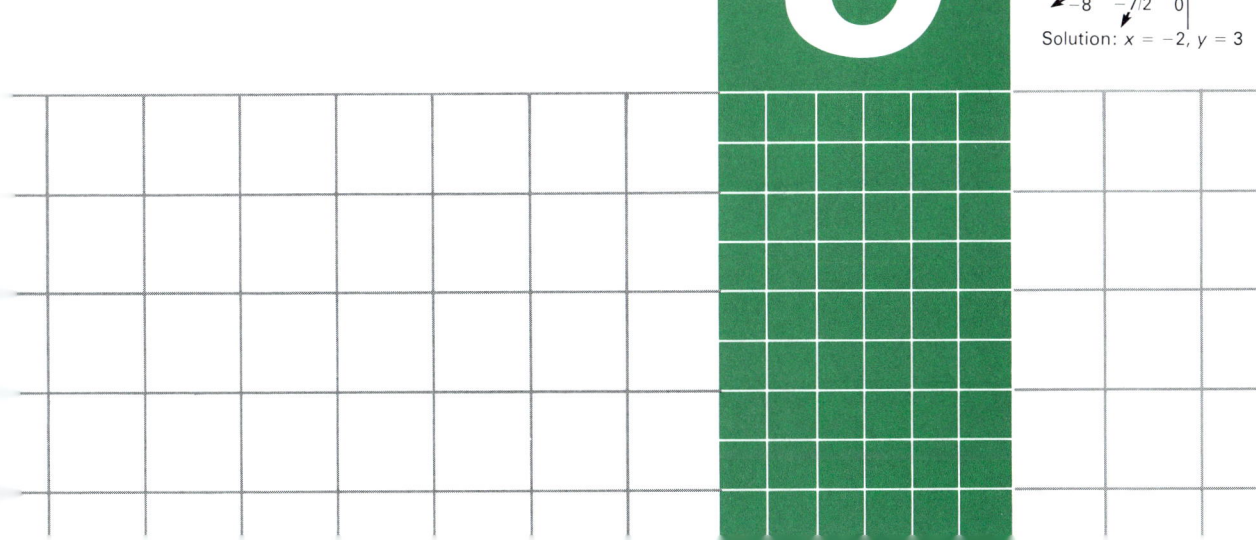

LINEAR SYSTEMS AND MATRICES

5

Solution: $x = -2, y = 3$

5–1

TWO EQUATIONS IN TWO VARIABLES

A set of linear equations in two or more unknowns such as

$$2x + 5y = 13$$
$$3x - 8y = 4$$

is called a **system** of linear equations. Observe that this system consists of two equations in two variables. A **solution** to a system is an ordered pair of numbers (x, y) which, when substituted into each equation, converts it into a true statement. Specifically, the ordered pair $(4, 1)$ is a solution to the preceding system since

$$2(4) + 5(1) = 13$$
$$3(4) - 8(1) = 4$$

Observe that the values $x = 4$ and $y = 1$ satisfy each equation of the system.

Solving a Linear System

We will discuss two methods for solving linear systems. The first method involves **adding** equations to eliminate variables. The second is a method of elimination by **substitution**.

To solve the preceding system by the method of addition, we multiply the first equation by 3 and the second by -2 to obtain

$$3(2x + 5y) = 3(13) \qquad\qquad 6x + 15y = 39$$
$$-2(3x - 8y) = -2(4) \qquad\longrightarrow\qquad -6x + 16y = -8$$

The resulting linear system has the same solution as the original linear system. Hence, it is said to be **equivalent** to the original system. Adding the two equations of the equivalent system eliminates the variable x. Thus,

$$6x + 15y = 39$$
$$-6x + 16y = -8$$
$$\overline{\qquad\qquad 31y = 31}$$
$$y = 1$$

We now substitute the value $y = 1$ into either of the two original equations. Choosing the first equation, we obtain

$$2x + 5y = 13$$
$$2x + 5(1) = 13$$
$$2x = 8$$
$$x = 4$$

Thus, the solution is $x = 4$ and $y = 1$, or $(4, 1)$. This can be checked

241 LINEAR SYSTEMS AND MATRICES

by substituting 4 for x and 1 for y into both equations of our original system. Note that we could have eliminated y instead of x by multiplying the first equation by 8, the second by 5, and adding the results.

We now solve our original system,

$$2x + 5y = 13$$
$$3x - 8y = 4$$

by the substitution method. Solving the first equation for x, we obtain

$$x = \frac{-5y + 13}{2}$$

Substituting $(-5y + 13)/2$ for x in the second equation gives us

$$3\left(\frac{-5y + 13}{2}\right) - 8y = 4$$

Simplifying this result yields

$$-\frac{15}{2}y + \frac{39}{2} - 8y = 4$$

$$-\frac{31}{2}y = -\frac{31}{2}$$

$$y = 1$$

We now substitute the value $y = 1$ into either of the original equations to obtain $x = 4$.

Graphical Interpretation When we solve a system of linear equations, we find an ordered pair (x, y) that satisfies both equations. Thus, (x, y) lies on both straight lines. In other words, the solution (x, y) to a system of linear equations is the **intersection point** of the straight lines. Figure 5–1 illustrates graphically the previous system and its solution.

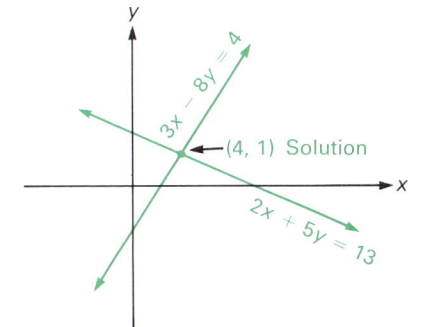

FIGURE 5–1

242 CHAPTER FIVE

A pair of straight lines can be drawn so that they either **intersect**, are **parallel**, or **coincide** (see Figure 5–2). Thus, when solving a system of equations, three possibilities exist:

1. The linear system has a *unique solution*. The straight lines *intersect* at one point (see Figure 5–2(A)).
2. The linear system has *no solution*. The straight lines are *parallel* and do not intersect (see Figure 5–2(B)).
3. The linear system has *infinitely many solutions*. The straight lines *coincide* (see Figure 5–2(C)).

(A) Intersecting lines (unique solution)

(B) Parallel lines (no solution)

(C) Coincident lines (infinitely many solutions)

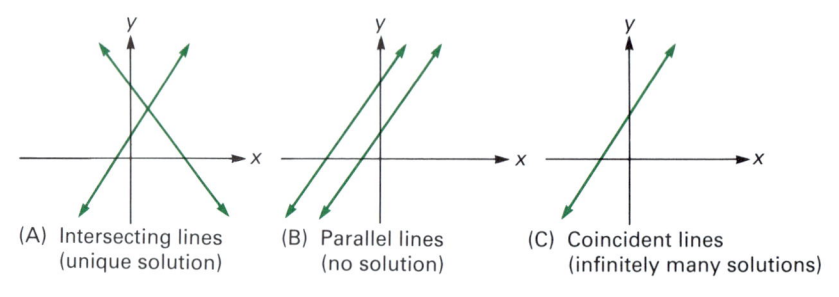

FIGURE 5–2

In this section, we have already shown a linear system having a unique solution. Examples 5–1 and 5–2 illustrate linear systems having no solutions and infinitely many solutions, respectively.

EXAMPLE 5–1 Solve the linear system

$$3x - 2y = 8$$
$$-6x + 4y = 10$$

Solution Multiplying the first equation by 2 yields the equivalent system

$$6x - 4y = 16$$
$$-6x + 4y = 10$$

Adding the two equations of the equivalent system, we obtain

$$6x - 4y = 16$$
$$\underline{-6x + 4y = 10}$$
$$0 = 26$$

The statement $0 = 26$ is a false statement, indicating that the linear system has no solution. Such a linear system (having

LINEAR SYSTEMS AND MATRICES

no solutions) is called **inconsistent**. The graph of this system appears in Figure 5–3.

Inconsistent (no solutions)

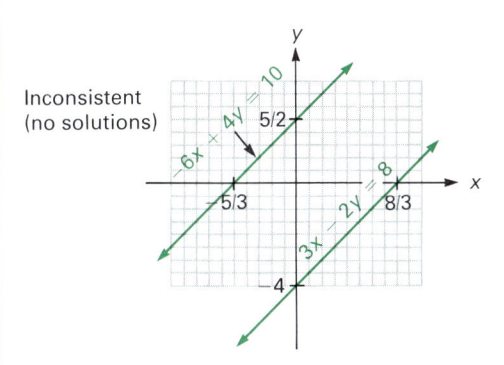

FIGURE 5–3

EXAMPLE 5–2 Solve the linear system

$$-5x + 3y = 30$$
$$15x - 9y = -90$$

Solution Multiplying the first equation by 3 yields the equivalent system

$$-15x + 9y = 90$$
$$15x - 9y = -90$$

Adding the two resulting equations, we obtain the true statement $0 = 0$. The statement $0 = 0$ indicates that both equations of the linear system have the same graph, and thus there are infinitely many solutions (see Figure 5–4).

FIGURE 5–4

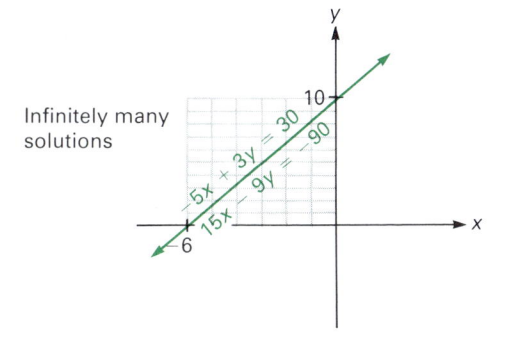

Infinitely many solutions

CHAPTER FIVE

The following examples provide some applications of linear systems to practical situations.

EXAMPLE 5-3 *(Product Mix)* The Ruszala Company manufactures bicycles and tricycles. Each bicycle and tricycle must pass through two departments: Department I (Assembly) and Department II (Finishing and Inspection). Each bicycle requires 3 hours in Department I and 5 hours in Department II. Each tricycle requires 4 hours in Department I and 2 hours in Department II. Each month, Departments I and II have available 450 and 400 hours, respectively. All of the time in both departments must be used (i.e., there must be no idle time or slack time). How many bicycles and tricycles should the company produce each month?

Solution The following table summarizes the information given in the problem:

	Bicycles	Tricycles	
Department I (Assembly)	3 hours for each bicycle	4 hours for each tricycle	450 hours for Department I
Department II (Finishing and Inspection)	5 hours for each bicycle	2 hours for each tricycle	400 hours for Department II

If

$$x = \text{number of bicycles manufactured}$$
$$y = \text{number of tricycles manufactured}$$

then the time equation for each department is

$$\text{Department I} \rightarrow 3x + 4y = 450$$
$$\text{Department II} \rightarrow 5x + 2y = 400$$

Solving this system by the addition method, we multiply the second equation by -2 and then add. Hence,

$$\begin{aligned} 3x + 4y &= 450 \\ -10x - 4y &= -800 \\ \hline -7x &= -350 \\ x &= 50 \end{aligned}$$

We may substitute $x = 50$ into either of the original equations. Choosing the second equation, we obtain

$$5x + 2y = 400$$
$$5(50) + 2y = 400$$
$$y = 75$$

Thus, the company should manufacture $x = 50$ bicycles and $y = 75$ tricycles.

EXAMPLE 5-4 C.P. Realty, Inc. is planning to build a housing development consisting of two- and three-bedroom ranch-style houses. Public demand indicates a need for three times as many three-bedroom houses as two-bedroom houses. Each two-bedroom house provides a net profit of $5000. Each three-bedroom house provides a net profit of $6000. If C.P. Realty, Inc. must net a total profit of $2,300,000 from this development, how many of each type ranch house should be built?

Solution Let

$$x = \text{number of 2-bedroom houses}$$
$$y = \text{number of 3-bedroom houses}$$

Then, we have

$$y = 3x$$
$$5000x + 6000y = 2,300,000$$

This system is most efficiently solved by the substitution method since the first equation expresses y in terms of x. Substituting $3x$ for y in the second equation yields

$$5000x + 6000(3x) = 2,300,000$$

Solving for x, we obtain

$$5000x + 18,000x = 2,300,000$$
$$23,000x = 2,300,000$$
$$x = 100$$

We may substitute $x = 100$ into either of the two equations. Choosing the first equation, we obtain

$$y = 3(100)$$
$$= 300$$

EXAMPLE 5-4 (continued)

Thus, C.P. Realty, Inc. should build $x = 100$ two-bedroom houses and $y = 300$ three-bedroom houses.

EXAMPLE 5-5 *(Equilibrium Point)*
The demand and supply equations for a given commodity are as follows:

$$\text{Demand equation} \longrightarrow 5q + 2p = 50$$
$$\text{Supply equation} \longrightarrow 4q - 3p = -52$$

where p is the unit price and q is the number of units. Find the equilibrium point.

Solution Solving this system by the addition method, we multiply the first equation by 3 and the second by 2 to obtain

$$15q + 6p = 150$$
$$8q - 6p = -104$$

Adding the resulting equations yields

$$23q = 46$$
$$q = 2$$

We may substitute $q = 2$ into either of the original equations. Choosing the first equation, we have

$$5q + 2p = 50$$
$$5(2) + 2p = 50$$
$$p = 20$$

Thus, at a unit price of $20, supply = demand = 2 units.

In this section, we have solved linear systems consisting of two equations in two variables. In Section 5-4, we will learn to solve linear systems with more than two equations and more than two variables.

EXERCISES

1. Solve each of the following linear systems and illustrate it graphically.

   (A) $x + 2y = 1$
           $3x - 5y = -8$

   (B) $2x - y = -7$
           $-x + 2y = 8$

LINEAR SYSTEMS AND MATRICES

    (C) $3x + 5y = 7$
          $2x - 6y = 11$

    (D) $x + y = 11$
          $x - y = 1$

2. Try to solve the linear system

$$5x - 7y = 70$$
$$-10x + 14y = 120$$

Note that your result is the contradictory statement $0 = 260$. To reconcile this result, show that this linear system has no solution by expressing each linear equation in slope–intercept form. What do you observe? What are your conclusions?

3. Try to solve the linear system

$$5x - 7y = 70$$
$$-10x + 14y = -140$$

Note that your result is the statement $0 = 0$. Show that this linear system has infinitely many solutions by expressing each linear equation in slope–intercept form. What do you observe? What are your conclusions?

4. Determine which of the following has no solutions and which has infinitely many solutions:

    (A) $3x - 8y = 10$
          $12x - 32y = 75$

    (B) $7x - 8y = -11$
          $-35x + 40y = 55$

5. Solve each of the following linear systems:

    (A) $2x - 3y = 6$
          $x - 7y = 25$

    (B) $4x - 5y = -2$
          $3x + 2y = -13$

    (C) $4x + y = 8$
          $6x - 2y = -9$

    (D) $2x + 3y = 3$
          $12x - 15y = -4$

    (E) $-3x + 10y = 5$
          $2x + 7y = 24$

    (F) $\frac{1}{2}x + 5y = 17$
          $3x + 2y = 18$

    (G) $\frac{1}{3}x - \frac{3}{2}y = -4$
          $5x - 4y = 14$

    (H) $1.5x + 2y = 20$
          $2.5x - 5y = -25$

6. The graph of Figure 5–5 appeared on a past Uniform CPA Examination. Find the intersection point.

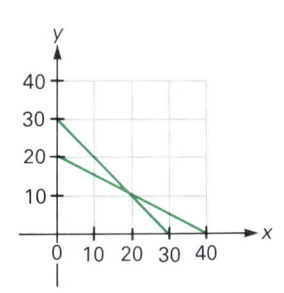

FIGURE 5–5

248 CHAPTER FIVE

7. A diet must provide exactly 1200 milligrams of protein and 1000 milligrams of iron. These nutrients will be obtained by eating meat and spinach. Each pound of meat contains 500 milligrams of protein and 100 milligrams of iron. Each pound of spinach contains 200 milligrams of protein and 800 milligrams of iron. How many pounds of meat and spinach should be eaten in order to provide the proper amounts of nutrients?

8. A farmer wants to plant a combination of two crops, cabbage and corn, on 100 acres. Cabbage requires 60 man-hours of labor per acre, and corn requires 80 man-hours of labor per acre. If the farmer has 6600 man-hours available, how many acres of each crop should be planted?

9. A toy company manufactures wagons and cars. The company usually sells four times as many wagons as cars. Each wagon provides a net profit of $6, and each car provides a net profit of $5. If the company wants a total profit of $29,000, how many wagons and cars must be produced?

10. The demand and supply equations for watches appear in the linear system

$$\text{Demand equation} \rightarrow 5p + 4q = 650$$
$$\text{Supply equation} \rightarrow 3p - 7q = -1020$$

where p is the unit price and q is the number of watches. Find the equilibrium point.

11. A company's sales revenue and cost equations appear in the linear system

$$\text{Revenue equation} \rightarrow y = 25x$$
$$\text{Cost equation} \longrightarrow y = 10x + 6000$$

where x is the number of units and y is the dollar amount. Find the break-even point.

5-2

MATRICES A **matrix** is a rectangular array of numbers. Each number is called an **element** of the matrix. Specifically,

$$\begin{array}{c} \phantom{\text{Row 1} \rightarrow} \text{Col 1} \quad \text{Col 2} \quad \text{Col 3} \\ \begin{array}{c} \text{Row 1} \rightarrow \\ \text{Row 2} \rightarrow \end{array} \begin{bmatrix} 3 & 1 & 2 \\ 8 & 0 & -5 \end{bmatrix} \end{array}$$

is a matrix with two rows and three columns. Thus, it is of **dimension** 2×3, read "two by three." In general, a matrix with m rows and n columns is of dimension $m \times n$. The following are matrices of various dimensions:

$$\begin{bmatrix} 3 \\ 4 \\ 2 \end{bmatrix} \quad [2 \quad 4 \quad -7] \quad \begin{bmatrix} 4 & 5 \\ 6 & -1 \end{bmatrix}$$

dimension 3×1 dimension 1×3 dimension 2×2

A matrix of only one column is called a **column matrix**. Column matrices are often called **column vectors**. The matrix

$$\begin{bmatrix} 3 \\ -1 \end{bmatrix}$$

is an example of a column matrix, or column vector. A matrix of only one row is called a **row matrix**. Row matrices are often called **row vectors**. The matrix

$$[5 \quad 0 \quad -1 \quad 7]$$

is an example of a row matrix, or row vector. A matrix with as many rows as columns is called a **square matrix**. The matrix

$$\begin{bmatrix} 3 & 4 \\ 6 & 0 \end{bmatrix}$$

is an example of a square matrix.

Matrices provide useful ways of presenting data. As a specific example, we consider the following product-mix problem. The departmental time requirements of three types of bolts are summarized by the 3×3 matrix

$$\begin{matrix} \text{Type A} & \text{Type B} & \text{Type C} & \\ \begin{bmatrix} 2 & 3 & 4 \\ 3 & 5 & 2 \\ 6 & 3 & 5 \end{bmatrix} & & & \begin{matrix} \text{Department I} \\ \text{Department II} \\ \text{Department III} \end{matrix} \end{matrix}$$

Studying this matrix, note that each Type A bolt requires 2 hours in Department I, 3 hours in Department II, and 6 hours in Department III. Each Type B bolt requires 3 hours in Department I, 5 hours in Department II, and 3 hours in Department III. And each Type C bolt requires 4 hours in Department I, 2 hours in Department II, and 5 hours in Department III.

Matrix Notation Matrices are usually denoted by capital letters. Thus, the matrix

$$B = \begin{bmatrix} 5 & 3 & 1 \\ 8 & 0 & 2 \end{bmatrix}$$

may be referred to by the letter B.

It is sometimes necessary to refer to a general matrix of a given dimension. For example, a general matrix A of dimension 2×3 is

$$A = \begin{bmatrix} a_{11} & a_{12} & a_{13} \\ a_{21} & a_{22} & a_{23} \end{bmatrix}$$

Note that the individual elements of the matrix A are denoted by a_{ij} where i denotes the row in which the element is located and j denotes the column. Thus, a_{11} denotes the element located in the first row and first column, a_{12} denotes the element in the first row and second column, . . ., a_{23} denotes the element in the second row and third column. In general, a matrix A of dimension $m \times n$ is denoted by

$$A = \begin{bmatrix} a_{11} & a_{12} & \cdots & a_{1n} \\ a_{21} & a_{22} & \cdots & a_{2n} \\ \vdots & \vdots & & \vdots \\ a_{m1} & a_{m2} & \cdots & a_{mn} \end{bmatrix}$$

Equality of Matrices Two matrices are **equal** if they are of the same dimension and if their corresponding elements are equal. Thus, if

$$A = \begin{bmatrix} 4 & 3 \\ -2 & 1 \end{bmatrix} \quad \text{and} \quad B = \begin{bmatrix} 8/2 & 3 \\ -2 & 6/6 \end{bmatrix}$$

then $A = B$. However, if $C = [1 \;\; 6]$ and $D = [6 \;\; 1]$, then $C \neq D$ since corresponding elements are not equal.

Adding and Subtracting Matrices If two or more matrices are of the same dimension, then they may be *added*. The **sum** of two or more matrices is a matrix where each element is the sum of the corresponding elements of the individual matrices. Similar statements hold for subtraction of matrices. Thus, if

$$A = \begin{bmatrix} 1 & 0 & -5 \\ 8 & -2 & 9 \end{bmatrix} \quad \text{and} \quad B = \begin{bmatrix} 4 & 2 & 3 \\ 5 & 1 & 7 \end{bmatrix}$$

LINEAR SYSTEMS AND MATRICES

then

$$A + B = \begin{bmatrix} 1+4 & 0+2 & -5+3 \\ 8+5 & -2+1 & 9+7 \end{bmatrix} = \begin{bmatrix} 5 & 2 & -2 \\ 13 & -1 & 16 \end{bmatrix}$$

$$A - B = \begin{bmatrix} 1-4 & 0-2 & -5-3 \\ 8-5 & -2-1 & 9-7 \end{bmatrix} = \begin{bmatrix} -3 & -2 & -8 \\ 3 & -3 & 2 \end{bmatrix}$$

EXAMPLE 5-6 Matrix N shows the number of dryers shipped from two plants, P_1 and P_2, to three warehouses, W_1, W_2, and W_3, during the month of November.

$$N = \begin{matrix} \\ P_1 \\ P_2 \end{matrix} \begin{matrix} W_1 & W_2 & W_3 \\ \begin{bmatrix} 100 & 50 & 70 \\ 300 & 20 & 80 \end{bmatrix} \end{matrix}$$

Matrix D shows the corresponding shipments made during December.

$$D = \begin{matrix} \\ P_1 \\ P_2 \end{matrix} \begin{matrix} W_1 & W_2 & W_3 \\ \begin{bmatrix} 200 & 150 & 80 \\ 400 & 90 & 100 \end{bmatrix} \end{matrix}$$

Find the matrix showing the combined shipment for both months.

Solution Here,

$$N + D = \begin{bmatrix} 100+200 & 50+150 & 70+80 \\ 300+400 & 20+90 & 80+100 \end{bmatrix}$$

$$= \begin{bmatrix} 300 & 200 & 150 \\ 700 & 110 & 180 \end{bmatrix}$$

Multiplying a Matrix by a Number

If a matrix A is multiplied by a number k, then the resulting matrix, kA, is determined by multiplying each element of matrix A by k. Specifically, if

$$A = \begin{bmatrix} 6 & 5 \\ 1 & 7 \end{bmatrix}$$

then

$$2A = \begin{bmatrix} 2 \cdot 6 & 2 \cdot 5 \\ 2 \cdot 1 & 2 \cdot 7 \end{bmatrix} = \begin{bmatrix} 12 & 10 \\ 2 & 14 \end{bmatrix}$$

252 CHAPTER FIVE

Observe that

$$A + A = \begin{bmatrix} 6 & 5 \\ 1 & 7 \end{bmatrix} + \begin{bmatrix} 6 & 5 \\ 1 & 7 \end{bmatrix} = \begin{bmatrix} 12 & 10 \\ 2 & 14 \end{bmatrix} = 2A$$

EXERCISES

1. State the dimension of each of the following matrices:

   (A) $\begin{bmatrix} 4 & 3 & -1 \\ 8 & 2 & 6 \end{bmatrix}$

   (B) $\begin{bmatrix} 6 & 8 \\ 5 & -4 \\ 2 & 0 \end{bmatrix}$

   (C) $\begin{bmatrix} 8 & 4 & 0 \\ 1 & 1 & 0 \\ 2 & 2 & 0 \end{bmatrix}$

   (D) $\begin{bmatrix} 8 & 4 \\ 6 & -10 \end{bmatrix}$

   (E) $[4 \ -1 \ 6]$

   (F) $\begin{bmatrix} 7 \\ -1 \end{bmatrix}$

   (G) $\begin{bmatrix} 5 \\ 0 \\ -1 \\ 4 \end{bmatrix}$

   (H) $[3 \ 0 \ -1 \ 5]$

2. Identify each of the following as either a row matrix or a column matrix:

   (A) $[4 \ 3 \ 0]$

   (B) $[7 \ -1]$

   (C) $\begin{bmatrix} 8 \\ 4 \end{bmatrix}$

   (D) $\begin{bmatrix} 9 \\ 2 \\ 0 \end{bmatrix}$

3. Which of the following are square matrices?

   (A) $\begin{bmatrix} 3 & 6 & 1 \\ 8 & 2 & 0 \end{bmatrix}$

   (B) $\begin{bmatrix} 4 & 3 \\ 2 & 0 \end{bmatrix}$

   (C) $\begin{bmatrix} 8 & 1 & 0 \\ 4 & 3 & 0 \\ 8 & 2 & 1 \end{bmatrix}$

   (D) $\begin{bmatrix} 8 & 6 & 1 & 0 \\ 2 & 3 & 0 & 0 \end{bmatrix}$

4. For each of the following, indicate whether the statement is true or false:

   (A) If

   $$A = \begin{bmatrix} 5 & 3 & 6 \\ 8 & 2 & 1 \end{bmatrix} \quad \text{and} \quad B = \begin{bmatrix} 10/2 & 12/4 & 6 \\ 8 & 14/7 & 9/9 \end{bmatrix}$$

   then $A = B$.

   (B) If

   $$C = \begin{bmatrix} 5 & 6 \\ 8 & 4 \end{bmatrix} \quad \text{and} \quad D = \begin{bmatrix} 5 & 6 & 0 \\ 8 & 4 & 0 \end{bmatrix}$$

   then $C = D$.

(C) If
$$E = \begin{bmatrix} 4 & 8 \\ 6 & 2 \end{bmatrix} \text{ and } F = \begin{bmatrix} 8 & 4 \\ 6 & 2 \end{bmatrix}$$
then $E = F$.

(D) If
$$H = \begin{bmatrix} 4 & 1 \\ 6 & -2 \end{bmatrix} \text{ and } K = \begin{bmatrix} 4 & 8/8 \\ 6 & -2 \end{bmatrix}$$
then $H = K$.

5. Let
$$A = \begin{bmatrix} x \\ y \end{bmatrix} \text{ and } B = \begin{bmatrix} 4 \\ -1 \end{bmatrix}$$
Given that $A = B$, what are the values of x and y?

6. Let
$$C = \begin{bmatrix} x_1 \\ x_2 \\ x_3 \end{bmatrix} \text{ and } D = \begin{bmatrix} 1 \\ 0 \\ -3 \end{bmatrix}$$
Given that $C = D$, what are the values of x_1, x_2, and x_3?

7. Let
$$H = \begin{bmatrix} x & y \\ z & w \end{bmatrix} \text{ and } K = \begin{bmatrix} 1 & -4 \\ 5 & -7 \end{bmatrix}$$
Given that $H = K$, what are the values of x, y, z, and w?

8. If
$$A = \begin{bmatrix} 3 & 1 & 2 \\ -1 & 5 & -2 \end{bmatrix} \quad B = \begin{bmatrix} 0 & 4 & 1 \\ 2 & -5 & 3 \end{bmatrix} \quad C = \begin{bmatrix} 4 & 3 & 0 \\ -2 & 5 & -1 \end{bmatrix}$$
compute each of the following:

(A) $A + B$ (B) $A - B$ (C) $B - A$
(D) $A + C$ (E) $A - C$ (F) $C - A$
(G) $B + C$ (H) $B - C$ (I) $C - B$
(J) $A + B + C$ (K) $A + B - C$ (L) $A + C - B$
(M) $2A$ (N) $3B$ (O) $5C$
(P) $-3A$ (Q) $-6B$ (R) $-2C$
(S) $C + 2A$ (T) $A - 3B$ (U) $B + 5C$
(V) $B - 3A$ (W) $A - 6B + 5C$ (X) $A + B - 2C$

9. If $A = [3 \quad -4 \quad 1]$ and $B = [2 \quad 0 \quad -3]$, compute each of the following:

(A) $A + B$ (B) $A - B$ (C) $B - A$
(D) $3A$ (E) $-2B$ (F) $A - 2B$
(G) $B + 3A$ (H) $B - 3A$ (I) $A + 2B$

10. If $C = \begin{bmatrix} 8 \\ 2 \end{bmatrix}$ and $D = \begin{bmatrix} -7 \\ 1 \end{bmatrix}$, compute each of the following:

(A) $C + D$ (B) $C - D$ (C) $D - C$
(D) $5C$ (E) $-3D$ (F) $C - 3D$
(G) $5C + D$ (H) $D - 5C$ (I) $C + 3D$

11. Let $X = \begin{bmatrix} x_1 \\ x_2 \end{bmatrix}$ and $B = \begin{bmatrix} 6 \\ 15 \end{bmatrix}$. Given that $3X = B$, find X.

12. Let $X = \begin{bmatrix} x_1 \\ x_2 \\ x_3 \end{bmatrix}$ and $C = \begin{bmatrix} 15 \\ 20 \\ 30 \end{bmatrix}$. Given that $5X = C$, find X.

13. If $A = \begin{bmatrix} 4 & 1 \\ 3 & 2 \end{bmatrix}$ and $B = \begin{bmatrix} -1 & 5 \\ 6 & 4 \end{bmatrix}$, verify that $A + B = B + A$.

14. Matrix J shows the number of sofas shipped from two plants, P_1 and P_2, to four warehouses, W_1, W_2, W_3, and W_4, during the month of July.

$$J = \begin{array}{c} \\ P_1 \\ P_2 \end{array} \begin{array}{cccc} W_1 & W_2 & W_3 & W_4 \\ \begin{bmatrix} 200 & 50 & 70 & 100 \\ 300 & 50 & 10 & 0 \end{bmatrix} \end{array}$$

Matrix A shows the corresponding shipments made during August.

$$A = \begin{array}{c} \\ P_1 \\ P_2 \end{array} \begin{array}{cccc} W_1 & W_2 & W_3 & W_4 \\ \begin{bmatrix} 100 & 30 & 10 & 50 \\ 70 & 400 & 200 & 80 \end{bmatrix} \end{array}$$

Find the matrix showing the combined shipment for both months.

15. Larry Merrick operates three fruit stores: S_1, S_2, and S_3. Each store stocks apples, oranges, grapes, and pears. Larry does his buying on Monday, Wednesday, and Friday. Matrix M shows the amounts spent by Larry on each item for each store on Monday.

$$M = \begin{array}{c} \\ \\ \\ \\ \end{array} \begin{array}{ccc} S_1 & S_2 & S_3 \\ \begin{bmatrix} \$200 & \$500 & \$300 \\ \$100 & \$400 & \$210 \\ \$500 & \$280 & \$80 \\ \$150 & \$350 & \$250 \end{bmatrix} \end{array} \begin{array}{l} \text{Apples} \\ \text{Oranges} \\ \text{Grapes} \\ \text{Pears} \end{array}$$

Matrices W and F show the corresponding expenditures for Wednesday and Friday respectively.

$$W = \begin{array}{ccc} S_1 & S_2 & S_3 \\ \begin{bmatrix} \$150 & \$80 & \$100 \\ \$250 & \$300 & \$150 \\ \$70 & \$50 & \$90 \\ \$120 & \$215 & \$160 \end{bmatrix} \end{array} \begin{array}{l} \text{Apples} \\ \text{Oranges} \\ \text{Grapes} \\ \text{Pears} \end{array}$$

$$F = \begin{array}{ccc} S_1 & S_2 & S_3 \\ \begin{bmatrix} \$209 & \$180 & \$120 \\ \$310 & \$140 & \$230 \\ \$80 & \$75 & \$55 \\ \$95 & \$90 & \$170 \end{bmatrix} \end{array} \begin{array}{l} \text{Apples} \\ \text{Oranges} \\ \text{Grapes} \\ \text{Pears} \end{array}$$

(A) Find the matrix showing the combined purchases for Monday and Wednesday.
(B) Find the matrix showing the combined purchases for Wednesday and Friday.
(C) Find the matrix showing the combined purchases for Monday, Wednesday, and Friday.

16. If Z is a matrix whose elements are all zeros, then given that

$$X = \begin{bmatrix} a & b \\ c & d \end{bmatrix}$$

verify the following:
(A) $X - X = Z$
(B) $X + Z = X$

17. If $A = \begin{bmatrix} 2 & -5 \\ 3 & 1 \end{bmatrix}$, $B = \begin{bmatrix} 2 & 1 \\ 5 & 0 \end{bmatrix}$, and $C = \begin{bmatrix} 2 & -6 \\ 7 & 1 \end{bmatrix}$, verify the following:

(A) $A + B = B + A$ (commutative property of addition)
(B) $A + (B + C) = (A + B) + C$ (associative property of addition)

5-3
MULTIPLYING MATRICES

Dot Product

Before we can multiply matrices, we must first be able to calculate the **dot product** of a row vector and a column vector. If

$$R = \begin{bmatrix} 1 & 3 & 5 \end{bmatrix} \quad \text{and} \quad C = \begin{bmatrix} 2 \\ 7 \\ 4 \end{bmatrix}$$

then the dot product of R and C, denoted $R \cdot C$, is the number determined by multiplying the first element of R by the first element of C, the second element of R by the second element of C, etc., then adding the resulting products. Hence,

$$\text{dot product } R \cdot C = \begin{bmatrix} 1 & 3 & 5 \end{bmatrix} \cdot \begin{bmatrix} 2 \\ 7 \\ 4 \end{bmatrix} = 1(2) + 3(7) + 5(4) = 43$$

Note that both vectors must have the same number of elements in order for a dot product to be calculated. Also notice that the left vector is a row vector and the right vector is a column vector. Although this is usually the case, it is possible to calculate the dot product of any two vectors as long as both have the same number of elements. In fact, we will encounter examples in Chapter 6 where the dot product of two column vectors will be calculated.

EXAMPLE 5–7 If

$$H = \begin{bmatrix} 2 & -1 & 0 & 3 \end{bmatrix} \quad \text{and} \quad K = \begin{bmatrix} 4 \\ -5 \\ 1 \\ 8 \end{bmatrix}$$

then find the dot product $H \cdot K$.
Solution The dot product is

$$H \cdot K = \begin{bmatrix} 2 & -1 & 0 & 3 \end{bmatrix} \cdot \begin{bmatrix} 4 \\ -5 \\ 1 \\ 8 \end{bmatrix}$$

$$= 2(4) + (-1)(-5) + 0(1) + 3(8)$$
$$= 37$$

EXAMPLE 5–8 If

$$D = \begin{bmatrix} 1 \\ 2 \\ -1 \end{bmatrix} \quad \text{and} \quad E = \begin{bmatrix} 3 \\ 0 \\ 4 \end{bmatrix}$$

find the dot product $D \cdot E$.
Solution Here,

$$D \cdot E = \begin{bmatrix} 1 \\ 2 \\ -1 \end{bmatrix} \cdot \begin{bmatrix} 3 \\ 0 \\ 4 \end{bmatrix} = 1(3) + 2(0) + (-1)(4) = -1$$

EXAMPLE 5–9 A grocery store carries three brands of detergent: Brand x, Brand y, and Brand z. Row matrix S represents the number of units of each of these brands sold during the month of February:

$$\begin{array}{cccc} & \text{Brand} & \text{Brand} & \text{Brand} \\ & x & y & z \\ S = [100 & 500 & 300] \end{array}$$

LINEAR SYSTEMS AND MATRICES

Column matrix P represents the unit selling price of each brand:

$$P = \begin{bmatrix} \$2.00 \\ \$1.00 \\ \$1.50 \end{bmatrix} \begin{matrix} \text{Brand x} \\ \text{Brand y} \\ \text{Brand z} \end{matrix}$$

Find the total sales revenue for all these products during the month of February.

Solution The calculation is as follows:

$$S \cdot P = \begin{bmatrix} 100 & 500 & 300 \end{bmatrix} \cdot \begin{bmatrix} \$2.00 \\ \$1.00 \\ \$1.50 \end{bmatrix}$$

$$= 100(\$2.00) + 500(\$1.00) + 300(\$1.50)$$

$$= \$1150$$

Product Matrices

We now discuss multiplying matrices in general. If

$$A = \begin{bmatrix} 4 & 3 & 2 \\ 5 & 1 & 6 \end{bmatrix} \quad \text{and} \quad B = \begin{bmatrix} 2 & 4 \\ 1 & 0 \\ -1 & 2 \end{bmatrix}$$

then the product matrix AB is determined by the following procedure:

Step 1 Partition the left matrix, A, into rows and the right matrix, B, into columns, as follows:

$$AB = \begin{bmatrix} \begin{array}{|ccc|} \hline 4 & 3 & 2 \\ \hline 5 & 1 & 6 \\ \hline \end{array} \end{bmatrix} \begin{bmatrix} \begin{array}{|c|c|} 2 & 4 \\ 1 & 0 \\ -1 & 2 \end{array} \end{bmatrix}$$

Step 2 Find the dot product of each row vector of the left matrix, A, and each column vector of the right matrix, B. Write these dot products in a matrix array such that the dot product resulting from row i of matrix A and column j of matrix B is located in row i and column j of the new matrix. The resulting new matrix is the product matrix AB.

$$\begin{bmatrix} 4 & 3 & 2 \end{bmatrix} \cdot \begin{bmatrix} 2 \\ 1 \\ -1 \end{bmatrix} = 4(2) + 3(1) + 2(-1) = 9 \quad \text{row 1, column 1}$$

258 CHAPTER FIVE

$$[4 \ 3 \ 2] \cdot \begin{bmatrix} 4 \\ 0 \\ 2 \end{bmatrix} = 4(4) + 3(0) + 2(2) = 20 \quad \text{row 1, column 2}$$

$$[5 \ 1 \ 6] \cdot \begin{bmatrix} 2 \\ 1 \\ -1 \end{bmatrix} = 5(2) + 1(1) + 6(-1) = 5 \quad \text{row 2, column 1}$$

$$[5 \ 1 \ 6] \cdot \begin{bmatrix} 4 \\ 0 \\ 2 \end{bmatrix} = 5(4) + 1(0) + 6(2) = 32 \quad \text{row 2, column 2}$$

Product Matrix
$$AB = \begin{bmatrix} 9 & 20 \\ 5 & 32 \end{bmatrix}$$

Observe that the product matrix AB is defined if and only if the number of columns of the left matrix, A, equals the number of rows of the right matrix, B. Note also that the product matrix has the same number of rows as the left matrix, A, and the same number of columns as the right matrix, B. These details are illustrated as follows:

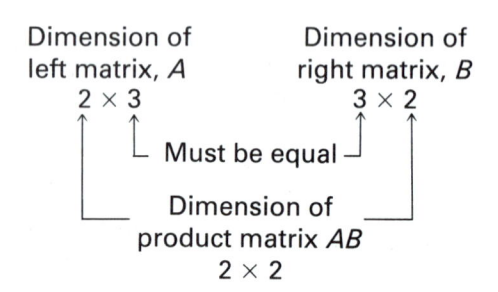

Dimension of left matrix, A
2×3

Dimension of right matrix, B
3×2

Must be equal

Dimension of product matrix AB
2×2

In general, if A is an $m \times n$ matrix and B is an $n \times k$ matrix, then the product AB is an $m \times k$ matrix, as shown here:

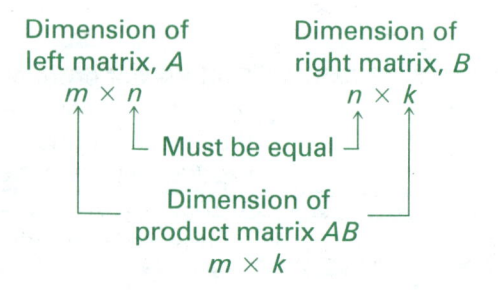

Dimension of left matrix, A
$m \times n$

Dimension of right matrix, B
$n \times k$

Must be equal

Dimension of product matrix AB
$m \times k$

Additionally, if A is a matrix of dimension $m \times n$, i.e.,

259 LINEAR SYSTEMS AND MATRICES

$$A = \begin{bmatrix} a_{11} & a_{12} & \cdots & a_{1n} \\ a_{21} & a_{22} & \cdots & a_{2n} \\ \vdots & \vdots & & \vdots \\ a_{i1} & a_{i2} & \cdots & a_{in} \\ \vdots & \vdots & & \vdots \\ a_{m1} & a_{m2} & & a_{mn} \end{bmatrix}$$

and B is a matrix of dimension $n \times k$, i.e.,

$$B = \begin{bmatrix} b_{11} & b_{12} & \cdots & b_{1j} & \cdots & b_{1k} \\ b_{21} & b_{22} & \cdots & b_{2j} & \cdots & b_{2k} \\ \vdots & \vdots & & \vdots & & \vdots \\ b_{n1} & b_{n2} & \cdots & b_{nj} & \cdots & b_{nk} \end{bmatrix}$$

then the product matrix AB is a matrix of dimension $m \times k$, i.e.,

$$AB = \begin{bmatrix} c_{11} & c_{12} & \cdots & c_{1j} & \cdots & c_{1k} \\ c_{21} & c_{22} & \cdots & c_{2j} & \cdots & c_{2k} \\ c_{i1} & c_{i2} & \cdots & c_{ij} & \cdots & c_{ik} \\ \vdots & \vdots & & \vdots & & \vdots \\ c_{m1} & c_{m2} & \cdots & c_{mj} & \cdots & c_{mk} \end{bmatrix}$$

If c_{ij} is the element in the ith row and the jth column of the product matrix AB, then

$$c_{ij} = a_{i1}b_{1j} + a_{i2}b_{2j} + \ldots + a_{in}b_{nj}$$

EXAMPLE 5–10 Let

$$F = \begin{bmatrix} 4 & 1 \\ 3 & 5 \\ 0 & 2 \end{bmatrix} \quad \text{and} \quad G = \begin{bmatrix} -3 & 7 \\ 1 & -2 \end{bmatrix}$$

Find FG.

EXAMPLE 5–10 (continued)

Solution Note the following:

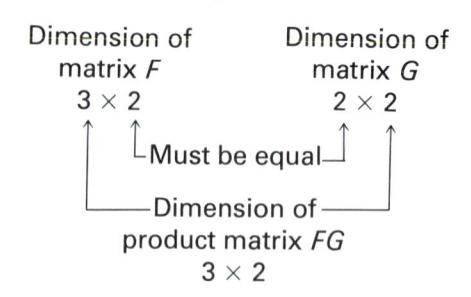

$$\underset{\substack{\text{Dimension of}\\ \text{matrix } F\\ 3 \times 2}}{\uparrow} \quad \underset{\substack{\text{Dimension of}\\ \text{matrix } G\\ 2 \times 2}}{\uparrow}$$

Must be equal

Dimension of product matrix FG
3×2

Thus,

$$FG = \begin{bmatrix} 4 & 1 \\ 3 & 5 \\ 0 & 2 \end{bmatrix} \begin{bmatrix} -3 & 7 \\ 1 & -2 \end{bmatrix} = \begin{bmatrix} p_{11} & p_{12} \\ p_{21} & p_{22} \\ p_{31} & p_{32} \end{bmatrix}$$

where

$$p_{11} = \begin{bmatrix} 4 & 1 \end{bmatrix} \begin{bmatrix} -3 \\ 1 \end{bmatrix} = 4(-3) + 1(1) = -11$$

$$p_{12} = \begin{bmatrix} 4 & 1 \end{bmatrix} \begin{bmatrix} 7 \\ -2 \end{bmatrix} = 4(7) + 1(-2) = 26$$

$$p_{21} = \begin{bmatrix} 3 & 5 \end{bmatrix} \begin{bmatrix} -3 \\ 1 \end{bmatrix} = 3(-3) + 5(1) = -4$$

$$p_{22} = \begin{bmatrix} 3 & 5 \end{bmatrix} \begin{bmatrix} 7 \\ -2 \end{bmatrix} = 3(7) + 5(-2) = 11$$

$$p_{31} = \begin{bmatrix} 0 & 2 \end{bmatrix} \begin{bmatrix} -3 \\ 1 \end{bmatrix} = 0(-3) + 2(1) = 2$$

$$p_{32} = \begin{bmatrix} 0 & 2 \end{bmatrix} \begin{bmatrix} 7 \\ -2 \end{bmatrix} = 0(7) + 2(-2) = -4$$

Hence,

$$FG = \begin{bmatrix} -11 & 26 \\ -4 & 11 \\ 2 & -4 \end{bmatrix}$$

EXAMPLE 5–11 The Saf-T-Flo Company manufactures two models of faucets: Model A and Model B. Each model must pass through Department I (Assembly) and Department II (Polishing). The unit time requirements (in hours) for each model in each department are given by matrix T:

$$T = \begin{bmatrix} 3 & 5 \\ 2 & 1 \end{bmatrix} \begin{matrix} \text{Department I} \\ \text{Department II} \end{matrix}$$

with columns labeled Model A and Model B.

The production requirements of each model are given by matrix P:

$$P = \begin{bmatrix} 500 \\ 700 \end{bmatrix} \begin{matrix} \text{Model A} \\ \text{Model B} \end{matrix}$$

Find the matrix that expresses the total time requirement for each department.

Solution

$$TP = \begin{bmatrix} 3 & 5 \\ 2 & 1 \end{bmatrix} \begin{bmatrix} 500 \\ 700 \end{bmatrix} = \begin{bmatrix} 5000 \\ 1700 \end{bmatrix} \begin{matrix} \text{Department I} \\ \text{Department II} \end{matrix}$$

Thus, Departments I and II need 5000 hours and 1700 hours, respectively, to satisfy production requirements.

EXAMPLE 5–12 Suppose that the Saf-T-Flo Company of Example 5–11 has two plants: Plant X and Plant Y. The unit time requirements (in hours) for each model in each department are the same for both plants and are given by matrix T:

$$T = \begin{bmatrix} 3 & 5 \\ 2 & 1 \end{bmatrix} \begin{matrix} \text{Department I} \\ \text{Department II} \end{matrix}$$

with columns labeled Model A and Model B.

The production requirements of each model in each plant are given by matrix R:

$$R = \begin{bmatrix} 500 & 800 \\ 700 & 300 \end{bmatrix} \begin{matrix} \text{Model A} \\ \text{Model B} \end{matrix}$$

with columns labeled Plant X and Plant Y.

Find the matrix that expresses the total time requirement for each department in each plant.

EXAMPLE 5–12 (continued)

Solution

$$TR = \begin{bmatrix} 3 & 5 \\ 2 & 1 \end{bmatrix} \begin{bmatrix} 500 \\ 700 \end{bmatrix} \begin{bmatrix} 800 \\ 300 \end{bmatrix}$$

$$= \begin{bmatrix} \text{Plant X} & \text{Plant Y} \\ 5000 & 3900 \\ 1700 & 1900 \end{bmatrix} \begin{matrix} \text{Department I} \\ \text{Department II} \end{matrix}$$

Thus, Departments I and II of Plant X need 5000 hours and 1700 hours, respectively, to satisfy production requirements. Also, Departments I and II of Plant Y need 3900 hours and 1900 hours, respectively, to satisfy production requirements.

Linear Systems and Matrix Equations

A linear system such as

$$3x + 5y = 1$$
$$-2x + 7y = -11$$

can be rewritten in matrix form as follows:

$$\begin{bmatrix} 3 & 5 \\ -2 & 7 \end{bmatrix} \begin{bmatrix} x \\ y \end{bmatrix} = \begin{bmatrix} 1 \\ -11 \end{bmatrix}$$

If

$$A = \begin{bmatrix} 3 & 5 \\ -2 & 7 \end{bmatrix} \quad X = \begin{bmatrix} x \\ y \end{bmatrix} \quad B = \begin{bmatrix} 1 \\ -11 \end{bmatrix}$$

then the preceding system may be recast as the matrix equation

$$AX = B$$

To verify that the matrix equation $AX = B$ is equivalent to the original system, we first find the product matrix AX. Hence,

$$AX = \begin{bmatrix} 3 & 5 \\ -2 & 7 \end{bmatrix} \begin{bmatrix} x \\ y \end{bmatrix} = \begin{bmatrix} 3x + 5y \\ -2x + 7y \end{bmatrix}$$

Since

$$AX = B$$

263 LINEAR SYSTEMS AND MATRICES

then
$$\begin{bmatrix} 3x + 5y \\ -2x + 7y \end{bmatrix} = \begin{bmatrix} 1 \\ -11 \end{bmatrix}$$

Matrix equality requires that corresponding elements of both matrices be equal. Thus,

$$3x + 5y = 1$$
$$-2x + 7y = -11$$

Observe that these are the equations of our original system.

EXAMPLE 5–13 The following is a linear system consisting of three equations and three variables. Rewrite this linear system in the matrix form $AX = B$.

$$4x_1 - 7x_2 + 3x_3 = 16$$
$$9x_1 + 3x_2 - 6x_3 = 5$$
$$-2x_1 - 5x_2 + 8x_3 = 11$$

Solution Here, we have

$$\underbrace{\begin{bmatrix} 4 & -7 & 3 \\ 9 & 3 & -6 \\ -2 & -5 & 8 \end{bmatrix} \begin{bmatrix} x_1 \\ x_2 \\ x_3 \end{bmatrix}}_{AX} = \underbrace{\begin{bmatrix} 16 \\ 5 \\ 11 \end{bmatrix}}_{B}$$

Identity Matrices The number 1 is called the *multiplicative identity* for real numbers because

$$a \cdot 1 = 1 \cdot a = a$$

for all real numbers a. In other words, the product of any real number a and 1 is the real number a. Similarly, an **identity matrix** is a square matrix I such that

$$AI = IA = A$$

where A is a square matrix of the same dimension as I.
The matrix

$$I = \begin{bmatrix} 1 & 0 \\ 0 & 1 \end{bmatrix}$$

is the multiplicative identity for square matrices of dimension 2 × 2. If

$$A = \begin{bmatrix} 3 & 5 \\ -2 & 7 \end{bmatrix}$$

then observe that

$$AI = A$$

$$\begin{array}{ccc} A & I & = A \\ \begin{bmatrix} 3 & 5 \\ -2 & 7 \end{bmatrix} \begin{bmatrix} 1 & 0 \\ 0 & 1 \end{bmatrix} &=& \begin{bmatrix} 3 & 5 \\ -2 & 7 \end{bmatrix} \end{array}$$

$$IA = A$$

$$\begin{array}{ccc} I & A & = A \\ \begin{bmatrix} 1 & 0 \\ 0 & 1 \end{bmatrix} \begin{bmatrix} 3 & 5 \\ -2 & 7 \end{bmatrix} &=& \begin{bmatrix} 3 & 5 \\ -2 & 7 \end{bmatrix} \end{array}$$

For square matrices of dimension 3 × 3, the multiplicative identity is

$$I = \begin{bmatrix} 1 & 0 & 0 \\ 0 & 1 & 0 \\ 0 & 0 & 1 \end{bmatrix}$$

In general, the $n \times n$ matrix

$$I = \begin{bmatrix} 1 & 0 & 0 & \ldots & 0 \\ 0 & 1 & 0 & \ldots & 0 \\ 0 & 0 & 1 & \ldots & 0 \\ \vdots & \vdots & \vdots & & \vdots \\ 0 & 0 & 0 & \ldots & 1 \end{bmatrix}$$

is the multiplicative identity for square matrices of dimension $n \times n$.

EXAMPLE 5–14 If

$$A = \begin{bmatrix} 1 & -4 & 7 \\ 3 & 2 & -5 \\ -1 & -6 & -8 \end{bmatrix} \text{ and } I = \begin{bmatrix} 1 & 0 & 0 \\ 0 & 1 & 0 \\ 0 & 0 & 1 \end{bmatrix}$$

verify that $AI = IA = A$.

LINEAR SYSTEMS AND MATRICES

Solution

$$AI = \begin{bmatrix} 1 & -4 & 7 \\ 3 & 2 & -5 \\ -1 & -6 & -8 \end{bmatrix} \begin{bmatrix} 1 & 0 & 0 \\ 0 & 1 & 0 \\ 0 & 0 & 1 \end{bmatrix}$$

$$= \begin{bmatrix} 1 & -4 & 7 \\ 3 & 2 & -5 \\ -1 & -6 & -8 \end{bmatrix} = A$$

$$IA = \begin{bmatrix} 1 & 0 & 0 \\ 0 & 1 & 0 \\ 0 & 0 & 1 \end{bmatrix} \begin{bmatrix} 1 & -4 & 7 \\ 3 & 2 & -5 \\ -1 & -6 & -8 \end{bmatrix}$$

$$= \begin{bmatrix} 1 & -4 & 7 \\ 3 & 2 & -5 \\ -1 & -6 & -8 \end{bmatrix} = A$$

EXERCISES

1. If $A = [1 \quad 2]$ and $B = \begin{bmatrix} -3 \\ 6 \end{bmatrix}$, calculate $A \cdot B$.

2. If $C = [1 \quad 4 \quad 0 \quad -3]$ and $D = \begin{bmatrix} 8 \\ -1 \\ 0 \\ 2 \end{bmatrix}$, calculate $C \cdot D$.

3. If $E = \begin{bmatrix} -4 \\ 3 \end{bmatrix}$ and $F = \begin{bmatrix} 1 \\ 5 \end{bmatrix}$, calculate $E \cdot F$.

4. (The following problem appeared on a past Uniform CPA Examination.) Dancy, Inc. is going to begin producing a new chemical cleaner containing alcohol, proxide, and cnzyme. Each quart of the new cleaner will require 1/2 quart of alcohol, 1/6 quart of proxide, and 1/3 quart of cnzyme. The costs per quart are $0.40 for alcohol, $0.60 for proxide, and $0.20 for cnzyme. If the requirements are listed in matrix R:

$$R = [1/2 \quad 1/6 \quad 1/3]$$

and their unit costs are listed in matrix C:

$$C = \begin{bmatrix} 0.40 \\ 0.60 \\ 0.20 \end{bmatrix}$$

then state and perform the matrix operation to determine the cost of producing 1 quart of cleaner.

5. Given that $A = \begin{bmatrix} 4 & 6 \\ -5 & 2 \end{bmatrix}$ and $B = \begin{bmatrix} 1 & -2 \\ -3 & 4 \end{bmatrix}$

   (A) Calculate AB.
   (B) Calculate BA.
   (C) Does $AB = BA$?

6. Given that

$$A = \begin{bmatrix} 1 & 3 & 7 \\ 2 & 4 & 0 \\ -1 & 5 & -2 \end{bmatrix} \quad B = \begin{bmatrix} 1 & 5 \\ 3 & 7 \\ -7 & 2 \end{bmatrix}$$

$$C = \begin{bmatrix} 2 & -1 & 0 & 6 \\ -1 & 4 & 3 & 2 \end{bmatrix} \quad D = \begin{bmatrix} 1 & 0 & -2 \\ 3 & -1 & 1 \end{bmatrix}$$

   calculate, if possible, each of the following:
   (A) AB (B) BA (C) BC
   (D) CB (E) BD (F) DB
   (G) DA (H) $(AB)C$ (I) $(DB)C$

7. Given that

$$A = \begin{bmatrix} 3 & 2 \\ -5 & -6 \end{bmatrix} \quad \text{and} \quad I = \begin{bmatrix} 1 & 0 \\ 0 & 1 \end{bmatrix}$$

   verify that $AI = IA = A$.

8. Given that

$$A = \begin{bmatrix} 4 & 3 & 6 \\ 8 & 2 & 7 \\ -1 & 1 & 4 \end{bmatrix} \quad \text{and} \quad I = \begin{bmatrix} 1 & 0 & 0 \\ 0 & 1 & 0 \\ 0 & 0 & 1 \end{bmatrix}$$

   verify that $AI = IA = A$.

9. For each of the following, determine the dimension of the product matrix AB:
   (A) A is a 2×2 matrix and B is a 2×4 matrix.
   (B) A is a 3×4 matrix and B is a 4×5 matrix.
   (C) A is a 2×5 matrix and B is a 5×3 matrix.
   (D) A is a 4×2 matrix and B is a 2×4 matrix.
   (E) A is a 4×4 matrix and B is a 4×4 matrix.

10. For each of the following, determine if it is possible to calculate the product matrix CD:
    (A) C is a 2×5 matrix and D is a 4×2 matrix.
    (B) C is a 3×4 matrix and D is a 2×5 matrix.
    (C) C is a 2×2 matrix and D is a 2×2 matrix.
    (D) C is a 2×3 matrix and D is a 3×7 matrix.

11. Repeat Exercise 10 for the product matrix DC.

12. Adam Apple operates a fruit store that stocks apples, oranges, pears, and limes. Adam does his buying on Monday, Wednesday, and Fri-

day. Matrix Q shows the quantities (in pounds) of each fruit bought by Adam during a given week:

$$Q = \begin{matrix} & \text{Apples} & \text{Oranges} & \text{Pears} & \text{Limes} & \\ & \begin{bmatrix} 300 & 100 & 200 & 50 \\ 200 & 300 & 100 & 20 \\ 500 & 400 & 50 & 100 \end{bmatrix} & \begin{matrix} \text{Monday} \\ \text{Wednesday} \\ \text{Friday} \end{matrix} \end{matrix}$$

Matrix P gives the price per pound (in dollars) of each fruit:

$$P = \begin{bmatrix} 0.20 \\ 0.10 \\ 0.30 \\ 0.40 \end{bmatrix} \begin{matrix} \text{Apples} \\ \text{Oranges} \\ \text{Pears} \\ \text{Limes} \end{matrix}$$

Find the matrix showing the dollar amount spent by Adam on each of his buying days.

13. The Tomorrow Transportation Company manufactures two models of mopeds: the F160 and the S130. Each F160 moped requires 2 hours in the Assembly Department and 1 hour in the Quality Control Department. Each S130 requires 3 hours in Assembly and 1.5 hours in Quality Control.

  (A) Determine the matrix A that shows the time required for each model in each department, i.e.,

$$A = \begin{matrix} & \text{F160} & \text{S130} & \\ & \begin{bmatrix} - & - \\ - & - \end{bmatrix} & \begin{matrix} \text{Assembly} \\ \text{Quality Control} \end{matrix} \end{matrix}$$

  (B) If the company must produce 100 F160 and 200 S130 mopeds, find the matrix B that lists these production requirements, i.e.,

$$B = \begin{bmatrix} - \\ - \end{bmatrix} \begin{matrix} \text{F160} \\ \text{S130} \end{matrix}$$

  (C) Find the matrix that shows the total time requirement for each department.

14. Suppose the Tomorrow Transportation Company of Exercise 13 has a second plant with product requirements of 250 F160 and 120 S130 mopeds.

  (A) Find the matrix C that shows the production requirements of both plants, i.e.,

$$C = \begin{matrix} & \text{Plant 1} & \text{Plant 2} & \\ & \begin{bmatrix} - & - \\ - & - \end{bmatrix} & \begin{matrix} \text{F160} \\ \text{S130} \end{matrix} \end{matrix}$$

  (B) Find the matrix that shows the total time requirements of each department in each plant.

15. Rewrite each of the following linear systems in the matrix form $AX = B$:

(A) $2x + 3y = 7$
       $-4x + 5y = 9$

(B) $x_1 + 5x_2 = 6$
       $4x_1 + 8x_2 = 11$

(C) $3x_1 - 7x_2 - 5x_3 = 11$
       $x_1 + 4x_2 - 2x_3 = 4$
       $5x_1 + 9x_2 + 8x_3 = 16$

(D) $2x + 3y + z = 11$
       $x + 2z = 9$
       $ 4y + 5z = 17$

16. Given that

$$A = \begin{bmatrix} 2 & -3 \\ 1 & 4 \end{bmatrix} \quad B = \begin{bmatrix} 2 & 0 \\ -1 & 5 \end{bmatrix} \quad C = \begin{bmatrix} 2 & 7 \\ -1 & -3 \end{bmatrix}$$

verify that
(A) $A(BC) = (AB)C$ (associative property of multiplication)
(B) $A(B + C) = AB + AC$ (distributive property)
(C) $(B + C)A = BA + CA$

17. Given that $X = [x_1 \ x_2 \ x_3]$ and $A = \begin{bmatrix} 1 & 2 & -1 \\ 4 & 0 & -3 \\ 5 & 1 & 2 \end{bmatrix}$ verify that

$X - XA = X(I - A)$.

18. If $A^2 = AA$, compute A^2 for each of the following matrices A:

(A) $\begin{bmatrix} 2 & 3 \\ -1 & 4 \end{bmatrix}$

(B) $\begin{bmatrix} -8 & 0 \\ 1 & 2 \end{bmatrix}$

(C) $\begin{bmatrix} 1 & 2 & 1 \\ 4 & -1 & 0 \\ 2 & 0 & 2 \end{bmatrix}$

(D) $\begin{bmatrix} -1 & 2 & 3 \\ 5 & -2 & 1 \\ 4 & -4 & 0 \end{bmatrix}$

19. If $A^3 = A^2A$, compute A^3 for each of the matrices of Exercise 18.

20. Compute A^3 for each of the following matrices A:

(A) $\begin{bmatrix} 4 & -7 \\ 0 & 1 \end{bmatrix}$

(B) $\begin{bmatrix} -2 & -3 \\ 1 & 4 \end{bmatrix}$

(C) $\begin{bmatrix} -1 & 0 & 2 \\ -3 & 1 & 1 \\ 2 & -1 & 3 \end{bmatrix}$

(D) $\begin{bmatrix} 4 & -1 & -3 \\ 1 & 2 & 1 \\ 0 & 1 & 0 \end{bmatrix}$

21. Using the results of Exercises 18 through 20, define A^n for positive integers n and square matrix A.

22. Given that $A = \begin{bmatrix} 4 & 3 \\ 8 & 0 \end{bmatrix}$ and $B = \begin{bmatrix} 7 & -1 \\ -2 & 4 \end{bmatrix}$, verify that $AB \neq BA$.

23. Given that $A = \begin{bmatrix} a & b \\ c & d \end{bmatrix}$ and $I = \begin{bmatrix} 1 & 0 \\ 0 & 1 \end{bmatrix}$, verify that $AI = IA = A$.

24. Given that $B = \begin{bmatrix} 2 & 3 \\ 7 & 4 \\ 5 & 7 \end{bmatrix}$ and $I = \begin{bmatrix} 1 & 0 \\ 0 & 1 \end{bmatrix}$, verify that $BI = B$.

LINEAR SYSTEMS AND MATRICES

25. Given that $A = \begin{bmatrix} 4 & 3 & 1 \\ 8 & 0 & 2 \\ -1 & 4 & 6 \end{bmatrix}$ and $I = \begin{bmatrix} 1 & 0 & 0 \\ 0 & 1 & 0 \\ 0 & 0 & 1 \end{bmatrix}$, verfiy that $AI = IA = A$.

26. Given that $D = \begin{bmatrix} 4 & 6 & -1 \\ 8 & 4 & 3 \end{bmatrix}$ and $I = \begin{bmatrix} 1 & 0 \\ 0 & 1 \end{bmatrix}$, verify that $ID = D$.

5–4

GAUSS–JORDAN METHOD OF SOLVING LINEAR SYSTEMS

In Section 5–1, we reviewed two methods of solving linear systems. Our discussion was limited to linear systems consisting of two equations and two variables. In this section, we will present a more structured method for solving linear systems of any size. The method consists of replacing an original linear system by a succession of **equivalent linear systems** (which have the same solution as the original linear system), the last of which is made up of equations that explicitly yield the solution values of the unknowns. An equivalent linear system is obtained by applying any of the following **row operations** to any equation* of a linear system.

> **Three Fundamental Row Operations**
>
> 1. Interchanging two rows (equations)
> 2. Multiplying a row (equation) by a nonzero constant
> 3. Multiplying a row (equation) by a nonzero constant and adding the result to another row (equation)

We illustrate by solving the linear system

$$2x_1 + 3x_2 = 9$$
$$x_1 + 4x_2 = 17$$

Recall that such a system may be expressed in the matrix form

$$AX = B$$

$$A \quad X \quad = \quad B$$
$$\begin{bmatrix} 2 & 3 \\ 1 & 4 \end{bmatrix} \begin{bmatrix} x_1 \\ x_2 \end{bmatrix} = \begin{bmatrix} 9 \\ 17 \end{bmatrix}$$

---

*The terms *equation* and *row* are used interchangeably.

If we write matrix A next to matrix B as illustrated here:

$$\begin{matrix} A & B \\ \begin{bmatrix} 2 & 3 & | & 9 \\ 1 & 4 & | & 17 \end{bmatrix} \end{matrix}$$

then the linear system is written in **tableau** form, denoted $[A \mid B]$. The tableau $[A \mid B]$ is also called an **augmented matrix**. Observe that each row of the tableau $[A \mid B]$ is composed of the coefficients of the corresponding equation of the linear system. As we present the process of solving a linear system by row operations, we will illustrate the corresponding tableau.

The original system and its corresponding tableau are as follows:

Initial Tableau

$$\begin{aligned} 2x_1 + 3x_2 &= 9 \\ x_1 + 4x_2 &= 17 \end{aligned} \qquad \begin{bmatrix} 2 & 3 & | & 9 \\ 1 & 4 & | & 17 \end{bmatrix}$$

We now begin the process of solving the system by row operations:

1. Using the first row operation, we interchange rows 1 and 2. This gives us a coefficient of 1 in the upper left-hand corner:

$$\begin{aligned} x_1 + 4x_2 &= 17 \\ 2x_1 + 3x_2 &= 9 \end{aligned} \qquad \begin{bmatrix} 1 & 4 & | & 17 \\ 2 & 3 & | & 9 \end{bmatrix}$$

2. Using the third row operation, we multiply row 1 by -2 and add the result to row 2. This eliminates variable x_1 in the second equation and thus gives us a coefficient of 0 in the lower left-hand corner of our tableau:

$$\begin{aligned} x_1 + 4x_2 &= 17 \\ -5x_2 &= -25 \end{aligned} \qquad \begin{bmatrix} 1 & 4 & | & 17 \\ 0 & -5 & | & -25 \end{bmatrix}$$

3. Using the second row operation, we multiply row 2 by $-1/5$. This gives us a solution for x_2 in the second equation and, correspondingly, a coefficient of 1 as the second element of row 2 in our tableau:

$$\begin{aligned} x_1 + 4x_2 &= 17 \\ x_2 &= 5 \end{aligned} \qquad \begin{bmatrix} 1 & 4 & | & 17 \\ 0 & 1 & | & 5 \end{bmatrix}$$

4. Using the third row operation, we multiply row 2 by -4 and add the result to row 1. This eliminates variable x_2 in the first

equation and thus gives us a solution for x_1. Correspondingly, a coefficient of 0 appears as the second element of row 1 in our tableau:

Final Tableau

$$x_1 = -3$$
$$x_2 = 5$$

$$\begin{bmatrix} 1 & 0 & | & -3 \\ 0 & 1 & | & 5 \end{bmatrix}$$

Note that matrix A of the initial tableau has been transformed by row operations into the identity matrix of the final tableau. Also, matrix B of the initial tableau has been transformed into the solution matrix. Additionally, note that the entire process of solving a system by row operations could have been carried out by writing only each tableau. This example has illustrated the **Gauss–Jordan method** of solving linear systems.

In general, when using the Gauss–Jordan method to solve a linear system

$$a_{11}x_1 + a_{12}x_2 + \ldots + a_{1n}x_n = b_1$$
$$a_{21}x_1 + a_{22}x_2 + \ldots + a_{2n}x_n = b_2$$
$$\vdots \vdots \vdots \vdots$$
$$a_{n1}x_1 + a_{n2}x_2 + \ldots + a_{nn}x_n = b_n$$

we begin with the augmented matrix $[A \mid B]$:

$$\begin{bmatrix} a_{11} & a_{12} & \ldots & a_{1n} & | & b_1 \\ a_{21} & a_{22} & \ldots & a_{2n} & | & b_2 \\ \vdots & \vdots & & \vdots & | & \vdots \\ a_{n1} & a_{n2} & \ldots & a_{nn} & | & b_n \end{bmatrix}$$

and use row operations to obtain the augmented matrix $[I \mid C]$, where C is the $n \times 1$ solution matrix:

$$\begin{bmatrix} 1 & 0 & \ldots & 0 & | & c_1 \\ 0 & 1 & \ldots & 0 & | & c_2 \\ \vdots & \vdots & & \vdots & | & \vdots \\ 0 & 0 & \ldots & 1 & | & c_n \end{bmatrix}$$

Hence, $x_1 = c_1, x_2 = c_2, \ldots, x_n = c_n$ is the solution to the linear system. If we cannot obtain the augmented matrix $[I \mid C]$, then the linear system has either no solutions or infinitely many solutions.

EXAMPLE 5–15 Solve the following linear system by the Gauss–Jordan method of row operations:

$$x_1 + 2x_2 - x_3 = 4$$
$$x_1 + x_2 - 2x_3 = 2$$
$$x_1 + 2x_2 + 3x_3 = 8$$

Solution We first write the augmented matrix $[A \mid B]$:

$$\begin{bmatrix} 1 & 2 & -1 & | & 4 \\ 1 & 1 & -2 & | & 2 \\ 1 & 2 & 3 & | & 8 \end{bmatrix}$$

Then, we use row operations to transform matrix A into the identity matrix, as follows:

1. We already have a 1 for the first element of row 1. We need a 0 for the first element in row 2. Multiplying row 1 by -1 and adding the result to row 2 gives us a 0 in this position:

$$\begin{bmatrix} 1 & 2 & -1 & | & 4 \\ 0 & -1 & -1 & | & -2 \\ 1 & 2 & 3 & | & 8 \end{bmatrix}$$

2. We need a 0 for the first element of row 3. Multiplying row 1 by -1 and adding the result to row 3 gives us the needed 0:

$$\begin{bmatrix} 1 & 2 & -1 & | & 4 \\ 0 & -1 & -1 & | & -2 \\ 0 & 0 & 4 & | & 4 \end{bmatrix}$$

The first column is now completed. We proceed to transform the second column to

$$\begin{bmatrix} 0 \\ 1 \\ 0 \end{bmatrix}$$

3. We need a 1 for the second element of row 2. Multiplying row 2 by -1 gives us the needed 1:

$$\begin{bmatrix} 1 & 2 & -1 & | & 4 \\ 0 & 1 & 1 & | & 2 \\ 0 & 0 & 4 & | & 4 \end{bmatrix}$$

LINEAR SYSTEMS AND MATRICES

4. We need a 0 for the second element of row 1. Multiplying row 2 by −2 and adding the result to row 1 gives us the 0:

$$\begin{bmatrix} 1 & 0 & -3 & | & 0 \\ 0 & 1 & 1 & | & 2 \\ 0 & 0 & 4 & | & 4 \end{bmatrix}$$

The second column is now completed. We proceed to transform the third column to

$$\begin{bmatrix} 0 \\ 0 \\ 1 \end{bmatrix}$$

5. We need a 1 for the third element of row 3. Multiplying row 3 by 1/4 gives us the 1:

$$\begin{bmatrix} 1 & 0 & -3 & | & 0 \\ 0 & 1 & 1 & | & 2 \\ 0 & 0 & 1 & | & 1 \end{bmatrix}$$

6. We need a 0 for the third element of row 2. Multiplying row 3 by −1 and adding the result to row 2 gives us the 0:

$$\begin{bmatrix} 1 & 0 & -3 & | & 0 \\ 0 & 1 & 0 & | & 1 \\ 0 & 0 & 1 & | & 1 \end{bmatrix}$$

7. We need a 0 for the third element of row 1. Multiplying row 3 by 3 and adding the result to row 1 gives us the 0:

$$\begin{bmatrix} 1 & 0 & 0 & | & 3 \\ 0 & 1 & 0 & | & 1 \\ 0 & 0 & 1 & | & 1 \end{bmatrix}$$

Thus, the solution to the linear system is $x_1 = 3$, $x_2 = 1$, $x_3 = 1$, or the ordered triple $(3, 1, 1)$.

EXAMPLE 5–16 Use the Gauss–Jordan method of row operations to solve the linear system

$$x_1 + 2x_2 = 3$$
$$2x_1 + 4x_2 = 8$$

EXAMPLE 5–16 (continued)

Solution We first write the augmented matrix $[A \mid B]$:

$$\begin{bmatrix} 1 & 2 & | & 3 \\ 2 & 4 & | & 8 \end{bmatrix}$$

Now, we use row operations to transform matrix A into the identity matrix:

1. We already have a 1 for the first element of row 1. We need a 0 for the first element of row 2. Multiplying row 1 by -2 and adding the result to row 2 gives us the desired 0:

$$\begin{bmatrix} 1 & 2 & | & 3 \\ 0 & 0 & | & 2 \end{bmatrix}$$

2. We now must get a 1 for the second element of row 2. Since this is impossible, we cannot proceed further.

    If we convert the preceding augmented matrix into equation form, we have the linear system

    $$x_1 + 2x_2 = 3$$
    $$0x_1 + 0x_2 = 2$$

Note that the second equation is $0 = 2$, an untrue statement *(inconsistency)*. Hence, the system has no solutions.

EXAMPLE 5–17 Use the Gauss–Jordan method of row operations to solve the linear system

$$x_1 + 2x_2 + x_3 = 3$$
$$2x_1 - 3x_2 - 2x_3 = 5$$
$$2x_1 + 4x_2 + 2x_3 = 6$$

Solution We first write the augmented matrix $[A \mid B]$:

$$\begin{bmatrix} 1 & 2 & 1 & | & 3 \\ 2 & -3 & -2 & | & 5 \\ 2 & 4 & 2 & | & 6 \end{bmatrix}$$

We need 0s for the second and third elements of column 1.

Multiplying row 1 by -2 and adding the result to row 2 gives us a 0 for the second element in column 1. Multiplying row 1 by -2 and adding the result to row 3 gives us a 0 for the third element in column 1:

$$\begin{bmatrix} 1 & 2 & 1 & | & 3 \\ 0 & -7 & -4 & | & -1 \\ 0 & 0 & 0 & | & 0 \end{bmatrix}$$

Since all the entries of row 3 are 0s, it will not be possible to get a 1 for the third element of row 3. Thus, if we convert the resulting augmented matrix into equation form, we have

$$x_1 + 2x_2 + x_3 = 3$$
$$-7x_2 - 4x_3 = -1$$
$$0x_1 + 0x_2 + 0x_3 = 0$$

Note that the third equation is $0 = 0$, a signal that the system has infinitely many solutions.

If we continue our attempt to obtain the identity matrix, we must get a 1 for the second element of row 2. Multiplying row 2 by $-1/7$ yields

$$\begin{bmatrix} 1 & 2 & 1 & | & 3 \\ 0 & 1 & 4/7 & | & 1/7 \\ 0 & 0 & 0 & | & 0 \end{bmatrix}$$

We now need a 0 for the first element of column 2. Multiplying row 2 by -2 and adding the result to row 1 yields

$$\begin{bmatrix} 1 & 0 & -1/7 & | & 19/7 \\ 0 & 1 & 4/7 & | & 1/7 \\ 0 & 0 & 0 & | & 0 \end{bmatrix}$$

Our next step is to get a 1 for the third element of row 3. However, since row 3 consists of all 0s, this is impossible. Thus, the preceding tableau is the final tableau matrix. Rewriting the equations corresponding to this final tableau matrix, we have

$$x_1 - \frac{1}{7}x_3 = \frac{19}{7}$$
$$x_2 + \frac{4}{7}x_3 = \frac{1}{7}$$

> **EXAMPLE 5–17** (continued)
>
> The infinitely many solutions of this linear system may be expressed by writing x_1 and x_2 in terms of x_3 to obtain
>
> $$x_1 = \frac{1}{7}x_3 + \frac{19}{7}$$
>
> $$x_2 = -\frac{4}{7}x_3 + \frac{1}{7}$$
>
> x_3 arbitrary
>
> We may generate any of the infinitely many solutions by choosing an arbitrary value for x_3. For example, if $x_3 = 2$, then our solution is $x_3 = 2$, $x_2 = -1$, $x_1 = 3$. If $x_3 = 1$, then another solution is $x_3 = 1$, $x_2 = -3/7$, $x_1 = 20/7$. The variable x_3 is called a **parameter**.

Note that the final tableau matrix of the linear system of Example 5–17 has two rows (rows 1 and 2) containing nonzero entries. Since there was no inconsistency and since the system had three variables, we were able to solve for two variables in terms of the remaining variable, the arbitrary parameter. In general, for a linear system containing n equations and n variables, the following three possibilities exist:

1. If the final tableau matrix contains an inconsistency, then there is *no solution.* Recall that an inconsistency appears as a row of 0s and a nonzero entry in B. The following tableau matrix gives an example of an inconsistency:

$$\begin{bmatrix} \cdot & \cdot & \cdot & \cdot & | & \cdot \\ \cdot & \cdot & \cdot & \cdot & | & \cdot \\ 0 & 0 & 0 & 0 & | & 5 \end{bmatrix}$$

2. If the final tableau matrix has k rows with nonzero entries and $k = n$ (and there is no inconsistency), then there is a *unique solution.*

3. If the final tableau matrix has k rows with nonzero entries, where $k < n$, and the remaining $n - k$ rows contain zero entries (and there is no inconsistency), then there are *infinitely many solutions.* Furthermore, it is possible to solve for k variables in terms of the remaining $n - k$ variables (or arbitrary parameters).

LINEAR SYSTEMS AND MATRICES

More Variables Than Equations Sometimes, we encounter a linear system with more variables than equations. Such systems have either an infinite number of solutions or no solutions. Example 5–18 considers such a system.

> **EXAMPLE 5–18** Using the Gauss–Jordan method, solve the linear system
>
> $$\begin{aligned} x_1 + 2x_2 \phantom{{}+x_3} + x_4 &= 5 \\ x_2 \phantom{{}+x_3} + 2x_4 &= 6 \\ 2x_1 + 4x_2 + x_3 + x_4 &= -5 \end{aligned}$$
>
> **Solution** We write the augmented matrix $[A \mid B]$:
>
> $$\begin{bmatrix} 1 & 2 & 0 & 1 & | & 5 \\ 0 & 1 & 0 & 2 & | & 6 \\ 2 & 4 & 1 & 1 & | & -5 \end{bmatrix}$$
>
> Using row operations, we transform $[A \mid B]$ into
>
> $$\begin{bmatrix} 1 & 0 & 0 & -3 & | & -7 \\ 0 & 1 & 0 & 2 & | & 6 \\ 0 & 0 & 1 & -1 & | & -15 \end{bmatrix}$$
>
> Converting this matrix into equation form, we have
>
> $$\begin{aligned} x_1 \phantom{{}+x_2} - 3x_4 &= -7 \\ x_2 \phantom{{}+x_3} + 2x_4 &= 6 \\ x_3 - x_4 &= -15 \end{aligned}$$
>
> If we choose x_4 as our arbitrary parameter, then x_1, x_2, and x_3 are expressed in terms of x_4 as follows:
>
> $$\begin{aligned} x_1 &= 3x_4 - 7 \\ x_2 &= -2x_4 + 6 \\ x_3 &= x_4 - 15 \\ x_4 &\text{ arbitrary} \end{aligned}$$
>
> Thus, we may generate any of our infinitely many solutions by choosing an arbitrary value for x_4.

Note that the final tableau matrix of the linear system of Example 5–18 has three rows containing nonzero entries. Since there was no inconsistency and since the system had four variables, we were able to solve for three variables in terms of a remaining variable, the arbitrary parameter. In general, for a linear system containing m equations and n variables, with $m < n$, the following hold:

1. If the final tableau matrix contains an inconsistency, then there is *no solution*.
2. If the final tableau matrix has k rows with nonzero entries, where $k \leq m$, and any remaining $m - k$ rows contain zero entries (and there is no inconsistency), then the linear system has *infinitely many solutions*. Furthermore, it is possible to solve for k variables in terms of the remaining $n - k$ variables (or arbitrary parameters).

More Equations Than Variables

Sometimes we encounter a linear system with more equations than variables. The following linear system is such an example:

$$x_1 - x_2 = 1$$
$$4x_1 - x_2 = 7$$
$$3x_1 - x_2 = 5$$

Note that each equation has a straight line as its graph. Thus, the three lines might intersect at a common point, in which case there would be a unique solution; the three lines might be parallel, in which case there would be no solution; the three lines might intersect at different points, in which case there would be no solution; two of the three lines might be the same so that the intersection with the third line would yield a unique solution; etc.

We will attempt to solve the preceding linear system by the Gauss–Jordan method of row operations. Writing the augmented matrix $[A\,|\,B]$, we have

$$\begin{bmatrix} 1 & -1 & | & 1 \\ 4 & -1 & | & 7 \\ 3 & -1 & | & 5 \end{bmatrix}$$

Since matrix A is of dimension 3×2, we try to convert it into the 3×2 matrix

$$\begin{bmatrix} 1 & 0 & | & \\ 0 & 1 & | & \\ 0 & 0 & | & \end{bmatrix}$$

Note that this 3×2 matrix consists of the first two columns of the 3×3 identity matrix.

Using row operations, we transform $[A \mid B]$ into

$$\begin{bmatrix} 1 & 0 & | & 2 \\ 0 & 1 & | & 1 \\ 0 & 0 & | & 0 \end{bmatrix}$$

Converting this tableau into equation form, we have

$$x_1 = 2$$
$$x_2 = 1$$

Ordinarily, we would convert the third row into the equation $0x_1 + 0x_2 = 0$, or $0 = 0$. However, since our linear system has one more equation than unknowns, we disregard a row consisting entirely of 0s. Hence, our linear system has the unique solution $x_1 = 2$, $x_2 = 1$, or the ordered pair $(2, 1)$.

EXAMPLE 5–19 Using the Gauss–Jordan method, solve

$$x_1 - x_2 = 4$$
$$2x_1 + 3x_2 = 8$$
$$5x_1 + x_2 = 7$$

Solution We write the augmented matrix $[A \mid B]$:

$$\begin{bmatrix} 1 & -1 & | & 4 \\ 2 & 3 & | & 8 \\ 5 & 1 & | & 7 \end{bmatrix}$$

Using row operations, we transform $[A \mid B]$ into

$$\begin{bmatrix} 1 & 0 & | & 4 \\ 0 & 1 & | & 0 \\ 0 & 0 & | & -13 \end{bmatrix}$$

Note that the third row results in the equation $0x_1 + 0x_2 = -13$, or $0 = -13$, an inconsistency. Hence, the system has no solution.

EXAMPLE 5–20 Using the Gauss–Jordan method, solve

$$x_1 + 2x_2 + x_3 = -3$$
$$2x_1 + 2x_2 + 4x_3 = 2$$
$$x_1 + x_2 + 2x_3 = 1$$
$$-4x_1 - 4x_2 - 8x_3 = -4$$

EXAMPLE 5–20 (continued)

Solution   We write the augmented matrix $[A \mid B]$:

$$\begin{bmatrix} 1 & 2 & 1 & | & -3 \\ 2 & 2 & 4 & | & 2 \\ 1 & 1 & 2 & | & 1 \\ -4 & -4 & -8 & | & -4 \end{bmatrix}$$

Using row operations, we transform $[A \mid B]$ into

$$\begin{bmatrix} 1 & 0 & 3 & | & 5 \\ 0 & 1 & -1 & | & -4 \\ 0 & 0 & 0 & | & 0 \\ 0 & 0 & 0 & | & 0 \end{bmatrix}$$

Note that we have two rows consisting entirely of 0s. Since our linear system has one more equation than variables, we may disregard one of the rows of 0s. However, the other row of 0s is a signal that the system has infinitely many solutions. Converting the matrix into equation form, we have

$$x_1 + 3x_3 = 5$$
$$ x_2 - x_3 = -4$$

If we choose x_3 as our arbitrary parameter, then the infinitely many solutions are expressed parametrically as

$$x_1 = -3x_3 + 5$$
$$x_2 = x_3 - 4$$
$$x_3 \text{ arbitrary}$$

For a linear system containing m equations and n unknowns, with $m > n$, the following hold:

1. If the final tableau matrix contains an inconsistency, then there is *no solution*.
2. If the final tableau matrix has k rows with nonzero entries and $k = n$ (and there is no inconsistency), the there is a *unique solution*.
3. If the final tableau matrix has k rows with nonzero entries, where $k < n$, and the remaining $m - k$ rows contain zero entries (and there is no inconsistency), then there are *infinitely many solutions*. Furthermore, it is possible to solve for k variables in terms of the remaining $n - k$ variables (or arbitrary parameters).

281 LINEAR SYSTEMS AND MATRICES

In this section, we have discussed the Gauss–Jordan method of solving linear systems. This method provides the structural framework for understanding important concepts such as the matrix inverse, to be discussed in Section 5–5, and the simplex method, to be covered in Chapter 6. Another method, **Gaussian elimination with back-substitution,** is a more efficient method, especially for solving larger linear systems. This method is outlined in Exercise 8 at the end of this section. Of course, today, computers are utilized to solve large linear systems.

EXERCISES

1. Solve each of the following linear systems by using the Gauss–Jordan method of row operations:

   (A) $2x - 3y = 6$
   $x - 7y = 25$

   (B) $2x_1 + 3x_2 = 3$
   $12x_1 - 15x_2 = -4$

   (C) $2x + 3y - 5z = -13$
   $-x + 2y + 3z = -7$
   $3x - 4y - 7z = 15$

   (D) $x_1 + 3x_2 + x_3 = -3$
   $2x_1 + 9x_2 + 2x_3 = -5$
   $5x_1 + 48x_2 + 7x_3 = -16$

   (E) $2x + y + 3z = 11$
   $4x + 3y - 2z = -1$
   $6x + 5y - 4z = -4$

   (F) $x_1 + x_2 - 5x_3 = -3$
   $2x_1 + x_2 + 10x_3 = 2$
   $3x_1 + 2x_2 + 25x_3 = 3$

2. Solve each of the following linear systems by using the Gauss–Jordan method of row operations:

   (A) $x_1 + 10x_2 = 34$
   $3x_1 + 2x_2 = 18$

   (B) $2x - y = -7$
   $-x + 2y = 8$

   (C) $5x_1 + 7x_2 + x_3 = 1$
   $3x_1 + 2x_2 + 3x_3 = 8$
   $2x_1 + 3x_2 + 5x_3 = 19$

   (D) $3x + 7y + 2z = 2$
   $4x + 3y + 3z = 8$
   $x + 2y + 4z = -9$

3. Solve the following linear system by using the Gauss–Jordan method of row operations:

$$x_1 + 2x_2 + x_3 - x_4 = 6$$
$$x_2 - x_3 + 2x_4 = 4$$
$$4x_1 + 5x_3 - 3x_4 = 2$$
$$ 2x_3 + 52x_4 = 8$$

4. Solve the following linear system by using the Gauss–Jordan method of row operations:

$$x_1 + 2x_2 - x_3 + x_4 = -7$$
$$2x_1 + x_2 + x_3 + 2x_4 = 1$$
$$-3x_1 - x_2 + x_3 - x_4 = 3$$
$$x_1 + 2x_4 = -1$$

5. Some of the following linear systems have no solution, and some have infinitely many solutions. Try to solve each by the Gauss–Jordan method of row operations. If the system has no solution, then state so. If the system has infinitely many solutions, express the solutions parametrically.

(A) $3x - 5y = 8$
$-6x + 10y = 30$

(B) $2x_1 + 3x_2 = 7$
$-x_1 - 1.5x_2 = -4.5$

(C) $8x_1 - 2x_2 = 10$
$-4x_1 + x_2 = -5$

(D) $4x_1 - x_2 = 9$
$-12x_1 + 3x_2 = 36$

(E) $x_1 - 2x_2 + x_3 = 3$
$3x_1 - 7x_2 + 2x_3 = 4$
$-2x_1 + 4x_2 - 2x_3 = 8$

(F) $x_1 + 4x_2 + x_3 = 6$
$2x_1 + 9x_2 + 2x_3 = 8$
$3x_1 + 12x_2 + 3x_3 = 18$

(G) $x_1 + x_2 + x_3 = 4$
$x_1 + 2x_2 + 3x_3 = 2$
$2x_1 + 4x_2 + 6x_3 = 5$

(H) $x_1 - 2x_2 - 2x_3 = 4$
$-2x_1 + 4x_2 + 4x_3 = -8$
$x_1 + 3x_2 + 2x_3 = 4$

6. Using the Gauss–Jordan method, solve or attempt to solve each of the following linear systems. If the system has no solutions, then state so. If the system has infinitely many solutions, express them parametrically.

(A) $x_1 + 2x_2 + x_3 = 4$
$2x_1 + x_2 + 5x_3 = 6$

(B) $x_1 - 3x_2 + x_3 = 6$
$-2x_1 + 6x_2 - 2x_3 = 9$

(C) $x_1 + 2x_2 + x_3 = 4$
$-x_1 - x_2 + x_4 = 5$
$3x_1 + 6x_2 + x_3 + x_4 = 6$

(D) $x_1 + 2x_2 + x_3 + x_4 = 5$
$x_1 + 3x_2 + 2x_3 + x_4 = 6$
$2x_1 + 5x_2 + 2x_3 + x_4 = 8$

(E) $2x_1 + x_2 + x_3 - x_4 = 6$
$x_1 + x_2 + x_3 - x_4 = 8$
$-2x_1 - x_2 - x_3 + x_4 = 9$

(F) $x_1 + x_2 - x_3 + x_4 = 8$
$-x_1 + 2x_2 + 2x_3 + x_4 = -4$

(G) $x_1 + 2x_2 - x_3 - x_4 = 5$
$2x_1 + 5x_2 + x_3 - 2x_4 = 8$

7. Using the Gauss–Jordan method, solve or attempt to solve each of the following linear systems. If the system has a unique solution, then state the solution. If the system has no solutions, then state so. If the system has infinitely many solutions, express them parametrically.

(A) $x_1 + x_2 = 1$
$2x_1 + x_2 = 4$
$4x_1 + 2x_2 = 8$

(B) $x_1 + 2x_2 = 3$
$2x_1 - x_2 = 1$
$2x_1 + x_2 = 4$

(C) $x_1 + 3x_2 = 4$
$2x_1 + 6x_2 = 8$
$-3x_1 - 9x_2 = -12$

(D) $x_1 + x_2 = 4$
$2x_1 + 3x_2 = 1$
$x_1 - x_2 = 8$

(E) $2x_1 + x_2 + 3x_3 = 11$
$6x_1 + 5x_2 - 4x_3 = -4$
$4x_1 + 3x_2 - 2x_3 = -1$
$-6x_1 - 5x_2 + 4x_3 = 4$

(F) $x_1 + x_2 + 2x_3 = 4$
$2x_1 + 3x_2 + x_3 = 8$
$2x_1 + 2x_2 + 4x_3 = 8$
$-x_1 - x_2 - 2x_3 = -4$

8. A very efficient method for solving linear systems is **Gaussian elimination with back-substitution**. We now use this method to solve the linear system

$$x_1 + 3x_2 + 7x_3 = -17$$
$$3x_1 + 2x_2 + 5x_3 = -8$$
$$4x_1 + 5x_2 + 6x_3 = -1$$

LINEAR SYSTEMS AND MATRICES

The initial tableau is given by the augmented matrix

$$\begin{bmatrix} 1 & 3 & 7 & | & -17 \\ 3 & 2 & 5 & | & -8 \\ 4 & 5 & 6 & | & -1 \end{bmatrix}$$

(A) Using row operations, show that this initial tableau matrix may be transformed into

$$\begin{bmatrix} 1 & 3 & 7 & | & -17 \\ 0 & -7 & -16 & | & 43 \\ 0 & 0 & -6 & | & 24 \end{bmatrix}$$

Note the *triangular* form of the 0s here.

(B) Write the equivalent linear system represented by the matrix of part (A).

(C) Solve the third equation of the linear system of part (B) for x_3. Then, substitute this result into the second equation and solve for x_2.

(D) Substitute the results of part (C) for x_2 and x_3 into the first equation and solve for x_1. The solution to the linear system is $x_1 = 2$, $x_2 = 3$, $x_3 = -4$.

9. Using Gaussian elimination with back-substitution, solve each of the following:

(A) $\quad x + 3y = 5$
$\quad\;\; 4x + 14y = 13$

(B) $\quad x_1 + 2x_2 = 3$
$\quad\;\; 2x_1 + 5x_2 = 8$

(C) $\quad x + 2y + 4z = -9$
$\quad\;\; 3x + 7y + 2z = 2$
$\quad\;\; 4x + 3y + 3z = 8$

(D) $\quad x_1 + 3x_2 + x_3 = -3$
$\quad\;\; 2x_1 + 9x_2 + 2x_3 = -5$
$\quad\;\; 5x_1 + 48x_2 + 7x_3 = -16$

(E) $\quad x_1 + x_2 - 5x_3 = -3$
$\quad\;\; 2x_1 + x_2 + 10x_3 = 2$
$\quad\;\; 3x_1 + 2x_2 + 25x_3 = 3$

(F) $\quad x_1 + x_2 + x_3 = 13$
$\quad\;\; 2x_1 - x_2 + 5x_3 = 3$
$\quad\;\; 4x_1 - 2x_2 + 10x_3 = 6$

5-5

INVERSE OF A SQUARE MATRIX

The multiplicative inverse of a real number a is that number $1/a$ which when multiplied by a, results in the multiplicative identity, 1. Thus,

$$a \cdot \frac{1}{a} = \frac{1}{a} \cdot a = 1$$

Analogously, the multiplicative inverse (if it exists) of a square matrix A is that square matrix A^{-1} which when multiplied by A results in the identity matrix, I. Thus,

$$A \cdot A^{-1} = A^{-1} \cdot A = I$$

EXAMPLE 5–21 If

$$A = \begin{bmatrix} 2 & 3 \\ 5 & 4 \end{bmatrix}$$

then verify that

$$A^{-1} = \begin{bmatrix} -4/7 & 3/7 \\ 5/7 & -2/7 \end{bmatrix}$$

Solution

$$A \cdot A^{-1} = \begin{bmatrix} 2 & 3 \\ 5 & 4 \end{bmatrix} \begin{bmatrix} -4/7 & 3/7 \\ 5/7 & -2/7 \end{bmatrix} = \begin{bmatrix} 1 & 0 \\ 0 & 1 \end{bmatrix} = I$$

$$A^{-1} \cdot A = \begin{bmatrix} -4/7 & 3/7 \\ 5/7 & -2/7 \end{bmatrix} \begin{bmatrix} 2 & 3 \\ 5 & 4 \end{bmatrix} = \begin{bmatrix} 1 & 0 \\ 0 & 1 \end{bmatrix} = I$$

Thus,

$$A \cdot A^{-1} = A^{-1} \cdot A = I$$

We note that because of the way matrix multiplication is defined, only square matrices can have inverses.

We will now learn to compute the multiplicative inverse of an $n \times n$ matrix by a method involving row operations. This method involves the three fundamental row operations discussed in the preceding section, which may be performed on any row of a given matrix. These three basic row operations are repeated here:

1. Interchanging two rows of a matrix
2. Multiplying a row of a matrix by a nonzero constant
3. Multiplying a row of a matrix by a nonzero constant and adding the result to another row

Since A^{-1} satisfies the matrix equation $AX = I$, where I is the identity matrix of the same dimension as matrix A, then we may compute A^{-1} using the Gauss-Jordan method to solve the matrix equation $AX = I$ for X. That is, we begin with the augmented matrix $[A \mid I]$ and use row operations to obtain the augmented matrix $[I \mid A^{-1}]$, where A^{-1} is the solution to the matrix equation $AX = I$.

To illustrate the computation of a matrix inverse, we begin with the matrix

LINEAR SYSTEMS AND MATRICES

$$A = \begin{bmatrix} 2 & 3 \\ 1 & 4 \end{bmatrix}$$

The computation of A^{-1} will be carried out in tableau form. The initial tableau is formed by writing the identity matrix, I, next to matrix A, as follows:

Initial Tableau

$$\left[\begin{array}{cc|cc} 2 & 3 & 1 & 0 \\ 1 & 4 & 0 & 1 \end{array}\right]$$

This resultant augmented matrix is denoted $[A \mid I]$. The procedure is to apply row operations so that matrix A is transformed to the identity matrix, I. Once this is accomplished, the corresponding final tableau will contain A^{-1} in place of I on the right-hand side, as shown here:

Final Tableau

$$\left[\begin{array}{cc|c} 1 & 0 & A^{-1} \\ 0 & 1 & \end{array}\right]$$

We now compute A^{-1}. We begin by writing the initial tableau $[A \mid I]$:

Initial Tableau $[A \mid I]$

$$\left[\begin{array}{cc|cc} 2 & 3 & 1 & 0 \\ 1 & 4 & 0 & 1 \end{array}\right]$$

1. We need a coefficient of 1 in the upper left-hand corner. Thus, we interchange rows 1 and 2:

$$\left[\begin{array}{cc|cc} 1 & 4 & 0 & 1 \\ 2 & 3 & 1 & 0 \end{array}\right]$$

2. We need a 0 for the first element in row 2. Multiplying row 1 by -2 and adding the result to row 2 gives us a 0 in the lower left-hand corner:

$$\left[\begin{array}{cc|cc} 1 & 4 & 0 & 1 \\ 0 & -5 & 1 & -2 \end{array}\right]$$

3. We need a 1 for the second element in row 2. Multiplying row 2 by $-1/5$ gives us a 1 in this position:

$$\left[\begin{array}{cc|cc} 1 & 4 & 0 & 1 \\ 0 & 1 & -1/5 & 2/5 \end{array}\right]$$

286 CHAPTER FIVE

4. We need a 0 for the second element in row 1. Multiplying row 2 by −4 and adding the result to row 1 gives us a 0 in this position. We have now transformed matrix A into the identity matrix. Thus,

Final Tableau
$$\begin{bmatrix} 1 & 0 & | & 4/5 & -3/5 \\ 0 & 1 & | & -1/5 & 2/5 \end{bmatrix}$$
$$\underbrace{}_{A^{-1}}$$

A^{-1} appears on the right-hand side of the final tableau.

In general, to find the multiplicative inverse of an $n \times n$ matrix A, we begin with the initial tableau or augmented matrix, $[A \mid I]$:

$$\begin{bmatrix} a_{11} & a_{12} & \ldots & a_{1n} & | & 1 & 0 & \ldots & 0 \\ a_{21} & a_{22} & \ldots & a_{2n} & | & 0 & 1 & \ldots & 0 \\ \vdots & \vdots & & \vdots & | & \vdots & \vdots & & \vdots \\ a_{n1} & a_{n2} & \ldots & a_{nn} & | & 0 & 0 & \ldots & 1 \end{bmatrix}$$

and use row operations to obtain the final tableau or augmented matrix, $[I \mid A^{-1}]$:

$$\begin{bmatrix} 1 & 0 & \ldots & 0 & | & & & \\ 0 & 1 & \ldots & 0 & | & & & \\ \vdots & \vdots & & \vdots & | & & A^{-1} & \\ 0 & 0 & \ldots & 1 & | & & & \end{bmatrix}$$

EXAMPLE 5–22 If

$$A = \begin{bmatrix} 3 & -1 & 1 \\ 2 & 2 & 0 \\ 0 & 1 & 2 \end{bmatrix}$$

then compute A^{-1}.
Solution Initial Tableau, $[A \mid I]$

$$\begin{bmatrix} 3 & -1 & 1 & | & 1 & 0 & 0 \\ 2 & 2 & 0 & | & 0 & 1 & 0 \\ 0 & 1 & 2 & | & 0 & 0 & 1 \end{bmatrix}$$

1. Multiplying row 2 by −1 and adding the result to row 1 gives us a 1 in the upper left-hand corner:

$$\begin{bmatrix} 1 & -3 & 1 & | & 1 & -1 & 0 \\ 2 & 2 & 0 & | & 0 & 1 & 0 \\ 0 & 1 & 2 & | & 0 & 0 & 1 \end{bmatrix}$$

(*Note:* Instead of using this row operation, we could have obtained a 1 in the upper left-hand corner by multiplying row 1 by 1/3. However, this would have resulted in fractions appearing in the transformed row 1. Since fractions usually slow down the computation process, we will try to prevent their appearance whenever possible. We will, however, reach a point in this problem where fractions must be used.)

2. Multiplying row 1 by −2 and adding the result to row 2 gives us a 0 in column 1 of row 2:

$$\begin{bmatrix} 1 & -3 & 1 & | & 1 & -1 & 0 \\ 0 & 8 & -2 & | & -2 & 3 & 0 \\ 0 & 1 & 2 & | & 0 & 0 & 1 \end{bmatrix}$$

3. Since a 0 already appears in column 1 of row 3, we now select a row operation that will result in a 1 appearing in column 2 of row 2. Multiplying row 3 by −7 and adding the result to row 2 accomplishes our goal:

$$\begin{bmatrix} 1 & -3 & 1 & | & 1 & -1 & 0 \\ 0 & 1 & -16 & | & -2 & 3 & -7 \\ 0 & 1 & 2 & | & 0 & 0 & 1 \end{bmatrix}$$

(*Note:* We could have obtained a 1 in column 2 of row 2 by multiplying row 2 by 1/8. However, this would have resulted in the appearance of fractions.)

4. Multiplying row 2 by 3 and adding the result to row 1 gives us a 0 in column 2 of row 1:

$$\begin{bmatrix} 1 & 0 & -47 & | & -5 & 8 & -21 \\ 0 & 1 & -16 & | & -2 & 3 & -7 \\ 0 & 1 & 2 & | & 0 & 0 & 1 \end{bmatrix}$$

5. Multiplying row 2 by −1 and adding the result to row 3 gives us a 0 in column 2 of row 3:

$$\begin{bmatrix} 1 & 0 & -47 & | & -5 & 8 & -21 \\ 0 & 1 & -16 & | & -2 & 3 & -7 \\ 0 & 0 & 18 & | & 2 & -3 & 8 \end{bmatrix}$$

6. We need a 1 in column 3 of row 3. The only feasible row

EXAMPLE 5-22 (continued)

operation is to multiply row 3 by 1/18. We can no longer prevent the appearance of fractions in our tableau:

$$\begin{bmatrix} 1 & 0 & -47 & | & -5 & 8 & -21 \\ 0 & 1 & -16 & | & -2 & 3 & -7 \\ 0 & 0 & 1 & | & 1/9 & -1/6 & 4/9 \end{bmatrix}$$

7. Multiplying row 3 by 16 and adding the result to row 2 gives us a 0 in column 3 of row 2:

$$\begin{bmatrix} 1 & 0 & -47 & | & -5 & 8 & -21 \\ 0 & 1 & 0 & | & -2/9 & 1/3 & 1/9 \\ 0 & 0 & 1 & | & 1/9 & -1/6 & 4/9 \end{bmatrix}$$

8. Multiplying row 3 by 47 and adding the result to row 1 gives us a 0 in column 3 of row 1. This results in the final tableau:

Final Tableau, $[I \mid A^{-1}]$

$$\begin{bmatrix} 1 & 0 & 0 & | & 2/9 & 1/6 & -1/9 \\ 0 & 1 & 0 & | & -2/9 & 1/3 & 1/9 \\ 0 & 0 & 1 & | & 1/9 & -1/6 & 4/9 \end{bmatrix}$$
$$\underbrace{}_{A^{-1}}$$

Matrix Inverse May Not Exist

Not all square matrices have inverses. If, during the process of computing the inverse of a matrix A, a row consisting entirely of 0s appears in the left-hand side of the tableau, then the matrix A has no inverse. To illustrate this case, we will attempt to compute the inverse of matrix A where

Initial Tableau, $[A \mid I]$

$$A = \begin{bmatrix} 1 & 2 \\ 4 & 8 \end{bmatrix} \qquad \begin{bmatrix} 1 & 2 & | & 1 & 0 \\ 4 & 8 & | & 0 & 1 \end{bmatrix}$$

Since we already have a 1 in the upper left-hand corner, we multiply row 1 by -4 and add the result to row 2. This gives us a 0 in the lower left-hand corner:

$$\begin{bmatrix} 1 & 2 & | & 1 & 0 \\ 0 & 0 & | & -4 & 1 \end{bmatrix}$$

Observe that the left-hand side of row 2 consists entirely of 0s.

289 LINEAR SYSTEMS AND MATRICES

Such a situation results when the matrix A has no inverse. Thus, for this example, A^{-1} does not exist.

EXERCISES

1. Determine whether or not the following matrices are inverses of each other:

   (A) $\begin{bmatrix} 1 & -3/2 \\ 1 & -2 \end{bmatrix}$ and $\begin{bmatrix} 4 & -3 \\ 2 & -2 \end{bmatrix}$

   (B) $\begin{bmatrix} 4 & 1 \\ 3 & 0 \end{bmatrix}$ and $\begin{bmatrix} 1 & 2 \\ -1 & 4 \end{bmatrix}$

   (C) $\begin{bmatrix} 7 & -8 \\ 3 & -3 \end{bmatrix}$ and $\begin{bmatrix} -1 & 8/3 \\ -1 & 7/3 \end{bmatrix}$

   (D) $\begin{bmatrix} 1 & 3 & 0 \\ 0 & 1 & 0 \\ 1 & 2 & 1 \end{bmatrix}$ and $\begin{bmatrix} 1 & -3 & 0 \\ 0 & 1 & 0 \\ 0 & -2 & 1 \end{bmatrix}$

2. Determine whether or not the following matrices are inverses of each other:

   (A) $\begin{bmatrix} 5 & 6 \\ 3 & 4 \end{bmatrix}$ and $\begin{bmatrix} 2 & -3 \\ -3/2 & 5/2 \end{bmatrix}$

   (B) $\begin{bmatrix} 1 & 0 & 0 \\ 0 & 1 & 0 \\ 2 & 3 & 1 \end{bmatrix}$ and $\begin{bmatrix} 1 & 0 & 0 \\ 0 & 1 & 0 \\ -2 & -3 & 1 \end{bmatrix}$

   (C) $\begin{bmatrix} 1 & 2 \\ 5 & -1 \end{bmatrix}$ and $\begin{bmatrix} 8 & 0 \\ 4 & 1 \end{bmatrix}$

   (D) $\begin{bmatrix} 1 & 3 & 2 \\ 0 & 1 & 4 \\ 0 & 0 & 1 \end{bmatrix}$ and $\begin{bmatrix} 1 & -3 & 10 \\ 0 & 1 & -4 \\ 0 & 0 & 1 \end{bmatrix}$

3. Given that $A = \begin{bmatrix} 2 & 5 \\ -8 & 11 \end{bmatrix}$.

   (A) Compute A^{-1}.
   (B) Verify that $AA^{-1} = A^{-1}A = I$.

4. Given that $B = \begin{bmatrix} 1 & 4 & 5 \\ 0 & 1 & 3 \\ 0 & 1 & 4 \end{bmatrix}$.

   (A) Compute B^{-1}.
   (B) Verify that $BB^{-1} = B^{-1}B = I$.

5. Find the inverse of each of the following matrices:

   (A) $\begin{bmatrix} 5 & -1 \\ -3 & 7 \end{bmatrix}$ (B) $\begin{bmatrix} 1 & 3 \\ 2 & -4 \end{bmatrix}$ (C) $\begin{bmatrix} 0 & 1 \\ 1 & 1 \end{bmatrix}$

   (D) $\begin{bmatrix} 2 & -1 & 3 \\ 3 & 2 & -4 \\ 4 & 2 & -5 \end{bmatrix}$ (E) $\begin{bmatrix} 3 & 2 & 1 \\ 4 & -3 & 2 \\ 2 & 4 & -3 \end{bmatrix}$ (F) $\begin{bmatrix} -1 & 1 & -2 \\ -2 & 0 & -4 \\ 6 & 2 & 10 \end{bmatrix}$

6. Find the inverse of each of the following matrices:

(A) $\begin{bmatrix} 3 & 4 \\ 2 & -7 \end{bmatrix}$ (B) $\begin{bmatrix} 1 & 3 \\ 2 & 4 \end{bmatrix}$ (C) $\begin{bmatrix} 2 & 3 \\ 4 & -1 \end{bmatrix}$

(D) $\begin{bmatrix} 1 & 1 & 2 \\ 3 & -1 & 3 \\ 2 & -5 & 2 \end{bmatrix}$ (E) $\begin{bmatrix} 1 & 2 & 1 \\ 4 & 1 & 0 \\ 0 & 0 & 1 \end{bmatrix}$ (F) $\begin{bmatrix} 1 & 2 & 2 \\ 8 & -6 & 2 \\ 8 & 4 & 4 \end{bmatrix}$

7. Given that $K = \begin{bmatrix} 2 & 3 \\ -10 & -15 \end{bmatrix}$, try to find K^{-1}. Does K^{-1} exist?

8. Given that $H = \begin{bmatrix} 1 & 2 & -1 \\ 2 & 4 & 3 \\ -2 & -4 & 2 \end{bmatrix}$, try to find H^{-1}. Does H^{-1} exist?

9. Find the inverse, if it exists, of each of the following matrices:

(A) $\begin{bmatrix} -1 & 1 & 1 & 0 \\ 0 & 0 & 0 & 2 \\ 3 & 3 & 0 & 0 \\ 4 & 2 & 2 & 0 \end{bmatrix}$ (B) $\begin{bmatrix} 1 & 2 & 1 & 0 \\ 2 & 2 & 0 & 4 \\ 6 & -3 & 3 & -3 \\ 1 & 1 & 0 & 2 \end{bmatrix}$

5–6

SOLVING SQUARE LINEAR SYSTEMS BY MATRIX INVERSES

In order to solve the linear equation

$$ax = b$$

for x, we multiply both sides by $1/a$, the multiplicative inverse of a, to obtain

$$\left(\frac{1}{a}\right)ax = \left(\frac{1}{a}\right)b$$

$$x = \left(\frac{1}{a}\right)b$$

Analogously, we may use the multiplicative inverse of a matrix to solve linear systems. Consider the linear system

$$2x_1 + 3x_2 = 9$$
$$5x_1 + 4x_2 = 26$$

Expressing this system in matrix form $AX = B$, we have

$$\begin{bmatrix} 2 & 3 \\ 5 & 4 \end{bmatrix} \begin{bmatrix} x_1 \\ x_2 \end{bmatrix} = \begin{bmatrix} 9 \\ 26 \end{bmatrix}$$

If we multiply both sides of the matrix equation $AX = B$ by A^{-1} (assuming that A^{-1} exists), we have

LINEAR SYSTEMS AND MATRICES

$$A^{-1}(AX) = A^{-1}B$$

Since $A^{-1}(AX) = (A^{-1}A)X$ by the associative property and $A^{-1}A = I$, the left-hand side becomes IX, and the equation reads

$$IX = A^{-1}B$$

Since $IX = X$, we obtain

$$X = A^{-1}B$$

Thus, the solution to a matrix equation $AX = B$ is

$$X = A^{-1}B$$

Returning to our example,

$$\begin{bmatrix} 2 & 3 \\ 5 & 4 \end{bmatrix} \begin{bmatrix} x_1 \\ x_2 \end{bmatrix} = \begin{bmatrix} 9 \\ 26 \end{bmatrix}$$

we must find A^{-1} by beginning with the augmented matrix or initial tableau, $[A \mid I]$:

$$\begin{bmatrix} 2 & 3 & | & 1 & 0 \\ 5 & 4 & | & 0 & 1 \end{bmatrix}$$

and using row operations to obtain the final tableau, $[I \mid A^{-1}]$:

$$\begin{bmatrix} 1 & 0 & | & -4/7 & 3/7 \\ 0 & 1 & | & 5/7 & -2/7 \end{bmatrix}$$

Hence,

$$A^{-1} = \begin{bmatrix} -4/7 & 3/7 \\ 5/7 & -2/7 \end{bmatrix}$$

Next, we find the product $A^{-1}B$:

$$A^{-1}B = \begin{bmatrix} -4/7 & 3/7 \\ 5/7 & -2/7 \end{bmatrix} \begin{bmatrix} 9 \\ 26 \end{bmatrix} = \begin{bmatrix} 6 \\ -1 \end{bmatrix}$$

Thus, the solution is

$$X = A^{-1}B$$
$$\begin{bmatrix} x_1 \\ x_2 \end{bmatrix} = \begin{bmatrix} 6 \\ -1 \end{bmatrix}$$

Therefore, $(6, -1)$ is the solution to our linear system.

EXAMPLE 5–23 Use the matrix inverse to solve the linear system

$$x_1 + 3x_2 + 3x_3 = 4$$
$$2x_1 + 7x_2 + 7x_3 = 9$$
$$2x_1 + 7x_2 + 6x_3 = 10$$

Solution Rewriting the linear system in matrix form $AX = B$, we have

$$\begin{bmatrix} 1 & 3 & 3 \\ 2 & 7 & 7 \\ 2 & 7 & 6 \end{bmatrix} \begin{bmatrix} x_1 \\ x_2 \\ x_3 \end{bmatrix} = \begin{bmatrix} 4 \\ 9 \\ 10 \end{bmatrix}$$

Hence,

$$A = \begin{bmatrix} 1 & 3 & 3 \\ 2 & 7 & 7 \\ 2 & 7 & 6 \end{bmatrix} \quad X = \begin{bmatrix} x_1 \\ x_2 \\ x_3 \end{bmatrix} \quad B = \begin{bmatrix} 4 \\ 9 \\ 10 \end{bmatrix}$$

We must find A^{-1} by beginning with the initial tableau, $[A \mid I]$:

$$\begin{bmatrix} 1 & 3 & 3 & | & 1 & 0 & 0 \\ 2 & 7 & 7 & | & 0 & 1 & 0 \\ 2 & 7 & 6 & | & 0 & 0 & 1 \end{bmatrix}$$

and using row operations to obtain the final tableau, $[I \mid A^{-1}]$:

$$\begin{bmatrix} 1 & 0 & 0 & | & 7 & -3 & 0 \\ 0 & 1 & 0 & | & -2 & 0 & 1 \\ 0 & 0 & 1 & | & 0 & 1 & -1 \end{bmatrix}$$

Hence,

$$A^{-1} = \begin{bmatrix} 7 & -3 & 0 \\ -2 & 0 & 1 \\ 0 & 1 & -1 \end{bmatrix}$$

We now find the product $A^{-1}B$:

$$A^{-1}B = \begin{bmatrix} 7 & -3 & 0 \\ -2 & 0 & 1 \\ 0 & 1 & -1 \end{bmatrix} \begin{bmatrix} 4 \\ 9 \\ 10 \end{bmatrix} = \begin{bmatrix} 1 \\ 2 \\ -1 \end{bmatrix}$$

LINEAR SYSTEMS AND MATRICES

Thus, the solution is

$$X = A^{-1}B$$

$$\begin{bmatrix} x_1 \\ x_2 \\ x_3 \end{bmatrix} = \begin{bmatrix} 1 \\ 2 \\ -1 \end{bmatrix}$$

Therefore, $(1, 2, -1)$ is the solution to our linear system.

The method of solving a square linear system $AX = B$ by using A^{-1} (if A^{-1} exists) is relatively inefficient for large square linear systems unless A^{-1} is known beforehand or can easily be determined by using a computer. However, in this age of "cheap computing," most students have access to a computer. Under these circumstances, this method is advantageous for solving square linear systems $AX = B$, especially when matrix B changes and matrix A does not.

EXERCISES

1. Express each of the following linear systems in matrix form $AX = B$. Then, compute A^{-1} and use it to solve the linear system.

   (A) $2x - 3y = 6$
   $x - 7y = 25$

   (B) $4x_1 - 5x_2 = -2$
   $3x_1 + 2x_2 = -13$

   (C) $2x + 3y - 5z = -13$
   $-x + 2y + 3z = -7$
   $3x - 4y - 7z = 15$

   (D) $2x_1 - 3x_2 + 4x_3 = 8$
   $3x_1 + x_2 - 2x_3 = 11$
   $5x_1 - 2x_2 + 3x_3 = 10$

   (E) $5x_1 + 7x_2 + x_3 = 1$
   $3x_1 + 2x_2 + 3x_3 = 8$
   $2x_1 + 3x_2 + 5x_3 = 19$

   (F) $3x + 7y + 2z = 2$
   $4x + 3y + 3z = 8$
   $x + 2y + 4z = -9$

2. Express each of the following linear systems in matrix form $AX = B$. Then, compute A^{-1} and use it to solve the linear system.

   (A) $4x_1 + x_2 = 8$
   $6x_1 - 2x_2 = -9$

   (B) $\frac{1}{2}x + 5y = 17$
   $3x + 2y = 18$

   (C) $2x + y + 3z = 11$
   $4x + 3y - 2z = -1$
   $6x + 5y - 4z = -4$

   (D) $x_1 + x_2 - 5x_3 = -3$
   $2x_1 + x_2 + 10x_3 = 2$
   $3x_1 + 2x_2 + 25x_3 = 3$

3. Express the following linear system in matrix form $AX = B$. Then, compute A^{-1} and use it to solve the linear system.

$$x_1 + 2x_2 + x_3 - x_4 = 6$$
$$x_2 - x_3 + 2x_4 = 4$$
$$4x_1 + 5x_3 - 3x_4 = 2$$
$$2x_3 + 52x_4 = 8$$

4. Express the following linear system in matrix form $AX = B$. Then, compute A^{-1} and use it to solve the linear system.

$$x_1 + 2x_2 - x_3 + x_4 = -7$$
$$2x_1 + x_2 + x_3 + 2x_4 = 1$$
$$-3x_1 - x_2 + x_3 - x_4 = 3$$
$$x_1 + 2x_4 = -1$$

5–7

APPLICATIONS

In this section, we will illustrate some useful applications of matrices and their inverses.

Product-Mix Problem

In Section 5–1, we discussed the time requirements for the manufacture of bicycles and tricycles by the Ruszala Company. We now reconsider this problem using matrices.

The time requirements (in hours) of each bicycle and tricycle in Departments I and II are summarized by matrix A, as shown here:

$$A = \begin{matrix} \text{Bicycles} & \text{Tricycles} \\ \begin{bmatrix} 3 & 4 \\ 5 & 2 \end{bmatrix} & \begin{matrix} \text{Department I} \\ \text{Department II} \end{matrix} \end{matrix}$$

Matrix B represents the time (in hours) available in each department. Remember that all this time must be used.

$$B = \begin{bmatrix} 450 \\ 400 \end{bmatrix} \begin{matrix} \text{Department I} \\ \text{Department II} \end{matrix}$$

Thus, we must determine how many bicycles and tricycles to produce in order to satisfy the time requirements of each department.

If the number of bicycles and tricycles to be manufactured is represented by the matrix

$$X = \begin{bmatrix} x_1 \\ x_2 \end{bmatrix} \begin{matrix} \text{Number of bicycles} \\ \text{Number of tricycles} \end{matrix}$$

LINEAR SYSTEMS AND MATRICES

then X satisfies the matrix equation

$$AX = B$$
$$\begin{bmatrix} 3 & 4 \\ 5 & 2 \end{bmatrix} \begin{bmatrix} x_1 \\ x_2 \end{bmatrix} = \begin{bmatrix} 450 \\ 400 \end{bmatrix}$$

Using row operations, it can be determined that

$$A^{-1} = \begin{bmatrix} -1/7 & 2/7 \\ 5/14 & -3/14 \end{bmatrix}$$

Solving for X yields

$$X = A^{-1}B$$
$$\begin{bmatrix} x_1 \\ x_2 \end{bmatrix} = \begin{bmatrix} -1/7 & 2/7 \\ 5/14 & -3/14 \end{bmatrix} \begin{bmatrix} 450 \\ 400 \end{bmatrix} = \begin{bmatrix} 50 \\ 75 \end{bmatrix}$$

Thus, the company should produce

$$x_1 = 50 \text{ bicycles}$$
$$x_2 = 75 \text{ tricycles}$$

The advantage of using A^{-1} to solve this problem becomes apparent when we consider the following change. Suppose matrix B, which represents the time available in each department, changes from

$$\begin{bmatrix} 450 \\ 400 \end{bmatrix} \quad \text{to} \quad \begin{bmatrix} 490 \\ 280 \end{bmatrix}$$

Then, since

$$X = A^{-1}B$$

the new solution X may be determined by multiplying A^{-1} times the new matrix B. Since matrix A is unchanged, it is not necessary to recalculate A^{-1}. For example, if matrix B is changed to

$$B = \begin{bmatrix} 490 \\ 280 \end{bmatrix}$$

then

$$X = A^{-1}B$$
$$\begin{bmatrix} x_1 \\ x_2 \end{bmatrix} = \begin{bmatrix} -1/7 & 2/7 \\ 5/14 & -3/14 \end{bmatrix} \begin{bmatrix} 490 \\ 280 \end{bmatrix} = \begin{bmatrix} 10 \\ 115 \end{bmatrix}$$

Thus, the company should produce

$$x_1 = 10 \text{ bicycles}$$
$$x_2 = 115 \text{ tricycles}$$

Leontief's Input–Output Model

An interesting application of matrices in economics is **Leontief's input–output model**. Consider a simplified economy which produces n commodities: $C_1, C_2, \ldots, C_n$. Each commodity is used in the production of the other commodities. Specifically, let us take an economy with $n = 3$ commodities: C_1, C_2, and C_3. Production of 1 unit of C_1 requires 0 units of C_1, 1/2 unit of C_2, and 1/4 unit of C_3. This is summarized by the following row vector:

$$\begin{array}{c} \phantom{\text{Producing 1 unit of}} \begin{array}{ccc} C_1 & C_2 & C_3 \end{array} \\ \text{Producing 1 unit of } \begin{bmatrix} 0 & 1/2 & 1/4 \end{bmatrix} \\ C_1 \text{ requires} \end{array}$$

The row vectors specifying production requirements for C_2 and C_3 are as follows:

$$\begin{array}{c} \phantom{\text{Producing 1 unit of}} \begin{array}{ccc} C_1 & C_2 & C_3 \end{array} \\ \text{Producing 1 unit of } \begin{bmatrix} 1/6 & 0 & 1/3 \end{bmatrix} \\ C_2 \text{ requires} \end{array}$$

$$\begin{array}{c} \phantom{\text{Producing 1 unit of}} \begin{array}{ccc} C_1 & C_2 & C_3 \end{array} \\ \text{Producing 1 unit of } \begin{bmatrix} 1/2 & 1/8 & 0 \end{bmatrix} \\ C_3 \text{ requires} \end{array}$$

If we write the second row vector beneath the first, and the third below the second, we obtain the matrix

$$A = \begin{matrix} & \begin{matrix} C_1 & C_2 & C_3 \end{matrix} & \\ & \begin{bmatrix} 0 & 1/2 & 1/4 \\ 1/6 & 0 & 1/3 \\ 1/2 & 1/8 & 0 \end{bmatrix} & \begin{matrix} \text{Requirements for 1 unit of } C_1 \\ \text{Requirements for 1 unit of } C_2 \\ \text{Requirements for 1 unit of } C_3 \end{matrix} \end{matrix}$$

Matrix A is called the **technological matrix** of the economy. If
$$x_1 = \text{number of units of } C_1 \text{ produced}$$
$$x_2 = \text{number of units of } C_2 \text{ produced}$$
$$x_3 = \text{number of units of } C_3 \text{ produced}$$

then this is summarized by the 1×3 matrix

$$X = [x_1 \quad x_2 \quad x_3]$$

297 LINEAR SYSTEMS AND MATRICES

Thus, X is the **gross production matrix**. However, a portion of each of these products is consumed in the production of the others. The product matrix XA represents these amounts consumed and is illustrated as follows:

$$XA = \begin{bmatrix} \underset{\underset{\text{of }C_1}{\downarrow}}{\text{Units}} & \underset{\underset{\text{of }C_2}{\downarrow}}{\text{Units}} & \underset{\underset{\text{of }C_3}{\downarrow}}{\text{Units}} \\ x_1 & x_2 & x_3 \end{bmatrix} \begin{bmatrix} 0 & 1/2 & 1/4 \\ 1/6 & 0 & 1/3 \\ 1/2 & 1/8 & 0 \end{bmatrix} \begin{matrix} \text{Requirements for 1 unit } C_1 \\ \text{Requirements for 1 unit } C_2 \\ \text{Requirements for 1 unit } C_3 \end{matrix}$$

$$= \begin{bmatrix} \underbrace{\frac{1}{6}x_2 + \frac{1}{2}x_3}_{\substack{\text{Units of} \\ C_1 \text{ consumed}}} & \underbrace{\frac{1}{2}x_1 + \frac{1}{8}x_3}_{\substack{\text{Units of} \\ C_2 \text{ consumed}}} & \underbrace{\frac{1}{4}x_1 + \frac{1}{3}x_2}_{\substack{\text{Units of} \\ C_3 \text{ consumed}}} \end{bmatrix}$$

Thus, XA is the **consumption matrix**. The **net production** (gross production minus consumption) is represented by the matrix

$$Y = \underset{\underset{\text{production}}{\text{gross}}}{X} - \underset{\text{consumption}}{XA}$$

Since

$$X - XA = X(I - A) \quad \text{(See Exercise 17, Section 5–3)}$$

then

$$Y = X(I - A)$$

The basic goal of input–output analysis is to determine the gross production, X, necessary for a given net production, Y. Thus, the preceding equation may be solved for X to yield

$$X = Y(I - A)^{-1}$$

if $(I - A)^{-1}$ exists.

Returning to our example,

$$I - A = \begin{bmatrix} 1 & 0 & 0 \\ 0 & 1 & 0 \\ 0 & 0 & 1 \end{bmatrix} - \begin{bmatrix} 0 & 1/2 & 1/4 \\ 1/6 & 0 & 1/3 \\ 1/2 & 1/8 & 0 \end{bmatrix} = \begin{bmatrix} 1 & -1/2 & -1/4 \\ -1/6 & 1 & -1/3 \\ -1/2 & -1/8 & 1 \end{bmatrix}$$

Using row operations, it may be determined that

$$(I - A)^{-1} = \begin{bmatrix} \frac{184}{127} & \frac{102}{127} & \frac{80}{127} \\ \frac{64}{127} & \frac{168}{127} & \frac{72}{127} \\ \frac{100}{127} & \frac{72}{127} & \frac{176}{127} \end{bmatrix}$$

Thus, if a net production of

$$Y = [\underset{\substack{\uparrow \\ \text{Units} \\ \text{of } C_1}}{254} \quad \underset{\substack{\uparrow \\ \text{Units} \\ \text{of } C_2}}{127} \quad \underset{\substack{\uparrow \\ \text{Units} \\ \text{of } C_3}}{381}]$$

is desired, then a gross production of

$$X = Y(I - A)^{-1}$$

$$= [254 \quad 127 \quad 381] \begin{bmatrix} \frac{184}{127} & \frac{102}{127} & \frac{80}{127} \\ \frac{64}{127} & \frac{168}{127} & \frac{72}{127} \\ \frac{100}{127} & \frac{72}{127} & \frac{176}{127} \end{bmatrix}$$

$$= [\underset{\substack{\uparrow \\ \text{Units} \\ \text{of } C_1}}{732} \quad \underset{\substack{\uparrow \\ \text{Units} \\ \text{of } C_2}}{588} \quad \underset{\substack{\uparrow \\ \text{Units} \\ \text{of } C_3}}{760}]$$

is required.

Fitting a Parabola to Given Points Another problem whose solution involves a linear system is that of finding the equation of a parabola passing through three given points. Example 5–24 illustrates this problem and its solution.

> **EXAMPLE 5–24** Find the equation of the parabola passing through the points (1, 9), (4, 6), and (6, 14).
> *Solution* The general form of the equation of the parabola is
>
> $$y = ax^2 + bx + c$$
>
> If we can determine the values of a, b, and c, then we will have the equation of the parabola. Substituting the coor-

299 LINEAR SYSTEMS AND MATRICES

dinates of each point into the general form, we have the linear system

$$9 = a(1)^2 + b(1) + c$$
$$6 = a(4)^2 + b(4) + c$$
$$14 = a(6)^2 + b(6) + c$$

Simplifying, we obtain

$$a + b + c = 9$$
$$16a + 4b + c = 6$$
$$36a + 6b + c = 14$$

Solving the system by the Gauss–Jordan method yields the final tableau,

$$\begin{bmatrix} 1 & 0 & 0 & | & 1 \\ 0 & 1 & 0 & | & -6 \\ 0 & 0 & 1 & | & 14 \end{bmatrix}$$

Hence, $a = 1$, $b = -6$, and $c = 14$, and the equation of the parabola is $y = x^2 - 6x + 14$.

EXERCISES

1. A diet must provide exactly 1200 milligrams of protein and 1000 milligrams of iron. These nutrients will be obtained by eating meat and spinach. Each pound of meat contains 500 milligrams of protein and 100 milligrams of iron. Each pound of spinach contains 200 milligrams of protein and 800 milligrams of iron.

   (A) If matrix A lists the number of milligrams of protein and iron obtained from a pound of meat and spinach, respectively, then fill in the elements of matrix A:

   $$A = \begin{bmatrix} \text{Meat} & \text{Spinach} \\ — & — \\ — & — \end{bmatrix} \begin{matrix} \text{Protein} \\ \text{Iron} \end{matrix}$$

   (B) If matrix B lists the amounts required of protein and iron, respectively, then fill in the elements of matrix B:

   $$B = \begin{bmatrix} — \\ — \end{bmatrix} \begin{matrix} \text{Protein} \\ \text{Iron} \end{matrix}$$

   (C) If

   $$X = \begin{bmatrix} x_1 \\ x_2 \end{bmatrix}$$

where x_1 and x_2 represent the number of pounds of meat and spinach, respectively, that should be eaten in order to provide the proper amounts of protein and iron, then write the matrix equation relating A, X, and B.

(D) Solve the matrix equation of part (C) for X and interpret the answer.

2. A company manufactures three products: A, B, and C. Each product must pass through three machines: I, II, and III. Each unit of Product A requires 3 hours on I, 2 hours on II, and 4 hours on III. Each unit of Product B requires 2 hours on I, 4 hours on II, and 6 hours on III. Each unit of Product C requires 3 hours on I, 5 hours on II, and 7 hours on III.

(A) If matrix T shows the time requirements of each product on each machine, then fill in the elements of matrix T:

$$T = \begin{bmatrix} & & \\ & & \\ & & \end{bmatrix} \begin{matrix} \text{I} \\ \text{II} \\ \text{III} \end{matrix} \quad \begin{matrix} \text{A} & \text{B} & \text{C} \end{matrix}$$

(B) If Machines I, II, and III have available 150, 240, and 360 hours, respectively, then express this information in matrix B:

$$B = \begin{bmatrix} \\ \\ \end{bmatrix} \begin{matrix} \text{I} \\ \text{II} \\ \text{III} \end{matrix}$$

(C) If

$$X = \begin{bmatrix} x_1 \\ x_2 \\ x_3 \end{bmatrix}$$

where x_1, x_2, and x_3 represent the numbers of units of Products A, B, and C, respectively, produced, then write the matrix equation relating T, X, and B.

(D) Solve the matrix equation of part (C) and interpret the answer.

3. A primitive economy has only two commodities: oil and coal. Production of 1 barrel of oil requires 1/2 ton of coal. Production of 1 ton of coal requires 1/4 barrel of oil.

(A) If A is the technological matrix of this economy, then fill in the elements of A:

$$A = \begin{bmatrix} & \\ & \end{bmatrix} \begin{matrix} \text{Requirements for 1 barrel of oil} \\ \text{Requirements for 1 ton of coal} \end{matrix} \quad \begin{matrix} \text{Oil} & \text{Coal} \end{matrix}$$

(B) If matrix Y shows the desired net production of oil and coal,

$$Y = [\ 210 \quad 490\ ]$$
$$\quad\quad \uparrow \quad\quad \uparrow$$
$$\quad\text{Barrels}\ \ \text{Tons}$$
$$\quad\text{of oil}\ \ \text{of coal}$$

then find the required gross production.

4. An economy has three commodities: C_1, C_2, and C_3. Production of 1 unit of C_1 requires 1/4 unit of C_2. Production of 1 unit of C_2 requires 1/2 unit of C_1 and 1/3 unit of C_3. Production of 1 unit of C_3 requires 1/4 unit of C_1 and 1/2 unit of C_2.
   (A) Write the technological matrix for this economy.
   (B) If matrix Y shows the desired net production of each commodity,

$$Y = [\ 231 \quad 462 \quad 924\ ]$$
$$\quad\quad \uparrow \quad\quad \uparrow \quad\quad \uparrow$$
$$\quad\text{Units}\ \ \text{Units}\ \ \text{Units}$$
$$\quad\text{of } C_1\ \ \text{of } C_2\ \ \text{of } C_3$$

then find the required gross production.

5. Find the equation of the parabola passing through the points (1, 4), (5, 40), and (3, 14).
6. Find the equation of the parabola passing through the points (2, −1), (3, −3), and (−2, 27).
7. A company manufacturing candy wishes to determine the quadratic equation expressing the relationship between profit and the number of boxes of candy produced. The company has accumulated the following data:

x (Number of Boxes)	y (Profit)
3	2
4	8
5	10

Find the quadratic equation.

CASE C OIL REFINERY SCHEDULING— MERCO OIL REFINERY

The Merco Oil Refinery owns three oil wells. Oil from the well in Saudi Arabia is refined into 0.2 million barrels of regular gasoline, 0.1 million barrels of unleaded gasoline, and 0.3 million barrels of kerosene each day. Oil from the well in Kuwait is refined into 0.3 million barrels of regular gasoline, 0.2 million barrels of unleaded gasoline, and 0.1 million barrels of kerosene each day. Oil from the well in Egypt is refined into 0.4 million barrels of regular gasoline, 0.1 million barrels of unleaded gasoline, and 0.4 million barrels of kerosene each day. The company needs to produce 19 million barrels of regular gasoline, 10 million barrels of unleaded gasoline, and 20 million barrels of kerosene to meet demand requirements. Assuming that adequate lead time for transportation assures a continual flow, how many days should each well be operated in order to meet the demand requirements?

EXERCISES

1. Complete the following table, which summarizes the preceding information:

	Well in Saudi Arabia	Well in Kuwait	Well in Egypt	Demand Requirements
Regular	(___)	(___)	(___)	(___)
Unleaded	(___)	(___)	(___)	(___)
Kerosene	(___)	(___)	(___)	(___)

2. Define each decision variable and write the linear system of equations for this problem.
3. Write the linear system in matrix form $AX = B$.
4. Determine A^{-1} and solve the linear system.
5. Suppose the demand requirements are changed to 20 million barrels of regular gasoline, 10 million barrels of unleaded gasoline, and 15 million barrels of kerosene. How many days should each well be operated in order to meet the demand requirements?

LINEAR PROGRAMMING

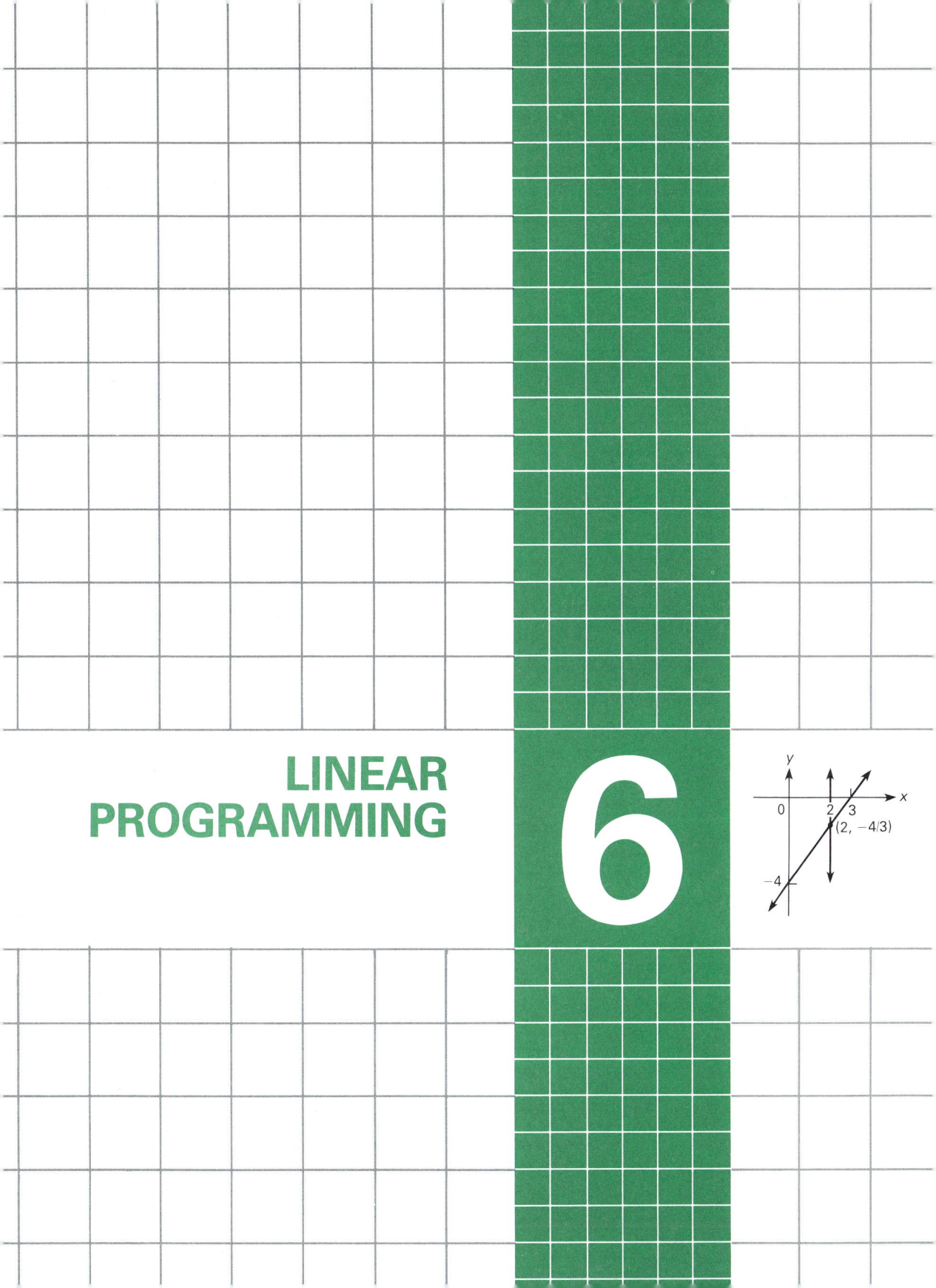

6

6-1

ALGEBRA REFRESHER

Linear Inequalities in One Variable

A statement such as

$$3x + 5 = 17$$

is a linear equation (equality) in one variable. To find its solution, we first subtract 5 from both sides to obtain

$$3x = 12$$

Then, we divide both sides by 3 to obtain the solution

$$x = 4$$

The solution is sketched on the real number line in Figure 6–1.

Solution: $x = 4$

FIGURE 6–1

When solving linear equations such as $3x + 5 = 17$, we use the following rules of equalities.

Rules of Equalities

Rule 1 If the same number is either added to or subtracted from both sides of an equality, the resulting equality remains true.

Rule 2 If both sides of an equality are either multiplied by or divided by the same nonzero number, the resulting equality remains true.

We now consider linear inequalities. In general, if the equal sign (=) of a linear equality such as

$$3x + 5 = 7$$

is replaced by an inequality sign ($<, >, \leq, \geq$), the resulting statement is a linear inequality. Thus, the statements

$$3x + 5 < 7$$
$$3x + 5 > 7$$
$$3x + 5 \leq 7$$
$$3x + 5 \geq 7$$

are examples of linear inequalities. To solve linear inequalities, we may use the following rules of inequalities.

Rules of Inequalities

Rule 1 If the same number is either added to or subtracted from both sides of an inequality, the resulting inequality is true.

Rule 2 (A) If both sides of an inequality are either multiplied by or divided by the same *positive* number, the resulting inequality is true.

(B) If both sides of an inequality are either multiplied by or divided by the same *negative* number, the original inequality sign must be *reversed* in order for the resulting inequality to be true.

Note that the rules of equalities also hold for inequalities with the exception involving either multiplication by or division by a negative number. Thus, if

$$2 < 5$$

then

$$-4(2) > -4(5)$$
$$-8 > -20$$

EXAMPLE 6-1 Solve the inequality $-5x + 3 \leq 13$ for x and sketch the solution on a real number line.

Solution We first subtract 3 from both sides (rule 1) to obtain

$$-5x \leq 10$$

Then we divide both sides by -5 (rule 2(B)) to get

$$x \geq -2$$

The solution is sketched in Figure 6-2.

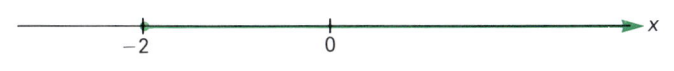

$x \geq -2$

FIGURE 6-2

EXAMPLE 6-2 Solve the inequality $3x + 5 < 17$ for x and sketch the solution on a real number line.

EXAMPLE 6–2 (continued)

Solution Subtracting 5 from both sides (rule 1), we obtain

$$3x < 12$$

Dividing both sides by 3 (rule 2(A)) yields

$$x < 4$$

The solution is sketched in Figure 6–3.

$$x < 4$$

FIGURE 6–3

EXAMPLE 6–3 Solve the inequality $(-1/2)x + 3 \geq -1$ for x and sketch the solution on a real number line.

Solution Subtracting 3 from both sides (rule 1) gives us

$$-\frac{1}{2}x \geq -4$$

Multiplying both sides by -2 (rule 2(B)) yields

$$x \leq 8$$

The solution is sketched in Figure 6–4.

$$x \leq 8$$

FIGURE 6–4

EXERCISES Solve each of the following inequalities and sketch its solution on a real number line:

1. $2x + 4 \leq 15$
2. $-3x + 5 \leq 32$
3. $4x - 5 < 25$
4. $5x + 3 > 17$
5. $-3x + 17 \geq -14$
6. $-6x + 5 > 23$
7. $-6x - 5 \geq -23$
8. $-3x - 2 \leq -14$
9. $3(x - 5) \geq 18$
10. $-4(x + 7) < 32$

LINEAR PROGRAMMING

11. $-\dfrac{1}{4}x + 11 \leq -20$

12. $\dfrac{1}{3}x - 3 \geq 5$

13. $-\dfrac{2}{5}x - 1 > -4$

14. $\dfrac{4}{3}x + 1 < 9$

6–2

LINEAR INEQUALITIES IN TWO VARIABLES

Linear Equalities

A statement such as

$$2x - 3y = 12$$

is a linear equation (equality) in two variables. When graphed, a linear equation results in a straight line. Recall that a linear equation is usually graphed by finding its x-intercept and y-intercept. Thus, our equation may be graphed as follows. First, we find the x-intercept (__, 0) by setting $y = 0$. Hence,

$$2x - 3(0) = 12$$
$$2x = 12$$
$$x = 6$$

Thus, the x-intercept is (6, 0). Next, we calculate the y-intercept (0, __) by setting $x = 0$. Therefore,

$$2(0) - 3y = 12$$
$$-3y = 12$$
$$y = -4$$

Hence, the y-intercept is (0, −4). The graph of this linear equation appears in Figure 6–5. It is important to remember that every point (x, y) on the straight line of Figure 6–5 satisfies the equation $2x - 3y = 12$.

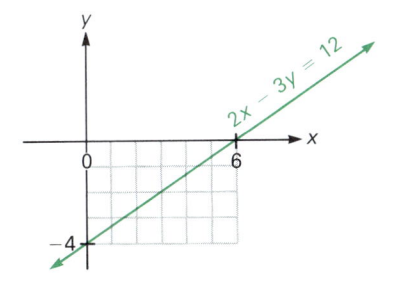

FIGURE 6–5

Linear Inequalities

If the equal sign (=) of a linear equation (equality) such as

$$2x - 3y = 12$$

is replaced by an inequality sign ($<, >, \leq, \geq$), the resulting statement is a linear inequality in two variables. Thus, the statements

$$2x - 3y < 12$$
$$2x - 3y > 12$$
$$2x - 3y \leq 12$$
$$2x - 3y \geq 12$$

are examples of linear inequalities in two variables.

To graph a linear inequality such as

$$2x - 3y \leq 12$$

we should observe that the corresponding straight line, $2x - 3y = 12$, divides the plane into two regions (see Figure 6–6). One region consists of all points satisfying the inequality $2x - 3y < 12$, whereas the other consists of all points satisfying the inequality $2x - 3y > 12$.

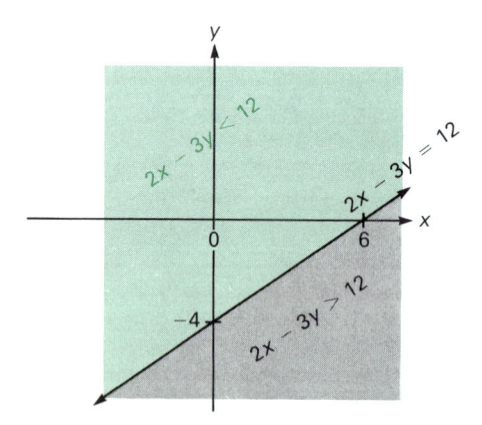

FIGURE 6–6

Observing Figure 6–6, note that points *above* the straight line satisfy the linear inequality $2x - 3y < 12$. This can be verified by selecting a few points (x, y) from this region and substituting their coordinates into the inequality $2x - 3y < 12$ (see Figure 6–7).

LINEAR PROGRAMMING

	$2x - 3y < 12$
(4, 1)	$\dfrac{2(4) - 3(1) < 12}{5}$
(0, 0)	$\dfrac{2(0) - 3(0) < 12}{0}$
(−1, 5)	$\dfrac{2(-1) - 3(5) < 12}{-17}$

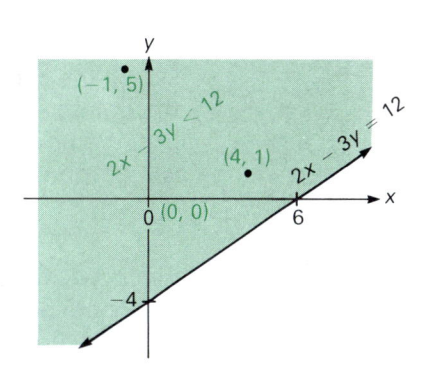

FIGURE 6–7

Points lying *below* the straight line satisfy the linear inequality $2x - 3y > 12$. This can be verified by selecting a few points (x, y) from this area below the straight line and substituting their coordinates into the inequality $2x - 3y > 12$ (see Figure 6–8). Thus, all points (x, y) satisfying the inequality $2x - 3y \leq 12$ are located either *on or above* the straight line $2x - 3y = 12$ (see Figure 6–9).

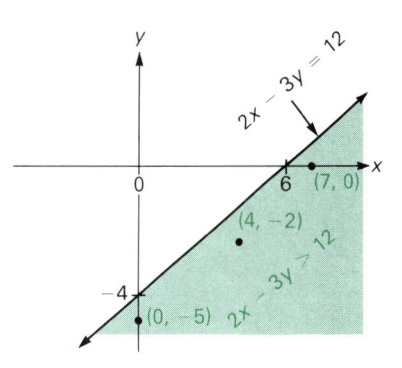

	$2x - 3y > 12$
(7, 0)	$\dfrac{2(7) - 3(0) > 12}{14}$
(4, −2)	$\dfrac{2(4) - 3(-2) > 12}{14}$
(0, −5)	$\dfrac{2(0) - 3(-5) > 12}{15}$

FIGURE 6–8

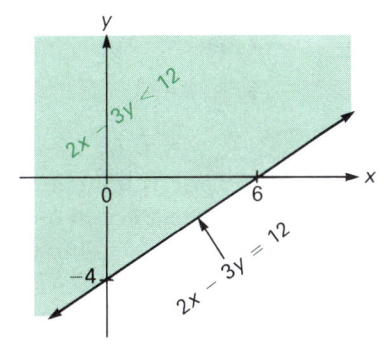

FIGURE 6–9

In general, when graphing linear inequalities, we should first sketch the corresponding straight line and then determine whether the points satisfying the inequality lie above or below the straight line. We now illustrate this graphing procedure. As an example, we will graph the inequality discussed in this section,

$$2x - 3y \leq 12$$

Since we have already illustrated its graph, let us pretend we have not seen Figure 6–9.

We first find the *y*-intercept of the corresponding straight line by setting $x = 0$ and solving for *y*. Hence,

$$2(0) - 3y \leq 12$$
$$-3y \leq 12$$
$$y \geq -4$$

Thus, the *y*-intercept is $(0, -4)$. The inequality $y \geq -4$ indicates that all points $(0, y)$ lie on the *y*-axis *at or above* $y = -4$. This result is graphed in Figure 6–10. We then find the *x*-intercept by setting $y = 0$ and solving for *x*. Hence,

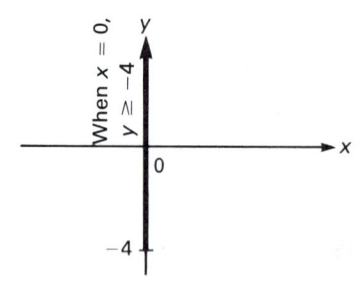

$$2x - 3(0) \leq 12$$
$$2x \leq 12$$
$$x \leq 6$$

FIGURE 6–10

Thus, the *x*-intercept is $(6, 0)$. The inequality $x \leq 6$ indicates that all points $(x, 0)$ lie on the *x*-axis *at or to the left of* $x = 6$. This result is graphed in Figure 6–11. Connecting the intercepts of Figure 6–11, we obtain the straight line $2x - 3y = 12$ along with the region represented by the inequality $2x - 3y < 12$ (see Figure 6–12). Observe that the direction of arrows indicates whether the associated region is either above or below the line.

LINEAR PROGRAMMING

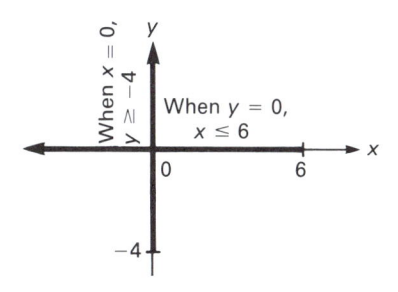

FIGURE 6–11

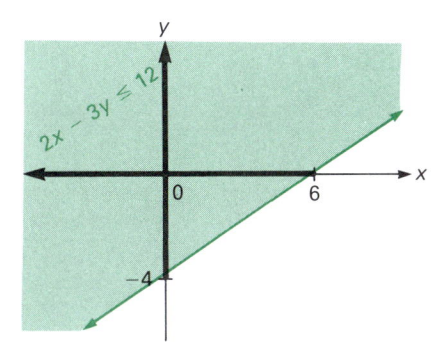

FIGURE 6–12

EXAMPLE 6–4 Graph $-4x + 5y \geq 40$.

Solution First, we find the x-intercept by setting $y = 0$ and using the rules of inequalities to solve for x. Hence,

$$-4x + 5(0) \geq 40$$
$$-4x \geq 40$$
$$x \leq -10$$

Thus, the x-intercept is $(-10, 0)$. We then find the y-intercept by setting $x = 0$ and using the rules of inequalities to solve for y. Here,

$$-4(0) + 5y \geq 40$$
$$5y \geq 40$$
$$y \geq 8$$

Therefore, the y-intercept is $(0, 8)$. Graphing these results, we obtain Figure 6–13.

EXAMPLE 6–4 (continued)

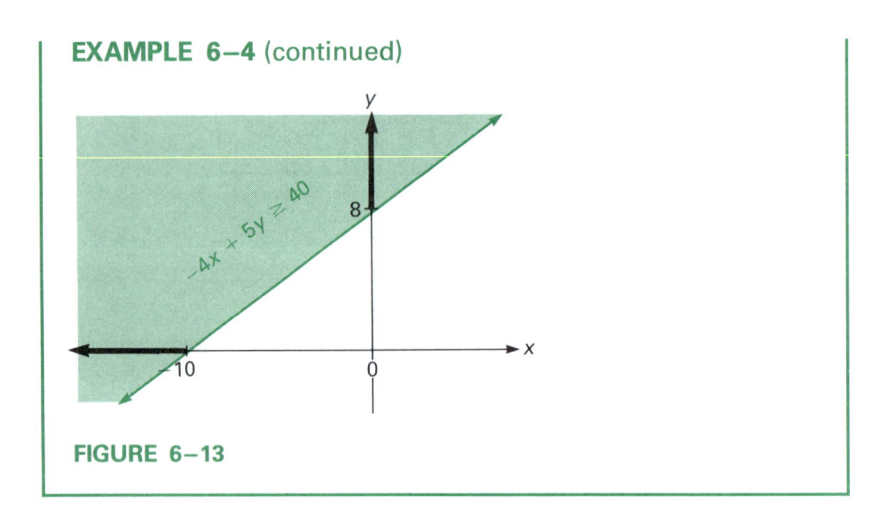

[Figure 6-13: Graph showing the inequality $-4x + 5y \geq 40$, with intercepts at -10 on the x-axis and 8 on the y-axis.]

FIGURE 6–13

Systems of Linear Inequalities

If we graph a set of points (x, y) satisfying *more than one* linear inequality, we are graphing a **system of linear inequalities**. Specifically, the graph of the system

$$3x + 5y \geq 30$$
$$4x - y \leq 8$$

consists of the set of points (x, y) satisfying *both* inequalities.

To graph a system of two linear inequalities, we graph each inequality on the same axis system and then determine the region common to both. Every point (x, y) in the common region will, of course, satisfy both inequalities. The preceding system of linear inequalities is graphed in Figures 6–14 through 6–16. Observe that we have first graphed each inequality on a separate axis system (Figures 6–14 and 6–15) before combining them (Figure 6–16).

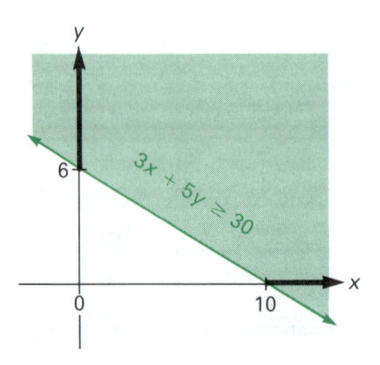

[Figure 6-14: Graph of $3x + 5y \geq 30$, with intercepts at 6 on the y-axis and 10 on the x-axis.]

FIGURE 6–14

313 LINEAR PROGRAMMING

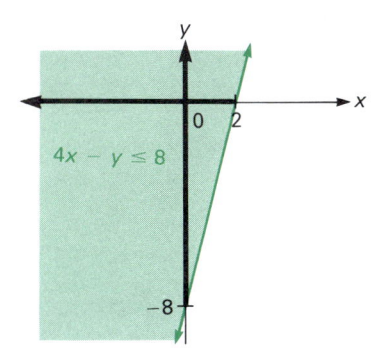

FIGURE 6–15

$4x - y \leq 8$

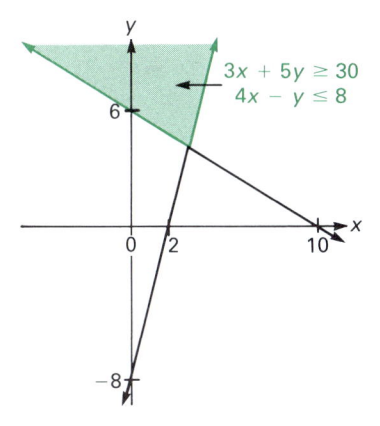

FIGURE 6–16

$3x + 5y \geq 30$
$4x - y \leq 8$

EXAMPLE 6–5 Graph the system

$$4x + 5y \leq 40$$
$$3x - 2y \leq 24$$
$$x \geq 0$$
$$y \geq 0$$

Solution Sketches of each inequality on a separate axis system appear in Figures 6–17 through 6–20. The region common to all of the inequalities is graphed in Figure 6–21.

CHAPTER SIX

EXAMPLE 6-5 (continued)

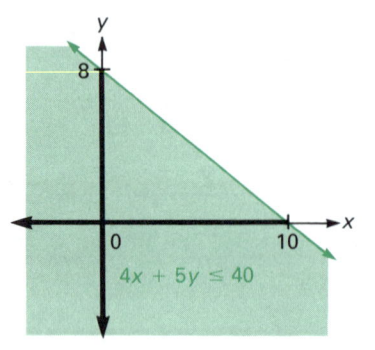

$4x + 5y \leq 40$

FIGURE 6-17

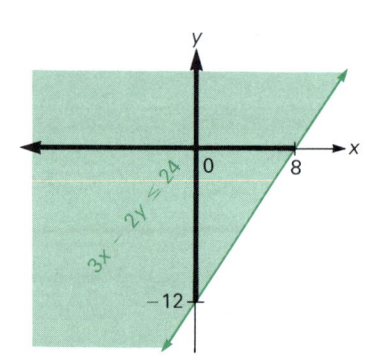

$3x - 2y \leq 24$

FIGURE 6-18

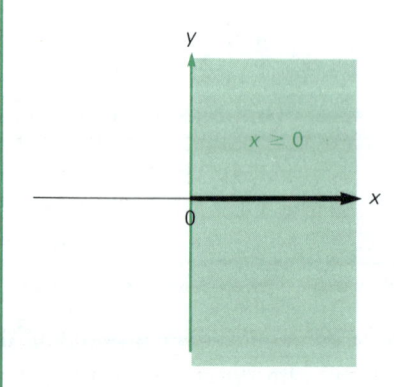

$x \geq 0$

FIGURE 6-19

315 LINEAR PROGRAMMING

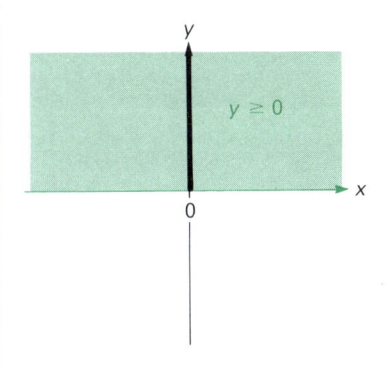

FIGURE 6-20

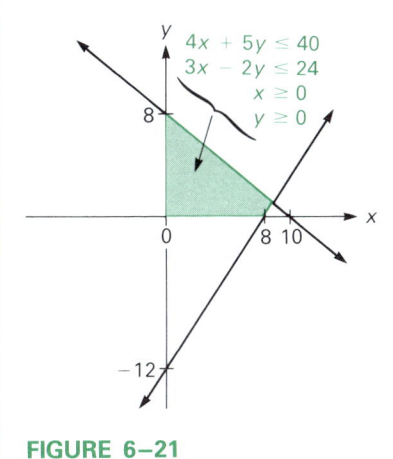

FIGURE 6-21

EXAMPLE 6-6 Graph the system

$$3x + 2y \geq 12$$
$$2x + 7y \geq 14$$
$$x \geq 0$$
$$y \geq 0$$

Solution We graph each of the first two inequalities on a separate axis system in Figures 6-22 and 6-23. Their common region is graphed in Figure 6-24. The common region and its boundaries, located in the first quadrant, are graphed in Figure 6-25.

EXAMPLE 6-6 (continued)

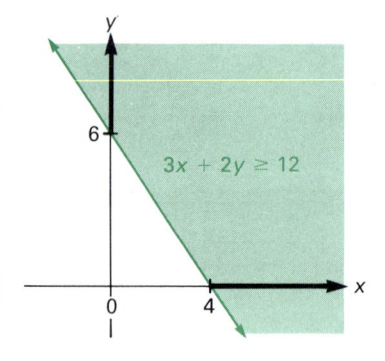

FIGURE 6-22
$3x + 2y \geq 12$

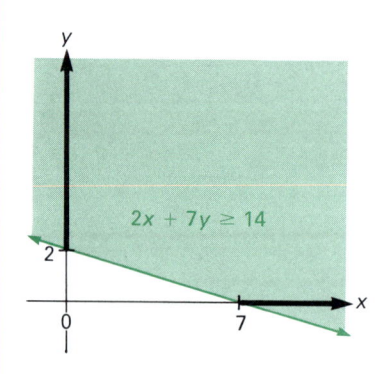

FIGURE 6-23
$2x + 7y \geq 14$

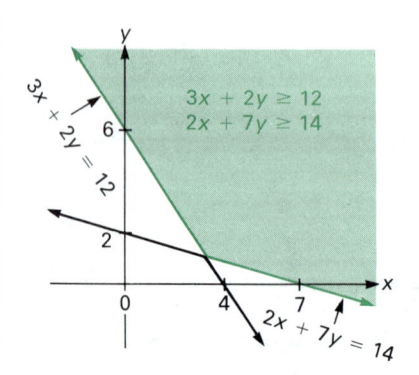

FIGURE 6-24
$3x + 2y \geq 12$
$2x + 7y \geq 14$
$3x + 2y = 12$
$2x + 7y = 14$

LINEAR PROGRAMMING

FIGURE 6–25

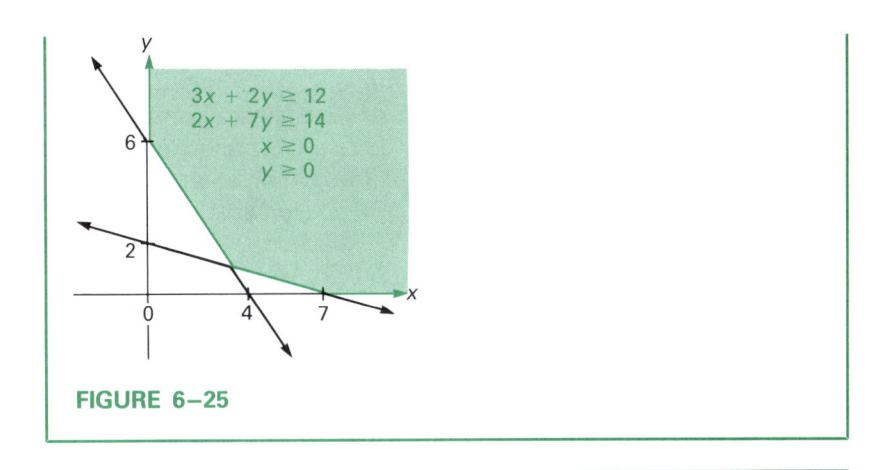

$3x + 2y \geq 12$
$2x + 7y \geq 14$
$x \geq 0$
$y \geq 0$

EXERCISES

1. Graph each of the following:
   - (A) $4x + 5y \geq 40$
   - (B) $-3x + 2y \leq 12$
   - (C) $x - y \geq 5$
   - (D) $-2x - 4y \geq 16$
   - (E) $2x + y \leq 12$
   - (F) $3x - y \leq 15$
   - (G) $x + 3y < 4$
   - (H) $-2x + y > 8$
   - (I) $9x - 2y \geq 36$
   - (J) $7x - 3y \leq -21$
   - (K) $-4x - 7y \geq -28$
   - (L) $x + y < 7$

2. Graph each of the following systems:
   - (A) $3x + 9y \geq 27$
     $2x - 3y \leq 12$
   - (B) $2x + 5y \leq 20$
     $x \geq 0$
   - (C) $-3x + 2y \geq -12$
     $y \geq -6$
   - (D) $4x - 3y \leq 12$
     $x \leq 2$
   - (E) $3x + 4y \leq 48$
     $5x + 3y \leq 30$
     $x \geq 0$
     $y \geq 0$
   - (F) $2x + 7y \geq 28$
     $4x + y \geq 16$
     $x \geq 0$
     $y \geq 0$
   - (G) $6x + 7y \leq 42$
     $3x + 5y \leq 25$
     $x \geq 0$
     $y \geq 0$
   - (H) $6x + 7y \leq 42$
     $3x + 5y \leq 25$
     $x + y \leq 6$
     $x \geq 0$
     $y \geq 0$
   - (I) $5x + 2y \geq 30$
     $3x + 4y \geq 24$
     $x + y \geq 7.6$
     $x \geq 0$
     $y \geq 0$
   - (J) $3x + y \leq 30$
     $2x + 3y \leq 24$
     $x \geq 1$
     $y \leq 7$

6–3

LINEAR PROGRAMMING

A common problem of most businesses is the allocation of limited resources among competing activities in an optimal way. Linear programming can be useful in solving such problems. Historically, linear programming was developed to solve resource allocation problems of the U.S. Air Force during World War II. Much of the development of linear programming is credited to George B. Dantzig, who gave a general formulation of a linear programming problem and a method of solving it. This method is called the *simplex method* and is discussed in Sections 6–4 through 6–7. In this section, we will discuss the *graphical method* for solving linear programming problems.

Profit Maximization: Production Scheduling

We now consider a problem involving the manufacture of bicycles and tricycles by the Ruszala Company. Each bicycle and tricycle must pass through Departments I and II. Each department has a limited number of hours available for the manufacture of bicycles and tricycles. The time requirements of each bicycle and tricycle in each department are listed in Table 6–1. Also included are the total number of hours available in each department as well as the unit profits of bicycles and tricycles.

TABLE 6–1

	Bicycles	Tricycles	
Unit Profits→	$6/bicycle	$4/tricycle	
Department I	3 hours/bicycle	4 hours/tricycle	At most 450 hours available
Department II	5 hours/bicycle	2 hours/tricycle	At most 400 hours available

Thus, we must determine how many bicycles and tricycles should be manufactured in order to satisfy departmental time constraints and *maximize total profit*. If

$$x = \text{number of bicycles manufactured}$$
$$y = \text{number of tricycles manufactured}$$

then the total profit, P, is expressed by

$$P = 6x + 4y$$

Since the objective is to maximize P, this profit equation is called the **objective function**.

The departmental time constraints are expressed by the inequalities

$$3x + 4y \leq 450$$
$$5x + 2y \leq 400$$

319 LINEAR PROGRAMMING

Note that since all the time in each department need not be used, then each of these constraints contains the inequality symbol ≤ instead of the equality symbol, =. Also, since x and y are nonnegative, the following inequalities are included:

$$x \geq 0$$
$$y \geq 0$$

Thus, our problem is formulated algebraically as

Maximize $P = 6x + 4y$
subject to
the constraints
$$3x + 4y \leq 450$$
$$5x + 2y \leq 400$$
$$x \geq 0$$
$$y \geq 0$$

Such a problem is called a **linear programming problem.** Note that a linear programming problem involves optimizing a linear objective function subject to linear constraints.

To solve our linear programming problem, we must realize that the constraints are a system of linear inequalities. Thus, we are initially seeking those points (x, y) that satisfy all the inequality constraints of the system. Hence, we graph the system (see Figure 6-26). Observe that the shaded region of Figure 6-26 represents those points (x, y) that satisfy all constraints of the problem. This region is called the **region of feasible solutions.** Also, note that we have identified the corner points on the boundary of the region of feasible solutions. These corner points are called **vertex points.** The vertex points (0, 0), (0, 112.5), and (80, 0) are easily identified since they lie on the axes. The vertex point (50, 75) was found by solving the linear system of equations

$$3x + 4y = 450$$
$$5x + 2y = 400$$

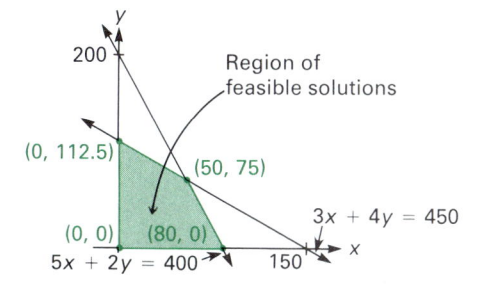

FIGURE 6-26

We must now determine which point(s) (x, y) of the region of feasible solutions maximize the profit function $P = 6x + 4y$. At first thought, it would appear that we must substitute each point (x, y) of the feasible region into the profit function. However, we will subsequently show the following.

> A maximum value of P, if it exists, will occur at one or more of the vertex points or on the boundary of the region of feasible solutions.

Thus, we may substitute the vertex points (x, y) into the profit function to determine which point yields the maximum profit. Hence,

$$P = 6x + 4y$$

(0, 0)	$P = 6(0) + 4(0) = \$0$	
(0, 112.5)	$P = 6(0) + 4(112.5) = \$450$	
(80, 0)	$P = 6(80) + 4(0) = \$480$	
(50, 75)	$P = 6(50) + 4(75) = \$600$	Maximum profit

Observe that the vertex point (50, 75) yields a maximum profit of $600. Therefore, the solution to the linear programming problem is

$x = 50$ bicycles should be manufactured

$y = 75$ tricycles should be manufactured

Why the Vertex Points?

Having solved a linear programming problem, we now show why an optimal value of an objective function, if it exists, will occur at one or more of the vertex points or on the boundary of the feasible region. Using the preceding linear programming problem as an example, we again illustrate the graph of its region of feasible solutions (see Figure 6–27). Observe that the objective function

$$P = 6x + 4y$$

with its maximum value, $P = 600$, also appears on the graph as

$$600 = 6x + 4y$$

Note that it passes through the optimal solution, (50, 75).

321 LINEAR PROGRAMMING

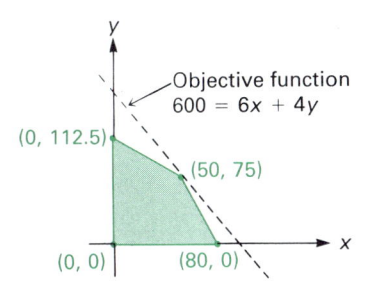

FIGURE 6–27

Now, suppose we substitute some other point of the feasible region into the objective function $P = 6x + 4y$. Choosing (20, 10), we obtain

$$P = 6(20) + 4(10) = \$160$$

Note that $P = \$160$ is less than the maximum value, $P = \$600$. If we include the graph of the objective function $160 = 6x + 4y$ with the graph of Figure 6–27, we obtain the graph of Figure 6–28. Observe that the objective function $160 = 6x + 4y$ is parallel to the objective function $600 = 6x + 4y$ but is located closer to the origin. Thus, we see that as the graph of the objective function moves *farther away* from the origin, the value of *P increases*. Therefore, the maximum value of *P* will occur at the point or points of the feasible region for which the graph of the objective function is farthest away from the origin.* Such points are either lone vertex points or all points on a boundary interval between two vertex points.

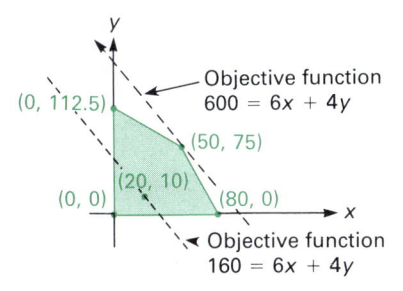

FIGURE 6–28

Note that the graph of the objective function must always pass through at least one point of the feasible region. Otherwise, the constraints of the linear programming problem would not be satisfied.

---

*This distance is measured in units of the objective function value.

CHAPTER SIX

Note also that if the unit profits were changed so that the slope of the objective function *increased* substantially in absolute value, then the vertex point (80, 0) would yield the maximum value of P (see Figure 6–29). On the other hand, if the unit profits were changed so that the slope of the objective function *decreased* substantially in absolute value, then the vertex point (0, 112.5) would yield the maximum value of P (see Figure 6–30).

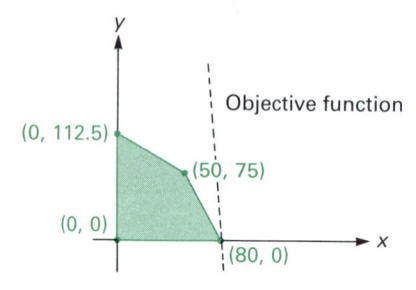

FIGURE 6–29

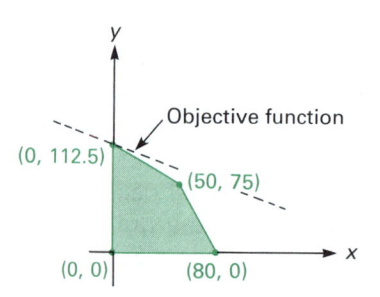

FIGURE 6–30

In addition, if the slope of the objective function *equaled* the slope of one of the constraints, then more than one vertex point would yield an optimal solution (see Figure 6–31). In fact, both vertex points and all points on the straight line between them yield optimal values for P. Thus, in the present example, the vertex points (0, 112.5) and (50, 75) and all points on the straight line between them yield optimal values of P (see Figure 6–31).

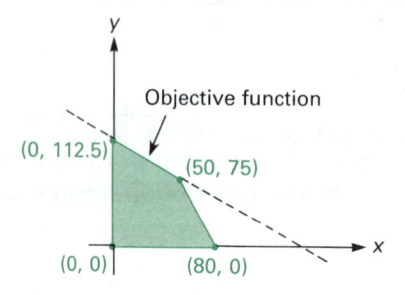

FIGURE 6–31

Cost Minimization: Vitamin Requirements

EXAMPLE 6–7 A diet is to include at least 140 milligrams of Vitamin A and at least 145 milligrams of Vitamin B. These requirements are to be obtained from two types of foods: Type I, which contains 10 milligrams of Vitamin A and 20 milligrams of Vitamin B per pound; and Type II, which contains 30 milligrams of Vitamin A and 15 milligrams of Vitamin B per pound. If Types I and II foods cost $2 and $8 per pound, respectively, how many pounds of each type should be purchased to satisfy the requirements at *minimum cost?*

Solution The preceding information is summarized in table form as follows:

	Type I	Type II	
Unit Costs→	$2/pound	$8/pound	
Vitamin A	10 milligrams/ pound	30 milligrams/ pound	At least 140 milligrams
Vitamin B	20 milligrams/ pound	15 milligrams/ pound	At least 145 milligrams

If

x = number of pounds of Type I food bought
y = number of pounds of Type II food bought
C = total cost

our problem is written algebraically as

Minimize $\quad C = 2x + 8y$

subject to $\qquad 10x + 30y \geq 140$
the constraints $\quad 20x + 15y \geq 145$
$\qquad\qquad\qquad\quad x \geq 0$
$\qquad\qquad\qquad\quad y \geq 0$

Note that the first two constraints contain the inequality symbol $\geq$ since 140 milligrams and 145 milligrams are *minimal* requirements of Vitamins A and B, respectively.

The region of feasible solutions is determined by graphing the linear inequality constraints (see Figure 6–32). The vertex points are $(0, 9\frac{2}{3})$, $(5, 3)$, and $(14, 0)$. Note that $(5, 3)$ was determined by solving the linear system of equations

$$10x + 30y = 140$$
$$20x + 15y = 145$$

EXAMPLE 6-7 (continued)

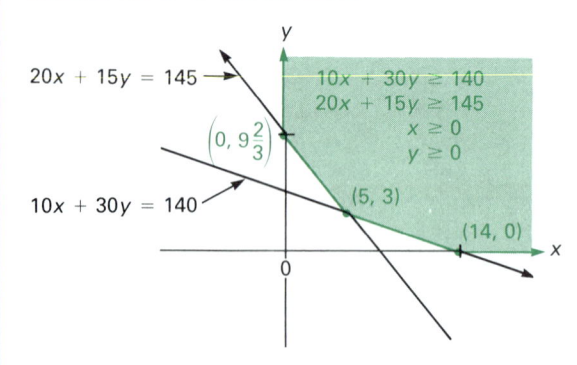

FIGURE 6-32

We now substitute the coordinates of each vertex point into the objective function $C = 2x + 8y$ to determine which yields the minimum cost. Hence,

$(0, 9\frac{2}{3})$	$C = 2(0) + 8\left(\frac{29}{3}\right) = \$77.33$
$(5, 3)$	$C = 2(5) + 8(3) = \$34.00$
$(14, 0)$	$C = 2(14) + 8(0) = \$28.00$ Minimum cost

Since the vertex point (14, 0) yields the minimum cost of $28, then

$x = 14$ pounds of Type I food should be bought
$y = 0$ pounds of Type II food should be bought

EXERCISES

1. A company manufactures motorcycles and mopeds, each of which must pass through two machines: Machine 1 and Machine 2. Each motorcycle requires 2 hours on Machine 1 and 5 hours on Machine 2. Each moped requires 3 hours on Machine 1 and 1 hour on Machine 2. Machines 1 and 2 have available 90 hours and 160 hours, respectively, for these two products.
   (A) If the company makes a profit of $120 on each motorcycle and $60 on each moped, how many of each should be produced in order to maximize total profit and satisfy the constraints of the problem?
   (B) If the company makes only $30 on each motorcycle and $90 on each moped, how many of each should be produced in order to maximize total profit and satisfy the constraints of the problem?

2. A diet must provide at least 1200 milligrams of protein and at least 1000 milligrams of iron. These nutrients are to be obtained from eating meat and spinach. Each pound of meat contains 500 milligrams of

protein and 100 milligrams of iron. Each pound of spinach contains 200 milligrams of protein and 800 milligrams of iron. If meat and spinach cost $3.00 and $1.50 per pound, respectively, how many pounds of each should be eaten in order to minimize total cost and satisfy the constraints of the problem?

3. A manufacturer produces two models of televisions, T140 and T240, each of which must pass through two departments, D1 and D2. Each unit of T140 requires 3 hours in D1 and 4 hours in D2. Each unit of T240 requires 6 hours in D1 and 4 hours in D2. Departments D1 and D2 each have 60 hours available. If the manufacturer makes a profit of $10 per unit on T140 and $30 per unit on T240, how many units of each should be manufactured in order to maximize total profit and satisfy the constraints of the problem?

4. A farmer owns a 100-acre farm and wants to plant a combination of two crops: A and B. Crop A requires 60 man-hours of labor per acre, and Crop B requires 80 man-hours of labor per acre. The farmer has 6600 man-hours of labor available.
   (A) If the farmer makes a profit of $400 per acre on Crop A and $500 per acre on Crop B, how many acres of each crop should be planted in order to maximize total profit and satisfy the constraints of the problem?
   (B) Suppose the farmer makes a profit of $500 per acre on each crop. How many acres of each crop should be planted in order to maximize total profit and satisfy the constraints of the problem?

5. A factory uses two types of fuel, F10 and F20, for heating and other purposes. At least 3800 gallons of fuel are needed each day. Some by-products are produced by the burning of the fuel. Each gallon of F10 leaves a residue of 0.02 pound of ash and 0.06 pound of soot. Each gallon of F20 leaves a residue of 0.05 pound of ash and 0.01 pound of soot. The factory needs at least 120 pounds of ash and at least 136 pounds of soot. If F10 and F20 cost $1.50 and $1.10 per gallon, respectively, how many gallons of each type should be purchased in order to minimize total cost and satisfy the constraints of the problem?

6. (The following problem appeared on a past Uniform CPA Examination.) A company markets two products: Alpha and Gamma. The marginal contributions per gallon are $5 for Alpha and $4 for Gamma. Both products consist of two ingredients: D and K. Alpha contains 80% D and 20% K, while the proportions of the same ingredients in Gamma are 40% and 60%, respectively. The current inventory is 16,000 gallons of D and 6000 gallons of K. The only company producing D and K is on strike and will neither deliver nor produce them in the foreseeable future. The company wishes to know the numbers of gallons of Alpha and Gamma that it should produce with its present stock of raw materials in order to maximize its total revenue.

7. (The following problem appeared on a past Uniform CPA Examination.) Patsy, Inc. manufactures two products: X and Y. Each product must be processed in each of three departments: Machining,

Assembling, and Finishing. The hours needed to produce 1 unit of product per department and the maximum possible hours per department are as follows:

Department	Production Hours per Unit X	Y	Maximum Capacity in Hours
Machining	2	1	420
Assembling	2	2	500
Finishing	2	3	600

In addition, $X \geq 50$ and $Y \geq 50$. The objective function is to maximize profits, where profit = $4X + $2Y. Given the objective and constraints, what is the most profitable number of units of X and Y to manufacture?

8. A pharmaceutical company plans to manufacture two new drugs: diopthelene and gramamine. Each case of diopthelene requires 3 hours of processing time and 1 hour of curing time per week. Each case of gramamine requires 5 hours of processing time and 5 hours of curing time per week. The company's time schedule allows 55 hours of processing time and 45 hours of curing time, weekly, for the two drugs. Additionally, the company must produce no more than 10 cases of diopthelene and no more than 9 cases of gramamine each week. If the company makes a profit of $400 on each case of diopthelene and $500 on each case of gramamine, how many cases of each should be produced in order to maximize profit and satisfy the constraints of the problem?

6–4

SIMPLEX METHOD FORMULATION

In the preceding sections of this chapter, we solved linear programming problems graphically. For each problem, our procedure was to graph the region of feasible solutions, determine its vertex points, and substitute the coordinates of each vertex point into the objective function to determine the optimal solution.

The graphical approach for solving linear programming problems is not viable when the number of variables exceeds two, and since most real-world linear programming problems have many variables, its applicability is very limited. Fortunately, there exists an *algebraic* method of determining the vertex points of the region of feasible solutions. This algebraic method is called the **simplex method.** It will enable us to determine the vertex points of the region of feasible solutions without having to graph the region. In fact, the simplex method will allow us to proceed systematically from one vertex point to another and improve the value of the objective function at each step until an optimal solution has been found. This systematic feature of the simplex method enhances its suitability for computer solution.

Maximization Problem: Slack Variables

The first step in applying the simplex method to a linear programming problem involves restating the linear *inequality* constraints into linear *equations*. This is accomplished by introducing either slack variables or surplus variables, depending upon the sense of the inequality.

To illustrate, we consider the linear programming maximization problem of Section 6–3. For reference purposes, this problem is summarized in the data box of Table 6–2. If

$$x_1 = \text{number of Product A to be produced}$$
$$x_2 = \text{number of Product B to be produced}$$
$$z = \text{total profit}$$

then the corresponding algebraic formulation is

Maximize $z = 6x_1 + 4x_2$ Objective function
subject to $3x_1 + 4x_2 \leq 450$
$5x_1 + 2x_2 \leq 400$
where x_1 and x_2 are nonnegative

TABLE 6–2

	x_1 Product A	x_2 Product B	
Unit Profits	$6/unit	$4/unit	
Department I	3 hours/unit	4 hours/unit	At most 450 hours available
Department II	5 hours/unit	2 hours/unit	At most 400 hours available

We now restate the linear inequalities as linear equations by adding a nonnegative quantity to the left-hand side of each inequality. Thus,

$$3x_1 + 4x_2 + x_3 = 450$$
$$5x_1 + 2x_2 + x_4 = 400$$

The nonnegative variables x_3 and x_4 are called **slack variables**. Hence, x_3 represents the number of hours of slack time in Department I, and x_4 represents the number of hours of slack time in Department II. Thus, if the number of hours of Department I time used—$3x_1 + 4x_2$—is less than 450, then x_3 represents the amount of unused Department I time. On the other hand, if all 450 hours of Department I time are used (i.e., $3x_1 + 4x_2 = 450$), then $x_3 = 0$. Similarly, if the number of hours of Department II time used—

$5x_1 + 2x_2$—is less than 400, then x_4 represents the amount of unused Department II time. On the other hand, if all 400 hours of Department II time are used (i.e., $5x_1 + 2x_2 = 400$), then $x_4 = 0$.

The entire problem is now formulated as follows:

$$\text{Maximize} \quad z = 6x_1 + 4x_2$$
$$\text{subject to} \quad 3x_1 + 4x_2 + x_3 = 450$$
$$\phantom{\text{subject to} \quad} 5x_1 + 2x_2 + x_4 = 400$$
$$\text{where} \quad x_1, x_2, x_3, \text{ and } x_4 \text{ are nonnegative}$$

Basic Feasible Solutions

We are now ready to begin our search for the vertex points of the region of feasible solutions. Finding the vertex points is equivalent to finding basic feasible solutions of our system. A **basic feasible solution** is a solution that contains exactly as many nonzero variables as the number of constraint equations; the remaining variables must equal 0.* In other words, if there are m constraint equations, then a basic feasible solution contains exactly m nonzero variables; the remaining variables must equal 0. In the preceding system, there are two constraint equations. Thus, a feasible solution with exactly two nonzero variables is a basic feasible solution. Hence, the solution

$$x_1 = 0, \quad x_2 = 0, \quad x_3 = 450, \quad x_4 = 400$$

is a basic feasible solution. Note that this corresponds to the vertex point (0, 0) of the region of feasible solutions.

The nonzero variables of a basic feasible solution are called **basic variables**, or **basis**, whereas the zero variables are called **nonbasic variables**. Hence, the variables of the preceding basic feasible solution are categorized as follows:

$$\underbrace{x_1 = 0, x_2 = 0}_{\text{nonbasic variables}} \qquad \underbrace{x_3 = 450, x_4 = 400}_{\substack{\text{basic variables} \\ \text{(or basis)}}}$$

The linear programming problem

$$\text{Maximize} \quad z = 6x_1 + 4x_2$$
$$\text{subject to} \quad 3x_1 + 4x_2 + x_3 = 450$$
$$\phantom{\text{subject to} \quad} 5x_1 + 2x_2 + x_4 = 400$$

---

*Actually, a basic feasible solution can contain fewer nonzero variables than the number of constraint equations. When this happens, the solution is termed **degenerate**. Since degeneracies do not occur very often, for practical purposes they are not considered in our definition of a basic feasible solution.

may be expressed in tableau form as follows:

Objective Function Coefficients of Basic Variables ↓

			x_1	x_2	x_3	x_4	
	Basis		6	4	0	0	← Objective Function of Coefficients
0	x_3		3	4	1	0	450 ←
0	x_4		5	2	0	1	400 ← Department Capacities

This tableau is called the **initial tableau** of the simplex method. Each column is identified by its respective variable. The objective function coefficients corresponding to the basic variables, x_3 and x_4, are identified by the column vectors $\begin{bmatrix}1\\0\end{bmatrix}$ and $\begin{bmatrix}0\\1\end{bmatrix}$, respectively. The basic variables are listed in the column labeled "Basis." The corresponding value of each basic variable is listed in the right-most column, entitled "Department Capacities." Specifically,

$$x_3 = 450$$
$$x_4 = 400$$

The corresponding objective function coefficients of the basic variables are listed in the leftmost column, next to the "Basis" column. Thus,

The objective function coefficient 0 corresponds to x_3.
The objective function coefficient 0 corresponds to x_4.

Having set up the initial tableau, we are now ready to search systematically for other basic feasible solutions in order to improve the value of the objective function.

Simplex Method (Maximization): Standard Form

Before going further, we will summarize the simplex method leading up to the formulation of the initial tableau and identifying the initial basic feasible solution. Thus, we consider the following general maximization problem:

Maximize $z = c_1x_1 + c_2x_2 + \ldots + c_kx_k$
subject to
the m constraints
$$a_{11}x_1 + a_{12}x_2 + \ldots + a_{1k}x_k \leq b_1$$
$$a_{21}x_1 + a_{22}x_2 + \ldots + a_{2k}x_k \leq b_2$$
$$\vdots \qquad \vdots \qquad \qquad \vdots \qquad \vdots$$
$$a_{m1}x_1 + a_{m2}x_2 + \ldots + a_{mk}x_k \leq b_m$$

where $x_1, x_2, \ldots, x_k$ are nonnegative variables
$a_{11}, a_{12}, \ldots, a_{mk}$ are constants
$b_1, b_2, \ldots, b_m$ are nonnegative constants
$c_1, c_2, \ldots, c_k$ are constants

This will be the standard form for maximization problems in this chapter.

Occasionally, linear programming problems that are not in standard form arise. For example, it is possible to encounter a linear programming problem where:

1. One or more of the variables is unrestricted in sign
2. One or more of the b_i's is negative

Techniques are available for handling these situations. We will not elaborate on such techniques, as they are outside the scope of this chapter.

To formulate the initial tableau for a linear programming problem in standard form, each linear inequality is restated as a linear equation by adding a nonnegative slack variable to the left-hand side of the inequality. Thus, the standard form becomes

Maximize $z = c_1 x_1 + c_2 x_2 + \ldots + c_k x_k$
subject to
$$a_{11} x_1 + a_{12} x_2 + \ldots + a_{1k} x_k + x_{k+1} = b_1$$
$$a_{21} x_1 + a_{22} x_2 + \ldots + a_{2k} x_k + x_{k+2} = b_2$$
$$\vdots$$
$$a_{m1} x_1 + a_{m2} x_2 + \ldots + a_{mk} x_k + \ldots + x_{k+m} = b_m$$

where $x_1, x_2, \ldots, x_k, x_{k+1}, \ldots, x_{k+m}$ are all nonnegative

Note that there are as many slack variables as there originally were inequalities.

Now we restate the following ideas upon which the simplex method depends.

An **optimum solution** (if one exists) to a linear objective function with m linear constraint equations is a **basic feasible solution**. A basic feasible solution contains exactly m non-zero variables; the remaining variables must equal 0. Basic feasible solutions correspond to vertex points of the region of feasible solutions.

LINEAR PROGRAMMING

Thus, in our general model, we set the variables $x_1, x_2, \ldots, x_k$ equal to 0. The basic variables and their respective values are

$$x_{k+1} = b_1$$
$$x_{k+2} = b_2$$
$$\vdots$$
$$x_{k+m} = b_m$$

This information is summarized by the following initial tableau:

Initial Tableau

Basic c_j	Basis	x_1 c_1	x_2 c_2	$\ldots$	x_k c_k	x_{k+1} 0	x_{k+2} 0	$\ldots$	x_{k+m} 0	$c_j \leftarrow$ $b_i \downarrow$
0	x_{k+1}	a_{11}	a_{12}	$\ldots$	a_{1k}	1	0	$\ldots$	0	b_1
0	x_{k+2}	a_{21}	a_{22}	$\ldots$	a_{2k}	0	1	$\ldots$	0	b_2
$\vdots$	$\vdots$	$\vdots$	$\vdots$		$\vdots$	$\vdots$	$\vdots$		$\vdots$	$\vdots$
0	x_{k+m}	a_{m1}	a_{m2}	$\ldots$	a_{mk}	0	0	$\ldots$	1	b_m

Note that three types of coefficients may be specified:

1. ***c*-Coefficients**—The coefficients $c_1, c_2, \ldots, c_k, 0, \ldots, 0$ of the objective function are termed ***c*-coefficients.**
2. ***b*-Coefficients**—The capacity constraints $b_1, b_2, \ldots, b_m$ of the m linear inequalities are called ***b*-coefficients.**
3. ***a*-Coefficients**—The coefficients

$$a_{11}, \ldots, a_{1k}, 1, 0, \ldots, 0$$
$$a_{21}, \ldots, a_{2k}, 0, 1, \ldots, 0$$
$$\vdots \qquad \vdots \qquad \vdots$$
$$a_{m1}, \ldots, a_{mk}, 0, 0, \ldots, 1$$

of the variables $x_1, x_2, \ldots, x_k, x_{k+1}, \ldots, x_{k+m}$ are known as ***a*-coefficients.**

Again, notice that the basic variables are listed in the column labeled "Basis." The objective function coefficients (*c*-coefficients) corresponding to the basic variables are listed in the leftmost column, labeled "Basic c_j."

EXAMPLE 6–8 Consider the following linear programming problem:

$$\text{Maximize} \quad z = 3x_1 + 5x_2 + 2x_3$$
$$\text{subject to} \quad x_1 + x_2 + 2x_3 \leq 440$$
$$x_1 + 4x_3 \leq 420$$
$$3x_1 + 2x_2 \leq 480$$
$$\text{where} \quad x_1 \geq 0, x_2 \geq 0, x_3 \geq 0$$

(A) Restate the linear inequalities as linear equations.
(B) Set up the initial tableau.
(C) State the initial basic feasible solution.

Solutions

(A) Introducing slack variables, we have

$$\text{Maximize} \quad z = 3x_1 + 5x_2 + 2x_3$$
$$\text{subject to} \quad x_1 + x_2 + 2x_3 + x_4 = 440$$
$$x_1 + 4x_3 + x_5 = 420$$
$$3x_1 + 2x_2 + x_6 = 480$$
$$\text{where} \quad x_1 \geq 0, x_2 \geq 0, x_3 \geq 0, x_4 \geq 0, x_5 \geq 0, x_6 \geq 0$$

(B)

Initial Tableau

Basic c_j	Basis	x_1	x_2	x_3	x_4	x_5	x_6	
		3	5	2	0	0	0	$\leftarrow c_j$
								$\leftarrow b_i$
0	x_4	1	1	2	1	0	0	440
0	x_5	1	0	4	0	1	0	420
0	x_6	3	2	0	0	0	1	480

(C) The initial basic feasible solution is as follows:

Nonbasic Variables	Basic Variables
$x_1 = 0$	$x_4 = 440$
$x_2 = 0$	$x_5 = 420$
$x_3 = 0$	$x_6 = 480$

EXAMPLE 6–9 Consider the linear programming problem

$$\text{Maximize} \quad z = 2x_1 + 5x_2 + 4x_3$$

subject to $x_1 + 3x_2 + 2x_3 \leq 60$
             $4x_1 + 2x_2 + x_3 \leq 100$
where $x_1 \geq 0, x_2 \geq 0, x_3 \geq 0$

(A) Restate the linear inequalities as linear equations.
(B) Construct the initial tableau.
(C) State the initial basic feasible solution.

Solutions

(A) Introducing slack variables, we have

Maximize $z = 2x_1 + 5x_2 + 4x_3$
subject to $x_1 + 3x_2 + 2x_3 + x_4 = 60$
             $4x_1 + 2x_2 + x_3 + x_5 = 100$
where $x_1 \geq 0, x_2 \geq 0, x_3 \geq 0, x_4 \geq 0, x_5 \geq 0$

(B)   Initial Tableau

		x_1	x_2	x_3	x_4	x_5	
Basic c_j	Basis	2	5	4	0	0	← c_j ↓ b_i
0	x_4	1	3	2	1	0	60
0	x_5	4	2	1	0	1	100

(C) The following is the initial basic feasible solution:

Nonbasic Variables   *Basic Variables*
   $x_1 = 0$ $x_4 = 60$
   $x_2 = 0$ $x_5 = 100$
   $x_3 = 0$

EXERCISES

1. Consider the following:

Maximize $z = 120x_1 + 60x_2$
subject to $2x_1 + 3x_2 \leq 90$
             $5x_1 + x_2 \leq 160$
where $x_1 \geq 0, x_2 \geq 0$

(A) Restate the linear inequalities as linear equations.
(B) Construct the initial tableau.
(C) State the initial basic feasible solution.

2. Consider the following:

$$\text{Maximize} \quad z = 10x_1 + 30x_2$$
$$\text{subject to} \quad 3x_1 + 6x_2 \le 60$$
$$4x_1 + 4x_2 \le 60$$
$$\text{where} \quad x_1 \ge 0, x_2 \ge 0$$

(A) Restate the linear inequalities as linear equations.
(B) Construct the initial tableau.
(C) State the initial basic feasible solution.

3. Consider the following:

$$\text{Maximize} \quad z = 400x_1 + 500x_2$$
$$\text{subject to} \quad x_1 + x_2 \le 100$$
$$60x_1 + 80x_2 \le 6600$$
$$\text{where} \quad x_1 \ge 0, x_2 \ge 0$$

(A) Construct the initial tableau.
(B) State the initial basic feasible solution.

4. Consider the following:

$$\text{Maximize} \quad z = 5x_1 + 6x_2 + 2x_3$$
$$\text{subject to} \quad 2x_1 + 3x_2 + 5x_3 \le 20$$
$$6x_1 + 2x_2 + 3x_3 \le 50$$
$$x_1 + 3x_2 + 4x_3 \le 24$$
$$\text{where} \quad x_1, x_2, \text{ and } x_3 \text{ are nonnegative}$$

(A) Restate the linear inequalities as linear equations.
(B) Construct the initial tableau.
(C) State the initial feasible solution.

5. Consider the following:

$$\text{Maximize} \quad z = 20x_1 + 42x_2 + 56x_3$$
$$\text{subject to} \quad 2x_1 + 3x_2 + x_3 \le 6$$
$$4x_1 + 2x_2 + 3x_3 \le 12$$
$$4x_1 + 2x_2 + x_3 \le 8$$
$$\text{where} \quad x_1 \ge 0, x_2 \ge 0, x_3 \ge 0$$

Prepare this problem for solution by the simplex method. Specifically,
(A) Construct the initial tableau.
(B) State the initial basic feasible solution.

6. Consider the following:

$$\text{Maximize} \quad z = 80x_1 + 10x_2 + 16x_3 + 12x_4$$
$$\text{subject to} \quad x_1 + x_2 + x_3 + x_4 \le 40$$
$$2x_1 + x_2 + 4x_3 + x_4 \le 80$$
$$\text{where} \quad x_1, x_2, x_3, \text{ and } x_4 \text{ are nonnegative}$$

6-5

SIMPLEX METHOD

We now return to the illustrative problem of Section 6–4. For reference purposes, we restate it here:

$$\text{Maximize} \quad z = 6x_1 + 4x_2$$
$$\text{subject to} \quad 3x_1 + 4x_2 + x_3 = 450$$
$$\phantom{\text{subject to} \quad} 5x_1 + 2x_2 + x_4 = 400$$

where x_1, x_2, x_3, and x_4 are nonnegative

The corresponding initial tableau is

Basic c_j	Basis	x_1 6	x_2 4	x_3 0	x_4 0	b_i
0	x_3	3	4	1	0	450
0	x_4	5	2	0	1	400
						$z = 0$

The initial basic feasible solution is

Nonbasic Variables	Basic Variables
$x_1 = 0$	$x_3 = 450$
$x_2 = 0$	$x_4 = 400$

The objective function value $z = 0$ is determined by calculating the dot product of the "Basic c_j" column vector and the "b_i" column vector. Hence,

$$\text{Basic } c_j \quad b_i$$
$$\begin{bmatrix} 0 \\ 0 \end{bmatrix} \begin{bmatrix} 450 \\ 400 \end{bmatrix} = 0(450) + 0(400) = 0$$

This has meaning if we realize that each unit of the basic variable x_3 (slack time in Department I) contributes 0 to the value of the objective function z and that each unit of the basic variable x_4 (slack time in Department II) contributes 0 to the value of z.

We must now search for a basic feasible solution that improves the value of the objective function. Specifically, we consider entering one of the nonbasic variables into the basis and at the same time exiting one of the basic variables from the basis. Consequently, we need to answer the following question:

Which of the nonbasic variables should enter the basis, and which of the basic variables should exit the basis, in order to provide the largest increase in the value of the objective function, z?

To answer this question, we must be able to interpret the c-coefficient and the a-coefficients of the column corresponding to the variable under consideration for entry into the basis. If we consider entering x_1 into the basis, then:

1. The corresponding objective function coefficient $c_1 = 6$ represents the amount by which the value of the objective function, z, is increased for each unit of x_1 entered. In other words, for each unit of x_1 entered into the basis, z will increase by 6.
2. The corresponding x_1 column of a-coefficients $\begin{bmatrix} 3 \\ 5 \end{bmatrix}$ (see Table 6–3) is interpreted as follows: For each unit of x_1 entered, 3 units of the basic variable x_3 will exit the basis and 5 units of the basic variable x_4 will exit the basis. This is illustrated by referring to the following equations:

$$3x_1 + 4x_2 + x_3 \qquad\qquad = 450$$
$$5x_1 + 2x_2 \qquad\quad + x_4 = 400$$

TABLE 6–3

Basis	x_1
	6
x_3	3 ⎫ a-coefficients
x_4	5 ⎭

The initial basic feasible solution is

Nonbasic Variables	Basic Variables
$x_1 = 0$	$x_3 = 450$
$x_2 = 0$	$x_4 = 400$

If we consider entering x_1 into the basis, then $x_1 > 0$, and $x_2 = 0$. Hence, the preceding equations become

$$3x_1 + x_3 \qquad\quad = 450$$
$$5x_1 \qquad + x_4 = 400$$

Solving for x_3 and x_4, we obtain

$x_3 = 450 - 3x_1$ (For each unit of x_1 to enter the basis, x_3 is decreased by 3.)

$x_4 = 400 - 5x_1$ (For each unit of x_1 to enter the basis, x_4 is decreased by 5.)

We must now determine what will happen to the value of the objective function, z, if x_1 is entered into the basis. Specifically, two things will occur:

1. Since $c_1 = 6$, then for each unit of x_1 entered into the basis, the value of the objective function will increase by 6.
2. For each unit of x_1 entered into the basis, 3 units of x_3 will exit, causing a decrease in the value of the objective function, z, by $0(3) = 0$; and 5 units of x_4 will exit, causing a decrease in the value of the objective function, z, by $0(5) = 0$. Thus, for each unit of x_1 entering the basis, the value of the objective function, z, is decreased by

$$z_1 = \begin{bmatrix} 0 \\ 0 \end{bmatrix} \begin{bmatrix} 3 \\ 5 \end{bmatrix} = 0(3) + 0(5) = 0$$

where the Basic c_j multiplies the left vector.

The value $z_1 = 0$ is entered in the tableau at the bottom of the x_1 column in the row labeled "z_j" (see Table 6–4). Hence, the net increase in the value of the objective function, z, will be $c_1 - z_1 = 6 - 0 = 6$. This result is entered in the tableau below $z_1 = 0$ in the row labeled "$c_j - z_j$." The $c_j - z_j$ values are called **indicators** since they indicate the net increase (or decrease, if negative) in the value of the objective function, z, if the corresponding variable x_j is entered into the basis.

TABLE 6–4

Basic c_j	Basis	x_1 6	x_2 4	x_3 0	x_4 0	b_i
0	x_3	3	4	1	0	450
0	x_4	5	2	0	1	400
	z_j	0				$z = 0$
	$c_j - z_j$	6				

If we consider entering x_2 into the basis, then:

1. The value of the objective function, z, is increased by $c_2 = 4$.
2. The value of the objective function, z, is decreased by

$$z_2 = \begin{bmatrix} 0 \\ 0 \end{bmatrix} \begin{bmatrix} 4 \\ 2 \end{bmatrix} = 0(4) + 0(2) = 0$$

The value $z_2 = 0$ is entered in the tableau at the bottom of the x_2 column (see Table 6–5). Hence, the net increase in the value of z is

$$c_2 - z_2 = 4 - 0 = 4$$

This is entered in Table 6–5 below $z_2 = 0$.

TABLE 6–5

Basic c_j	Basis	x_1	x_2	x_3	x_4	c_j b_i
		6	4	0	0	
0	x_3	3	4	1	0	450
0	x_4	5	2	0	1	400
	z_j	0	0	0	0	$z = 0$
	$c_j - z_j$	6	4	0	0	

We now calculate the indicators for x_3 and x_4, as follows:

$$z_3 = \begin{bmatrix} 0 \\ 0 \end{bmatrix}\begin{bmatrix} 1 \\ 0 \end{bmatrix} = 0(1) + 0(0) = 0 \qquad z_4 = \begin{bmatrix} 0 \\ 0 \end{bmatrix}\begin{bmatrix} 0 \\ 1 \end{bmatrix} = 0(0) + 0(1) = 0$$

$$c_3 - z_3 = 0 - 0 = 0 \qquad\qquad\qquad c_4 - z_4 = 0 - 0 = 0$$

These values are entered in the appropriate positions at the bottom of the x_3 and x_4 columns, respectively (see Table 6–5).

In answer to the question regarding which variable should enter the basis in order to improve the value of the objective function, z, by the largest amount, we choose the variable x_j with the largest positive indicator $c_j - z_j$. Hence, x_1 will be the entering variable.

Having decided which variable to enter into the basis, we now must learn how to enter it. Referring to Table 6–5, we must use matrix row operations to transform the x_1 column of a-coefficients $\begin{bmatrix} 3 \\ 5 \end{bmatrix}$ into either $\begin{bmatrix} 1 \\ 0 \end{bmatrix}$ or $\begin{bmatrix} 0 \\ 1 \end{bmatrix}$, the column vectors that identify basic variables. We must now decide which a-coefficient of the x_1 column vector $\begin{bmatrix} 3 \\ 5 \end{bmatrix}$ becomes 1. This coefficient will be called the **pivot** element. Determining the pivot element is equivalent to determining which of the present basic variables will exit the basis. Thus, we again consider the following equations:

$$3x_1 + 4x_2 + x_3 \qquad\quad = 450$$
$$5x_1 + 2x_2 \qquad\quad + x_4 = 400$$

LINEAR PROGRAMMING

The initial basic feasible solution is

Nonbasic Variables	Basic Variables
$x_1 = 0$	$x_3 = 450$
$x_2 = 0$	$x_4 = 400$

Since x_1 will be entering the basis, then $x_1 > 0$, and $x_2 = 0$. Hence, the equations become

$$3x_1 + x_3 = 450$$
$$5x_1 + x_4 = 400$$

Solving for x_3, we obtain

$$x_3 = 450 - 3x_1 \quad \text{(For each unit of } x_1 \text{ to enter the basis,}$$
$$\text{3 units of } x_3 \text{ will exit the basis.)}$$

Thus, since $x_3 \geq 0$, the maximum number of units of x_1 that can be entered will be that number at which x_3 diminishes to 0. Hence,

$$0 = 450 - 3x_1$$
$$3x_1 = 450$$
$$x_1 = \frac{450}{3}$$
$$= 150$$

Therefore, 150 units of x_1 may be entered before x_3 diminishes to 0. Solving for x_4 yields

$$x_4 = 400 - 5x_1 \quad \text{(For each unit of } x_1 \text{ to enter the basis,}$$
$$\text{5 units of } x_4 \text{ will exit the basis.)}$$

Thus, since $x_4 \geq 0$, the maximum number of units of x_1 that can be entered will be that number at which x_4 diminishes to 0. Hence,

$$0 = 400 - 5x_1$$
$$5x_1 = 400$$
$$x_1 = \frac{400}{5}$$
$$= 80$$

Therefore, 80 units of x_1 may be entered before x_4 diminishes to 0.

Logically, we cannot enter more units of x_1 than the *smaller* of the two ratios $450/3 = 150$ and $400/5 = 80$, because if we did enter 150 units (the larger ratio) of x_1, more units of x_4 than are available would exist the basis. Hence, 80 units (the smaller ratio) of x_1 will enter the basis, and x_4 will exit (diminish to a value of 0). Thus, the pivot element of the entering variable x_1 column (see Table 6–6) is determined as follows:

TABLE 6–6

x_1

6	$\leftarrow c_j$
	$\downarrow b_i$
3	450
5	400

1. Compute the ratios:

$$\frac{450}{3} = 150 \qquad \frac{400}{5} = 80$$

Note that only ratios for which the a-coefficients are positive should be considered.

2. Choose the pivot element to be the a-coefficient corresponding to the minimum ratio.

Hence, the pivot element is 5. The pivot element is circled in Table 6–7.

TABLE 6–7

Basic c_j	Basis	x_1	x_2	x_3	x_4	$\leftarrow c_j$
		6	4	0	0	$\downarrow b_i$
0	x_3	3	4	1	0	450
0	x_4	⑤	2	0	1	400
	z_j	0	0	0	0	$z = 0$
	$c_j - z_j$	6	4	0	0	

In general (see Table 6–8), if variable x_j is entering the basis, the pivot element is determined as follows:

1. Compute all ratios b_i/a_{ij} such that $a_{ij} > 0$.
2. Choose the pivot element to be the a-coefficient corresponding to the minimum ratio.

LINEAR PROGRAMMING

TABLE 6-8

$$
\begin{array}{c|c}
x_j & \\
\hline
c_j & \\
\hline
a_{1j} & b_1 \\
\vdots & \vdots \\
a_{ij} & b_i \\
\vdots & \vdots \\
a_{mj} & b_m \\
\end{array}
$$

Returning to our problem, we now enter x_1 using 5 as our pivot element:

	x_1	x_2	x_3	x_4	$\leftarrow c_j$
	6	4	0	0	$\swarrow b_i$
Row 1	3	4	1	0	450
Row 2	⑤	2	0	1	400

We transform the 5 to 1 by the following row operation. Multiply row 2 by 1/5. Rows 1 and 2 of our tableau then become

					b_i
Row 1	3	4	1	0	450
Row 2	1	2/5	0	1/5	80

Next, we transform the 3 in column 1 to 0 by multiplying row 2 by -3 and adding the result to row 1. Our tableau now becomes

		x_1	x_2	x_3	x_4	
Basic c_j	Basis	6	4	0	0	$\leftarrow c_j$
						$\swarrow b_i$
0	x_3	0	14/5	1	$-3/5$	210
6	x_1	1	2/5	0	1/5	80
	z_j	6	12/5	0	6/5	$z = 480$
	$c_j - z_j$	0	8/5	0	$-6/5$	

Note that x_1 has replaced x_4 in the basis. The c-coefficient corresponding to x_1 ($c_1 = 6$) has replaced $c_4 = 0$ in the "Basic c_j" column. Also, the z_j and $c_j - z_j$ values for each column have been calculated and placed in their appropriate positions. Note that the value of the objective function, z, has increased from 0 to 480. The z_j and $c_j - z_j$ computations are shown in Table 6–9.

TABLE 6–9

$$z_1 = \begin{bmatrix} 0 \\ 6 \end{bmatrix} \begin{bmatrix} 0 \\ 1 \end{bmatrix} = 0(0) + 6(1) = 6; \quad c_1 - z_1 = 6 - 6 = 0$$

$$z_2 = \begin{bmatrix} 0 \\ 6 \end{bmatrix} \begin{bmatrix} 14/5 \\ 2/5 \end{bmatrix} = 0(14/5) + 6(2/5) = 12/5; \quad c_2 - z_2 = 4 - 12/5 = 8/5$$

$$z_3 = \begin{bmatrix} 0 \\ 6 \end{bmatrix} \begin{bmatrix} 1 \\ 0 \end{bmatrix} = 0(1) + 6(0) = 0; \quad c_3 - z_3 = 0 - 0 = 0$$

$$z_4 = \begin{bmatrix} 0 \\ 6 \end{bmatrix} \begin{bmatrix} -3/5 \\ 1/5 \end{bmatrix} = 0(-3/5) + 6(1/5) = 6/5; \quad c_4 - z_4 = 0 - 6/5 = -6/5$$

Objective function value, $z = \begin{bmatrix} 0 \\ 6 \end{bmatrix} \begin{bmatrix} 210 \\ 80 \end{bmatrix} = 0(210) + 6(80) = 480$

Looking at the tableau, note that the present basic feasible solution is

Nonbasic Variables	Basic Variables
$x_2 = 0$	$x_3 = 210$
$x_4 = 0$	$x_1 = 80$

However, this basic feasible solution is not optimal since a scanning of the $c_j - z_j$ indicator row reveals a positive indicator of $c_2 - z_2 = 8/5$. Thus, for each unit of x_2 entered into the basis, the objective function value, z, will be increased by 8/5. We restate the following idea.

> As long as there exists a positive indicator $c_j - z_j$, the present basic feasible solution is not optimal. The value of the objective function may be increased by entering the variable corresponding to a positive $c_j - z_j$. If there is more than one positive indicator $c_j - z_j$, then enter the variable corresponding to the largest positive indicator $c_j - z_j$.

LINEAR PROGRAMMING

Thus, x_2 will enter the basis. We must now determine the pivot element of the x_2 column. Hence, we compute the ratios b_i/a_{ij}, where a_{ij} represents the a-coefficients of the x_2 column. Again, we only consider such ratios for positive a_{ij}'s. Observing Table 6–10, the ratios are

$$\frac{210}{\frac{14}{5}} = \frac{1050}{14} = 75$$

$$\frac{80}{\frac{2}{5}} = \frac{400}{2} = 200$$

TABLE 6–10

x_2

c_j	
4	b_i
14/5	210
2/5	80

The minimum ratio is 75. Its corresponding a-coefficient, 14/5, is our pivot element (see Table 6–11). We now transform 14/5 to 1 by the following row operation.

TABLE 6–11

	x_1	x_2	x_3	x_4	c_j	
	6	4	0	0		b_i
Row 1	0	(14/5)	1	−3/5	210
Row 2	1	2/5	0	1/5	80

Multiply row 1 by 5/14. Thus, rows 1 and 2 of our tableau become

					b_i
Row 1	0	1	5/14	−3/14	75
Row 2	1	2/5	0	1/5	80

We then transform 2/5 in column 2 to 0 by multiplying row 1 by −2/5 and adding the result to row 2. Our tableau now becomes Table 6–12.

TABLE 6–12

Basic c_j	Basis	x_1 6	x_2 4	x_3 0	x_4 0	c_j ← b_i ↓
4	x_2	0	1	5/14	–3/14	75
6	x_1	1	0	–1/7	2/7	50
	z_j	6	4	4/7	6/7	z = 600
	$c_j - z_j$	0	0	–4/7	–6/7	

Note that x_2 has replaced x_3 in the basis. The c-coefficient corresponding to x_2 ($c_2 = 4$) has replaced $c_3 = 0$ in the "Basic c_j" column. The z_j and $c_j - z_j$ values for each column have been calculated and placed in their positions. Note that the value of the objective function, z, has increased from 480 to 600. These computations are shown in Table 6–13.

TABLE 6–13

$z_1 = \begin{bmatrix} 4 \\ 6 \end{bmatrix} \begin{bmatrix} 0 \\ 1 \end{bmatrix} = 4(0) + 6(1) = 6;\ c_1 - z_1 = 6 - 6 = 0$

$z_2 = \begin{bmatrix} 4 \\ 6 \end{bmatrix} \begin{bmatrix} 1 \\ 0 \end{bmatrix} = 4(1) + 6(0) = 4;\ c_2 - z_2 = 4 - 4 = 0$

$z_3 = \begin{bmatrix} 4 \\ 6 \end{bmatrix} \begin{bmatrix} 5/14 \\ -1/7 \end{bmatrix} = 4(5/14) + 6(-1/7) = 4/7;\ c_3 - z_3 = 0 - 4/7 = -4/7$

$z_4 = \begin{bmatrix} 4 \\ 6 \end{bmatrix} \begin{bmatrix} -3/14 \\ 2/7 \end{bmatrix} = 4(-3/14) + 6(2/7) = 6/7;\ c_4 - z_4 = 0 - 6/7 = -6/7$

Objective function value, $z = \begin{bmatrix} 4 \\ 6 \end{bmatrix} \begin{bmatrix} 75 \\ 50 \end{bmatrix} = 4(75) + 6(50) = 600$

Scanning the $c_j - z_j$ indicator row of the tableau of Table 6–12, we note that there are no positive indicators. Hence, the following basic feasible solution is optimal:

Nonbasic Variables	Basic Variables
$x_3 = 0$	$x_2 = 75$
$x_4 = 0$	$x_1 = 50$

The tableau of Table 6–12 is called the **final tableau**.

EXAMPLE 6–10 Using the simplex method, solve the following:

$$\text{Maximize} \quad z = 2x_1 + 5x_2 + 4x_3$$
$$\text{subject to} \quad x_1 + 3x_2 + 2x_3 \leq 60$$
$$4x_1 + 2x_2 + x_3 \leq 100$$

where x_1, x_2, and x_3 are nonnegative

Solution Introducing slack variables x_4 and x_5, formulating the initial tableau, and computing z_j and $c_j - z_j$ values, we have

Initial Tableau

Basic c_j	Basis	x_1	x_2	x_3	x_4	x_5	c_j
		2	5	4	0	0	b_i
0	x_4	1	③	2	1	0	60
0	x_5	4	2	1	0	1	100
	z_j	0	0	0	0	0	$z = 0$
	$c_j - z_j$	2	5	4	0	0	

Scanning the indicators, $c_j - z_j$, of the initial tableau, we note that variable x_2 must enter the basis since it has the largest positive indicator, 5. Computing the b_i/a_{ij} ratios corresponding to the x_2 column, we have

$$\frac{60}{3} = 20 \qquad \frac{100}{2} = 50$$

Choosing the a-coefficient corresponding to the minimum ratio, 20, we have a pivot element of 3. The pivot element is circled in our initial tableau. We transform the 3 to 1 by multiplying row 1 by 1/3. Rows 1 and 2 of our tableau then become

Row 1	1/3	1	2/3	1/3	0	20
Row 2	4	2	1	0	1	100

Next, we transform the 2 in column 2 to 0 by multiplying row

EXAMPLE 6–10 (continued)

1 by −2 and adding the result to row 2. Our tableau then becomes

Basic c_j	Basis	x_1 2	x_2 5	x_3 4	x_4 0	x_5 0	← c_j ↓ b_i
5	x_2	1/3	1	2/3	1/3	0	20
0	x_5	10/3	0	−1/3	−2/3	1	60
	z_j	5/3	5	10/3	5/3	0	z = 100
	$c_j - z_j$	1/3	0	2/3	−5/3	0	

Observe that z_j and $c_j - z_j$ values have been computed. Since the largest positive indicator is 2/3, we must enter x_3 into the basis. There is only one b_i/a_{ij} value to compute since there is only one positive a-coefficient—2/3—in the x_3 column. Thus, 2/3 is the pivot element. We transform 2/3 to 1 by multiplying row 1 by 3/2. Rows 1 and 2 of our tableau are now

						b_i
Row 1	1/2	3/2	1	1/2	0	30
Row 2	10/3	0	−1/3	−2/3	1	60

Then, we transform −1/3 of the x_3 column to 0 by multiplying row 1 by 1/3 and adding the result to row 2. This gives us the following tableau:

Basic c_j	Basis	x_1 2	x_2 5	x_3 4	x_4 0	x_5 0	← c_j ↓ b_i
4	x_3	1/2	3/2	1	1/2	0	30
0	x_5	7/2	1/2	0	−1/2	1	70
	z_j	2	6	4	2	0	z = 120
	$c_j - z_j$	0	−1	0	−2	0	

Observe that z_j and $c_j - z_j$ values have been calculated.

Since there are no positive indicators, this tableau is the final tableau. Thus, the optimal solution is

Nonbasic Variables	Basic Variables
$x_1 = 0$	$x_3 = 30$
$x_2 = 0$	$x_5 = 70$
$x_4 = 0$	

The objective function value is $z = 120$.

EXERCISES

1. Using the simplex method, solve the following:

$$\text{Maximize} \quad z = 120x_1 + 60x_2$$
$$\text{subject to} \quad 2x_1 + 3x_2 \leq 90$$
$$5x_1 + x_2 \leq 160$$
$$\text{where} \quad x_1 \geq 0, x_2 \geq 0$$

2. Using the simplex method, solve the following:

$$\text{Maximize} \quad z = 10x_1 + 30x_2$$
$$\text{subject to} \quad 3x_1 + 6x_2 \leq 60$$
$$4x_1 + 4x_2 \leq 60$$
$$\text{where} \quad x_1 \geq 0, x_2 \geq 0$$

3. Using the simplex method, solve the following:

$$\text{Maximize} \quad z = 400x_1 + 500x_2$$
$$\text{subject to} \quad x_1 + x_2 \leq 100$$
$$60x_1 + 80x_2 \leq 6600$$
$$\text{where} \quad x_1 \geq 0, x_2 \geq 0$$

4. Using the simplex method, solve the following:

$$\text{Maximize} \quad z = 5x_1 + 6x_2 + x_3$$
$$\text{subject to} \quad 2x_1 + 3x_2 + 5x_3 \leq 20$$
$$6x_1 + 2x_2 + 3x_3 \leq 50$$
$$x_1 + 3x_2 + 4x_3 \leq 24$$
$$\text{where} \quad x_1 \geq 0, x_2 \geq 0, x_3 \geq 0$$

5. Using the simplex method, solve the following:

$$\text{Maximize} \quad z = 20x_1 + 42x_2 + 56x_3$$
$$\text{subject to} \quad 2x_1 + 3x_2 + x_3 \leq 6$$
$$4x_1 + 2x_2 + 3x_3 \leq 12$$
$$4x_1 + 2x_2 + x_3 \leq 8$$
$$\text{where} \quad x_1 \geq 0, x_2 \geq 0, x_3 \geq 0$$

6. Using the simplex method, solve the following:

$$\text{Maximize} \quad z = 80x_1 + 10x_2 + 16x_3 + 12x_4$$
$$\text{subject to} \quad x_1 + x_2 + x_3 + x_4 \leq 40$$
$$2x_1 + x_2 + 4x_3 + x_4 \leq 90$$

where $x_1, x_2, x_3,$ and x_4 are nonnegative

7. (The following problem appeared in a Uniform CPA Examination.) Beekley, Inc. manufactures widgets, gadgets, and trinkets and has asked for advice in determining the best production mix for its three products. Demand for the company's products is excellent, and management finds that it is unable to meet potential sales with existing plant capacity.

Each product goes through three operations: milling, grinding, and painting. The effective weekly departmental capacities in minutes are: milling—10,000; grinding—11,000; and painting—10,000. The following data are available on the three products:

Product	Selling Price per Unit	Variable Cost per Unit	Milling	Grinding	Painting
Widgets	$5.25	$4.45	4	8	4
Gadgets	5.00	3.90	10	4	2
Trinkets	4.50	3.30	4	8	2

Per-Unit Production Time (minutes)

Solve this problem by the simplex method.

8. *(Alternate Optima)* If, in a final tableau, a nonbasic variable has a zero indicator, the variable may be entered into the basis without changing the optimal value of the objective function. Thus, an alternate optimal solution exists. Refer to the final tableau of Example 6–10 of this section. Observe that the nonbasic variable x_1 has a zero indicator.
   (A) Enter x_1 into the basis.
   (B) State the alternate optimal solution.
   (C) Has the optimal value of z changed?

9. Solve the following by the simplex method:

$$\text{Maximize} \quad z = 3x_1 + 6x_2$$
$$\text{subject to} \quad x_1 + x_2 \leq 55$$
$$2x_1 + 5x_2 \leq 200$$
$$x_1 \leq 40$$
$$x_2 \leq 38$$

where $x_1 \geq 0, x_2 \geq 0$

10. *(Degeneracy)* This exercise is a step-by-step introduction to the concept of degeneracy. Consider the problem below and follow the indicated steps.

$$\text{Maximize} \quad z = 400x_1 + 500x_2$$
$$\text{subject to} \quad 3x_1 + 5x_2 \le 55$$
$$x_1 + 5x_2 \le 45$$
$$x_1 \le 10$$
$$x_2 \le 9$$
$$\text{where} \quad x_1 \ge 0, x_2 \ge 0$$

Step 1 Formulate the initial tableau and determine the entering variable.

Step 2 Determine the pivot element. Note that there is more than one minimum ratio b_i/a_{ij}. Such a situation is called a **degeneracy**. Arbitrarily select either of the two candidates as the pivot element and enter the variable into the basis.

Step 3 Observe the resulting solution. Note that one of the basic variables has a value of 0. Such a solution is called a **degenerate solution**.

Step 4 Continue the simplex method until an optimal solution is determined. Notice that the degeneracy is resolved and that the basic variables of the optimal solution have nonzero values.

Comments

a. Sometimes, an optimal solution may be degenerate. In this case, one or more of the basic variables would have zero values. Such an optimal solution is called an **optimal degenerate solution**.

b. Intuitively, one characteristic of degeneracy is **redundant constraints**. This may be observed for this exercise by returning to the graphical solution of this linear programming problem, found in Exercise 8, Section 6–3. Observe that the constraint $y \le 9$ (or $x_2 \le 9$) is redundant.

6–6

SHADOW PRICES

We return to the illustrative problem of Sections 6–4 and 6–5. For reference purposes, we restate it here:

$$\text{Maximize} \quad z = 6x_1 + 4x_2$$
$$\text{subject to} \quad 3x_1 + 4x_2 + x_3 = 450 \quad \text{Department I}$$
$$5x_1 + 2x_2 + x_4 = 400 \quad \text{Department II}$$
$$\text{where} \quad x_1 \ge 0, x_2 \ge 0, x_3 \ge 0, x_4 \ge 0$$

CHAPTER SIX

The initial tableau is

Basic c_j	Basis	x_1 6	x_2 4	x_3 0	x_4 0	$\leftarrow c_j$
0	x_3	3	4	1	0	450
0	x_4	5	2	0	1	400

($\downarrow b_i$)

The final tableau is

Basic c_j	Basis	x_1 6	x_2 4	x_3 0	x_4 0	$\leftarrow c_j$
4	x_2	0	1	5/14	−3/14	75
6	x_1	1	0	−1/7	2/7	50
	z_j	6	4	4/7	6/7	$z = 600$
	$c_j - z_j$	0	0	−4/7	−6/7	

↑ ↑
Shadow price indicators

Referring to the final tableau, we note that the absolute value of the indicators corresponding to the nonbasic slack variables are called **shadow prices**. Thus, $|c_3 - z_3| = 4/7$ and $|c_4 - z_4| = 6/7$ are shadow prices. In order to interpret a shadow price, we must identify its associated slack variable and corresponding constraint. Specifically, the shadow price $|c_3 - z_3| = 4/7$ is associated with slack variable x_3. Recall that x_3 represents the amount of slack in Department I (which is constrained by a capacity of 450 hours). Since x_3 is nonbasic, its value is 0. Thus, there is no slack in Department I (its capacity of 450 hours is used up). Note that the Department I constraint, $3x_1 + 4x_2 \leq 450$, is of the "less than or equal to" type. For a "less than or equal to" constraint, decreasing its b-coefficient (or capacity) by one unit is equivalent to entering one unit of slack into the constraint. Thus, for each unit *decrease* in the value of the b-coefficient $b_1 = 450$ of the Department I constraint, one unit of slack, x_3, is entered into the basis of the final tableau, resulting in a decrease ($c_3 - z_3 = -4/7$) in the value of the objective function by 4/7. Analogously, for each unit *increase* in the value of the b-coefficient $b_1 = 450$, the value of the objective function increases by 4/7, or $0.57. In summary, we have the following.

> A **shadow price** represents the dollar increase in the value of the objective function z resulting from a unit increase in the associated constraint capacity or, equivalently, the maximum amount one is willing to pay for a unit increase in the associated constraint capacity.

The shadow price $|c_4 - z_4| = 6/7$ is associated with slack variable x_4. Recall that x_4 represents the amount of slack in Department II (which is constrained by a capacity of 400 hours). Since x_4 is nonbasic, its value is 0. Thus, there is no slack in Department II (its capacity of 400 hours is used up). The shadow price 6/7 indicates that for each unit increase in the value of the b-coefficient $b_2 = 400$, the value of the objective function increases by 6/7, or $0.86. Accordingly, one is willing to pay a maximum of $0.86 for each additional hour of Department II time.

EXERCISES

1. Refer to the final tableau of Example 6–10.
   (A) Identify the shadow prices.
   (B) Interpret the shadow prices.
2. Refer to the final tableau of Exercise 1, Section 6–5.
   (A) Identify the shadow prices.
   (B) Interpret the shadow prices.
3. Refer to the final tableau of Exercise 2, Section 6–5.
   (A) Identify the shadow prices.
   (B) Interpret the shadow prices.
4. Refer to the final tableau of Exercise 3, Section 6–5.
   (A) Identify the shadow prices.
   (B Interpret the shadow prices.
5. Refer to the final tableau of Exercise 4, Section 6–5.
   (A) Identify the shadow prices.
   (B) Interpret the shadow prices.
6. Refer to the final tableau of Exercise 5, Section 6–5.
   (A) Identify the shadow prices.
   (B) Interpret the shadow prices.
7. Refer to the final tableau of Exercise 6, Section 6–5.
   (A) Identify the shadow prices.
   (B) Interpret the shadow prices.

6-7
SIMPLEX METHOD: MINIMIZATION

Up to this point, we have used the simplex method to solve maximization problems. The typical maximization problem (see Table 6–14) involves:

1. An objective function whose value, z, is *maximized*
2. "Less than or equal to" constraints

TABLE 6–14

> Maximize $z = c_1x_1 + c_2x_2 + \ldots + c_kx_k$
> subject to $a_{11}x_1 + a_{12}x_2 + \ldots + a_{1k}x_k \leq b_1$
> $\qquad \vdots \qquad \vdots \qquad \vdots \qquad \vdots$
> $\qquad a_{m1}x_1 + a_{m2}x_2 + \ldots + a_{mk}x_k \leq b_m$
> where $x_1 \geq 0, x_2 \geq 0, \ldots, x_k \geq 0; b_1 \geq 0, b_2 \geq 0, \ldots, b_m \geq 0$

We will now discuss minimization problems. The typical minimization problem (see Table 6–15) involves:

1. An objective function whose value, z, is *minimized*
2. "Greater than or equal to" constraints

TABLE 6–15

> Minimize $z = c_1x_1 + c_2x_2 + \ldots + c_kx_k$
> subject to $a_{11}x_1 + a_{12}x_2 + \ldots + a_{1k}x_k \geq b_1$
> $\qquad \vdots \qquad \vdots \qquad \vdots \qquad \vdots$
> $\qquad a_{m1}x_1 + a_{m2}x_2 + \ldots + a_{mk}x_k \geq b_m$
> where $x_1 \geq 0, x_2 \geq 0, \ldots, x_k \geq 0; b_1 \geq 0, b_2 \geq 0, \ldots, b_m \geq 0$

We now consider the minimization problem of Section 6–3. For reference purposes, the problem is summarized in the data box of Table 6–16.

TABLE 6–16

	x_1 Type I Food	x_2 Type II Food	
Unit Costs →	$2/pound	$8/pound	
Vitamin A	10 milligrams/pound	30 milligrams/pound	At least 140 milligrams
Vitamin B	20 milligrams/pound	15 milligrams/pound	At least 145 milligrams

LINEAR PROGRAMMING

If

x_1 = number of pounds of Type I food purchased
x_2 = number of pounds of Type II food purchased
z = total cost

then the corresponding algebraic formulation is

Minimize $z = 2x_1 + 8x_2$
subject to $10x_1 + 30x_2 \geq 140$
$20x_1 + 15x_2 \geq 145$
where $x_1 \geq 0, x_2 \geq 0$

Since the inequalities are of the "greater than or equal to" type, we restate them as linear equations by subtracting nonnegative quantities from their left-hand sides. Thus, we obtain

$10x_1 + 30x_2 - x_3 = 140$
$20x_1 + 15x_2 - x_4 = 145$
where $x_1, x_2, x_3,$ and x_4 are nonnegative

The nonnegative variables x_3 and x_4 are called **surplus variables**: x_3 represents the amount of surplus of Vitamin A, and x_4 represents the amount of surplus of Vitamin B.

For the preceding system, the initial basic solution would be the following:

Nonbasic Variables	Basic Variables
$x_1 = 0$	$-x_3 = 140$ or $x_3 = -140$
$x_2 = 0$	$-x_4 = 145$ or $x_4 = -145$

Unfortunately, this presents a problem since x_3 and x_4 have violated the nonnegativity condition. Therefore, this basic solution is not feasible. This problem is remedied by the addition of a nonnegative **artificial variable** to the left-hand side of each inequality. Thus, we get

$10x_1 + 30x_2 - x_3 + x_5 = 140$
$20x_1 + 15x_2 - x_4 + x_6 = 145$
where $x_1, x_2, \underbrace{x_3, x_4,}_{\text{surplus variables}} \underbrace{x_5, \text{ and } x_6}_{\text{artificial variables}}$ are nonnegative

Since the variables x_5 and x_6 are artificial, the simplex method must be made to drive their values to 0. This is accomplished by assigning them arbitrarily large objective function coefficients.

354 CHAPTER SIX

These coefficients are designated by M. Thus, the minimization process will force these artificial variables to 0. Our problem is now stated as

$$\text{Minimize} \quad z = 2x_1 + 8x_2 + 0x_3 + 0x_4 + Mx_5 + Mx_6$$
$$\text{subject to} \quad 10x_1 + 30x_2 - x_3 \quad\quad + x_5 \quad\quad = 140$$
$$20x_1 + 15x_2 \quad\quad - x_4 \quad\quad + x_6 = 145$$
$$\text{where} \quad x_1, x_2, x_3, x_4, x_5, \text{ and } x_6 \text{ are nonnegative}$$

The initial basic feasible solution is

Nonbasic Variables	Basic Variables
$x_1 = 0$	$x_5 = 140$
$x_2 = 0$	$x_6 = 145$
$x_3 = 0$	
$x_4 = 0$	

The initial tableau is

Basic c_j	Basis	x_1 2	x_2 8	x_3 0	x_4 0	x_5 M	x_6 M	c_j b_i
M	x_5	10	30	-1	0	1	0	140
M	x_6	20	15	0	-1	0	1	145
	z_j	$30M$	$45M$	$-M$	$-M$	M	M	$z = 285M$
	$c_j - z_j$	$2 - 30M$	$8 - 45M$	M	M	0	0	

The calculations for z_j and $c_j - z_j$ values are illustrated in Table 6–17.

TABLE 6–17

$$z_1 = \begin{bmatrix} M \\ M \end{bmatrix} \begin{bmatrix} 10 \\ 20 \end{bmatrix} = 10M + 20M = 30M; \; c_1 - z_1 = 2 - 30M$$

$$z_2 = \begin{bmatrix} M \\ M \end{bmatrix} \begin{bmatrix} 30 \\ 15 \end{bmatrix} = 30M + 15M = 45M; \; c_2 - z_2 = 8 - 45M$$

$$z_3 = \begin{bmatrix} M \\ M \end{bmatrix} \begin{bmatrix} -1 \\ 0 \end{bmatrix} = -1(M) + 0(M) = -M; \; c_3 - z_3 = 0 - (-M) = M$$

$$z_4 = \begin{bmatrix} M \\ M \end{bmatrix} \begin{bmatrix} 0 \\ -1 \end{bmatrix} = 0(M) + (-1)(M) = -M; \; c_4 - z_4 = 0 - (-M) = M$$

$$z_5 = \begin{bmatrix} M \\ M \end{bmatrix} \begin{bmatrix} 1 \\ 0 \end{bmatrix} = 1(M) + 0(M) = M; \; c_5 - z_5 = M - M = 0$$

$$z_6 = \begin{bmatrix} M \\ M \end{bmatrix} \begin{bmatrix} 0 \\ 1 \end{bmatrix} = 0(M) + 1(M) = M; \; c_6 - z_6 = M - M = 0$$

LINEAR PROGRAMMING

$$\text{Objective function value, } z = \begin{bmatrix} M \\ M \end{bmatrix} \begin{bmatrix} 140 \\ 145 \end{bmatrix} = 140M + 145M = 285M$$

We must now choose an entering variable. Since we are *minimizing* the value of the objective function, z, we must *choose the variable with the most negative indicator $c_j - z_j$*. Recall that an indicator value $c_j - z_j$ designates the net increase (or decrease, if negative) in the value of the objective function if the corresponding variable x_j is entered into the basis. Hence, x_2 will enter the basis since its indicator, $8 - 45M$, is the most negative, as M is arbitrarily large.

We must now determine the pivot element of the x_2 column. Thus, we compute the positive ratios b_i/a_{ij}, where a_{ij} represents the a-coefficients of the x_2 column (see Table 6–18). Hence, the ratios are

$$\frac{140}{30} = 4\frac{2}{3} \qquad \frac{145}{15} = 9\frac{2}{3}$$

TABLE 6–18

x_2

8	$\leftarrow c_j$
	$\downarrow b_i$
30	140
15	145

The minimum ratio is $4\frac{2}{3}$. Its corresponding a-coefficient, 30, is the pivot element (see Table 6–19).

TABLE 6–19

	x_1	x_2	x_3	x_4	x_5	x_6	
	2	8	0	0	M	M	$\leftarrow c_j$
							$\downarrow b_i$
Row 1	10	㉚	−1	0	1	0	140
Row 2	20	15	0	−1	0	1	145

We now transform 30 to 1 by multiplying row 1 by 1/30. Rows 1 and 2 of the tableau then become

							b_i
Row 1	1/3	1	−1/30	0	1/30	0	14/3
Row 2	20	15	0	−1	0	1	145

Next, we transform the 15 in column 2 to 0 by multiplying row 1 by −15 and adding the result to row 2. The tableau is illustrated in

Table 6–20. The calculations for z_j and $c_j - z_j$ values are illustrated in Table 6–21.

TABLE 6–20

Basic c_j	Basis	x_1	x_2	x_3	x_4	x_5	x_6	b_i
		2	8	0	0	M	M ← c_j	
8	x_2	1/3	1	−1/30	0	1/30	0	14/3
M	x_6	15	0	1/2	−1	−1/2	1	75
	z_j	8/3 + 15M	8	−8/30 + (1/2)M	−M	8/30 − (1/2)M	M	112/3 + 75M = z
	$c_j - z_j$	−2/3 − 15M	0	8/30 − (1/2)M	M	−8/30 + (3/2)M	0	

TABLE 6–21

$z_1 = \begin{bmatrix} 8 \\ M \end{bmatrix} \begin{bmatrix} 1/3 \\ 15 \end{bmatrix} = 8/3 + 15M$; $c_1 - z_1 = 2 - (8/3 + 15M) = -2/3 - 15M$

$z_2 = \begin{bmatrix} 8 \\ M \end{bmatrix} \begin{bmatrix} 1 \\ 0 \end{bmatrix} = 1(8) + 0(M) = 8$; $c_2 - z_2 = 8 - 8 = 0$

$z_3 = \begin{bmatrix} 8 \\ M \end{bmatrix} \begin{bmatrix} -1/30 \\ 1/2 \end{bmatrix} = -8/30 + (1/2)M$; $c_3 - z_3 = 0 - [-8/30 + (1/2)M] = 8/30 - (1/2)M$

$z_4 = \begin{bmatrix} 8 \\ M \end{bmatrix} \begin{bmatrix} 0 \\ -1 \end{bmatrix} = 0(8) - 1(M) = -M$; $c_4 - z_4 = 0 - (-M) = M$

$z_5 = \begin{bmatrix} 8 \\ M \end{bmatrix} \begin{bmatrix} 1/30 \\ -1/2 \end{bmatrix} = 8/30 - (1/2)M$; $c_5 - z_5 = M - [8/30 - (1/2)M] = -8/30 + (3/2)M$

$z_6 = \begin{bmatrix} 8 \\ M \end{bmatrix} \begin{bmatrix} 0 \\ 1 \end{bmatrix} = 0(8) + 1(M) = M$; $c_6 - z_6 = M - M = 0$

Objective function value, $z = \begin{bmatrix} 8 \\ M \end{bmatrix} \begin{bmatrix} 14/3 \\ 75 \end{bmatrix} = 8(14/3) + 75M = 112/3 + 75M$

Scanning the $c_j - z_j$ indicator row of the tableau of Table 6–20, we note that the most negative indicator is $-2/3 - 15M$ since M is arbitrarily large. Thus, its associated variable, x_1, will now enter the basis. Next, we must determine the pivot element of the x_1 column. Therefore, we compute the positive ratios b_i/a_{ij}, where a_{ij} represents the a-coefficients of the x_1 column (see Table 6–22).

TABLE 6–22

x_1	
2 ← c_j	
	b_i
1/3	14/3
15	75

LINEAR PROGRAMMING

Hence,

$$\frac{\frac{14}{3}}{\frac{1}{3}} = \frac{14}{3} \cdot \frac{3}{1} = 14 \qquad \frac{75}{15} = 5$$

The minimum ratio is 5. Its corresponding a-coefficient, 15, is the pivot element (see Table 6–23).

TABLE 6–23

	x_1	x_2	x_3	x_4	x_5	x_6	c_i	b_i
	2	8	0	0	M	M		
Row 1	1/3	1	−1/30	0	1/30	0		14/3
Row 2	⑮	0	1/2	−1	−1/2	1		75

We now transform 15 to 1 by multiplying row 2 by 1/15. Rows 1 and 2 of our tableau then become

							b_i
Row 1	1/3	1	−1/30	0	1/30	0	14/3
Row 2	1	0	1/30	−1/15	−1/30	1/15	5

We now transform the 1/3 in column 1 to 0 by multiplying row 2 by −1/3 and adding the result to row 1. Our tableau is shown in Table 6–24. The calculations for z_j and $c_j - z_j$ are illustrated in Table 6–25.

TABLE 6–24

Basic c_j	Basis	x_1	x_2	x_3	x_4	x_5	x_6	c_j	b_i
		2	8	0	0	M	M		
8	x_2	0	1	−4/90	1/45	4/90	−1/45		3
2	x_1	1	0	1/30	−1/15	−1/30	1/15		5
	z_j	2	8	−26/90	2/45	26/90	−2/45		$z = 34$
	$c_j - z_j$	0	0	26/90	−2/45	M − 26/90	M + 2/45		

TABLE 6–25

$$z_1 = \begin{bmatrix} 8 \\ 2 \end{bmatrix}\begin{bmatrix} 0 \\ 1 \end{bmatrix} = 8(0) + 2(1) = 2; \quad c_1 - z_1 = 2 - 2 = 0$$

$$z_2 = \begin{bmatrix} 8 \\ 2 \end{bmatrix}\begin{bmatrix} 1 \\ 0 \end{bmatrix} = 8(1) + 2(0) = 8; \quad c_2 - z_2 = 8 - 8 = 0$$

$$z_3 = \begin{bmatrix} 8 \\ 2 \end{bmatrix}\begin{bmatrix} -4/90 \\ 1/30 \end{bmatrix} = 8(-4/90) + 2(1/30) = -26/90; \quad c_3 - z_3 = 0 - (-26/90) = 26/90$$

$$z_4 = \begin{bmatrix} 8 \\ 2 \end{bmatrix}\begin{bmatrix} 1/45 \\ -1/15 \end{bmatrix} = 8(1/45) + 2(-1/15) = 2/45; \quad c_4 - z_4 = 0 - 2/45 = -2/45$$

$$z_5 = \begin{bmatrix} 8 \\ 2 \end{bmatrix}\begin{bmatrix} 4/90 \\ -1/30 \end{bmatrix} = 8(4/90) + 2(-1/30) = 26/90; \quad c_5 - z_5 = M - 26/90$$

$$z_6 = \begin{bmatrix} 8 \\ 2 \end{bmatrix}\begin{bmatrix} -1/45 \\ 1/15 \end{bmatrix} = 8(-1/45) + 2(1/15) = -2/45; \quad c_6 - z_6 = M - (-2/45) = M + 2/45$$

Objective function value, $z = \begin{bmatrix} 8 \\ 2 \end{bmatrix}\begin{bmatrix} 3 \\ 5 \end{bmatrix} = 8(3) + 2(5) = 34$

Scanning the $c_j - z_j$ indicator row of Table 6–24, we note that the most negative indicator is $-2/45$ since M is arbitrarily large. Thus, its associated variable, x_4, will now enter the basis. Next, we must determine the pivot element of the x_4 column. Thus, we compute the positive ratios b_i/a_{ij}, where a_{ij} represents the a-coefficients of the x_4 column (see Table 6–26).

TABLE 6–26

x_4

c_j	b_i
0	
1/45	3
−1/15	5

Since the a-coefficient $-1/15$ is negative, we must disregard the ratio $5/(-1/15)$. Thus, $-1/15$ cannot be a pivot element. The only alternative is $1/45$. Therefore, $1/45$ is the pivot element (see Table 6–27).

TABLE 6–27

	x_1	x_2	x_3	x_4	x_5	x_6	c_j	b_i
	2	8	0	0	M	M		
Row 1	0	1	−4/90	⓵1/45	4/90	−1/45		3
Row 2	1	0	1/30	−1/15	−1/30	1/15		5

LINEAR PROGRAMMING

We now change 1/45 to 1 by multiplying row 1 by 45. Rows 1 and 2 of our tableau become

Row 1	0	45	−2	1	2	−1	135
Row 2	1	0	1/30	−1/15	−1/30	1/15	5

We then change −1/15 in column 4 to 0 by multiplying row 1 by 1/15 and adding the result to row 2. The resulting tableau is shown in Table 6–28.

TABLE 6–28

Basic c_j	Basis	x_1	x_2	x_3	x_4	x_5	x_6	$c_j \leftarrow$ / $b_i \downarrow$
		2	8	0	0	M	M	
0	x_4	0	45	−2	1	2	−1	135
2	x_1	1	3	−1/10	0	1/10	0	14
	z_j	2	6	−1/5	0	1/5	0	z = 28
	$c_j − z_j$	0	2	1/5	0	M − 1/5	M	

The calculations for z_j and $c_j − z_j$ values are illustrated in Table 6–29.

TABLE 6–29

$$z_1 = \begin{bmatrix} 0 \\ 2 \end{bmatrix} \begin{bmatrix} 0 \\ 1 \end{bmatrix} = 0(0) + 2(1) = 2; \; c_1 − z_1 = 2 − 2 = 0$$

$$z_2 = \begin{bmatrix} 0 \\ 2 \end{bmatrix} \begin{bmatrix} 45 \\ 3 \end{bmatrix} = 0(45) + 2(3) = 6; \; c_2 − z_2 = 8 − 6 = 2$$

$$z_3 = \begin{bmatrix} 0 \\ 2 \end{bmatrix} \begin{bmatrix} -2 \\ -1/10 \end{bmatrix} = 0(-2) + 2(-1/10) = -1/5; \; c_3 − z_3 = 0 − (−1/5) = 1/5$$

$$z_4 = \begin{bmatrix} 0 \\ 2 \end{bmatrix} \begin{bmatrix} 1 \\ 0 \end{bmatrix} = 0(1) + 2(0) = 0; \; c_4 − z_4 = 0 − 0 = 0$$

$$z_5 = \begin{bmatrix} 0 \\ 2 \end{bmatrix} \begin{bmatrix} 2 \\ 1/10 \end{bmatrix} = 0(2) + 2(1/10) = 1/5; \; c_5 − z_5 = M − 1/5$$

$$z_6 = \begin{bmatrix} 0 \\ 2 \end{bmatrix} \begin{bmatrix} -1 \\ 0 \end{bmatrix} = 0(-1) + 2(0) = 0; \; c_6 − z_6 = M − 0 = M$$

Objective function value, $z = \begin{bmatrix} 0 \\ 2 \end{bmatrix} \begin{bmatrix} 135 \\ 14 \end{bmatrix} = 0(135) + 2(14) = 28$

Scanning the $c_j - z_j$ indicator row of Table 6–28, we note that there are no negative indicators. Hence, the following basic feasible solution is optimal:

Nonbasic Variables	Basic Variables
$x_2 = 0$	$x_4 = 135$
$x_3 = 0$	$x_1 = 14$
$x_5 = 0$	
$x_6 = 0$	

The objective function value is $z = 28$.
Comparing the initial tableau,

Initial Tableau

	x_1	x_2	x_3	x_4	x_5	x_6	
c_j	2	8	0	0	M	M	
Vitamin A	10	30	−1	0	1	0	140 ←Minimum requirement
Vitamin B	20	15	0	−1	0	1	145 ←Minimum requirement

with the final tableau,

Final Tableau

Basic c_j	Basis	x_1	x_2	x_3	x_4	x_5	x_6	
c_j		2	8	0	0	M	M	
0	x_4	0	45	−2	1	2	−1	135
2	x_1	1	3	−1/10	0	1/10	0	14
	z_j	2	6	−1/5	0	1/5	0	$z = 28$
	$c_j - z_j$	0	2	1/5	0	M − 1/5	M	

↑
Shadow price indicator

note the shadow price 1/5 (indicator $c_3 - z_3$) corresponding to the nonbasic surplus variable x_3. Observe that x_3 is the surplus variable for the "greater than or equal to" Vitamin A constraint. For a "greater than or equal to" constraint, increasing the right-hand side b-coefficient by one unit is equivalent to entering one unit of surplus into that constraint. Thus, for each unit increase in the value of the b-coefficient $b_1 = 140$ of the Vitamin A constraint, one unit of surplus, x_3, is entered into the basis, resulting in an increase of $c_3 - z_3 = 1/5$ in the value of the objective function. Analogously, for each unit decrease in the value of the b-coefficient $b_1 = 140$ of the Vitamin A constraint, the objective function value decreases by $c_3 - z_3 = 1/5$.

EXERCISES

1. The linear programming problem of Exercise 2, Section 6–3 is expressed algebraically as

$$\text{Minimize} \quad z = 3x_1 + 1.5x_2$$
$$\text{subject to} \quad 500x_1 + 200x_2 \geq 1200$$
$$100x_1 + 800x_2 \geq 1000$$
$$\text{where} \quad x_1 \geq 0, x_2 \geq 0$$

  (A) Solve this problem by the simplex method.
  (B) Identify the shadow prices.
  (C) Interpret the shadow prices.

2. Refer to the linear programming problem of Exercise 5, Section 6–3.
  (A) Solve this problem by the simplex method.
  (B) Identify the shadow prices.
  (C) Interpret the shadow prices.

3. Consider the following:

$$\text{Minimize} \quad z = 5x_1 + 2x_2 + x_3$$
$$\text{subject to} \quad x_1 + 2x_2 + x_3 \geq 20$$
$$3x_1 + 3x_2 + 6x_3 \geq 24$$
$$\text{where} \quad x_1 \geq 0, x_2 \geq 0, x_3 \geq 0$$

  (A) Solve this problem by the simplex method.
  (B) Identify the shadow prices.
  (C) Interpret the shadow prices.

4. Consider the following:

$$\text{Minimize} \quad z = 3x_1 + 4x_2$$
$$\text{subject to} \quad x_1 + 4x_2 \geq 12$$
$$2x_1 + x_2 \geq 10$$
$$x_1 \geq 2$$
$$x_2 \geq 1$$

  (A) Solve this problem by the simplex method.
  (B) Identify the shadow prices.
  (C) Interpret the shadow prices.

6-8

LINEAR PROGRAMMING: FORMULATION AND APPLICATIONS

In this section, we will consider problems solvable by linear programming. We will concentrate upon formulating problems from various areas of application.

Investment Allocation

An investor has a maximum amount of $600,000 to allocate among the following investment alternatives:

Investment Alternative	Rate of Return	Risk Factor
Government bonds	7.5%	0
Money market funds	8.6	2
Hi-tech stock	15.5	7
Municipal bonds	7.8	1

The investor will tolerate an average risk of no more than 4.5 and will invest no more than $100,000 in hi-tech stock. If the objective is to maximize yield subject to the stated restrictions, how much should be allocated among the various investment alternatives? We will only find the algebraic formulation for this problem.

We let

x_1 = amount invested in government bonds
x_2 = amount invested in money market funds
x_3 = amount invested in hi-tech stock
x_4 = amount invested in municipal bonds

Since the objective is to maximize yield, the objective function is defined by

Maximize $z = 0.075x_1 + 0.086x_2 + 0.155x_3 + 0.078x_4$

Since the maximum amount invested is $600,000, one constraint is

$$x_1 + x_2 + x_3 + x_4 \leq 600,000$$

And since an average risk factor of no more than 4.5 is to be tolerated, we have

$$\frac{0x_1 + 2x_2 + 7x_3 + x_4}{x_1 + x_2 + x_3 + x_4} \leq 4.5$$

Multiplying both sides of this statement by $x_1 + x_2 + x_3 + x_4$ and then simplifying the result yields the constraint

$$-4.5x_1 - 2.5x_2 + 2.5x_3 - 3.5x_4 \leq 0$$

Finally, since the investor will invest no more than $100,000 in hi-tech stock, another constraint is

$$x_3 \leq 100{,}000$$

Thus, the complete algebraic formulation is

$$\begin{aligned}
\text{Maximize} \quad & z = 0.075x_1 + 0.086x_2 + 0.155x_3 + 0.078x_4 \\
\text{subject to} \quad & x_1 + x_2 + x_3 + x_4 \leq 600{,}000 \\
& -4.5x_1 - 2.5x_2 + 2.5x_3 - 3.5x_4 \leq 0 \\
& x_3 \leq 100{,}000
\end{aligned}$$

where $x_1, x_2, x_3,$ and x_4 are nonnegative

Blending Problem

A company manufactures two types of plant food, Type A and Type B, packaged in 50-pound bags. Each type is a blend of two raw materials, Raw Material 1 and Raw Material 2. Raw Material 1 contains 70% nitrogen and 10% phosphorus. Raw Material 2 contains 20% nitrogen and 40% phosphorus. Type A must contain at least 55% nitrogen. Type B must contain at least 15% phosphorus. The season's demand is for at least 50,000 bags for Type A and at least 20,000 bags for Type B. If each pound of Raw Material 1 costs $0.60 and each pound of Raw Material 2 costs $0.45, how many pounds of each raw material should be ordered to satisfy demand at minimum cost?

Here, we let

x_1 = number of pounds of Raw Material 1 used for Type A food
x_2 = number of pounds of Raw Material 1 used for Type B food
x_3 = number of pounds of Raw Material 2 used for Type A food
x_4 = number of pounds of Raw Material 2 used for Type B food

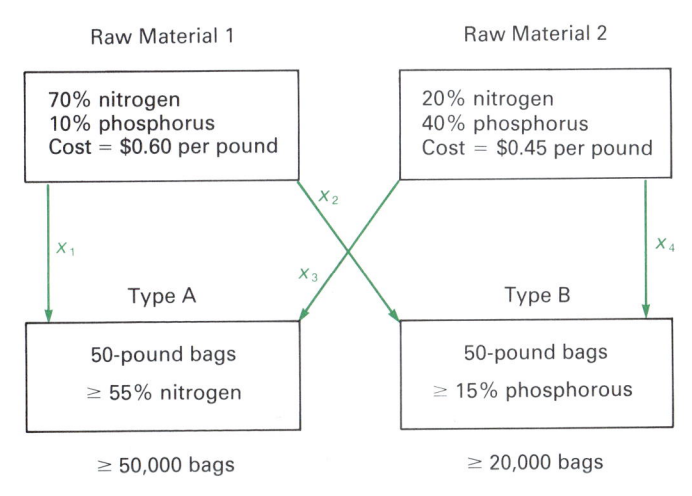

FIGURE 6–33

(See Figure 6–33.) Since the objective is to minimize the total cost of the raw materials, the objective function is defined by

$$\text{Minimize} \quad z = 0.60x_1 + 0.60x_2 + 0.45x_3 + 0.45x_4$$

Since we must have at least 50,000 fifty-pound bags of Type A, then

$$x_1 + x_3 \geq 50,000(50)$$

Since we must have at least 20,000 fifty-pound bags of Type B, then

$$x_2 + x_4 \geq 20,000(50)$$

Since Type A must contain at least 55% nitrogen, then

$$0.70x_1 + 0.20x_3 \geq 0.55(x_1 + x_3)$$

or

$$0.15x_1 - 0.35x_3 \geq 0$$

Since Type B must contain at least 15% phosphorus, then

$$0.10x_2 + 0.40x_4 \geq 0.15(x_2 + x_4)$$

or

$$-0.05x_2 + 0.25x_4 \geq 0$$

Thus, the linear programming problem is stated as

$$\begin{aligned}
\text{Minimize} \quad & z = 0.60x_1 + 0.60x_2 + 0.45x_3 + 0.45x_4 \\
\text{subject to} \quad & x_1 \phantom{{}+ x_2} + x_3 \phantom{{}+ x_4} \geq 2{,}500{,}000 \\
& \phantom{x_1 + {}} x_2 \phantom{{}+ x_3} + x_4 \geq 1{,}000{,}000 \\
& 0.15x_1 \phantom{{}+ x_2} - 0.35x_3 \phantom{{}+ x_4} \geq 0 \\
& \phantom{0.15x_1 + {}} -0.05x_2 \phantom{{}- 0.35x_3} + 0.25x_4 \geq 0
\end{aligned}$$

where $x_1, x_2, x_3,$ and x_4 are nonnegative

Project Scheduling   Figure 6–34 contains a pictorial description or network of a project showing the precedence relationships between the activities (or jobs) of the project. The project involves installation of a heating system in an office building. Table 6–30 lists each activity (or job), its immediate predecessor(s), and its estimated completion time. Studying Table 6–30, note that Activity A is listed as an immediate predecessor of Activity B. This means that Activity B cannot begin until Activity A is completed. Notice how this precedence relationship is illustrated by the network of Figure 6–34. Specifically, Activity A is represented by an arrow beginning at Node 1 and termi-

LINEAR PROGRAMMING

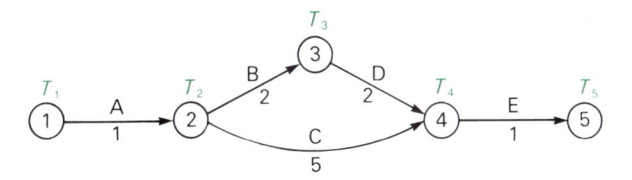

FIGURE 6-34

nating at Node 2; Activity B is represented by an arrow that begins at the same node at which Activity A's arrow terminates, Node 2. This indicates that Activity A is an immediate predecessor of Activity B.

TABLE 6-30
Project: Installation of Heating System

Activity	Description	Immediate Predecessors	Estimated Time
A	Remove old furnace	—	1 day
B	Set new furnace	A	2 days
C	Install piping	A	5 days
D	Install controls	B	2 days
E	Clean up	C, D	1 day

Studying the network of Figure 6-34, note that the estimated completion time of an activity is listed below its respective arrow. Observe that the nodes are labeled T_1, T_2, T_3, T_4, and T_5. Specifically, T_4 represents the completion time of Node 4 or, in other words, the earliest time that all activities with arrows terminating at Node 4 are completed. Analogously, T_1, T_2, T_3, and T_5 represent the completion times of Nodes 1, 2, 3, and 5, respectively.

Our goal is to determine the completion times of the nodes so that the project completion time, T_5, is minimized. Thus, the objective function is defined by

$$\text{Minimize} \quad z = T_5$$

Since T_2 is the earliest time that all activities with arrows terminating at Node 2 are completed, then

$$T_2 \geq T_1 + 1$$

is a constraint. Since T_3 is the earliest time that all activities with arrows terminating at Node 3 are completed, then

$$T_3 \geq T_2 + 2$$

is a constraint. Since T_4 is the earliest time that all activities with arrows terminating at Node 4 are completed, then

$$T_4 \geq T_3 + 2 \quad \text{and} \quad T_4 \geq T_2 + 5$$

are constraints. Since T_5 is the earliest time that all activities with arrows terminating at Node 5 are completed, then

$$T_5 \geq T_4 + 1$$

is a constraint. Thus, our problem is formulated as

Minimize $z = T_5$
subject to Activity A constraint $\longrightarrow T_2 \geq T_1 + 1$
 Activity B constraint $\longrightarrow T_3 \geq T_2 + 2$
 Activity C constraint $\longrightarrow T_4 \geq T_2 + 5$
 Activity D constraint $\longrightarrow T_4 \geq T_3 + 2$
 Activity E constraint $\longrightarrow T_5 \geq T_4 + 1$
where $T_1, T_2, T_3, T_4,$ and T_5 are nonnegative

This formulation may be rewritten as

Minimize $z = 0T_1 + 0T_2 + 0T_3 + 0T_4 + T_5$
subject to
$$\begin{aligned} -T_1 + T_2 &\geq 1 \\ -T_2 + T_3 &\geq 2 \\ -T_2 + T_4 &\geq 5 \\ -T_3 + T_4 &\geq 2 \\ -T_4 + T_5 &\geq 1 \end{aligned}$$
where $T_1, T_2, T_3, T_4,$ and T_5 are nonnegative

EXERCISES

1. A trust officer has a maximum amount of $800,000 to allocate among the following investment alternatives:

Investment Alternative	Rate of Return	Risk Factor
Treasury bills	8.3%	0
Municipal bonds	9.8	1
Real estate	15.9	3
Mutual fund	16.3	4
Energy stocks	18.4	6

The investor will tolerate an average risk of no more than 5.7 and will invest no more than $200,000 in energy stocks. If the objective is to maximize yield subject to the stated restrictions, how much should be allocated among the various investment alternatives? Formulate this problem. Do not solve.

2. Suppose the trust officer of Exercise 1 imposes the restriction that no more than 40% of the invested money is to be allocated among treasury bills and municipal bonds. Reformulate the problem with this added restriction.

3. A company manufactures two types of liquid fertilizer: Type A and Type B. Each type is a blend of two raw materials: Raw Material 1 and Raw Material 2. Raw Material 1 contains 60% nitrogen and 20% phosphorus. Raw Material 2 contains 30% nitrogen and 40% phosphorus. Type A must contain at least 40% nitrogen, and Type B must contain at least 18% phosphorus. The season's demand is for at least 100,000 gallons of Type A and at least 70,000 gallons of Type B. If Raw Material 1 costs $0.80 per gallon and Raw Material 2 costs $0.55 per gallon, how many pounds of each raw material should be ordered to satisfy demand at minimum cost?

4. The network of Figure 6–35 shows the precedence relationships between the activities of some project. The goal is to determine the completion times of the nodes so that the project completion time is minimized. Formulate this problem.

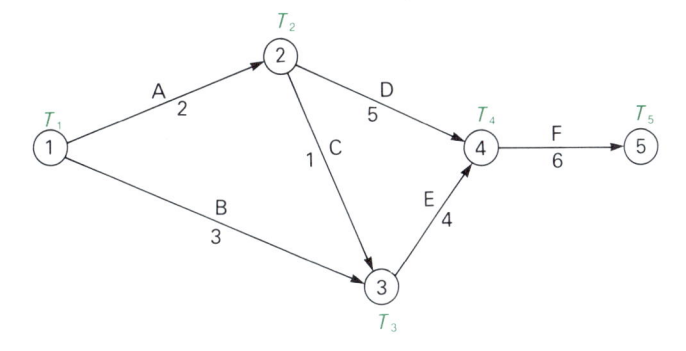

FIGURE 6–35

5. *(Optional)* Run the linear programming problem formulated in Exercise 1 on a computer. Interpret the results.

6. *(Optional)* Run the linear programming problem formulated in Exercise 2 on a computer. Interpret the results.

7. *(Optional)* Run the linear programming problem formulated in Exercise 3 on a computer. Interpret the results.

8. *(Optional)* Run the linear programming problem formulated in Exercise 4 on a computer. Interpret the results.

CASE D PROFIT MAXIMIZATION—COMFORT HOUSE FURNITURE COMPANY

The Comfort House Furniture Company manufactures three products: chairs, desks, and tables. The production process involves passing each product through each of three departments: Cutting, Assembly, and Finishing and Inspection. The available time (per month) for each department and the time requirements for each type of furniture are as follows:

Department	Time Requirements (hours) Chairs	Desks	Tables	Time Availability (hours per month)
Cutting	30	20	10	240
Assembly	20	60	20	320
Finishing and Inspection	10	40	90	260

The profit per chair is $15, the profit per desk is $50, and the profit per table is $80. Management wishes to determine the quantity of each type of furniture to be produced in order to maximize profit and satisfy the departmental time constraints.

EXERCISES
1. Define each decision variable.
2. Formulate as a linear programming problem.
3. Solve the problem by the simplex method.

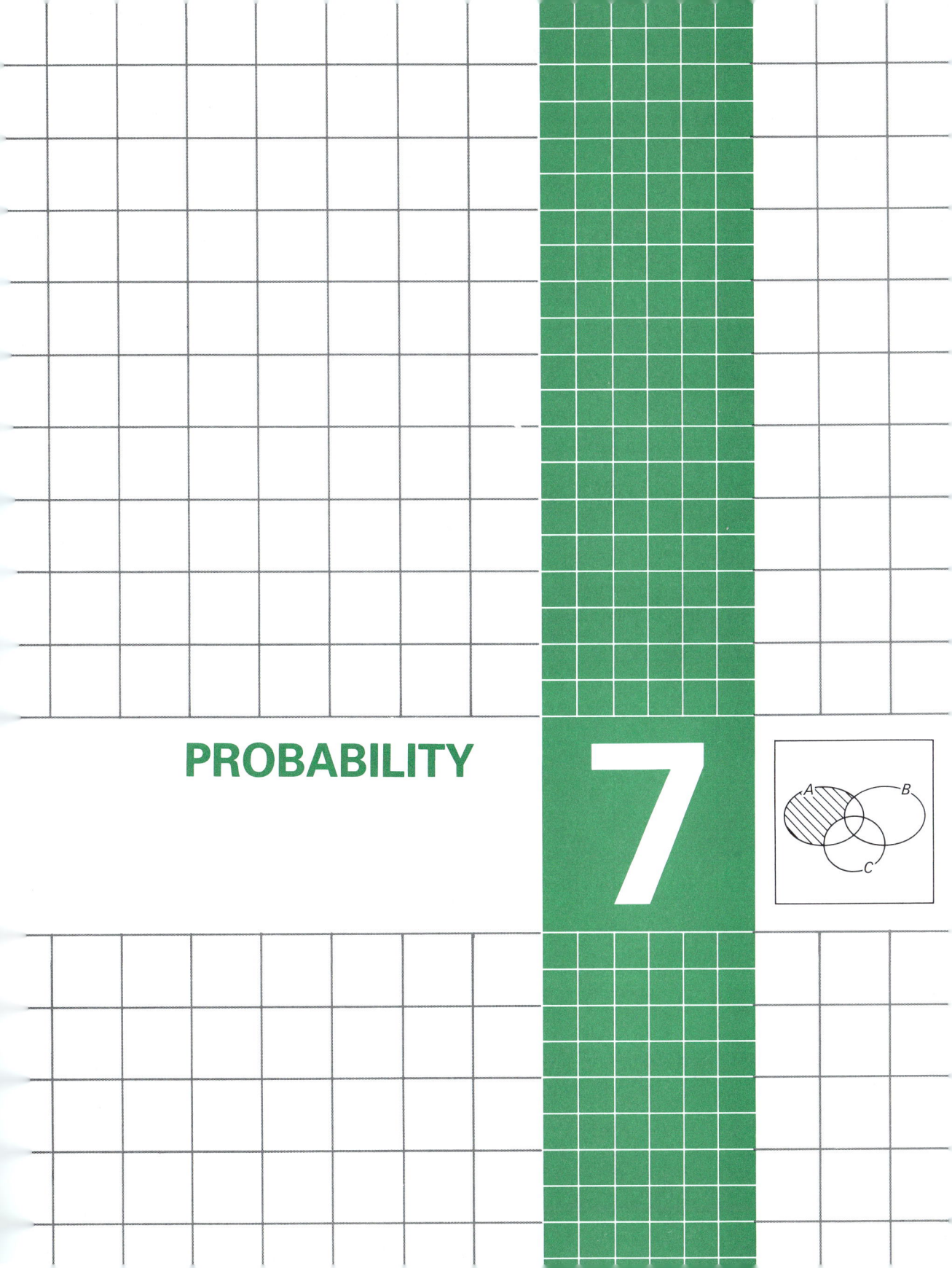

PROBABILITY

7

CHAPTER SEVEN

In this chapter, we will discuss *probability*—the likelihood of occurrence of chance events. Since probability will be presented by using sets, we begin the chapter with a review of sets.

7-1

SETS

A **set** is a collection of things. Examples of sets exist everywhere in our daily lives. Some specific sets are the following:

>The collection of letters of the English alphabet
>The collection of vowels in the English alphabet
>The collection of integers 1 through 9

Sets are usually written using set braces, { }. Thus, the preceding sets may be expressed as

>{letters in the English alphabet}
>{a, e, i, o, u}
>{1, 2, 3, 4, 5, 6, 7, 8, 9}

Sets are often named by capital letters. The objects belonging to a set are called **elements** of that set. The phrase "is an element of" is written in shorthand by the symbol $\in$. Thus, if

$$A = \{a, e, i, o, u\}$$

then

$$a \in A \quad e \in A \quad i \in A \quad o \in A \quad u \in A$$

The fact that s "is not an element of" A is written in shorthand as

$$s \notin A$$

It is frequently necessary to count the number of elements in a set. Again, consider set A:

$$A = \{a, e, i, o, u\}$$

The statement "the number of elements in set A is 5" may be written in shorthand notation as

$$n(A) = 5$$

EXAMPLE 7-1 If

$$B = \{4, 3, 0, 2, 8, 1\}$$

then $n(B) = 6$.

Empty Set

The **empty set**, or **null set**, is the set containing no elements. The empty set is commonly represented by either of the following symbols: $\emptyset$, { }. Hence, $n(\emptyset) = 0$.

Subset

If there exist two sets A and B such that every element of A is also an element of B, then A is a **subset** of B. The statement "A is a subset of B" is written in shorthand as $A \subset B$.

> **EXAMPLE 7-2** If
> $$A = \{1, 4, 6\} \quad B = \{1, 2, 4, 6, 8\}$$
> then $A \subset B$, but $B \not\subset A$.

The concept of subset is pictorially illustrated by the **Venn diagram** of Figure 7-1. Note that set A is completely contained inside set B. Hence, each element of A is also in B.

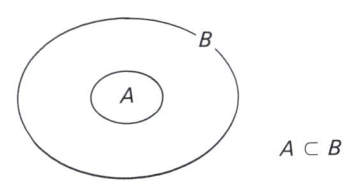

$A \subset B$

FIGURE 7-1

> **EXAMPLE 7-3** If
>
> A = {members of the U.S. Congress}
> B = {members of the U.S. Senate}
> C = {members of the U.S. House of Representatives}
>
> then $B \subset A$ and $C \subset A$.

Equality of Sets

Two sets are **equal** if they contain exactly the same elements. If A and B represent equal sets, then we may write $A = B$.

> **EXAMPLE 7-4** If
> $$A = \{1, 3, 4\} \quad B = \{3, 4, 1\}$$
> then $A = B$.

Note that if two sets are equal, then each is a subset of the other.

Universal Set Usually when discussing a specific problem involving sets, a universal set is identified. A **universal set** contains all elements of all sets of a given problem. In other words, every set of the given problem is a subset of the universal set for that problem. The letter U will represent the universal set.

EXAMPLE 7–5 Let

$$U = \{1, 2, 3, 4, 5, 6\} \quad A = \{1, 3, 4\} \quad B = \{6\}$$

Note that $A \subset U$ and $B \subset U$. The Venn diagram of Figure 7–2 pictorially illustrates these sets.

U

A: 1, 3, 4

B: 6

2, 5

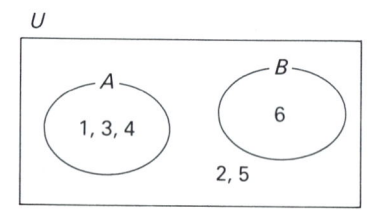

FIGURE 7–2

Intersection of Sets Let A and B represent two sets. The **intersection** of the two sets, written $A \cap B$, is the set of elements belonging to *both* A and B. The intersection of two sets consists of elements common to both.

EXAMPLE 7–6 If

$$A = \{1, 3, 5, 7\} \quad B = \{2, 3, 4, 5, 9\}$$

then $A \cap B = \{3, 5\}$. Note that in order for an element to belong to $A \cap B$, it must belong to both A and B.

EXAMPLE 7–7 If

$$A = \{2, 3, 4\} \quad B = \{1, 5, 6, 9\}$$

then $A \cap B = \emptyset$. Since the two sets A and B have no elements in common, their intersection is empty. Such sets are called **disjoint sets**.

373 PROBABILITY

The Venn diagrams of Figure 7–3 pictorially illustrate the concept of intersection.

$A \cap B$ = shaded area

$A \cap B = \emptyset$
Disjoint sets

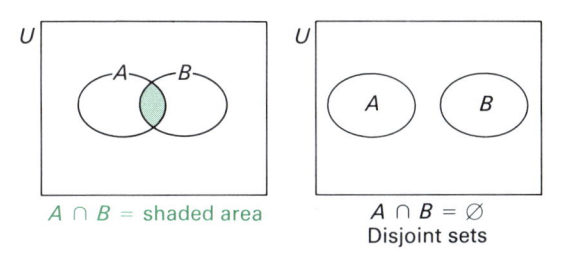

FIGURE 7–3

Union of Sets Let A and B represent two sets. The **union** of the two sets, written $A \cup B$, is the set of elements belonging to *either* A *or* B *or both*. When we "union" two sets, we "join them together."

EXAMPLE 7–8 If

$$A = \{1, 3, 5, 7\} \quad B = \{2, 3, 4, 5, 9\}$$

then $A \cup B = \{1, 2, 3, 4, 5, 7, 9\}$. Note that in order for an element to belong to $A \cup B$, it must belong to either A or B or both.

EXAMPLE 7–9 If

$$A = \{2, 3, 4\} \quad B = \{1, 5, 6, 9\}$$

then $A \cup B = \{1, 2, 3, 4, 5, 6, 9\}$.

The Venn diagrams of Figure 7–4 pictorially illustrate the concept of union.

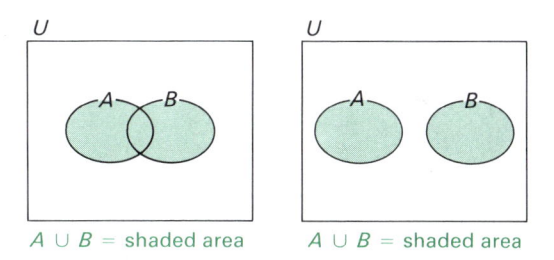

$A \cup B$ = shaded area $A \cup B$ = shaded area

FIGURE 7–4

Complement of a Set

Let A represent a set and U represent a universal set. The **complement** of A, written A', is the set of all elements of U that *do not belong to A*.

EXAMPLE 7–10 If

$$U = \{0, 1, 2, 3, 4, 5\} \qquad A = \{1, 3, 5\}$$

then $A' = \{0, 2, 4\}$.

The Venn diagram of Figure 7–5 pictorially illustrates the concept of complement.

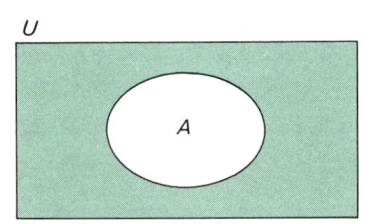

A' = shaded area

FIGURE 7–5

EXAMPLE 7–11 If

$$U = \{0, 1, 2, 3, 4, 5, 6, 7, 8, 9\}$$
$$A = \{0, 1, 2, 3\}$$
$$B = \{2, 3, 6\}$$
$$C = \{3, 7, 8, 9\}$$

find each of the following:
(A) $A \cup B$ (B) $A \cap B$ (C) $A \cap B \cap C$
(D) $n(A \cap B \cap C)$ (E) $A \cup B \cup C$ (F) $n(A \cup B \cup C)$
(G) A' (H) $(A \cap B)'$

Solutions
(A) $A \cup B = \{0, 1, 2, 3, 6\}$
(B) $A \cap B = \{2, 3\}$
(C) $A \cap B \cap C = \{3\}$
(D) $n(A \cap B \cap C) = 1$
(E) $A \cup B \cup C = \{0, 1, 2, 3, 6, 7, 8, 9\}$
(F) $n(A \cup B \cup C) = 8$
(G) $A' = \{4, 5, 6, 7, 8, 9\}$
(H) $(A \cap B)' = \{0, 1, 4, 5, 6, 7, 8, 9\}$

Sets are applied to many areas. We now give some examples of applications.

EXAMPLE 7–12 A marketing research group conducted a survey of 70 automobile buyers. The results indicated the following buyer preferences:

45 bought air-conditioned cars
20 bought 2-door models
9 bought 2-door, air-conditioned models

(A) How many bought two-door models without air-conditioning?
(B) How many bought four-door, air-conditioned cars?
(C) How many bought neither air-conditioned cars nor two-door models?
(D) Of those who bought air-conditioned cars, what percentage bought two-door models?

Solutions If

U = {all buyers surveyed}
A = {buyers of air-conditioned cars}
B = {buyers of 2-door models}
$A \cap B$ = {buyers of 2-door, air-conditioned cars}

then $n(U) = 70$, $n(A) = 45$, $n(B) = 20$, and $n(A \cap B) = 9$ (see Figure 7–6).

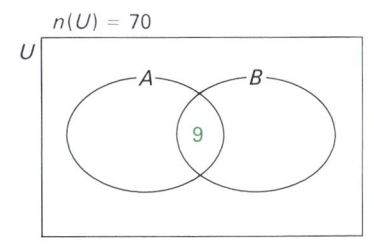

FIGURE 7–6

(A) Buyers of two-door models without air-conditioning are represented by the region outside set A but inside set B, i.e., $A' \cap B$. Since $n(B) = 20$ and $n(A \cap B) = 9$, then $n(A' \cap B) = n(B) - n(A \cap B) = 20 - 9 = 11$ (see Figure 7–7). Thus, 11 people bought two-door models without air-conditioning.

EXAMPLE 7–12 (continued)

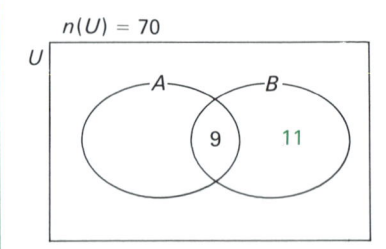

FIGURE 7–7

(B) Buyers of four-door, air-conditioned models are represented by the region outside set *B* but inside set *A*, i.e., $A \cap B'$. Since $n(A) = 45$ and $n(A \cap B) = 9$, then $n(A \cap B') = n(A) - n(A \cap B) = 45 - 9 = 36$ (see Figure 7–8). Thus, 36 people bought four-door, air-conditioned cars.

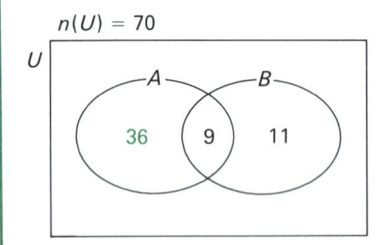

FIGURE 7–8

(C) Buyers who bought neither air-conditioned nor two-door cars are represented by the region outside set $A \cup B$, i.e., $(A \cup B)'$. Since $n(U) = 70$ and $n(A \cup B) = 36 + 9 + 11 = 56$, then $n(A \cup B)' = n(U) - n(A \cup B) = 70 - 56 = 14$. Thus, 14 people bought neither air-conditioned nor two-door cars (see Figure 7–9).

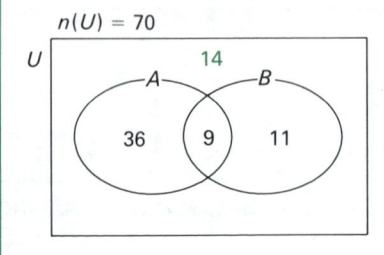

FIGURE 7–9

(D) Since $n(A) = 45$ and $n(A \cap B) = 9$, then $9/45 = 20\%$ of those who bought air-conditioned cars bought two-door models (see Figure 7–10).

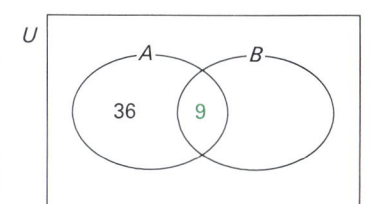

FIGURE 7–10

EXERCISES

1. If $A = \{0, 1, 2, 3, 4\}$, state whether each of the following is true or false:
   - (A) $0 \in A$
   - (B) $1 \in A$
   - (C) $2 \in A$
   - (D) $3 \in A$
   - (E) $4 \in A$
   - (F) $7 \in A$
   - (G) $n(A) = 4$
   - (H) $8 \in A$

2. If $A = \{5, 6, 7\}$ and $B = \{5, 6, 7, 8, 9\}$, answer each of the following:
   - (A) Is $A \subset B$?
   - (B) Is $B \subset A$?
   - (C) Does $A = B$?

3. If $A = \{7, 9, 10\}$ and $B = \{7, 10, 9\}$, answer each of the following:
   - (A) Is $A \subset B$?
   - (B) Is $B \subset A$?
   - (C) Does $A = B$?

4. If $U = \{1, 2, 3, 4, 5, 6, 7, 8, 9\}$, $A = \{2, 3, 4, 5\}$, and $B = \{4, 5, 6, 7\}$, find each of the following:
   - (A) $A \cup B$
   - (B) $A \cap B$
   - (C) A'
   - (D) B'
   - (E) $A' \cap B$
   - (F) $A \cap B'$
   - (G) $(A \cup B)'$
   - (H) $(A \cap B)'$
   - (I) $A' \cap B'$
   - (J) $A' \cup B'$
   - (K) $n(A \cap B)$
   - (L) $n(B')$

5. If $U = \{3, 4, 5, 10, 11, 12, a, b, c\}$, $A = \{3, 5, 12\}$, $B = \{11, 12, 3, a\}$, and $C = \{12, a\}$, find each of the following:
   - (A) $A \cup B$
   - (B) $B \cup C$
   - (C) $n(A \cup B)$
   - (D) $A \cup C$
   - (E) $n(A \cup C)$
   - (F) $A \cup B \cup C$
   - (G) $A \cap B$
   - (H) $n(A \cap B)$
   - (I) $A \cap B \cap C$
   - (J) $A \cap C$
   - (K) $(A \cap C)'$
   - (L) $(A \cup B)'$
   - (M) A'
   - (N) $A' \cap C$
   - (O) $(B \cup C) \cap A$
   - (P) $(A \cup B)' \cap C$
   - (Q) C'
   - (R) $C' \cup A$
   - (S) $C' \cap A$
   - (T) $(A \cap B) \cup C$
   - (U) $A' \cap B'$

6. For the Venn diagram of Figure 7–11 on the next page, shade the region corresponding to each of the following:
   - (A) $A \cup B$
   - (B) $(A \cup B)'$
   - (C) $A \cap B$
   - (D) $(A \cap B)'$
   - (E) A'
   - (F) $A' \cap B$
   - (G) B'
   - (H) $B' \cap A$
   - (I) $A' \cap B'$
   - (J) $A' \cup B'$

378 CHAPTER SEVEN

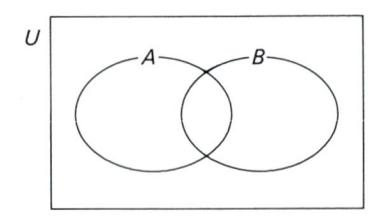

FIGURE 7–11

7. For each of the following Venn diagrams, shade the region corresponding to $A \cap B$:

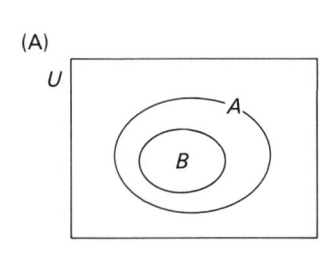

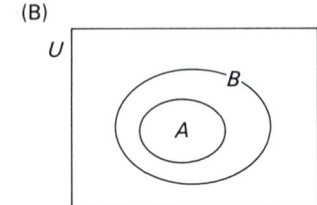

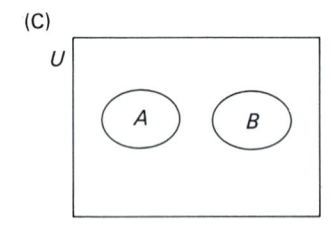

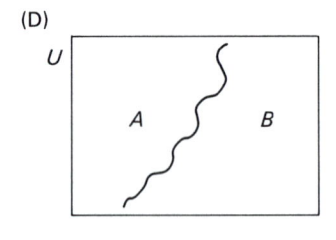

(A)

(B)

(C)

(D)

8. For each of the Venn diagrams of Exercise 7, shade the region corresponding to $A \cup B$.
9. For each of the Venn diagrams of Exercise 7, shade the region corresponding to A'.
10. For each of the Venn diagrams of Exercise 7, shade the region corresponding to B'.
11. For the Venn diagram of Figure 7–12, shade the region corresponding to each of the following:

(A)	$A \cup B$	(B)	$A \cup C$	(C)	$B \cup C$
(D)	$A \cap B$	(E)	$A \cap C$	(F)	$B \cap C$
(G)	$A \cap B \cap C$	(H)	$A \cup B \cup C$	(I)	$A \cap B'$
(J)	$(A \cup B)'$	(K)	$(A \cap B)'$	(L)	$A' \cup B'$
(M)	$A' \cap B'$	(N)	$A \cap (B \cup C)'$	(O)	$A \cap (B \cap C)'$
(P)	$A' \cap (B \cap C)$	(Q)	$C' \cap (A \cap B)$	(R)	$B' \cap (A \cap C)$
(S)	$(A \cup B)' \cap C$	(T)	$(A \cup C)' \cap B$	(U)	$(A \cup B \cup C)'$

379 PROBABILITY

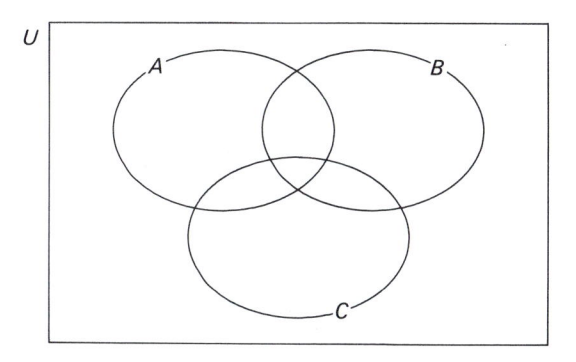

Venn diagram showing three overlapping circles A, B, C within universal set U

FIGURE 7–12

12. Referring to the Venn diagram of Figure 7–13, show that $(A \cap B)' = A' \cup B'$ by: *first*, shading the region corresponding to $(A \cap B)'$; *second*, shading the region corresponding to $A' \cup B'$; *third*, comparing the shaded regions. They should be identical.

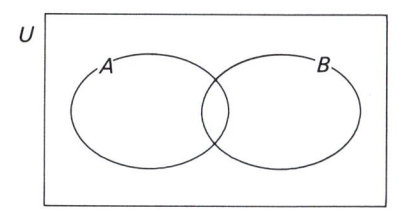

Venn diagram showing two overlapping circles A and B within universal set U

FIGURE 7–13

13. Referring to the Venn diagram of Figure 7–14, show that $(A \cup B)' = A' \cap B'$. The statements

$$(A \cap B)' = A' \cup B'$$
$$(A \cup B)' = A' \cap B'$$

are called **De Morgan's laws**.

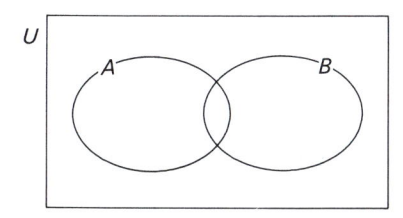

Venn diagram showing two overlapping circles A and B within universal set U

FIGURE 7–14

380 CHAPTER SEVEN

14. A survey of 60 executives of Montcalf Corporation revealed the following results:

    40 read *The Wall Street Journal*
    30 read *Barron's*
    20 read both *Barron's* and *The Wall Street Journal*

    (A) How many read only *The Wall Street Journal?*
    (B) How many read only *Barron's?*
    (C) How many read neither *Barron's* nor *The Wall Street Journal?*
    (D) Of those who read *Barron's,* what percentage also read *The Wall Street Journal?*
    (E) Of those who read *The Wall Street Journal,* what percentage also read *Barron's?*

15. The Heavenly Flavor Ice Cream Co. surveyed 100 families to determine flavor preferences. The results were as follows:

    40 families liked vanilla
    34 families liked strawberry
    52 families liked coffee
    21 families liked both vanilla and strawberry
    15 families liked both coffee and strawberry
    17 families liked both vanilla and coffee
    10 families liked all three flavors

    (A) How many families liked both vanilla and strawberry but not coffee?
    (B) How many families liked both strawberry and coffee but not vanilla?
    (C) How many families liked both vanilla and coffee but not strawberry?
    (D) How many families liked only vanilla?
    (E) How many families liked only strawberry?
    (F) How many families liked only coffee?
    (G) How many families liked none of the three flavors?

16. The results of a survey of 50 single residents of Hilltop Manor Exclusive Apartments are expressed in the following table:

	Doctors	Lawyers	Business People	Totals
Jaguar Owners	6	3	1	10
Mercedes Owners	18	7	15	40
Totals	24	10	16	50

Let D = {doctors}, L = {lawyers}, B = {business people}, J = {Jaguar owners}, and M = {Mercedes owners}.
(A) Find $n(D)$. (B) Find $n(L)$. (C) Find $n(B)$.
(D) Find $n(J)$. (E) Find $n(M)$. (F) Find $n(D \cap J)$.
(G) Find $n(D \cap M)$. (H) Find $n(L \cap J)$. (I) Find $n(L \cap M)$.
(J) Find $n(B \cap J)$. (K) Find $n(B \cap M)$. (L) Find $n(D \cup J)$.
(M) Find $n(B \cup M)$. (N) Find $n(L \cup J)$. (O) Find $n(B')$.
(P) Find $n(J')$. (G) Find $n(B' \cap M)$. (R) Find $n(D' \cap J)$.
(S) Of those residents who are Jaguar owners, what percentage are doctors?
(T) Of those residents who are doctors, what percentage are Jaguar owners?
(U) Of those residents who are lawyers, what percentage are Mercedes owners?
(V) Of those residents who are Mercedes owners, what percentage are lawyers?

17. Referring to the Venn diagram of Figure 7–15, explain why

$$n(A \cup B) = n(A) + n(B) - n(A \cap B)$$

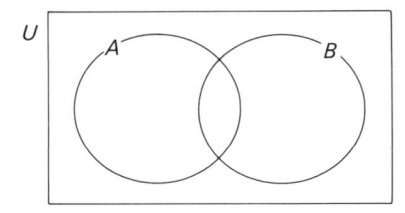

FIGURE 7–15

18. Referring to the Venn diagram of Figure 7–16, explain why

$$n(A \cup B) = n(A) + n(B)$$

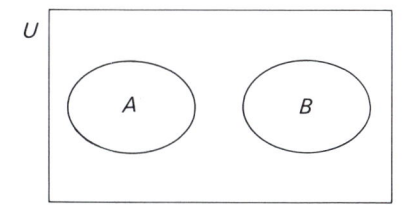

FIGURE 7–16

7-2

BASIC PROBABILITY FORMULA

In this chapter, we will encounter happenings whose outcomes are uncertain. Happenings with uncertain outcomes are called **chance experiments**. The start of a new day with its unknown weather, the rolling of a pair of dice, the flipping of a coin, and the selection of a card from a deck are some examples of chance experiments. In this chapter, we will encounter many chance experiments. Given a chance experiment, we will determine the likelihood of occurrence of certain events. The likelihood of occurrence of a chance event is called its **probability**.

The probability of a chance event is a number between 0 and 1. The closer the number is to 1, the more likely it is that the event will occur. The closer the number is to 0, the less likely it is that the event will occur. If an event is certain to occur, then its probability is 1. If an event is certain not to occur, then its probability is 0.

To determine probabilities of events, we must define the following terms:

A **sample space**, *S*, is the set of all possible outcomes of a chance experiment.

An **event**, *E*, is a subset of a sample space.

EXAMPLE 7-13 A fair coin is flipped.* H = heads; T = tails.
(A) Write the sample space, S, for this chance experiment.
(B) Write the set E that represents the event of a head occurring.

Solutions
(A) $S = \{H, T\}$
(B) $E = \{H\}$

---
*A coin is fair if, when flipped, either of its two sides is equally likely to come up.

EXAMPLE 7-14 Two fair dice are rolled.*
(A) Write the sample space, S, for this chance experiment.
(B) Write the set E that represents the event of getting a 1 on the first die.
(C) Write the set E that represents the event of getting a sum of 5.

Solutions
(A) Table 7-1 lists the 36 elements of S.

TABLE 7-1

		Die 2					
		1	2	3	4	5	6
	1	(1, 1)	(1, 2)	(1, 3)	(1, 4)	(1, 5)	(1, 6)
	2	(2, 1)	(2, 2)	(2, 3)	(2, 4)	(2, 5)	(2, 6)
Die 1	3	(3, 1)	(3, 2)	(3, 3)	(3, 4)	(3, 5)	(3, 6)
	4	(4, 1)	(4, 2)	(4, 3)	(4, 4)	(4, 5)	(4, 6)
	5	(5, 1)	(5, 2)	(5, 3)	(5, 4)	(5, 5)	(5, 6)
	6	(6, 1)	(6, 2)	(6, 3)	(6, 4)	(6, 5)	(6, 6)

(B) The event E of getting a 1 on the first die is represented by

$$E = \{(1, 1), (1, 2), (1, 3), (1, 4), (1, 5), (1, 6)\}$$

(C) The event E of getting a sum of 5 is represented by

$$E = \{(1, 4), (4, 1), (2, 3), (3, 2)\}$$

---

*A die is fair if, when rolled, any one of its six faces is equally likely to come up.

In general, the relationship between an event E and its sample space, S, is illustrated in Figure 7-17. Observe that a sample space S is a universal set for a given experiment. An event E is simply a subset of a sample space.

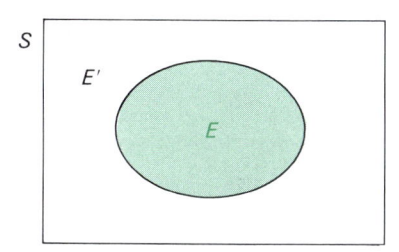

FIGURE 7-17

Probabilities of Events

The probability of an event E occurring will be denoted by the symbol

$$P(E)$$

If $n(S)$ is the number of elements of a finite sample space whose outcomes are equally likely to occur, and if $n(E)$ is the number of elements of an event set E, then $P(E)$, the probability of event E occurring, is given by

$$P(E) = \frac{n(E)}{n(S)}$$

This expression is called the **basic probability formula.**

EXAMPLE 7–15 If a fair coin is flipped once, find the probability of a head occurring.
Solution In Example 7–13, we determined that $S = \{H, T\}$ and that $E = \{H\}$. Since $n(S) = 2$ and $n(E) = 1$, then

$$P(E) = \frac{n(E)}{n(S)} = \frac{1}{2}, \text{ or } .50$$

EXAMPLE 7–16 If two fair dice are rolled, find the probability of getting
(A) A 1 on the first die.
(B) A sum of 5.
Solutions In Example 7–14(A), we determined that $n(S) = 36$.
(A) In Example 7–14(B), we determined that

$$E = \{(1, 1), (1, 2), (1, 3), (1, 4), (1, 5), (1, 6)\}$$

Since $n(E) = 6$, then

$$P(E) = \frac{n(E)}{n(S)} = \frac{6}{36} = \frac{1}{6}, \text{ or } 16\tfrac{2}{3}\%$$

(B) In Example 7–14(C), we determined that

$$E = \{(1, 4), (4, 1), (2, 3), (3, 2)\}$$

Since $n(E) = 4$, then

$$P(E) = \frac{n(E)}{n(S)} = \frac{4}{36} = \frac{1}{9}, \text{ or } 11\frac{1}{9}\%$$

EXAMPLE 7–17 Two fair coins are flipped.
(A) Determine the sample space.
(B) Determine the probability that exactly 1 head occurs.
(C) Determine the probability that at most 1 head occurs.

Solutions
(A) Since this experiment consists of flipping more than one coin, we will determine the sample space by using a **tree diagram** (see Figure 7–18). From the tree diagram of Figure 7–18, we obtain

$$S = \{HH, HT, TH, TT\}$$

Possible Outcomes

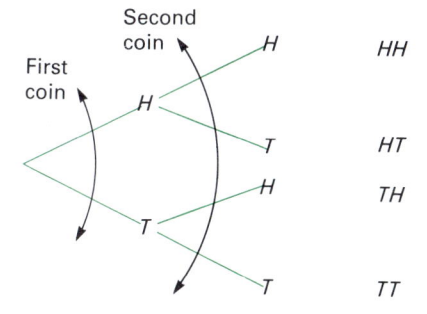

```
Second
coin      H         HH
First
coin   H
          T         HT
                    
          H         TH

       T
          T         TT
```

FIGURE 7–18

(B) The event of getting *exactly 1 head* is represented by the event set

$$E = \{HT, TH\}$$

Since $n(E) = 2$ and $n(S) = 4$, then

$$P(E) = \frac{n(E)}{n(S)} = \frac{2}{4}, \text{ or } 50\%$$

(C) The event of getting *at most 1 head* is represented by the event set

$$E = \{TT, HT, TH\}$$

EXAMPLE 7-17 (continued)

Since $n(E) = 3$ and $n(S) = 4$, then

$$P(E) = \frac{n(E)}{n(S)} = \frac{3}{4}, \text{ or } 75\%$$

EXAMPLE 7-18 A chance experiment consists of flipping three fair coins.
(A) Determine the sample space.
(B) Determine the probability that exactly 2 heads occur.
(C) Determine the probability that at least 2 heads occur.

Solutions
(A) The sample space is determined by the tree diagram of Figure 7-19. From this diagram, we obtain

$$S = \{HHH, HHT, HTH, HTT, THH, THT, TTH, TTT\}$$

Possible Outcomes

HHH
HHT
HTH
HTT
THH
THT
TTH
TTT

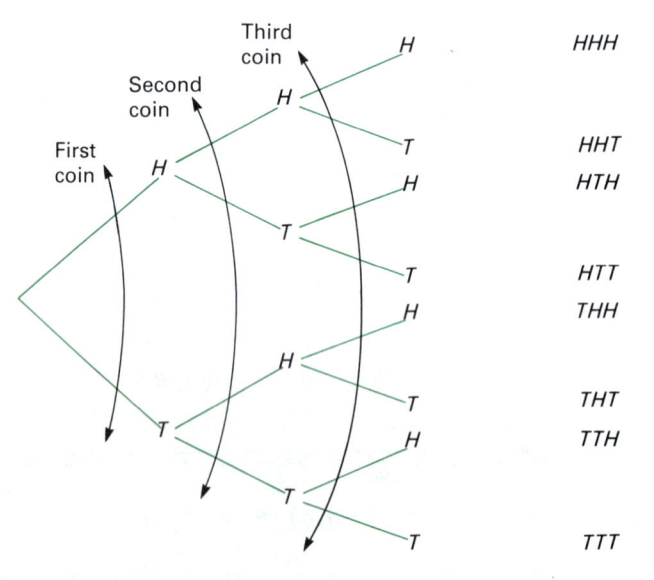

FIGURE 7-19

(B) The event of getting *exactly 2 heads* is represented by the event set

$$E = \{HHT, HTH, THH\}$$

PROBABILITY

Since $n(E) = 3$ and $n(S) = 8$, then

$$P(E) = \frac{n(E)}{n(S)} = \frac{3}{8}, \text{ or } 37\tfrac{1}{2}\%$$

(C) The event of getting *at least 2 heads* is represented by the event set

$$E = \{HHH, HHT, HTH, THH\}$$

Since $n(E) = 4$ and $n(S) = 8$, then

$$P(E) = \frac{n(E)}{n(S)} = \frac{4}{8}, \text{ or } 50\%$$

Complement Law The Venn diagram of Figure 7–20 illustrates the relationship between an event E and its sample space, S. The complement of event E is E', read "not E." We now derive a formula for $P(E')$.

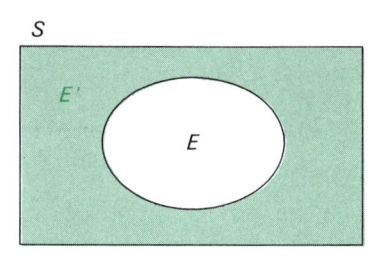

FIGURE 7–20

Beginning with

$$P(E') = \frac{n(E')}{n(S)}$$

and observing in Figure 7–20 that

$$n(E') = n(S) - n(E)$$

we have

$$P(E') = \frac{n(E')}{n(S)} = \frac{n(S) - n(E)}{n(S)}$$
$$= \frac{n(S)}{n(S)} - \frac{n(E)}{n(S)}$$
$$= 1 - P(E)$$

Thus, the formula

$$P(E') = 1 - P(E)$$

is called the **complement law**.

> **EXAMPLE 7–19** If the probability of a supermarket succeeding at a given location is .70, find the probability of its failing at that location.
> *Solution* By the complement law, we have
> $$P(\text{failing}) = 1 - P(\text{succeeding})$$
> $$= 1 - .70$$
> $$= .30$$

Interpreting Probabilities

In this section, we have calculated probabilities of various events. In Example 7–15, we determined that if a coin is flipped, the probability of a head occurring is 1/2, or 50%. This may be interpreted *objectively* as follows: If a coin is repeatedly flipped, then the ratio

$$\frac{\text{cumulative number of heads}}{\text{cumulative number of flips}}$$

will approach 1/2 as the number of flips increases (see Figure 7–21). This interpretation is based on the **law of large numbers**.

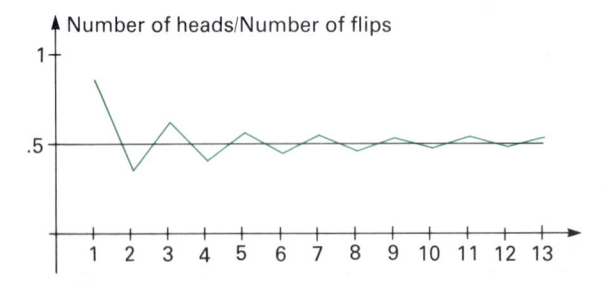

FIGURE 7–21

Sometimes, probabilities of certain events must be assigned *subjectively*. Such probabilities are called **subjective probabilities**. For example, suppose a supermarket chain is planning to open a new market in a developing community. Management has no past record of success or failure at this location. However, using personal judgment, hunches, or perhaps past experience with a similar location, management assigns an 80% probability of success to the new market. The 80% probability is a subjective probability.

389 PROBABILITY

In this section, the basic probability formula $P(E) = n(E)/n(S)$ was stated for finite sample spaces with equally likely outcomes. Sometimes, we encounter situations where these conditions are not met and must therefore use other ways of determining probabilities. In the next three sections, we will discuss laws of probabilities that will enable us to determine probabilities under more general situations.

EXERCISES

1. Each letter of the word *money* is written on a separate slip of paper, and the slips are placed in a box. A chance experiment consists of selecting two slips, without replacement, from the box.
   (A) Determine the sample space.
   (B) What is the probability of getting *o* on the first selection?
   (C) What is the probability of getting *n* on the second selection?
   (D) What is the probability of getting the same letter for both selections?

2. Repeat Exercise 1 under the assumption that the first slip is replaced before the second is selected.

3. A boy has a penny, a nickel, a dime, and a quarter in his pocket. Upon entering a candy store, he takes two coins out of his pocket—first one, then another.
   (A) Determine the sample space.
   (B) What is the probability the boy selects a total of 35 cents?
   (C) What is the probability of getting the same coin for both selections?

4. Repeat Exercise 3 under the assumption that the first coin is replaced before the second is selected.

5. A coin is tossed four times.
   (A) Determine the sample space.
   (B) Find the probability of getting exactly 3 heads.
   (C) Find the probability of getting at most 3 heads.
   (D) Find the probability of getting at least 3 heads.
   (E) Find the probability of getting the same result on all four trials.

6. A placement office in a large company received applications from recent college graduates with the following racial origins: 65 whites, 15 blacks, 10 Hispanics, 5 Asians, and 5 American Indians. If one applicant is screened out, find the probability that the selected applicant's racial origin is
   (A) White.   (B) Non-white.
   (C) Black.   (D) Non-black.
   (E) Hispanic or Asian. (F) Asian or American Indian.

7. If there is a 30% chance of a stock price increase tomorrow, what is the probability that it will not increase tomorrow?

8. If two dice are rolled, find the probability of getting
   (A) A 3 on the second die.
   (B) A sum of 10.

(C) An even number on the first die and an odd number on the second die.
(D) Even numbers on both.
(E) A sum of 13.
(F) A sum of 7.
(G) A sum not equal to 7.

9. A card is selected from an ordinary deck of 52 cards. Find the probability that the selected card is
   (A) A king.
   (B) A queen.
   (C) Not a queen.
   (D) A red card.
   (E) A jack.
   (F) Not a jack.
   (G) A black jack.
   (H) The ace of spades.
   (I) Not a red king.
   (J) Either a jack or a queen.

7-3

LAWS OF ADDITION

In Section 7-2, we used the formula $P(E) = n(E)/n(S)$ to determine the probabilities of various events. Sometimes, an event is the union of two events. Consider the survey group of 70 automobile buyers discussed in Example 7-12. Recall that

$$A = \{\text{buyers of air-conditioned cars}\}$$
$$B = \{\text{buyers of 2-door models}\}$$
$$A \cap B = \{\text{buyers of 2-door, air-conditioned cars}\}$$

These sets are illustrated in the Venn diagram of Figure 7-22. If a chance experiment consists of selecting one person from this group, then the sample space, S, is the universal set of Example 7-12. Hence,

$$S = \{\text{all buyers surveyed}\}$$

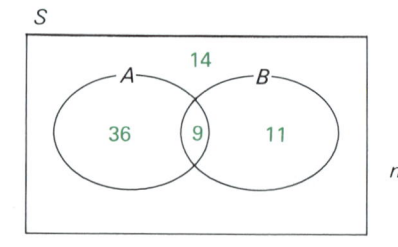

$n(S) = 70$
$n(A) = 45$
$n(B) = 20$
$n(A \cap B) = 9$

FIGURE 7-22

We now derive a formula for $P(A \cup B)$, the probability of selecting a buyer of either an air-conditioned car *or* a two-door model. Beginning with the basic probability formula, we have

$$P(A \cup B) = \frac{n(A \cup B)}{n(S)}$$

PROBABILITY

Since, as discussed in Exercise 17, Section 7-1,

$$n(A \cup B) = n(A) + n(B) - n(A \cap B)$$

then

$$P(A \cup B) = \frac{n(A) + n(B) - n(A \cap B)}{n(S)}$$
$$= \frac{n(A)}{n(S)} + \frac{n(B)}{n(S)} - \frac{n(A \cap B)}{n(S)}$$
$$= P(A) + P(B) - P(A \cap B)$$

The formula

$$P(A \cup B) = P(A) + P(B) - P(A \cap B)$$

is called the **general law of addition**. The general law of addition defines $P(A \cup B)$ for any events A and B. From Figure 7-22, we note that

$$P(A) = \frac{n(A)}{n(S)} = \frac{45}{70} \qquad P(B) = \frac{n(B)}{n(S)} = \frac{20}{70}$$
$$P(A \cap B) = \frac{n(A \cap B)}{n(S)} = \frac{9}{70}$$

Thus, we calculate

$$P(A \cup B) = P(A) + P(B) - P(A \cap B)$$
$$= \frac{45}{70} + \frac{20}{70} - \frac{9}{70}$$
$$= \frac{56}{70}$$

EXAMPLE 7-20 A card is selected from an ordinary deck. Find the probability of selecting either a red card *or* a king.
Solution If

$S = \{\text{cards from an ordinary deck}\}$
$R = \{\text{red cards}\}$
$K = \{\text{kings}\}$

then the Venn diagram of Figure 7-23 illustrates these sets.

EXAMPLE 7-20 (continued)

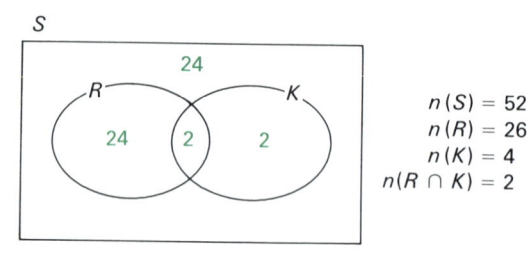

[Venn diagram showing sets R and K within sample space S, with 24 in R only, 2 in intersection, 2 in K only, and 24 outside both]

$n(S) = 52$
$n(R) = 26$
$n(K) = 4$
$n(R \cap K) = 2$

FIGURE 7-23

We seek to determine $P(R \text{ or } K)$. Since $P(R \text{ or } K)$ is equivalent to $P(R \cup K)$, we use the general law of addition to obtain

$$P(R \cup K) = P(R) + P(K) - P(R \cap K)$$
$$= \frac{26}{52} + \frac{4}{52} - \frac{2}{52}$$
$$= \frac{28}{52}$$

EXAMPLE 7-21 According to a weather forecaster, the probability of rain tomorrow is .60, the probability of hail is .10, and the probability of both rain and hail is .08. Find the probability of either rain or hail tomorrow.
Solution Let

R = event of rain tomorrow
H = event of hail tomorrow
R and H = event of both rain and hail tomorrow

Thus, $P(R) = .60$ and $P(H) = .10$. Since $P(R \text{ and } H)$ is equivalent to $P(R \cap H)$, then $P(R \cap H) = .08$. We seek $P(R \text{ or } H)$. Since $P(R \text{ or } H)$ is equivalent to $P(R \cup H)$, we use the general law of addition to obtain

$$P(R \cup H) = P(R) + P(H) - P(R \cap H)$$
$$= .60 + .10 - .08$$
$$= .62$$

If two event sets A and B are disjoint, then they are called **mutually exclusive** (see Figure 7-24). Since $A \cap B = \emptyset$, then

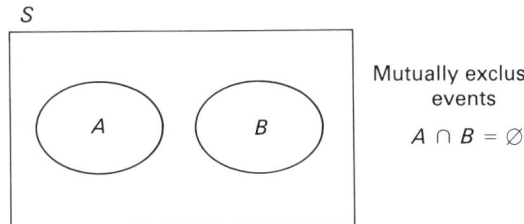

PROBABILITY

FIGURE 7-24

Mutually exclusive events
$A \cap B = \emptyset$

$n(A \cap B) = 0$ and $P(A \cap B) = 0$. Thus, if events A and B are mutually exclusive, the general law of addition

$$P(A \cup B) = P(A) + P(B) - P(A \cap B)$$

becomes

$$P(A \cup B) = P(A) + P(B)$$

This result is called the **special law of addition.** It must be remembered that the special law of addition applies only to mutually exclusive events. This law defines $P(A \cup B)$ for mutually exclusive events A and B.

Since mutually exclusive events have no common outcomes, they cannot occur at the same time. Thus, the occurrence of one of them prohibits the occurrence of the other. Consider again the chance experiment of selecting one card from an ordinary deck. This time, let us find the probability of either a king or queen occurring. If

S = {cards from an ordinary deck}
K = {kings}
Q = {queens}

then the Venn diagram of Figure 7-25 illustrates these sets. Observe that $Q \cap K = \emptyset$. Thus, events Q and K are mutually exclusive. Note that since only one card is to be selected, both the king and queen cannot occur at the same time. Therefore, using the special law of addition, we obtain $P(K$ or $Q)$. Hence,

$$P(K \cup Q) = P(K) + P(Q)$$
$$= \frac{4}{52} + \frac{4}{52}$$
$$= \frac{8}{52}$$

394 CHAPTER SEVEN

S

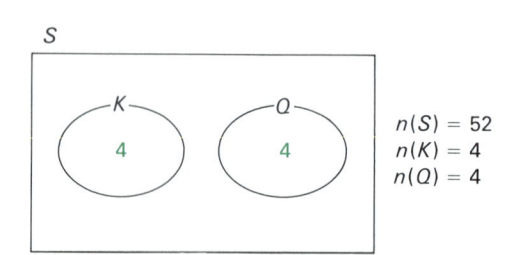

$n(S) = 52$
$n(K) = 4$
$n(Q) = 4$

FIGURE 7–25

EXAMPLE 7–22 The sales manager of an appliance store has found that only washers, dryers, and refrigerators have been sold during the past week. A survey of this past week's buyers reveals the information presented in Table 7–2.

TABLE 7–2

	C Paid Cash	N Used Credit Card	Totals
W: Bought Washers	20	53	73
D: Bought Dryers	7	13	20
R: Bought Refrigerators	3	4	7
Totals	30	70	100

Observe that each customer bought only one appliance. Let

C = {customers who paid cash}
N = {customers who used a credit card}
W = {customers who bought washers}
D = {customers who bought dryers}
R = {customers who bought refrigerators}

If a person is to be selected from this group, find each of the following:
(A) $P(W$ and $N)$ (B) $P(W)$ (C) $P(N)$
(D) $P(W$ or $N)$ (E) $P(W$ or $R)$

Solutions

(A) $P(W$ and $N) = \dfrac{n(W \cap N)}{n(S)} = \dfrac{53}{100} = .53$

(B) $P(W) = \dfrac{n(W)}{n(S)} = \dfrac{73}{100} = .73$

(C) $P(N) = \dfrac{n(N)}{n(S)} = \dfrac{70}{100} = .70$

PROBABILITY

(D) $P(W \text{ or } N) = P(W) + P(N) - P(W \cap N)$
$\qquad = .73 + .70 - .53$
$\qquad = .90$

(E) $P(W \text{ or } R) = P(W) + P(R)$
$\qquad = .73 + .07$
$\qquad = .80$

EXERCISES

1. A die is tossed. If A is the event of getting an even number and B is the event of getting a number less than or equal to 4, find each of the following:
   (A) $P(A)$
   (B) $P(B)$
   (C) $P(A \cap B)$
   (D) $P(A \cup B)$

2. The probability that a student passes mathematics is .60; the probability that he or she passes sociology is .50; and the probability that he or she passes both subjects is .30. Find the probability that the student passes either mathematics or sociology.

3. A box contains 100 marbles: 20 red, 30 green, 40 orange, and 10 blue. A person is to select one marble from the box. Find the probability of getting
   (A) A red marble.
   (B) A blue marble.
   (C) Either a red or a blue marble.
   (D) Either a green or an orange marble.

4. A survey of 60 executives of Montcalf Corporation revealed the following results:

   40 read *The Wall Street Journal*
   30 read *Barron's*
   20 read both *The Wall Street Journal* and *Barron's*

   If a person is selected from this group, find the probability that he or she
   (A) Reads *The Wall Street Journal*.
   (B) Reads *Barron's*.
   (C) Reads both.
   (D) Reads either *The Wall Street Journal* or *Barron's*.

5. The results of a survey of 50 car buyers selected at random in Franklin County, Ohio, are listed in the following table:

	Doctors	Lawyers	Business People	Totals
Jaguar Owners	6	3	1	10
Mercedes Owners	18	7	15	40
Totals	24	10	16	50

If a person is selected from this group, find the probability that he or she is

(A) A doctor.
(B) A Jaguar owner.
(C) Both a doctor and a Jaguar owner.
(D) Either a doctor or a Jaguar owner.
(E) Either a doctor or a lawyer.

6. Using a Venn diagram, verify that for mutually exclusive events A, B, and C,

$$P(A \text{ or } B \text{ or } C) = P(A) + P(B) + P(C)$$

7. During her lunch break, a teacher can do one of the following: go to lunch, buy a dress, visit a friend, or relax in the lounge, with probabilities of .50, .10, .25, and .15, respectively. Find the probability that the teacher either goes to lunch, visits a friend, or buys a dress.

8. Of the employees of the ABK Company, 30% earn $4 per hour, 10% earn $5 per hour, 40% earn $6 per hour, and 20% earn $8 per hour. If a person is selected from this group, find the probability that he or she earns

(A) Less than $6 per hour.
(B) At most $6 per hour.

9. Using the Venn diagram of Figure 7–26, verify that

$$P(A \cup B \cup C) = P(A) + P(B) + P(C) - P(A \cap B) - P(A \cap C) \\ - P(B \cap C) + P(A \cap B \cap C)$$

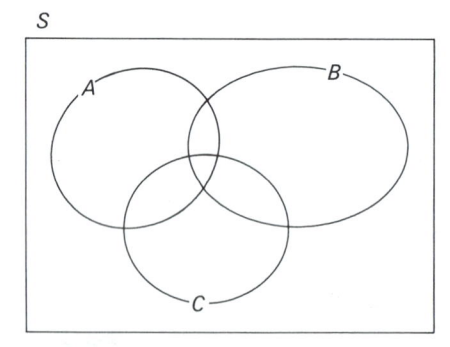

FIGURE 7–26

10. The Heavenly Flavor Ice Cream Company surveyed 100 families to determine flavor preferences. The results were as follows:

    40 families liked vanilla
    34 families liked strawberry
    52 families liked coffee
    21 families liked both vanilla and strawberry
    15 families liked both coffee and strawberry

17 families liked both vanilla and coffee
10 families liked all three flavors

If a family is selected from this group, find the probability that it likes
(A) Vanilla.
(B) Both vanilla and strawberry.
(C) Either vanilla or strawberry.
(D) Either vanilla or strawberry or coffee.

11. Three individuals are applying for the position of manager in a large company. Candidates A and B have the same chance of being hired. The probability of Candidate C being hired is twice that of Candidate A. Find the probability that
(A) Candidate A is hired.
(B) Candidate B is hired.
(C) Candidate C is hired.

7-4

CONDITIONAL PROBABILITY

Consider again the group of buyers of Example 7-22. For convenience, we repeat this set data in Table 7-3.

TABLE 7-3

	C Paid Cash	N Used Credit Card	Totals
W: Bought Washers	20	53	73
D: Bought Dryers	7	13	20
R: Bought Refrigerators	3	4	7
Totals	30	70	100

A chance experiment consists of selecting one person from the set of washer buyers. Note that the person is not to be selected from the full sample space. The person is to be selected from W, where

$$W = \{\text{customers who bought washers}\}$$

Thus, W is called the **restricted sample space** for this chance experiment. Suppose we want to determine the probability of selecting a credit card user knowing that he or she bought a washer. Such a probability is called a **conditional probability**. Since

$$N = \{\text{customers who used credit cards}\}$$

then this conditional probability is denoted as

$$P(N|W)$$

Note that we seek the probability of selecting a person from $N \cap W$. Since the restricted sample space is W, then

$$P(N|W) = \frac{n(N \cap W)}{n(W)} = \frac{53}{73}$$

In general, the formula

$$P(A|B) = \frac{n(A \cap B)}{n(B)}$$

is used to calculate $P(A|B)$ if $n(A \cap B)$ and $n(B)$ are known.

If we begin with this formula and divide both numerator and denominator by $n(S)$, the number of elements in the sample space, we obtain

$$P(A|B) = \frac{\frac{n(A \cap B)}{n(S)}}{\frac{n(B)}{n(S)}}$$
$$= \frac{P(A \cap B)}{P(B)}$$

Thus, if $P(A \cap B)$ and $P(B)$ are known, then $P(A|B)$ may be calculated by the formula

$$P(A|B) = \frac{P(A \cap B)}{P(B)}$$

This formula defines the conditional probability, $P(A|B)$, for any events A and B. Returning to Table 7–3, observe that

$$P(N|W) = \frac{P(N \cap W)}{P(W)} = \frac{\frac{53}{100}}{\frac{73}{100}} = \frac{53}{73}$$

EXAMPLE 7–23 Before placing a person in a highly skilled position, a printing company gives each applicant a physical examination. The results of the examination are presented in Table 7–4.

TABLE 7–4

	A Satisfactory	B Unsatisfactory	Totals
M: Male	30%	10%	40%
F: Female	18%	42%	60%
Totals	48%	52%	100%

Determine each of the following conditional probabilities:

(A) $P(A|M)$ (B) $P(M|A)$
(C) $P(F|B)$ (D) $P(B|M)$

Solutions

(A) $P(A|M) = \dfrac{P(A \cap M)}{P(M)} = \dfrac{.30}{.40} = \dfrac{3}{4}$

(B) $P(M|A) = \dfrac{P(M \cap A)}{P(A)} = \dfrac{.30}{.48} = \dfrac{5}{8}$

(C) $P(F|B) = \dfrac{P(F \cap B)}{P(B)} = \dfrac{.42}{.52} = \dfrac{21}{26}$

(D) $P(B|M) = \dfrac{P(B \cap M)}{P(M)} = \dfrac{.10}{.40} = \dfrac{1}{4}$

Sometimes, conditional probabilities must be calculated intuitively, without using the formula $P(A|B) = P(A \cap B)/P(B)$. Example 7–24 involves such a case.

EXAMPLE 7–24 A chance experiment consists of two selections, without replacement, from the box of marbles illustrated here:

Box of Marbles
30 red
70 blue

(A) Find the probability of selecting a red marble on the first selection.
(B) Find the probability of selecting a red marble on the second selection given that a red marble was obtained on the first selection.

EXAMPLE 7-24 (continued)

Solutions
(A) If R_1 represents the event of selecting a red marble on the first selection, then since there are 30 red marbles out of a total of 100 marbles, we have

$$P(R_1) = \frac{30}{100}$$

(B) If R_2 represents the event of getting a red marble on the second selection, then we seek $P(R_2|R_1)$. If a red marble is obtained on the first selection, then, for the second selection, there are only 29 red marbles out of a total of 99 marbles. Thus, we obtain

$$P(R_2|R_1) = \frac{29}{99}$$

EXERCISES

1. A survey of recent buyers of automobiles revealed the information presented in the following table:

	Bought Compact Car	Bought Full-size Car	Totals
Paid Cash	50	30	80
Paid Credit	100	20	120
Totals	150	50	200

If a person is selected from this group, find the probability of getting
(A) A full-size car buyer given that he or she used credit.
(B) A credit user given that he or she bought a full-size car.
(C) A full-size car buyer and a credit user.
(D) Either a full-size car buyer or a credit user.

2. A survey of employees at a large company reveals the following information:

	M Male	F Female	Totals
W: White-Collar Workers	15%	25%	40%
B: Blue-Collar Workers	50%	10%	60%
Totals	65%	35%	100%

Determine each of the following probabilities:
(A) $P(M|W)$ (B) $P(W|M)$ (C) $P(F|B)$

(D) $P(B|F)$ (E) $P(B$ and $F)$ (F) $P(B$ or $F)$
(G) $P(M|B)$ (H) $P(B|M)$ (I) $P(M$ and $B)$
(J) $P(M$ or $B)$ (K) $P(M$ and $W)$ (L) $P(M$ or $W)$

3. Referring to the Venn diagram of Figure 7–27, determine each of the following probabilities:
(A) $P(A)$ (B) $P(B)$ (C) $P(A \cup B)$
(D) $P(A|B)$ (E) $P(B|A)$ (F) $P(A \cap B)$
(G) $P(A')$ (H) $P(B')$ (I) $P((A \cap B)')$

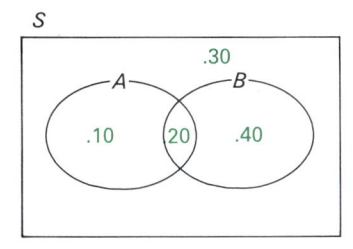

FIGURE 7–27

4. A bond broker is considering using a list of stockholders for direct mail advertising. She knows that 60% of the investors hold stocks, that 30% hold bonds, and that 20% hold both.
   (A) If an investor is a stockholder, what is the probability that he or she is also a bondholder?
   (B) If an investor is a bondholder, what is the probability that he or she is also a stockholder?

5. In a certain store, 40% of entering customers buy washers, 30% buy dryers, and 20% buy both washers and dryers.
   (A) If a customer has bought a washer, find the probability that he or she also buys a dryer.
   (B) If a customer has bought a dryer, find the probability that he or she also buys a washer.

6. In a given small town, 40% of the drivers buy Mercurys, 70% buy Oldsmobiles, and 30% buy both.
   (A) If a driver buys a Mercury, find the probability that he or she also buys an Oldsmobile.
   (B) If a driver buys an Oldsmobile, find the probability that he or she also buys a Mercury.
   (C) What percentage of the drivers buy either an Oldsmobile or a Mercury?

7. A box of marbles contains 40 red and 60 green marbles. A chance experiment consists of two selections, without replacement, from the box. Find the probability of selecting
   (A) A red marble on the first selection.
   (B) A red marble on the second selection given that a red marble was selected on the first selection.
   (C) A green marble on the second selection given that a red marble was selected on the first selection.

8. Repeat Exercise 7 under the assumption that the two selections are made *with replacement*.
9. A production lot of 200 spark plugs contains 20 defectives. A quality control check consists of selecting 3 spark plugs, without replacement, from the lot. Find the probability of selecting
   (A) A defective spark plug on the first selection.
   (B) A defective spark plug on the second selection given that a defective was obtained on the first selection.
   (C) A defective spark plug on the third selection given that defectives were obtained on the first two selections.
   (D) A defective spark plug on the third selection given that nondefectives were obtained on the first two selections.
10. Repeat Exercise 9 under the assumption that the three selections are made *with replacement*.
11. Given that events A and B are mutually exclusive, find $P(A|B)$ and $P(B|A)$. (*Hint:* A Venn diagram will be helpful.)
12. At a given college, 35% of the students are under 20. Also, 80% of the students are male, and 25% of the males are under 20. Using a cross-classification table similar to Table 7–4, answer each of the following:
    (A) What percentage of the students are female and 20 or over?
    (B) Given that a student is a female, what is the probability that she is under 20?

7–5

LAWS OF MULTIPLICATION

In Section 7–4, we derived the formula

$$P(A|B) = \frac{P(A \cap B)}{P(B)}$$

Since $A \cap B$ is equivalent to the expression (A and B), this formula may be restated as

$$P(A|B) = \frac{P(A \text{ and } B)}{P(B)}$$

Multiplying both sides of this equation by $P(B)$ yields

$$P(A \text{ and } B) = P(B)P(A|B)$$

Had we started with the formula

$$P(B|A) = \frac{P(A \text{ and } B)}{P(A)}$$

and multiplied both sides by $P(A)$, we would have obtained
$$P(A \text{ and } B) = P(A)P(B|A)$$

The two formulas

$$P(A \text{ and } B) = P(B)P(A|B)$$
$$P(A \text{ and } B) = P(A)P(B|A)$$

where A and B are any events, are called the **general laws of multiplication**. They are usually used to determine $P(A \text{ and } B)$ for chance experiments consisting of *multiple trials*. These formulas define $P(A \text{ and } B)$ for any events A and B.

EXAMPLE 7–25 A chance experiment consists of selecting two marbles, without replacement, from the following box:

Box of Marbles

| 40 red |
| 60 blue |

Find the probability of selecting
(A) Two red marbles.
(B) Two blue marbles.
(C) A red on the first selection and a blue on the second.
(D) A blue on the first selection and a red on the second.
(E) One of each color.

Solutions

(A) If R_1 represents the event of getting a red on the first selection and R_2 represents the event of getting a red on the second, then we seek $P(R_1 \text{ and } R_2)$. Using the general law of multiplication, we have

$$P(R_1 \text{ and } R_2) = P(R_1)P(R_2|R_1)$$
$$= \frac{40}{100} \cdot \frac{39}{99}$$
$$= \frac{26}{165}$$

EXAMPLE 7–25 (continued)

(B) If B_1 and B_2 represent the events of getting a blue on the first and second selections, respectively, then we calculate

$$P(B_1 \text{ and } B_2) = P(B_1)P(B_2|B_1)$$
$$= \frac{60}{100} \cdot \frac{59}{99}$$
$$= \frac{59}{165}$$

(C) Using the preceding notation, we seek $P(R_1 \text{ and } B_2)$. By the general law of multiplication, we have

$$P(R_1 \text{ and } B_2) = P(R_1)P(B_2|R_1)$$
$$= \frac{40}{100} \cdot \frac{60}{99}$$
$$= \frac{40}{165} = \frac{8}{33}$$

(D) $P(B_1 \text{ and } R_2) = P(B_1)P(R_2|B_1)$
$$= \frac{60}{100} \cdot \frac{40}{99}$$
$$= \frac{40}{165} = \frac{8}{33}$$

(E) Here, we seek $P((R_1 \text{ and } B_2) \text{ or } (B_1 \text{ and } R_2))$. Since the events $(R_1 \text{ and } B_2)$ and $(B_1 \text{ and } R_2)$ are mutually exclusive, then by the special law of addition,

$$P((R_1 \text{ and } B_2) \text{ or } (B_1 \text{ and } R_2)) = P(R_1 \text{ and } B_2) + P(B_1 \text{ and } R_2)$$

In parts (C) and (D) of this example, we determined that

$$P(R_1 \text{ and } B_2) = \frac{8}{33}$$
$$P(B_1 \text{ and } R_2) = \frac{8}{33}$$

Thus,

$$P((R_1 \text{ and } B_2) \text{ or } (B_1 \text{ and } R_2)) = P(R_1 \text{ and } B_2) + P(B_1 \text{ and } R_2)$$
$$= \frac{8}{33} + \frac{8}{33}$$
$$= \frac{16}{33}$$

The possible outcomes of this chance experiment are illustrated by the tree diagram of Figure 7–28. Since the events are not equiprobable, the probability of each event is written along its respective branch of the diagram. The probability of each outcome is calculated at the far right. Note that the sum of these probabilities is 1. Thus, all possible outcomes are included in the tree diagram.

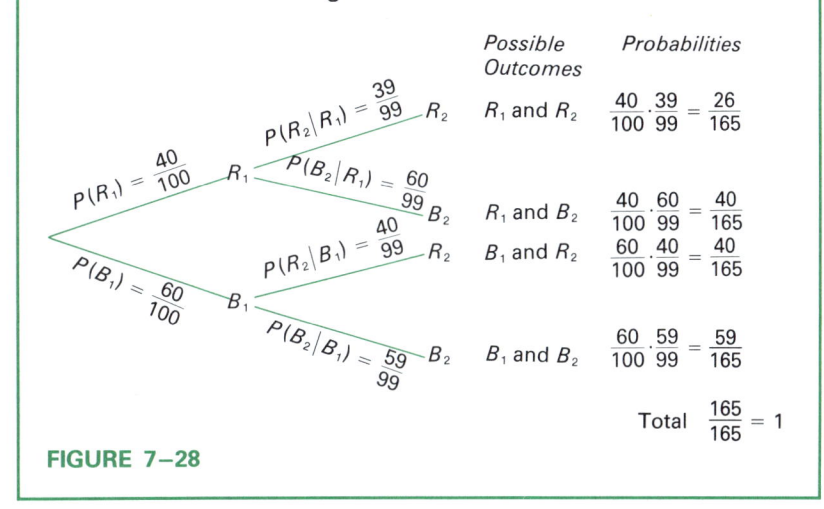

	Possible Outcomes	Probabilities
$P(R_1) = \frac{40}{100}$, $P(R_2\|R_1) = \frac{39}{99} \to R_2$	R_1 and R_2	$\frac{40}{100} \cdot \frac{39}{99} = \frac{26}{165}$
$P(B_2\|R_1) = \frac{60}{99} \to B_2$	R_1 and B_2	$\frac{40}{100} \cdot \frac{60}{99} = \frac{40}{165}$
$P(B_1) = \frac{60}{100}$, $P(R_2\|B_1) = \frac{40}{99} \to R_2$	B_1 and R_2	$\frac{60}{100} \cdot \frac{40}{99} = \frac{40}{165}$
$P(B_2\|B_1) = \frac{59}{99} \to B_2$	B_1 and B_2	$\frac{60}{100} \cdot \frac{59}{99} = \frac{59}{165}$
	Total	$\frac{165}{165} = 1$

FIGURE 7–28

Independent Events

Two events A and B are **independent** if

$$P(A|B) = P(A)$$

Intuitively, two events are independent if the occurrence of one does not change the probability of the occurrence (or nonoccurrence) of the other. If two events are not independent, then they are **dependent**. As a specific example, consider the chance experiment of selecting, with replacement, two marbles from the box of Example 7–25, which is repeated here for reference:

Box of Marbles

40 red
60 blue

Recall from Example 7–25 that the probability of selecting a red marble on the second selection given that a red occurred on the first selection is

$$P(R_2|R_1) = \frac{39}{99}$$

Since we are now selecting the two marbles *with replacement*, the first marble selected is replaced before selecting the second marble. Thus,

$$P(R_2|R_1) = P(R_2) = \frac{40}{100}$$

and the events of getting a red marble on the second selection and getting a red marble on the first selection are independent.

If two events A and B are independent, the general law of multiplication

$$P(A \text{ and } B) = P(A)P(B|A)$$

becomes

$$P(A \text{ and } B) = P(A)P(B)$$

since $P(B|A) = P(B)$. The result

$$P(A \text{ and } B) = P(A)P(B)$$

is called the **special law of multiplication**. It must be remembered that this law applies only to independent events. The special law of multiplication defines $P(A \text{ and } B)$ for independent events A and B.

EXAMPLE 7-26 A chance experiment consists of selecting two marbles, with replacement, from the following box:

Box of Marbles

| 40 red |
| 60 blue |

Find the probability of selecting
(A) Two red marbles.
(B) Two blue marbles.
(C) A red on the first selection and a blue on the second.
(D) A blue on the first selection and a red on the second.
(E) One of each color.

Solutions Using the special law of multiplication, we have

(A) $P(R_1 \text{ and } R_2) = P(R_1)P(R_2)$
$$= \frac{40}{100} \cdot \frac{40}{100}$$
$$= .16$$

(B) $P(B_1 \text{ and } B_2) = P(B_1)P(B_2)$
$$= \frac{60}{100} \cdot \frac{60}{100}$$
$$= .36$$

(C) $P(R_1 \text{ and } B_2) = P(R_1)P(B_2)$
$$= \frac{40}{100} \cdot \frac{60}{100}$$
$$= .24$$

(D) $P(B_1 \text{ and } R_2) = P(B_1)P(R_2)$
$$= \frac{60}{100} \cdot \frac{40}{100}$$
$$= .24$$

(E) Using the special law of addition, we have
$$P((R_1 \text{ and } B_2) \text{ or } (B_1 \text{ and } R_2)) = P(R_1 \text{ and } B_2) + P(B_1 \text{ and } R_2)$$
$$= .24 + .24$$
$$= .48$$

Sometimes, we must use the laws of addition and multiplication and conditional probability formulas to solve certain problems. Example 7–27 illustrates such a case.

EXAMPLE 7–27 In a certain hospital, 70% of the patients are smokers, 60% of the patients have lung cancer, and 80% of those who are smokers have lung cancer. If one patient is to be selected from this group, find the probability that he or she
(A) Is a smoker and has lung cancer.
(B) Is a smoker given that he or she has lung cancer.
(C) Is a smoker or has lung cancer.
Solutions If

S = event of selecting a smoker
L = event of selecting a patient with lung cancer

EXAMPLE 7–27 (continued)

then

$$P(S) = .70 \quad P(L) = .60 \quad P(L|S) = .80$$

(A) $P(S \text{ and } L) = P(S)P(L|S)$
$= (.70)(.80)$
$= .56$

(B) $P(S|L) = \dfrac{P(S \text{ and } L)}{P(L)}$
$= \dfrac{.56}{.60} = \dfrac{14}{15}$
$\approx .93$

(C) $P(S \text{ or } L) = P(S) + P(L) - P(S \text{ and } L)$
$= .70 + .60 - .56$
$= .74$

EXERCISES

1. A box contains 50 white balls, 110 green balls, and 40 red balls. A chance experiment consists of selecting two balls, without replacement, from the box. Find the probability of getting
   (A) Two red balls.
   (B) Two white balls.
   (C) First a red, then a green.
   (D) First a green, then a red.
   (E) A red and a green.

2. Repeat Exercise 1 under the assumption that the selections are made with replacement.

3. Of the households of a given city, 30% have electric dryers, 40% have electric stoves, and 25% of those that have electric stoves also have electric dryers. Find the probability that
   (A) A household has both an electric dryer and an electric stove.
   (B) A household with an electric dryer also has an electric stove.

4. Analyzing the sales of a given product in a retail store, we discover that 10% of the purchases were made by men and 20% of the purchases were over $10 in value. Given that 80% of male customers make purchases over $10, find what percentages of the purchases
   (A) Were over $10 and were made by men.
   (B) Were over $10 or were made by men.
   (C) Over $10 were made by men.

5. The probability of a husband not voting in an election is .30. The probability of a wife voting given that her husband votes is .80. Find the probability that both a husband and wife vote.

PROBABILITY

6. Sixty percent of the patients at a medical center are female and 55 percent are over 40. Ten percent are male and over 40. If a patient is selected from this group, find the probability that he or she is
   (A) Male and 40 or under.
   (B) Female and over 40.
   (C) Female and 40 or under.
   (D) Over 40.
   (*Hint:* Use a cross-classification table.)

7. A private ambulance service keeps two vehicles in readiness for emergencies. The probability that a given vehicle is available when needed is .90. If the availability of one vehicle is independent of the availability of the other, find the probability that
   (A) Both vehicles will be available in the event of two emergencies occurring at the same time.
   (B) A vehicle will be available when a call is received for service (i.e., the probability that either vehicle will be available for service).
   (C) Neither vehicle is available for service. (*Hint:* Use one of De Morgan's laws. See Exercise 13, Section 7–1.)

8. If events A and B are independent and $P(A) = .3$ and $P(B) = .6$, find each of the following:
   (A) $P(A$ and $B)$　　　(B) $P(A$ or $B)$
   (C) $P(A|B)$　　　(D) $P(B|A)$

9. Given that $P(A) = .6$, $P(B) = .7$, and $P(A \cap B) = .4$.
   (A) Find $P(A \cup B)$　　　(B) Find $P(A|B)$
   (C) Find $P(B|A)$　　　(D) Are events A and B independent? Why or why not?

7–6

BAYES' FORMULA

In the previous two sections, we encountered conditional probabilities such as $P(A|B)$, $P(B|A)$, etc. Sometimes, we come across situations where we know, for example, $P(A|B)$ and wish to determine $P(B|A)$. Such a problem is usually solved by using **Bayes' formula**. We now proceed to develop this famous useful expression.

Consider two boxes containing marbles. Box 1 contains 4 red marbles and 6 green marbles. Box 2 has 2 red, 1 white, and 5 green marbles. A chance experiment consists of first selecting a box and then selecting a marble from that box. Here, we will find the probability that a red marble is chosen.

We let B_1 and B_2 represent the events of choosing Box 1 and Box 2, respectively, and R, G, and W represent the events of getting a red, green, and white marble, respectively. Hence, we have

$$P(B_1) = \frac{1}{2} \qquad P(B_2) = \frac{1}{2}$$

$$P(R|B_1) = \frac{4}{10} \qquad P(R|B_2) = \frac{2}{8}$$

$$P(G|B_1) = \frac{6}{10} \qquad P(W|B_2) = \frac{1}{8}$$

$$P(G|B_2) = \frac{5}{8}$$

The tree diagram of Figure 7–29 illustrates all possible outcomes for this experiment. Observing Figure 7–29, note that event R—getting a red marble—can be written as

$$R = (B_1 \text{ and } R) \quad \text{or} \quad (B_2 \text{ and } R)$$

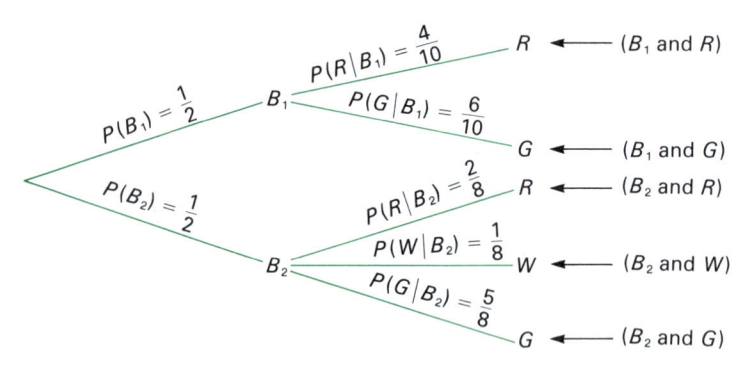

FIGURE 7–29

In other words, we could have chosen Box 1 and then selected a red marble from Box 1 or we could have chosen Box 2 and then selected a red marble from Box 2. Since the events $(B_1$ and $R)$ and $(B_2$ and $R)$ are mutually exclusive, then by the special law of addition, we have

$$P(R) = P(B_1 \text{ and } R) + P(B_2 \text{ and } R)$$

And since

$$P(B_1 \text{ and } R) = P(B_1)P(R|B_1)$$
$$P(B_2 \text{ and } R) = P(B_2)P(R|B_2)$$

the formula for $P(R)$ may be written as

$$P(R) = P(B_1)P(R|B_1) + P(B_2)P(R|B_2)$$
$$= \frac{1}{2} \cdot \frac{4}{10} + \frac{1}{2} \cdot \frac{2}{8}$$
$$= \frac{13}{40}$$

We now consider another question regarding this experiment of selecting a box and then selecting a marble from that box:

Suppose the experiment is performed and a red marble selected. What is the probability that the red marble came from Box 1?

Here, we seek $P(B_1|R)$. Using the formula for conditional probability, we have

$$P(B_1|R) = \frac{P(B_1 \text{ and } R)}{P(R)}$$

Since $P(B_1 \text{ and } R) = P(B_1)P(R|B_1)$, this equation becomes

$$P(B_1|R) = \frac{P(B_1)P(R|B_1)}{P(R)}$$

From the discussion of the preceding paragraphs,

$$P(R) = P(B_1)P(R|B_1) + P(B_2)P(R|B_2)$$

Substituting this result into the formula for $P(B_1|R)$, we have

$$P(B_1|R) = \frac{P(B_1)P(R|B_1)}{P(B_1)P(R|B_1) + P(B_2)P(R|B_2)}$$

Thus, the probability that the red marble came from Box 1 is

$$P(B_1|R) = \frac{\frac{1}{2} \cdot \frac{4}{10}}{\frac{1}{2} \cdot \frac{4}{10} + \frac{1}{2} \cdot \frac{2}{8}}$$
$$= \frac{8}{13}$$

The preceding formula for $P(B_1|R)$ is a special case of Bayes' formula.

The general expression of Bayes' formula is as follows.

Bayes' Formula

If $B_1, B_2, \ldots, B_k$ are mutually exclusive events, one of which must occur, and if A is an event such that $P(A) > 0$, then

$$P(B_i|A) = \frac{P(B_i)P(A|B_i)}{P(B_1)P(A|B_1) + P(B_2)P(A|B_2) + \ldots + P(B_k)P(A|B_k)}$$

for $i = 1, 2, \ldots, k$.

Figure 7–30 contains a tree diagram illustration of Bayes' formula. Observe that $B_1, B_2, \ldots, B_k$ are mutually exclusive events, one of which must occur. Also, $P(A|B_1), P(A|B_2), \ldots, P(A|B_k)$ are known probabilities. Bayes' formula calculates the probabilities $P(B_i|A)$ for $i = 1, 2, \ldots, k$.

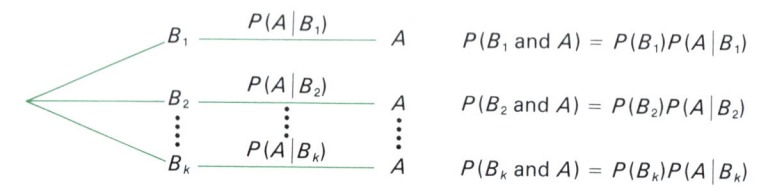

```
B₁ ——P(A|B₁)—— A    P(B₁ and A) = P(B₁)P(A|B₁)
B₂ ——P(A|B₂)—— A    P(B₂ and A) = P(B₂)P(A|B₂)
 ⋮
Bₖ ——P(A|Bₖ)—— A    P(Bₖ and A) = P(Bₖ)P(A|Bₖ)
```

Bayes' formula
$$P(B_i|A) = \frac{P(B_i)P(A|B_i)}{P(B_1)P(A|B_1) + P(B_2)P(A|B_2) + \cdots + P(B_k)P(A|B_k)}$$
for $i = 1, 2, \ldots, k$

FIGURE 7–30

EXAMPLE 7–28 Referring to the illustrative problem of this section, suppose a red marble was selected. What is the probability that it came from Box 2?

Solution We seek $P(B_2|R)$. Using Bayes' formula, we obtain

$$P(B_2|R) = \frac{P(B_2)P(R|B_2)}{P(B_1)P(R|B_1) + P(B_2)P(R|B_2)}$$

$$= \frac{\frac{1}{2} \cdot \frac{2}{8}}{\frac{1}{2} \cdot \frac{4}{10} + \frac{1}{2} \cdot \frac{2}{8}}$$

$$= \frac{5}{13}$$

A Posteriori versus A Priori Probabilities

Consider the probabilities $P(B_1)$ and $P(B_2)$ of the marble selection problem of this section. The probabilities $P(B_1)$ and $P(B_2)$ are the probabilities of choosing Box 1 and Box 2, respectively, before the experiment is performed. Such probabilities are called **before the fact** or **a priori probabilities**. The probability $P(B_1|R)$ is the probability that a selected red marble came from Box 1 after the experiment was performed. Such a probability is called an **after the fact** or **a posteriori probability**. Bayes' formula enables us to calculate a posteriori probabilities.

EXAMPLE 7–29 A distributor carries three brands of tires: Brand B_1, Brand B_2, and Brand B_3. Of all the tires, 50% are Brand B_1, 30% are Brand B_2, and 20% are Brand B_3. Percentages of defectives are as follows: 10% of Brand B_1, 5% of Brand B_2, and 8% of Brand B_3.
(A) If the distributor selects a tire from this inventory, find the probability that the tire is a defective.
(B) If the distributor selects a tire and gets a defective, find the probability that the tire is a Brand B_2 tire.

Solutions If

D = event of getting a defective tire
B_1 = event of getting a Brand B_1 tire
B_2 = event of getting a Brand B_2 tire
B_3 = event of getting a Brand B_3 tire

then

$P(B_1) = .50$ $P(B_2) = .30$ $P(B_3) = .20$
$P(D|B_1) = .10$ $P(D|B_2) = .05$ $P(D|B_3) = .08$

EXAMPLE 7-29 (continued)

(A) Here, we seek the probability of getting a defective tire, $P(D)$. Since a defective tire can be either a Brand B_1, a Brand B_2, or a Brand B_3 tire, we write

$$D = (B_1 \text{ and } D) \text{ or } (B_2 \text{ and } D) \text{ or } (B_3 \text{ and } D)$$

(See Figure 7-31.) Since $(B_1 \text{ and } D)$, $(B_2 \text{ and } D)$, and $(B_3 \text{ and } D)$ are mutually exclusive events, we have

$$P(D) = P(B_1 \text{ and } D) + P(B_2 \text{ and } D) + P(B_3 \text{ and } D)$$

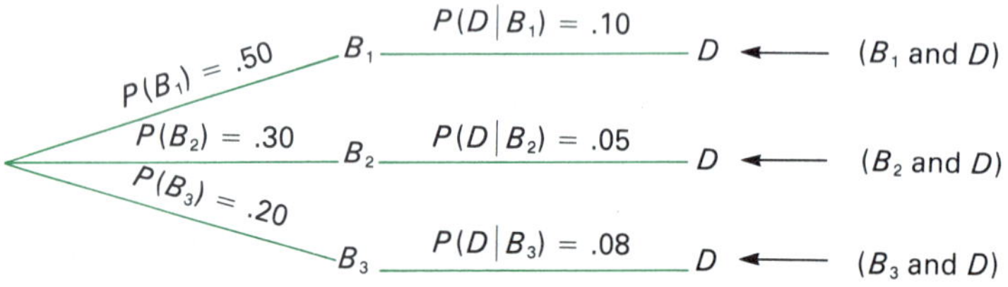

FIGURE 7-31

$P(B_1) = .50$, B_1, $P(D|B_1) = .10$, $D \leftarrow (B_1 \text{ and } D)$
$P(B_2) = .30$, B_2, $P(D|B_2) = .05$, $D \leftarrow (B_2 \text{ and } D)$
$P(B_3) = .20$, B_3, $P(D|B_3) = .08$, $D \leftarrow (B_3 \text{ and } D)$

By the general law of multiplication,

$$P(B_1 \text{ and } D) = P(B_1)P(D|B_1)$$
$$P(B_2 \text{ and } D) = P(B_2)P(D|B_2)$$
$$P(B_3 \text{ and } D) = P(B_3)P(D|B_3)$$

Substituting these results into the formula for $P(D)$ gives us

$$P(D) = P(B_1)P(D|B_1) + P(B_2)P(D|B_2) + P(B_3)P(D|B_3)$$
$$= (.50)(.10) + (.30)(.05) + (.20)(.08)$$
$$= .081$$

(B) In this case, we seek the a posteriori probability $P(B_2|D)$. Using Bayes' formula, we write

$$P(B_2|D) = \frac{P(B_2)P(D|B_2)}{P(B_1)P(D|B_1) + P(B_2)P(D|B_2) + P(B_3)P(D|B_3)}$$

Note that the denominator of this formula is $P(D)$, which we calculated in part (A). Hence,

PROBABILITY

$$P(B_2|D) = \frac{(.30)(.05)}{.081}$$
$$= .185$$

EXERCISES

1. Box 1 contains 5 red and 3 green marbles, and Box 2 contains 1 yellow, 7 red, and 2 green marbles. A chance experiment consists of first selecting a box and then selecting a marble from that box.
   (A) Find the probability of getting a red marble.
   (B) Find the probability of getting a green marble.
   (C) Find the probability of getting a yellow marble.
   (D) If a red marble is selected, find the probability that it came from Box 1.
   (E) If a green marble is selected, find the probability that it came from Box 1.

2. Refer to the chance experiment of Exercise 1.
   (A) If a red marble is selected, find the probability that it came from Box 2.
   (B) If a green marble is selected, find the probability that it came from Box 2.
   (C) If a yellow marble is selected, find the probability that it came from Box 2.

3. A company that manufactures mopeds has three plants. Plant 1 produces 40% of the output, Plant 2 produces 35%, and Plant 3 produces 25%. Of the mopeds, 2% produced by Plant 1 are defective, 1% produced by Plant 2 are defective, and 3% produced by Plant 3 are defective. If a moped is selected from this company's output and found to be defective, find the probability that it came from
   (a) Plant 1.   (B) Plant 2.   (C) Plant 3.

4. If a person with a certain disease is given a screening, there is a 90% probability that the disease will be detected. If a person without this disease is given the same screening, there is a 20% probability that the person will be diagnosed incorrectly as having this disease. In a given community, 15% of the residents have this disease. If a resident of this community is diagnosed as having this disease, what is the probability that he or she actually has the disease?

5. A given city is divided into submarkets for consumer behavior. Submarket 1 contains 30% of the consumers, Submarket 2 contains 25%, Submarket 3 contains 21%, and Submarket 4 contains 24%. Of the consumers, 40% from Submarket 1 usually favor foreign-made cars, as do 20% from Submarket 2, 35% from Submarket 3, and 50% from Submarket 4.

(A) If a consumer is chosen at random, what is the probability that he or she will favor foreign-made cars?

(B) If a chosen consumer has favored foreign-made cars, what is the probability that he or she came from Submarket 1?

6. The following table shows the accident rate for various age groups insured with a particular insurance company:

Age Group	Proportion of Total Insured	Accident Rate
Under 21	.10	.08
21–30	.15	.05
31–40	.35	.03
41–50	.25	.02
Over 50	.15	.04

If a policyholder reports an accident, what is the probability that he or she is

(A) Under 21? (B) In the age group 31–40?

7. A local department store has classified 90% of its credit customers as "Good Payers" and 10% as "Poor Payers." From past experience, it has been determined that 98% of the "Good Payers" do not return any items purchased whereas only 30% of the "Poor Payers" do not return any items purchased. If a credit customer returns a purchased item, what is the probability that he or she is a "Poor Payer"?

8. A mail-order house has two workers filling orders. Worker 1 usually fills 70% of all orders, and Worker 2 fills the remaining 30%. Worker 1 has an error rate of 5%, and Worker 2 has an error rate of 3%. If an order is returned due to an error, what is the probability that it was filled by Worker 1?

7-7

COUNTING, PERMUTATIONS, COMBINATIONS, AND PROBABILITY

Many times, we must determine how many possible outcomes there are for a certain chance experiment. We often do this by using principles of counting. The following experiment will illustrate a basic principle of counting called the multiplication rule.

Counting

Consider two sets, A_1 and A_2, containing 4 and 2 elements, respectively. In how many different ways can we first select an element from A_1 and then one from A_2?

PROBABILITY

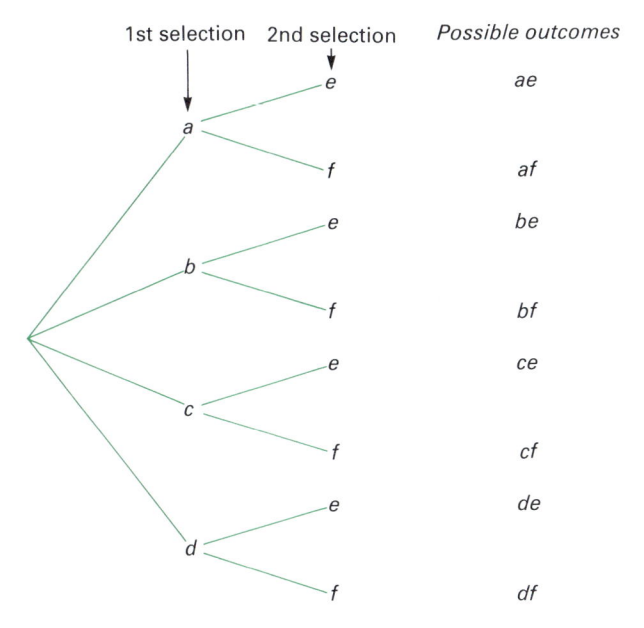

```
1st selection   2nd selection    Possible outcomes
                      e               ae
          a
                      f               af
                      e               be
          b
                      f               bf
                      e               ce
          c
                      f               cf
                      e               de
          d
                      f               df
```

FIGURE 7–32

The tree diagram of Figure 7–32 shows all the possible outcomes for this experiment. Note that for the sake of illustration, we let $A_1 = \{a, b, c, d\}$ and $A_2 = \{e, f\}$. Observe that there are $4 \cdot 2 = 8$ possible ways of first selecting an element from A_1 and then selecting an element from A_2. This result can be generalized to any number of sets to yield the **multiplication rule of counting.**

Multiplication Rule of Counting

If sets $A_1, A_2, \ldots, A_k$ contain $n_1, n_2, \ldots, n_k$ elements, respectively, then there are $n_1 \cdot n_2 \cdot \ldots \cdot n_k$ ways in which one can first select an element from A_1, then an element from A_2, and finally an element from A_k.

EXAMPLE 7–30 A student must arrange 3 books on a shelf. How many possible arrangements are there?
Solution This situation may be perceived as selecting one element from each of 3 sets. Since the student may select any of the 3 books for the first selection, then the first selection is made from a set containing 3 elements. Having selected the first book, the student will make the second selection from a

EXAMPLE 7–30 (continued)

set containing 2 elements. And the third selection is made from a set containing 1 element. Thus, by the multiplication rule of counting, there are 3 · 2 · 1 = 6 possible arrangements. The tree diagram of Figure 7–33 illustrates all possible arrangements. In this figure, the 3 books have been labeled as a, b, and c.

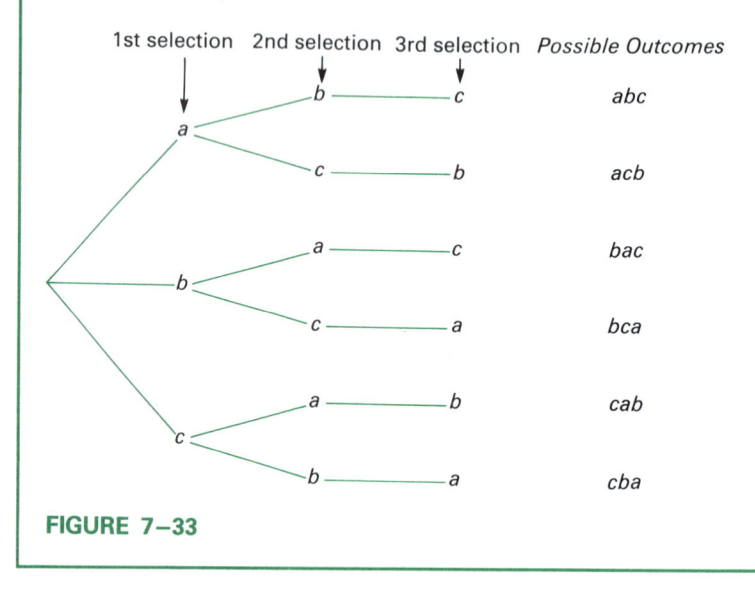

1st selection   2nd selection   3rd selection   *Possible Outcomes*

```
         b ——————— c           abc
      a
         c ——————— b           acb

         a ——————— c           bac
      b
         c ——————— a           bca

         a ——————— b           cab
      c
         b ——————— a           cba
```

FIGURE 7–33

Factorial In Example 7–30, we determined that there are 3 · 2 · 1 = 6 different arrangements of 3 books. Note that 3 · 2 · 1 is a product of all consecutive integers from 1 to 3. There are many chance experiments where the number of possible outcomes is a product of all consecutive integers from 1 to n. Such a product is called ***n* factorial** and is denoted by the symbol ***n*!** Thus, for any counting number n

$$n! = n(n-1)(n-2) \ldots (3)(2)(1)$$

Hence, the statement "5 factorial" is denoted by the symbol 5! and implies the product

$$5! = 5 \cdot 4 \cdot 3 \cdot 2 \cdot 1$$

And the statement "7 factorial" is denoted by 7! and implies the product

$$7! = 7 \cdot 6 \cdot 5 \cdot 4 \cdot 3 \cdot 2 \cdot 1$$

By definition, 0! = 1.

PROBABILITY

> **EXAMPLE 7–31** How many different batting orders does the manager of a baseball team (9 players) have to choose from?
>
> *Solution* The first selection is made from a set of 9 players, the second from a set of 8 players, the third from a set of 7 players, etc., and so by the multiplication rule of counting, there are
>
> $$9 \cdot 8 \cdot 7 \cdot 6 \cdot 5 \cdot 4 \cdot 3 \cdot 2 \cdot 1 = 9! = 362,880$$
>
> possible batting orders.

> **EXAMPLE 7–32** Referring to Example 7–31, suppose the star hitter must be third in the batting order and the pitcher must be sixth in the batting order. How many different batting orders are there under these restrictions?
>
> *Solution* Since the star hitter must bat third and the pitcher must bat sixth, these two players are fixed in their respective batting orders. Thus, the first selection is made from a set of 7 players, the second from a set of 6 players, the third from a set of 1 player (the star hitter), the fourth from a set of 5 players, the fifth from a set of 4 players, the sixth from a set of 1 player (the pitcher), the seventh from a set of 3 players, the eighth from a set of 2 players, and the ninth from a set of 1 player. Thus, by the multiplication rule of counting, there are
>
> $$7 \cdot 6 \cdot 1 \cdot 5 \cdot 4 \cdot 1 \cdot 3 \cdot 2 \cdot 1 = 5040$$
>
> possible batting orders.

Permutations Suppose we want to select 2 letters from the set $\{a, b, c, d, e\}$. Since there are 5 possibilities for the first selection and 4 possibilities for the second selection, then by the multiplication rule of counting, there are $5 \cdot 4 = 20$ possible outcomes. The tree diagram of Figure 7–34 illustrates all the possible outcomes for this experiment.

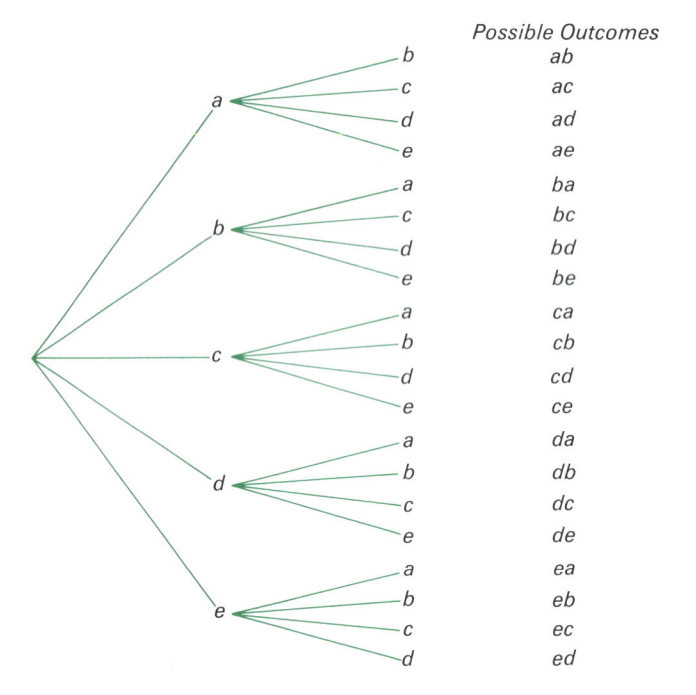

```
                    Possible Outcomes
             b          ab
            c           ac
       a    d           ad
            e           ae
            a           ba
            c           bc
       b    d           bd
            e           be
            a           ca
            b           cb
       c    d           cd
            e           ce
            a           da
            b           db
       d    c           dc
            e           de
            a           ea
            b           eb
       e    c           ec
            d           ed
```

FIGURE 7–34

Observing the outcomes of Figure 7–34, we note two important characteristics of this counting problem:

1. No letter is repeated. In other words, a selected letter is not replaced before a subsequent selection.
2. The order of selection makes a difference. That is, despite the fact that *ab* and *ba* consist of the same 2 letters, they are considered different outcomes because the order of the 2 letters is different.

Counting problems that consist of selecting r elements out of a set of n elements and satisfy these two characteristics are called **permutations of n elements taken r at a time.**

> The number of **permutations** of n elements taken r at a time ($r \leq n$) is the number of all possible ways of selecting r elements from a set of n elements where order of selection is important and replacement between selections is not allowed. This quantity is denoted by the symbol $P(n, r)$.

We now derive a formula for $P(n, r)$, the number of possible ways of selecting r elements from a set of n elements where order

421 PROBABILITY

of selection is important and replacement between selections is not allowed. Since the first selection has n possibilities, the second selection has $n - 1$ possibilities, the third selection has $n - 2$ possibilities, etc., by the multiplication rule of counting, there will be

$$\underbrace{n(n - 1)(n - 2) \ldots (n - r + 1)}_{r \text{ factors}}$$

possible arrangements. Thus,

$$P(n, r) = \underbrace{n(n - 1)(n - 2) \ldots (n - r + 1)}_{r \text{ factors}}$$

For example, referring to the counting problem of Figure 7–34, where $n = 5$ and $r = 2$, we saw that there were

$$P(5, 2) = \underbrace{5 \cdot 4}_{r = 2 \text{ factors}}$$

arrangements of 5 elements taken 2 at a time. Note that the product $5 \cdot 4$ can be written in factorial notation as follows:

$$5 \cdot 4 = \frac{5 \cdot 4 \cdot 3 \cdot 2 \cdot 1}{3 \cdot 2 \cdot 1}$$

$$= \frac{5!}{3!}$$

$$= \frac{5!}{(5 - 2)!}$$

In general, the product

$$P(n, r) = \underbrace{n(n - 1)(n - 2) \ldots (n - r + 1)}_{r \text{ factors}}$$

may be written in factorial notation as

$$P(n, r) = \frac{n!}{(n - r)!}$$

Thus, we conclude the following.

> The number of permutations of n elements taken r at a time $(r \leq n)$ is denoted by $P(n, r)$, where
>
> $$P(n, r) = \frac{n!}{(n - r)!}$$

EXAMPLE 7-33 In how many ways can 4 people be seated in a row of 7 seats?

Solution This problem is equivalent to selecting 4 seats out of 7 where order is important and replacement is not permitted. Thus, there are

$$P(7, 4) = \frac{7!}{(7 - 4)!}$$
$$= \frac{7!}{3!}$$
$$= \frac{7 \cdot 6 \cdot 5 \cdot 4 \cdot 3!}{3!}$$
$$= 840$$

possible seating arrangements.

Combinations

Consider all possible ways of selecting 3 letters from the 4-element set $\{a, b, c, d\}$ where order makes a difference and replacement is not allowed. Thus, we are considering all possible permutations of 4 elements taken 3 at a time, or

$$P(4, 3) = \frac{4!}{(4 - 3)!} = 24$$

possible outcomes. Figure 7-35 illustrates the 24 permutations.

abc	acb	bac	bca	cab	cba
abd	adb	bad	bda	dab	dba
acd	adc	cad	cda	dac	dca
bcd	bdc	cbd	cdb	dbc	dcb

24 permutations of 4 elements taken 3 at a time

FIGURE 7-35

Suppose we want to determine the number of possible ways of selecting 3 elements out of 4 elements where order makes *no* difference and replacement is not allowed. Here, we are seeking the number of combinations of 4 elements taken 3 at a time. This quantity is denoted by the symbol $\binom{4}{3}$. In general, counting problems that consist of selecting r elements out of a set of n elements where the order of selection makes no difference and replacement is not allowed are called **combinations of n elements taken r at a time**.

PROBABILITY

> The number of **combinations** of n elements taken r at a time ($r \leq n$) is the number of all possible ways of selecting r elements from a set of n elements where order of selection makes no difference and replacement between selections is not allowed. This quantity is denoted by the symbol $\binom{n}{r}$.

We now derive a formula for $\binom{n}{r}$. Observing the 24 permutations of 4 elements taken 3 at a time of Figure 7–35, note that the outcomes of each row contain the same 3 letters. Since 3 letters can be arranged in 3! different orders, each row has 3! = 6 outcomes which are now considered equal since order makes no difference. Thus, the number of combinations of 4 elements taken 3 at a time, or $\binom{4}{3}$, is found by dividing the number of permutations by 3!. Hence, we have

$$\binom{4}{3} = \frac{P(4, 3)}{3!}$$

In general, the number of combinations of n elements taken r at a time is

$$\binom{n}{r} = \frac{P(n, r)}{r!}$$

$$= \frac{n!}{r!(n - r)!}$$

Thus, we conclude the following.

> The number of combinations of n elements taken r at a time ($r \leq n$) is denoted by $\binom{n}{r}$, where
>
> $$\binom{n}{r} = \frac{n!}{r!(n - r)!}$$

EXAMPLE 7–34 How many 3-element subsets can be selected from a set containing 5 elements?

EXAMPLE 7-34 (continued)

Solution This situation is equivalent to selecting 3 elements out of 5 where order makes no difference and replacement is not allowed. Thus, there are

$$\binom{5}{3} = \frac{5!}{3!(5-3)!}$$
$$= \frac{5!}{3!2!}$$
$$= \frac{5 \cdot 4 \cdot 3!}{3!2!}$$
$$= 10$$

possible subsets or combinations of 5 elements taken 3 at a time.

EXAMPLE 7-35
In how many ways can a committee of 5 be chosen from a group of 9 persons?

Solution We seek the number of combinations of 9 elements taken 5 at a time. There are

$$\binom{9}{5} = \frac{9!}{5!(9-5)!}$$
$$= \frac{9!}{5!4!}$$
$$= \frac{9 \cdot 8 \cdot 7 \cdot 6 \cdot 5!}{5! \cdot 4 \cdot 3 \cdot 2 \cdot 1}$$
$$= 126$$

such combinations.

EXAMPLE 7-36
A production lot contains 20 radios, 4 of which are defective. A sample of 5 radios is to be selected from the lot.
(A) How many possible samples are there?
(B) Suppose we want the sample of 5 radios to contain all nondefective radios. In how many ways can this happen?
(C) What is the probability that the sample contains 5 nondefective radios?

Solutions

(A) Since we are selecting 5 elements out of a set of 20 elements where order makes no difference and replacement is not allowed, there are

$$\binom{20}{5} = \frac{20!}{5!(20-5)!}$$
$$= \frac{20 \cdot 19 \cdot 18 \cdot 17 \cdot 16 \cdot 15!}{5 \cdot 4 \cdot 3 \cdot 2 \cdot 1 \cdot 15!}$$
$$= 15,504$$

possible samples.

(B) Since the production lot contains 4 defectives, we want the sample of 5 radios to be selected from 16 nondefective radios. There are

$$\binom{16}{5} = \frac{16!}{5!11!}$$
$$= 4368$$

possible ways of doing this.

(C) In part (B), we determined that there are $\binom{16}{5} = 4368$ different ways of selecting a sample of 5 nondefective radios. Thus, the event set, E, contains $\binom{16}{5} = 4368$ elements. In part (A), we determined that there are $\binom{20}{5} = 15,504$ possible 5-element samples that can be selected from the production lot of 20 radios. Thus, the sample space, S, contains $\binom{20}{5} = 15,504$ possible outcomes. Hence, the probability of selecting 5 nondefectives is

$$P(E) = \frac{n(E)}{n(S)} = \frac{\binom{16}{5}}{\binom{20}{5}}$$
$$= \frac{4368}{15,504}$$
$$= 0.28$$

EXAMPLE 7-37 Referring to the situation of Example 7-36, find the probability that the 5-element sample contains exactly 3 defectives.

Solution We want the sample to contain 3 defectives and 2 nondefectives. Since the 3 defectives must be chosen from

426 CHAPTER SEVEN

EXAMPLE 7–37 (continued)

the 4 defectives in the production lot, there are $\binom{4}{3} = 4$ possible ways of doing this. The 2 nondefectives must be chosen from the 16 nondefectives of the production lot. There are $\binom{16}{2} = 120$ ways of doing this. Thus, by the multiplication rule of counting, there are $\binom{4}{3} \cdot \binom{16}{2} = 480$ possible ways of selecting a sample of 3 defectives and 2 nondefectives. Hence, the event set, E, contains $\binom{4}{3} \cdot \binom{16}{2} = 480$ elements. In part (A) of Example 7–36, we determined that the sample space has $\binom{20}{5} = 15{,}504$ possible outcomes. Thus, the probability that the 5-element sample contains exactly 3 defectives is

$$P(E) = \frac{n(E)}{n(S)} = \frac{\binom{4}{3} \cdot \binom{16}{2}}{\binom{20}{5}}$$

$$= \frac{480}{15{,}504}$$

$$= .031$$

EXAMPLE 7–38 Referring to the situation in Examples 7–36 and 7–37, find the probability that the 5-element sample contains at least 3 defectives.

Solution Since the production lot contains only 4 defectives, we want the 5-element sample to contain either 3 defectives or 4 defectives. In Example 7–37, we determined the probability that the sample contains exactly 3 defectives to be

$$\frac{\binom{4}{3} \cdot \binom{16}{2}}{\binom{20}{5}} = .031$$

In a similar manner, we determine the probability that the sample contains exactly 4 defectives to be

$$\frac{\binom{4}{4} \cdot \binom{16}{1}}{\binom{20}{5}} = .001$$

By the special law of addition, the probability that the 5-element sample contains at least 3 defectives is

$$\frac{\binom{4}{3} \cdot \binom{16}{2}}{\binom{20}{5}} + \frac{\binom{4}{4} \cdot \binom{16}{1}}{\binom{20}{5}} = .031 + .001 = .032$$

EXAMPLE 7-39 What is the probability that a bridge hand (13 cards) will contain exactly 3 queens?

Solution Since we are selecting 13 cards from a deck of 52 cards, there are $\binom{52}{13}$ possible ways of doing this. Thus, our sample space, S, has $\binom{52}{13}$ possible outcomes. We want our hand (13 cards) to contain 3 queens and 10 nonqueens. Since the 3 queens must be chosen from the 4 queens in the deck, there are $\binom{4}{3}$ ways of doing this. The 10 nonqueens must be chosen from the 48 nonqueens of the deck. There are $\binom{48}{10}$ ways of doing this. Thus, by the multiplication rule of counting, there are $\binom{4}{3} \cdot \binom{48}{10}$ ways of selecting exactly 3 queens. Hence, the event set, E, contains $\binom{4}{3} \cdot \binom{48}{10}$ elements. Therefore, the probability of getting a bridge hand containing exactly 3 queens is

$$P(E) = \frac{n(E)}{n(S)} = \frac{\binom{4}{3} \cdot \binom{48}{10}}{\binom{52}{13}}$$

$$= \frac{858}{20{,}825}$$

$$\approx 0.041$$

EXERCISES

1. Evaluate each of the following:
   (A) 4!   (B) 5!   (C) 0!   (D) 1!   (E) 6!

2. Evaluate each of the following:
   (A) $P(5, 3)$ (B) $P(3, 3)$ (C) $P(3, 0)$
   (D) $P(4, 1)$ (E) $P(8, 7)$ (F) $P(5, 3)$

3. Evaluate each of the following:
   (A) $\binom{7}{4}$ (B) $\binom{4}{4}$ (C) $\binom{4}{2}$
   (D) $\binom{5}{3}$ (E) $\binom{5}{2}$ (F) $\binom{6}{1}$

428 CHAPTER SEVEN

4. After evaluating $\binom{6}{4}$ and $\binom{6}{2}$, explain why, in general,
$$\binom{n}{r} = \binom{n}{n-r}$$

5. In how many ways can 5 letters be selected from $\{a, b, c, d, e\}$ if the order of selection makes a difference?

6. In how many ways can 5 people be seated in a row of 5 seats?

7. In how many ways can 3 letters be chosen from $\{a, b, c, d, e\}$ if the order of selection makes a difference?

8. In how many ways can 3 people be seated in a row of 5 seats?

9. In how many ways can 3 letters be selected from $\{a, b, c, d, e\}$ if the order of selection makes no difference?

10. In how many ways can a committee of 3 be chosen from a group of 5 people?

11. Given that 2 points determine a straight line, how many straight lines are determined by a set of 5 points, no 3 of which are collinear?

12. A multiple-choice quiz has 10 questions. Each question has 5 possible choices. How many different sets of 10 answers are possible?

13. A restaurant offers a choice of 7 different types of sandwiches, 3 different types of desserts, and 5 different types of beverages. If a lunch consists of a sandwich, a dessert, and a beverage, how many different lunches are available?

14. A placement officer analyzes a job applicant's personality by rating that applicant in each of 4 different characteristics. If each characteristic has 6 different ratings, how many different personality types are possible?

15. What is the probability that a bridge hand (13 cards) will contain
    (A) Exactly 2 queens?
    (B) Exactly 4 queens?
    (C) At least 2 queens?

16. What is the probability that a bridge hand (13 cards) will contain
    (A) No queens?
    (B) Exactly 1 queen?
    (C) At most 1 queen?

17. A shipment of 24 televisions contains 5 defectives. A sample of 6 televisions is to be selected from this shipment.
    (A) How many possible samples are there?
    (B) Suppose we want the sample to contain all nondefective televisions. In how many ways can this happen?
    (C) What is the probability that the sample contains 6 nondefective televisions?
    (D) Find the probability that the sample contains exactly 1 defective.
    (E) Find the probability that the sample contains exactly 2 defectives.
    (F) Find the probability that the sample contains at most 2 defectives.

(G) Find the probability that the sample contains more than 2 defectives.

18. An auditor is selecting a sample of 8 accounts from a set of 30 accounts receivable for the purpose of balance confirmation. Of the 30 accounts receivable, 4 are in error.
    (A) How many possible samples are there?
    (B) What is the probability that the sample contains all correct accounts?
    (C) What is the probability that the sample contains exactly 1 account in error?
    (D) What is the probability that the sample contains exactly 2 accounts in error?
    (E) What is the probability that the sample contains at most 2 accounts in error?
    (F) What is the probability that the sample contains at least 3 accounts in error?

19. *(Birthday Problem)* A group of 6 people is to be selected at random. Find the probability that at least 2 of them have the same birthday. Assume that there are 365 days to a year. Parts (A) through (C) offer a step-by-step solution to this problem.
    (A) Since each person has 365 possible days for a birthday, we are selecting 1 element from each of 6 sets where each set has 365 elements. Thus, by the multiplication rule of counting, there are _____ possible outcomes in the sample space, S.
    (B) Since it is difficult to calculate $n(E)$, the number of ways in which at least 2 selected people have the same birthday, we calculate $n(E')$, the number of ways in which all 6 birthdays are different. If all 6 birthdays are to be different, the first selection has 365 possibilities, the second has 364, the third has _____, the fourth has _____, the fifth has _____, and the sixth has _____. Thus, by the multiplication rule of counting, there are _____ ways in which all 6 selected birthdays are different.
    (C) Hence, $P(E') = n(E')/n(S) =$ _____, and $P(E) =$ _____. So, the probability that at least 2 selected people have the same birthday is _____.

20. Repeat Exercise 19 for 5 selected people.

7–8

PROBABILITY DISTRIBUTIONS AND RANDOM VARIABLES

In business and industry, decisions must often be made on the basis of expected profits, revenue, costs, losses, etc. Crucial to these decisions is the concept of expected value of a random variable. In this section, we will discuss this concept.

Consider a simple type of gambling machine—a box full of numbers (see Table 7–5). Observing Table 7–5, note that the box contains the numbers 1, 3, 5, 8, and 10. Additionally, 20% of all the numbers in the box are 1s, 50% are 3s, 15% are 5s, 10% are 8s, and 5% are 10s.

TABLE 7–5
Box Full of Numbers

```
1 (20%)
3 (50%)
5 (15%)
8 (10%)
10 (5%)
```

A chance experiment consists of selecting one number from the box. The number selected may be taken to a cashier who will pay the player a dollar amount equal to the number. Observing the percentages in Table 7–5, it is obvious that the player has a 20% probability of winning $1, a 50% probability of winning $3, a 15% probability of winning $5, a 10% probability of winning $8, and a 5% probability of winning $10.

Since the outcomes of this chance experiment are numerical values, they may be listed in tabular form as shown in Table 7–6.

TABLE 7–6

x	P(x)
1	.20
3	.50
5	.15
8	.10
10	.05
	1.00

Observing Table 7–6, note that the numerical values 1, 3, 5, 8, and 10 are denoted by a letter—in this case, x. Since the values of x are chance outcomes, x is called a **random variable**. Observe that the probability of occurrence of each x-value is listed in the "$P(x)$" column (i.e., the "probability of x" column). Such a display of values of a random variable and their corresponding probabilities is called a **probability distribution**. As shown in Table 7–6, the sum of the probabilities of a probability distribution is 1. Figure 7–36 graphically illustrates this probability distribution. Since the random variable x takes on a finite number of values, it is called a **discrete random variable**. And its probability distribution is termed a **discrete probability distribution**. If a random variable can take on values within an interval of numbers, it is termed a **continuous random variable** and its probability distribution is called a **continuous**

probability distribution. In this section, we will limit our discussion to discrete random variables.

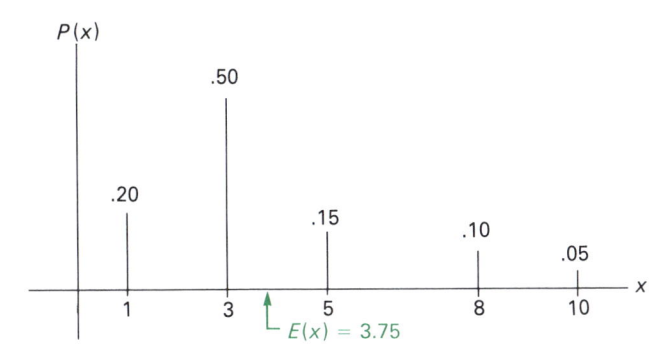

FIGURE 7–36

Expected Value of a Discrete Random Variable

If a discrete random variable x takes on the values $x_1, x_2, \ldots, x_n$, having respective probabilities of occurrence $p_1, p_2, \ldots, p_n$, then the **expected value** or **mean** of x, which is denoted by $E(x)$, is determined by

$$E(x) = x_1 p_1 + x_2 p_2 + \ldots + x_n p_n$$

Table 7–7 illustrates the calculation of the expected value of the random variable x of this section.

TABLE 7–7

x	$P(x)$	$xP(x)$
1	.20	.20
3	.50	1.50
5	.15	.75
8	.10	.80
10	.05	.50
	1.00	3.75

$E(x) =$ ⤴

Thus, the expected value is 3.75 and is interpreted as follows. If one were to repeat the chance experiment (i.e., selecting one number from the box) over and over again, then one would win, on the average, $3.75 per repetition. In other words, the average winnings would be approximately $3.75, in the long run. Thus, if the gambling machine operator charges $4.00 per play, then the operator's profit is, on the average, $4.00 − $3.75 = $0.25 per play, in the long run. One can see how gambling houses make money on games of chance.

EXAMPLE 7–40 The manager of Howie's Hamburger Stand has kept a record of daily demand for hamburgers during the past 400 days. The results are shown in Table 7–8. Such a display of data is called a **frequency (f) distribution**.

TABLE 7–8

x Demand	f (# days)
100	20
110	80
120	200
130	80
140	20
	400

(A) Convert this frequency distribution into a probability distribution by finding the percent frequency of occurrence of each x-value.
(B) Find the probability that daily demand is 130 hamburgers.
(C) Find the probability that daily demand is at most 130 hamburgers.
(D) How many hamburgers should be kept on hand in order to satisfy daily demand 95% of the time?

Solutions

(A) Dividing each frequency value by the total frequency, 400, we have the probability distribution of Table 7–9.

TABLE 7–9

x	P(x)	
100	.05	← 20/400
110	.20	← 80/400
120	.50	← 200/400
130	.20	← 80/400
140	.05	← 20/400
	1.00	

(B) Observing Table 7–9, we note that the probability that daily demand is 130 hamburgers is .20. This is expressed symbolically as

$$P(x = 130) = P(130) = .20$$

(C) The probability that daily demand is at most 130 is

$$P(x \leq 130) = P(130) + P(120) + P(110) + P(100)$$
$$= .20 + .50 + .20 + .05$$
$$= .95$$

(D) Since there is a 95% probability (see part (C)) that daily demand will be at most 130 hamburgers, then 130 hamburgers should be kept on hand in order to satisfy demand 95% of the time.

EXAMPLE 7-41 Referring to the probability distribution of demand for Howie's Hamburgers in Example 7-40 (see Table 7-9), calculate the expected value of demand, $E(x)$.

Solution

x	P(x)	xP(x)
100	.05	5.00
110	.20	22.00
120	.50	60.00
130	.20	26.00
140	.05	7.00
	1.00	120.00

$E(x) =$ ⤴

Thus, in the long run, the average daily demand for hamburgers is approximately 120 hamburgers.

EXERCISES

1. A person playing a gambling machine at a given casino has a chance of "winning" either $1, $2, $5, or $10. If the random variable x denotes these dollar value outcomes, the probability distribution of x is as follows:

x	P(x)
1	.40
2	.50
5	.06
10	.04
	1.00

(A) Graph this probability distribution.
(B) Find $P(x < 5)$ and interpret the result.

(C) Find $P(x \leq 5)$ and interpret the result.
(D) Calculate $E(x)$ and interpret the result.
(E) If a gambler must pay $2.50 each time he or she plays the machine, find the house's average profit per play.

2. For the following probability distribution, the random variable x represents the possible closing prices of a given stock as determined by an investment analyst:

x	P(x)
20	.30
25	.40
30	.20
35	.10
	1.00

(A) Calculate $E(x)$.
(B) Interpret $E(x)$.

3. A company is considering building a retail outlet at a given location. If successful at this location, the outlet should net $6 million. If it fails, there will be a loss of $2 million. Assume that the probability of success at this location is .70.
   (A) Construct a probability distribution indicating the dollar value outcomes and their probabilities.
   (B) Find the expected value of the probability distribution of part (A).

4. Suppose the company in Exercise 3 is considering another location for which the probability of success is .80. If successful at this location, the company will net $5 million. If it fails, there will be a loss of $100,000.
   (A) Construct a probability distribution indicating the dollar value outcomes and their probabilities.
   (B) Find the expected value of the probability distribution of part (A).
   (C) If the company will choose a location based on its expected value, which location should be chosen?

5. The manager of Wally's Weiner Hut has kept a record of daily demand for weiners during the past 500 days. The results are as follows:

x Demand	f (# days)
200	50
250	100
300	200
350	100
400	50
	500

(A) Convert this frequency distribution into a probability distribution.
(B) Find the probability that daily demand is 250 weiners.

(C) Find the probability that daily demand is at most 250 weiners.
(D) How many weiners should be kept on hand in order to satisfy daily demand 90% of the time?
(E) Calculate $E(x)$, the expected value of daily demand for hamburgers.

7-9

BINOMIAL EXPERIMENTS

A **binomial experiment** is a chance experiment possessing the following characteristics:

1. The experiment consists of a sequence of n independent trials.
2. Each trial has two possible outcomes—*success* or *failure.*
3. The probability of getting a "success" on an individual trial is denoted by p and does not change from one trial to the next.

As an example of a binomial experiment, consider the chance experiment of selecting 3 marbles from the following box:

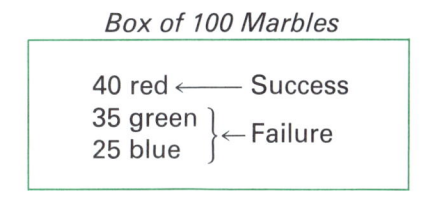

Box of 100 Marbles

40 red ⟵ Success
35 green ⎫
25 blue ⎭ ⟵ Failure

If we define the selection of a red marble as a "success," then the selection of either a green or a blue marble is termed a "failure." For this experiment, $n = 3$ independent trials, and the probability of getting a success (red) on an individual trial is $p = 40/100 = .4$.

Typically, with a binomial experiment, we must determine the probability of getting x successes out of the n trials. Such a probability is called a **binomial probability.** Referring to our chance experiment with marbles, we will now determine the probability of selecting 2 successes out of the 3 trials. The tree diagram of Figure 7-37 illustrates all possible outcomes and their probabilities for this chance experiment. Observe that there are 3 ways of getting 2 successes out of the 3 trials. This should not surprise us since there are

$$\binom{3}{2} = \frac{3!}{2!(3-2)!} = 3$$

possible ways in which 2 successes (S) and 1 failure (F) can occur. These 3 outcomes and their probabilities are as follows:

$$SSF \qquad SFS \qquad FSS$$
$$(.4)^2(.6) \qquad (.4)^2(.6) \qquad (.4)^2(.6)$$

Possible Outcomes	Probabilities
SSS	$(.4)(.4)(.4) = (.4)^3$
SSF	$(.4)(.4)(.6) = (.4)^2(.6)$
SFS	$(.4)(.6)(.4) = (.4)^2(.6)$
SFF	$(.4)(.6)(.6) = (.4)(.6)^2$
FSS	$(.6)(.4)(.4) = (.6)(.4)^2$
FSF	$(.6)(.4)(.6) = (.4)(.6)^2$
FFS	$(.6)(.6)(.4) = (.6)^2(.4)$
FFF	$(.6)(.6)(.6) = (.6)^3$

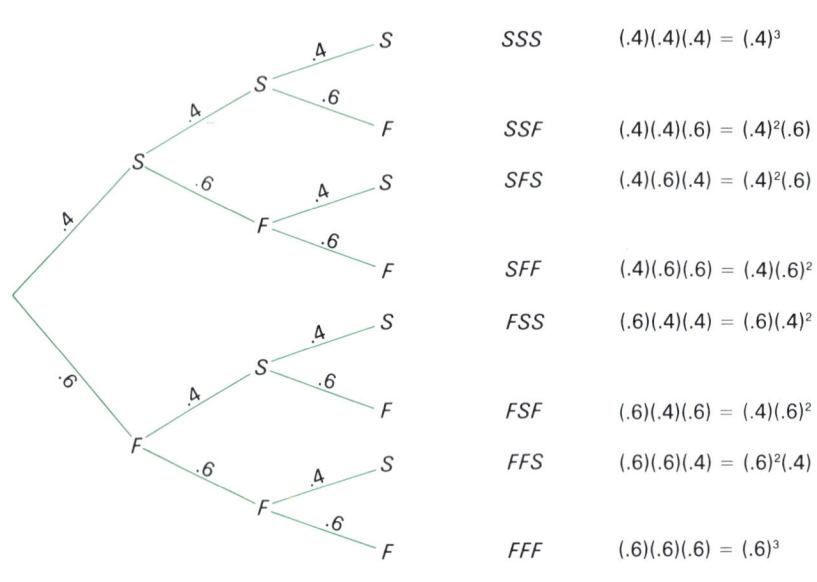

FIGURE 7-37

Thus, the probability of selecting 2 successes (reds) out of 3 trials is

$$(.4)^2(.6) + (.4)^2(.6) + (.4)^2(.6)$$

or

$$\binom{3}{2}(.4)^2(.6) = .288$$

We now generalize this concept.

For a binomial experiment, the probability of getting x successes out of n independent trials ($x \leq n$) is

$$\binom{n}{x} p^x (1-p)^{n-x}$$

where p is the *probability of success* per trial.

PROBABILITY

Since p is the probability of success on an individual trial, $1 - p$ is the *probability of failure* on an individual trial. The expression $1 - p$ is usually denoted by q. Hence,

$$q = 1 - p$$

and the binomial probability formula

$$\binom{n}{x} p^x (1-p)^{n-x}$$

is often written as

$$\binom{n}{x} p^x q^{n-x}$$

EXAMPLE 7–42 Again referring to the chance experiment of selecting 3 marbles from the box of this section, find the probability of getting
(A) 0 successes out of 3 trials.
(B) 1 success out of 3 trials.
(C) 3 successes out of 3 trials.
Solutions
(A) Since $n = 3$, $x = 0$, $p = .4$, and $q = 1 - p = 1 - .4 = .6$, then

$$\binom{n}{x} p^x q^{n-x} = \binom{3}{0}(.4)^0(.6)^3$$

$$= \frac{3!}{0!3!}(.4)^0(.6)^3$$

$$= .216$$

(B) Since $n = 3$, $x = 1$, $p = .4$, and $q = .6$, then

$$\binom{n}{x} p^x q^{n-x} = \binom{3}{1}(.4)^1(.6)^2$$

$$= \frac{3!}{1!2!}(.4)^1(.6)^2$$

$$= .432$$

438 CHAPTER SEVEN

EXAMPLE 7–42 (continued)

(C) Since $n = 3$, $x = 3$, $p = .4$, and $q = .6$, then

$$\binom{n}{x} p^x q^{n-x} = \binom{3}{3}(.4)^3(.6)^0$$

$$= \frac{3!}{3!0!}(.4)^3(.6)^0$$

$$= .064$$

EXAMPLE 7–43 Each question of a 10-question multiple-choice exam has 5 choices, only 1 of which is correct. If a student guesses on each question, find the probability that he or she gets
(A) Exactly 9 questions correct.
(B) All 10 questions correct.
(C) At least 9 questions correct.

Solutions
(A) Since $n = 10$, $x = 9$, $p = 1/5 = .2$, and $q = 1 - p = 1 - .2 = .8$, then

$$\binom{n}{x} p^x q^{n-x} = \binom{10}{9}(.2)^9(.8)^1$$

$$= .000004096$$

(B) Since $n = 10$, $x = 10$, $p = .2$, and $q = .8$, then

$$\binom{n}{x} p^x q^{n-x} = \binom{10}{10}(.2)^{10}(.8)^0$$

$$= .0000001024$$

(C) Since x is either 9 or 10, then

$$\binom{10}{9}(.2)^9(.8)^1 + \binom{10}{10}(.2)^{10}(.8)^0 = .000004096 + .0000001024$$

$$= .0000041984$$

Binomial Distribution

In the preceding paragraphs, a binomial experiment was defined as a chance experiment consisting of n independent trials. We also derived the formula

$$\binom{n}{x} p^x (1-p)^{n-x}$$

for the probability of getting x successes out of the n trials, where p is the probability of success per trial. Since x denotes the numerical outcomes (i.e., number of successes) of a chance experiment, the x is really a random variable. Thus, for any binomial experiment, a probability distribution may be generated by listing, in tabular form, all possible values of x and their corresponding probabilities, $P(x)$, where

$$P(x) = \binom{n}{x} p^x (1-p)^{n-x}$$

Specifically, consider the chance experiment of taking a 4-question, multiple-choice quiz wherein each question has 5 choices, only 1 of which is correct. If a student guesses at each question, the probability of getting an individual question correct is 1/5, or .2. Thus, this is a binomial experiment with $n = 4$ trials and $p = .2$. If $x =$ number of correctly guessed questions, the probability distribution of x is given in Table 7–10. A graph of this binomial distribution appears in Figure 7–38 on page 440.

TABLE 7–10

$$P(x) = \binom{n}{x} p^x (1-p)^{n-x}$$

x	$P(x)$	
0	.4096	$P(0) = \binom{4}{0}(.2)^0(.8)^4 = .4096$
1	.4096	$P(1) = \binom{4}{1}(.2)^1(.8)^3 = .4096$
2	.1536	$P(2) = \binom{4}{2}(.2)^2(.8)^2 = .1536$
3	.0256	$P(3) = \binom{4}{3}(.2)^3(.8)^1 = .0256$
4	.0016	$P(4) = \binom{4}{4}(.2)^4(.8)^0 = .0016$
	1.0000	

440 CHAPTER SEVEN

Binomial distribution
$n = 4, p = .2$

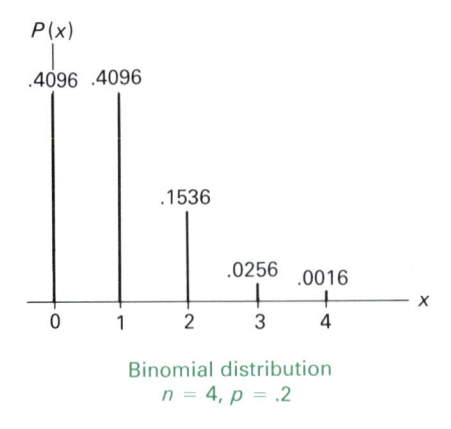

FIGURE 7–38

EXAMPLE 7–44 *(Drug Testing)* Medical researchers claim that a new drug is 80% effective in curing flu-related illnesses. A chance experiment consists of testing the drug on 6 patients with flu-related illnesses. Let x = number of patients cured.

(A) Determine the probability distribution of x.
(B) Graph the distribution of part (A).

Solutions

(A) Note that this is a binomial experiment with $n = 6$ and $p = .8$. Its distribution is determined as follows:

x	$P(x)$	$P(x) = \binom{n}{x}p^x(1-p)^{n-x}$
0	.0001	$P(0) = \binom{6}{0}(.8)^0(.2)^6 = .0001$
1	.0015	$P(1) = \binom{6}{1}(.8)^1(.2)^5 = .0015$
2	.0154	$P(2) = \binom{6}{2}(.8)^2(.2)^4 = .0154$
3	.0819	$P(3) = \binom{6}{3}(.8)^3(.2)^3 = .0819$
4	.2458	$P(4) = \binom{6}{4}(.8)^4(.2)^2 = .2458$
5	.3932	$P(5) = \binom{6}{5}(.8)^5(.2)^1 = .3932$
6	.2621	$P(6) = \binom{6}{6}(.8)^6(.2)^0 = .2621$
	1.0000	

(B) Figure 7-39 illustrates the graph of the distribution of part (A).

P(x)

.0001 .0015 .0154 .0819 .2458 .3932 .2621

0 1 2 3 4 5 6 — *x*

Binomial distribution
$n = 6$, $p = .8$

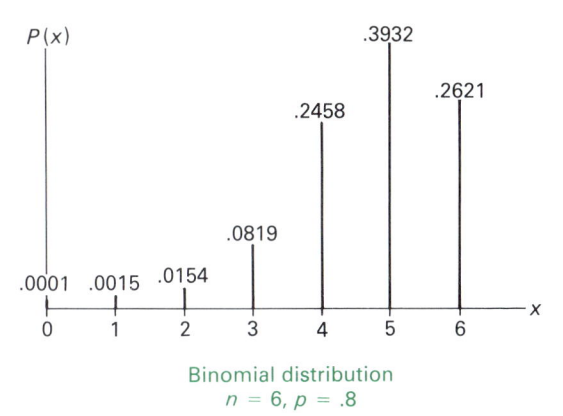

FIGURE 7-39

EXAMPLE 7-45 Referring to Example 7-44, suppose only 2 patients out of the 6 were cured by the drug. What might we conclude about the claim that $p = .8$?

Solution Looking at the graph of the probability distribution of Figure 7-39, note that the probability of getting 2 successes or fewer is

$$P(x \leq 2) = P(2) + P(1) + P(0)$$
$$= .0154 + .0015 + .0001$$
$$= .0170$$

Thus, there is very little likelihood (only 1.7%, to be exact) of getting 2 successes or fewer if the claim $p = .8$ is true. Since the actual experiment produced only 2 successes, we might question whether the claim $p = .8$ is true.

Expected Value of a Binomially Distributed Random Variable

In Section 7-8, we stated that the expected value, $E(x)$, of a discrete random variable x, which takes on the values $x_1, x_2 \ldots, x_n$, with respective probabilities $p_1, p_2, \ldots, p_n$, is given by

$$E(x) = x_1 p_1 + x_2 p_2 + \ldots + x_n p_n$$

Using this formula, we calculate the expected value of the binomially distributed random variable of Figure 7-38 (see Table 7-11).

CHAPTER SEVEN

TABLE 7–11

x	P(x)	xP(x)
0	.4096	.0000
1	.4096	.4096
2	.1536	.3072
3	.0256	.0768
4	.0016	.0064
		.8000

$$E(x) = x_1 p_1 + x_2 p_2 + \ldots + x_n p_n$$
$$= .8$$

The expected value, $E(x)$, of a binomially distributed random variable x can be more easily calculated by the formula

$$E(x) = np$$

Note that for the random variable of Table 7–11, $n = 4$ and $p = .2$, therefore

$$E(x) = np = 4(.2) = .8$$

Obviously, the formula $E(x) = np$ is considerably quicker and easier to use. However, we must remember that this formula is valid only for a binomially distributed random variable.

EXAMPLE 7–46 Of the 6 patients treated by the drug of Example 7–44, how many can we expect to be cured?
Solution We seek the expected value of the binomially distributed random variable x. Since $n = 6$ and $p = .8$, we calculate

$$E(x) = np = 6(.8) = 4.8$$

Thus, in the long run, we expect 4.8 cures out of every 6 patients.

EXERCISES

1. A box contains 30 green, 50 red, and 20 blue marbles. If 4 marbles are selected, with replacement, from the box, find the probability of getting
   (A) Exactly 2 green marbles out of the 4 trials.
   (B) Exactly 1 green marble out of the 4 trials.
   (C) No green marbles out of the 4 trials.
   (D) At most 2 green marbles out of the 4 trials.
   (E) Fewer than 2 green marbles out of the 4 trials.

2. Each question of a 12-question multiple-choice exam has 4 choices, only 1 of which is correct. If a student guesses on each question, find the probability that he or she gets
   (A) Exactly 10 questions correct.
   (B) Exactly 11 questions correct.
   (C) All 12 questions correct.
   (D) At least 10 questions correct.
   (E) More than 10 questions correct.

3. A true-false examination consists of 10 questions. If a student guesses at each question, find the probability that he or she gets
   (A) Exactly 2 questions correct.
   (B) No questions correct.
   (C) At most 2 questions correct.
   (D) Fewer than 2 questions correct.
   (E) Exactly 9 questions correct.
   (F) All 10 questions correct.
   (G) At least 9 questions correct.
   (H) More than 9 questions correct.

4. A production lot of small electrical components contains 10% defectives. A quality control procedure consists of selecting 5 components from the lot. Despite the fact that there is no replacement between selections, we will assume independence since there are so many components in the production lot that the proportion of defectives (10%) remains nearly constant as the selections begin. Find the probability of getting
   (A) Exactly 3 defectives out of the 5 selections.
   (B) Exactly 2 nondefectives out of the 5 selections.
   (C) Exactly 1 defective out of the 5 selections.
   (D) Exactly 4 nondefectives out of the 5 selections.
   (E) Exactly 3 nondefectives out of the 5 selections.

5. A new drug for treating a given disease has been proven to be effective 70% of the time. If the drug is administered to 9 patients, find the probability that it cures
   (A) Exactly 7 of the 9 patients.
   (B) Exactly 8 of the 9 patients.
   (C) All 9 patients.
   (D) At least 7 of the 9 patients.

6. On a given journey, a driver who always exceeds the speed limit must pass 3 radar traps. If there is a 40% chance of getting caught by any given radar trap, find the probability that the driver
   (A) Passes all 3 radar traps without getting caught.
   (B) Gets caught at least once.

7. Each time a salesperson makes a presentation, there is a 20% chance of a sale. If this salesperson makes 4 presentations during a day, find the probability that
   (A) No sales are made.
   (B) Exactly 2 sales are made.
   (C) At least 2 sales are made.
   (D) More than 2 sales are made.

444 CHAPTER SEVEN

8. A random variable x is binomially distributed with $n = 3$ and $p = .4$.
   (A) Calculate $E(x)$.
   (B) Determine the probability distribution of x.
   (C) Graph the distribution of part (B).

9. The probability that a tomato seed will germinate is .6. A gardener plants 5 tomato seeds. Let x represent the number of seeds that germinate.
   (A) Determine the probability distribution of x.
   (B) Graph the distribution of part (A).
   (C) On the average, how many seeds should the gardener expect to germinate?

10. Each time a particular salesperson makes a presentation, there is a 20% chance that the customer will buy the product. The salesperson usually makes 5 presentations per day. Let x represent the number of customers that buy the product during an ordinary day.
    (A) Determine the probability distribution of x.
    (B) Graph the distribution of part (A).
    (C) On the average, this salesperson should expect to make how many sales per day?

11. An advertising agency claims that 60% of the viewers of a given television commercial will buy the product. A chance experiment consists of surveying 10 viewers. Let x represent the number of viewers who bought the product.
    (A) Determine the probability distribution of x.
    (B) Graph the distribution of part (A).
    (C) Calculate $E(x)$.
    (D) Suppose only 2 out of the 10 viewers bought the product. What might we conclude about the agency's claim that $p = .6$?

12. Blue Hills Winery manufactures a domestic substitute for an expensive imported champagne. Blue Hills conducts a taste test. In the test, 12 judges (assumed to be unbiased) are each given 2 unidentified glasses of wine. One glass contains the expensive imported wine, and the other contains the domestic substitute. Let x represent the number of judges preferring the domestic substitute. If there is a 50 percent chance that a judge prefers the domestic substitute
    (A) Determine the probability distribution of x.
    (B) Graph the distribution of part (A).
    (C) Calculate $E(x)$.
    (D) Find $P(x = 10)$, $P(x = 11)$, and $P(x = 12)$.
    (E) Find the probability that 10 or more judges prefer the domestic substitute. If 10 or more judges actually do prefer the domestic substitute, what might we conclude?
    (F) Find the probability that 2 or fewer judges prefer the domestic substitute. If 2 or fewer judges actually do prefer the domestic substitute, what might we conclude?

13. The Computer Card Company manufactures data-processing cards for computers. A quality control engineer claims that each production lot of cards contains 10% defectives. A chance experiment con-

PROBABILITY

sists of selecting 9 cards from a production lot. Let x be the number of defective cards selected.
(A) Determine the probability distribution of x.
(B) Graph the distribution of part (A).
(C) Calculate $E(x)$.

14. *(Quality Control)* The Computer Card Company plans to use the chance experiment of Exercise 13 in the following manner. If more than 2 defective cards are selected, then the entire production lot will be rejected. If 2 or fewer defective cards are selected, then the entire production lot will be accepted.
(A) Find the probability that a production lot is accepted.
(B) Find the probability that a production lot is rejected.

7–10 MARKOV CHAINS

In this section, we will study chance experiments consisting of a sequence of trials. Such an experiment is called a **Markov chain**. We now define a Markov chain.

> A **Markov chain** is a sequence of trials such that:
>
> 1. The outcome of each trial is one of a finite number of possible states 1, 2, . . . , r.
> 2. The probability of a particular outcome at a given trial depends only upon the outcome of the preceding trial.

As an example, let us consider a machine producing trivets. During a given day, the machine can be either *in adjustment (state 1)* or *out of adjustment (state 2)*. Thus, each trial is the birth of a new day with the machine being in one of two possible states during that day:

State 1: Machine is in adjustment
State 2: Machine is out of adjustment

TABLE 7–12

From \ To	In Adjustment	Out of Adjustment
In Adjustment	.8	.2
Out of Adjustment	.6	.4

Table 7–12 lists the probabilities of the machine changing from one state to another state during two successive days. Such probabilities are called **probabilities of change,** or **transition probabilities.** A table composed of transition probabilities such as Table 7–12 is usually expressed as a matrix. Thus, Table 7–12 is expressed as the matrix

$$P = \begin{bmatrix} .8 & .2 \\ .6 & .4 \end{bmatrix}$$

Such a matrix is called a **transition matrix.** Note that the sum of the probabilities in each row of the transition matrix P equals 1.

In general, if each trial of a Markov chain has r possible states, then the transition matrix is the matrix

$$P = \begin{bmatrix} p_{11} & p_{12} & \cdots & p_{1r} \\ p_{21} & p_{22} & \cdots & p_{2r} \\ \vdots & \vdots & & \vdots \\ p_{r1} & p_{r2} & & p_{rr} \end{bmatrix}$$

where p_{ij} is the probability of changing from state i to state j. Also, the sum of the probabilities in each row equals 1. Thus, p_{11} is the probability of remaining in state 1 from one trial to the next trial, p_{12} is the probability of changing from state 1 to state 2 from one trial to the next trial, etc.

We now return to our machine producing trivets with transition matrix

$$P = \begin{bmatrix} .8 & .2 \\ .6 & .4 \end{bmatrix}$$

Additionally, we are given the following information:

1. The probability of the machine initially being in state 1 is .7.
2. The probability of the machine initially being in state 2 is .3.

We now seek the probabilities of the machine being in either state after 1 day, 2 days, 3 days, . . . , n days. Figure 7–40 gives a tree diagram illustration of this Markov chain. Observing Figure 7–40, note the following:

1. The probability of the machine being in state 1 during day 1 is

$$(.7)(.8) + (.3)(.6)$$

2. The probability of the machine being in state 2 during day 1 is

$$(.7)(.2) + (.3)(.4)$$

447 PROBABILITY

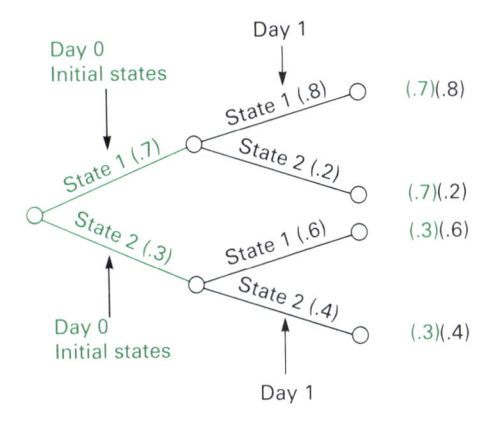

FIGURE 7-40

These probabilities can be expressed as the matrix product

$$[.7 \quad .3]\begin{bmatrix} .8 & .2 \\ .6 & .4 \end{bmatrix}$$

In general, if

$p_1^{(n)}$ = probability of being in state 1 during day n
$p_2^{(n)}$ = probability of being in state 2 during day n

then the matrix

$$p^{(n)} = [p_1^{(n)} \quad p_2^{(n)}]$$

gives the probabilities of the machine being in either state during day n. Thus, the probabilities of the machine being in either state during day 0 are given by

$$p^{(0)} = [p_1^{(0)} \quad p_2^{(0)}]$$
$$= [.7 \quad .3]$$

And the probabilities of the machine being in either state during day 1 are given by

$$p^{(1)} = [.7 \quad .3]\begin{bmatrix} .8 & .2 \\ .6 & .4 \end{bmatrix}$$
$$= [(.7)(.8) + (.3)(.6) \quad (.7)(.2) + (.3)(.4)]$$
$$= [.74 \quad .26]$$

Therefore,

$$p^{(1)} = [p_1^{(1)} \quad p_2^{(1)}] = [.74 \quad .26]$$

Since

$$p^{(0)} = [.7 \quad .3] \quad \text{and} \quad P = \begin{bmatrix} .8 & .2 \\ .6 & .4 \end{bmatrix}$$

then

$$p^{(1)} = p^{(0)}P$$

Observing Figure 7–41, we see that the probabilities of the machine being in either state during day 2 are given by

$$p^{(2)} = [p_1^{(2)} \quad p_2^{(2)}]$$
$$= [(.74)(.8) + (.26)(.6) \quad (.74)(.2) + (.26)(.4)]$$
$$= [.748 \quad .252]$$

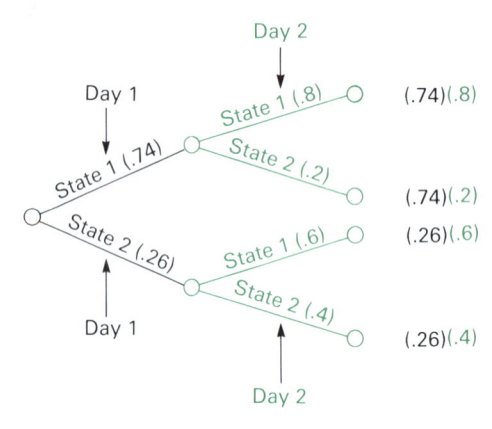

FIGURE 7–41

Thus, the probability of the machine being in state 1 during day 2 is .748, and the probability of the machine being in state 2 during day 2 is .252.

Note that $p^{(2)}$ can be written as

$$p^{(2)} = [.74 \quad .26] \begin{bmatrix} .8 & .2 \\ .6 & .4 \end{bmatrix}$$
$$= p^{(1)}P$$

In a similar manner, we can determine the probabilities of the machine being in either state during day 3 to be

$$p^{(3)} = p^{(2)}P$$

In general, the probabilities of the machine being in either state during day n are given by

PROBABILITY

$$p^{(n)} = p^{(n-1)}P$$

We will now develop an alternate formula for $p^{(n)}$. Recall that

$$p^{(1)} = p^{(0)}P$$

and

$$p^{(2)} = p^{(1)}P$$

If we substitute $p^{(0)}P$ in place of $p^{(1)}$ in the formula for $p^{(2)}$, we obtain

$$p^{(2)} = (p^{(0)}P)P$$
$$= p^{(0)}P^2$$

where $P^2 = P \cdot P$. Thus, $p^{(2)}$ may be written as

$$p^{(2)} = p^{(0)}P^2$$

Now, consider the formula

$$p^{(3)} = p^{(2)}P$$

Substituting $p^{(0)}P^2$ in place of $p^{(2)}$ yields

$$p^{(3)} = (p^{(0)}P^2)P$$
$$= p^{(0)}P^3$$

where $P^3 = P \cdot P \cdot P$. Thus, $p^{(3)}$ may be written as

$$p^{(3)} = p^{(0)}P^3$$

In general, $p^{(n)}$ may be written as

$$p^{(n)} = p^{(0)}P^n$$

where

$$P^n = \underbrace{P \cdot P \ldots P}_{nP's}$$

EXAMPLE 7–47 Companies A, B, and C have 60%, 30%, and 10%, respectively, of the cosmetic market in a given state. Experience has shown that customers change companies on a monthly basis. The transition matrix P shows the transition probabilities from one month to the next.

EXAMPLE 7–47 (continued)

$$\begin{array}{c}\text{From}\end{array}\begin{array}{c}\text{To}\\\begin{array}{ccc}A & B & C\end{array}\end{array}$$
$$\begin{array}{c}A\\B\\C\end{array}\begin{bmatrix}.5 & .4 & .1\\.3 & .3 & .4\\.2 & .5 & .3\end{bmatrix}$$

(A) Find the probability of a customer changing from Company A to Company B.
(B) Find the market share of each company at the end of the first month.
(C) Find the market share of each company at the end of the second month.

Solutions

(A) We seek $p_{12} = .4$.

(B) $p^{(1)} = p^{(0)}P$

$$= [.6 \quad .3 \quad .1]\begin{bmatrix}.5 & .4 & .1\\.3 & .3 & .4\\.2 & .5 & .3\end{bmatrix}$$

$$= [.41 \quad .38 \quad .21]$$

Thus, at the end of the first month, the market shares of Companies A, B, and C are 41%, 38%, and 21%, respectively.

(C) $p^{(2)} = p^{(1)}P$

$$= [.41 \quad .38 \quad .21]\begin{bmatrix}.5 & .4 & .1\\.3 & .3 & .4\\.2 & .5 & .3\end{bmatrix}$$

$$= [.361 \quad .383 \quad .256]$$

Thus, at the end of the second month, the market shares of Companies A, B, and C are 36.1%, 38.3%, and 25.6%, respectively.

EXERCISES

1. In a given state, a voter is either a Democrat, a Republican, or an Independent. The transition matrix P

$$\begin{array}{c}\text{From}\end{array}\begin{array}{c}\text{To}\\\begin{array}{ccc}\text{Democrats} & \text{Republicans} & \text{Independents}\end{array}\end{array}$$
$$\begin{array}{c}\text{Democrats}\\\text{Republicans}\\\text{Independents}\end{array}\begin{bmatrix}.6 & .2 & .2\\.1 & .8 & .1\\.3 & .3 & .4\end{bmatrix}$$

shows the transition probabilities of voters switching parties from one election to the next. If, initially, there are 60% Democrats, 30% Republicans, and 10% Independents, find the percentage of voters in each party
(A) After the first election.
(B) After the second election.
(C) After the third election.

2. An insurance company annually classifies its drivers into Good, Fair, and Poor risk groups. The transition matrix P is as follows:

$$\begin{array}{c|ccc} & \multicolumn{3}{c}{\text{To}} \\ \text{From} & \text{Good} & \text{Fair} & \text{Poor} \\ \text{Good} & .8 & .1 & .1 \\ \text{Fair} & .1 & .7 & .2 \\ \text{Poor} & 0 & .1 & .9 \end{array}$$

If, initially, the percentages of Good, Fair, and Poor drivers are 60%, 30%, and 10%, respectively, find the percentage in each category
(A) After the first year.
(B) After the second year.
(C) After the third year.

3. A communication system is designed to transmit messages by passing the digits 0 and 1 through several stages. The probability that a digit entering a given stage remains unchanged as it leaves that stage is .6.
(A) Write the transition matrix for transmission of 0s and 1s for this system.
(B) If, initially, 50% of the digits entering this system are 0s, find the percentages of 0s and 1s after two stages.

4. A sample of voters in a given state is polled monthly and asked the following question: "Do you think that the governor is doing a good job? Answer yes or no." Experience has shown that the probability of a person changing his or her opinion from one month to the next is 40%.
(A) Write the transition matrix for this problem.
(B) If, initially, 80% of the voters in the sample answered yes, find the percentages of yes's and no's after 3 months.

7–11

MARKOV CHAINS IN EQUILIBRIUM

In Section 7–10, we determined the probabilities of a Markov chain being in various states after a finite number of trials. If the Markov process stabilizes as n, the number of trials, gets arbitrarily large, then the Markov chain is said to be in **equilibrium** or **steady-state**. The probabilities of a Markov chain in equilibrium being in various states are called **steady-state probabilities**. In this section, we will determine steady-state probabilities.

Not all Markov processes stabilize as n, the number of trials, gets arbitrarily large. It can be shown that a Markov chain will reach a steady-state or equilibrium condition if for some positive integer n, the matrix P^n (where P is the transition matrix) consists of all positive probabilities. Such a transition matrix P is said to be **regular**.

Let us return to the machine producing trivets of Section 7–10. Recall that on any given day, the machine could be in either of the following states:

State 1: Machine is in adjustment
State 2: Machine is out of adjustment

The transition matrix P is

From \ To	In Adjustment	Out of Adjustment
In Adjustment	.8	.2
Out of Adjustment	.6	.4

Recall from Section 7–10 that the probability of the machine being in states 1 and 2, respectively, during day 0 was given by the matrix

$$p^{(0)} = [.7 \quad .3]$$

The probability of the machine being in states 1 and 2, respectively, during day 1 was determined to be

$$p^{(1)} = [.74 \quad .26]$$

The probability of the machine being in states 1 and 2, respectively, during day 2 was determined to be

$$p^{(2)} = [.748 \quad .252]$$

In general, the probability of the machine being in states 1 and 2, respectively, during day n is given by

$$p^{(n)} = [p_1^{(n)} \quad p_2^{(n)}]$$

Intuitively, the steady-state probabilities are those values that $p_1^{(n)}$ and $p_2^{(n)}$ are approaching as n, the number of trials, gets arbitrarily large. The steady-state probabilities are denoted by

$$p = [p_1 \quad p_2]$$

where p_1 and p_2 are the steady-state probabilities of the machine being in states 1 and 2, respectively.

If the transition matrix P is regular, then the Markov chain will eventually reach equilibrium and the steady-state probabilities can be determined. To determine the steady-state probabilities, we begin with the matrix equation for $p^{(n)}$, the probabilities of a Markov chain being in its respective states during the nth trial:

$$p^{(n)} = p^{(n-1)}P$$

As the process approaches equilibrium or steady-state, $p^{(n-1)}$ and $p^{(n)}$ will approach the steady-state probabilities p. Thus, if we replace $p^{(n)}$ and $p^{(n-1)}$ in the preceding matrix equation with p, we obtain the matrix equation

$$p = pP$$

Solving this equation for p (and including the fact that the sum of the steady-state probabilities equals 1) yields the steady-state probability matrix.

We now return to our illustrative example with

$$P = \begin{bmatrix} .8 & .2 \\ .6 & .4 \end{bmatrix}$$

and solve for the steady-state probabilities. Here,

$$p = pP$$

$$[p_1 \quad p_2] = [p_1 \quad p_2]\begin{bmatrix} .8 & .2 \\ .6 & .4 \end{bmatrix}$$

Rewriting this in equation form, we have

$$p_1 = .8p_1 + .6p_2$$
$$p_2 = .2p_1 + .4p_2$$

This results in the system

$$-.2p_1 + .6p_2 = 0$$
$$.2p_1 - .6p_2 = 0$$

Note that these two equations are really the same. Thus, we choose one and include the equation $p_1 + p_2 = 1$, which states that the sum of the steady-state probabilities equals 1. Our system then becomes

$$-.2p_1 + .6p_2 = 0$$
$$p_1 + p_2 = 1$$

Solving for p_1 and p_2 yields $p_1 = .75$ and $p_2 = .25$. Thus, the steady-state probability matrix is

$$p = [p_1 \ p_2]$$
$$= [.75 \ .25]$$

Therefore, in the long run, when the Markov process reaches equilibrium, it will be in state 1 (machine is in adjustment) 75% of the time and in state 2 (machine is out of adjustment) 25% of the time.

Successive powers of the regular transition matrix P are as follows:

$$P = \begin{bmatrix} .8 & .2 \\ .6 & .4 \end{bmatrix} \qquad P^2 = \begin{bmatrix} .76 & .24 \\ .72 & .28 \end{bmatrix}$$

$$P^3 = \begin{bmatrix} .752 & .248 \\ .744 & .256 \end{bmatrix} \qquad P^4 = \begin{bmatrix} .7504 & .2496 \\ .7488 & .2512 \end{bmatrix}$$

$$P^5 = \begin{bmatrix} .75008 & .24992 \\ .74976 & .25024 \end{bmatrix}$$

Note that the probabilities of each row are approaching the steady-state probabilities $p = [.75 \ .25]$ as the power of P increases. This is why the Markov chain approaches equilibrium or steady-state.

EXAMPLE 7–48 In Example 7–47, we gave the transition matrix P

$$\begin{array}{c} \text{From} \end{array} \begin{array}{c} \text{To} \\ \begin{array}{cccc} & A & B & C \end{array} \\ \begin{array}{c} A \\ B \\ C \end{array} \begin{bmatrix} .5 & .4 & .1 \\ .3 & .3 & .4 \\ .2 & .5 & .3 \end{bmatrix} \end{array}$$

showing the probabilities of customers changing cosmetic companies from one month to the next. Find the steady-state probabilities of this Markov chain.
Solution
The steady-rate probability matrix is $p = [p_1 \ p_2 \ p_3]$, where

$$p = pP$$

or

$$[p_1 \ p_2 \ p_3] = [p_1 \ p_2 \ p_3] \begin{bmatrix} .5 & .4 & .1 \\ .3 & .3 & .4 \\ .2 & .5 & .3 \end{bmatrix}$$

Rewriting this last expression in equation form, we have

$$p_1 = .5p_1 + .3p_2 + .2p_3$$
$$p_2 = .4p_1 + .3p_2 + .5p_3$$
$$p_3 = .1p_1 + .4p_2 + .3p_3$$

This results in the system

$$-.5p_1 + .3p_2 + .2p_3 = 0$$
$$.4p_1 - .7p_2 + .5p_3 = 0$$
$$.1p_1 + .4p_2 - .7p_3 = 0$$

Since one of the equations is redundant, we choose any two equations and include the equation $p_1 + p_2 + p_3 = 1$, which states that the sum of the steady-state probabilities equals 1. Choosing the last two equations and including $p_1 + p_2 + p_3 = 1$, we have the system

$$.4p_1 - .7p_2 + .5p_3 = 0$$
$$.1p_1 + .4p_2 - .7p_3 = 0$$
$$p_1 + p_2 + p_3 = 1$$

Solving this system, we obtain

$$p_1 = \frac{29}{85} \qquad p_2 = \frac{33}{85} \qquad p_3 = \frac{23}{85}$$

Thus, the steady-state probability matrix is

$$p = [29/85 \quad 33/85 \quad 23/85]$$
$$= [.34 \quad .39 \quad .27]$$

Therefore, in the long run, the market shares of Companies A, B, and C are 34%, 39%, and 27%, respectively.

EXERCISES

1. Which of the following transition matrices are regular?

   (A) $\begin{bmatrix} 1/2 & 1/2 \\ 1/3 & 2/3 \end{bmatrix}$
   (B) $\begin{bmatrix} 0 & 1 \\ 1/8 & 7/8 \end{bmatrix}$

   (C) $\begin{bmatrix} 0 & 1 \\ 1 & 0 \end{bmatrix}$
   (D) $\begin{bmatrix} 1 & 0 \\ 1/4 & 3/4 \end{bmatrix}$

2. Which of the following transition matrices are regular?

   (A) $\begin{bmatrix} 1 & 0 \\ 0 & 1 \end{bmatrix}$
   (B) $\begin{bmatrix} 1/4 & 3/4 \\ 1/3 & 2/3 \end{bmatrix}$

(C) $\begin{bmatrix} 0 & 1 \\ 1/5 & 4/5 \end{bmatrix}$ (D) $\begin{bmatrix} 1 & 0 \\ 1/2 & 1/2 \end{bmatrix}$

3. Determine the steady-state probabilities for each of the following transition matrices P:

   (A) $\begin{bmatrix} .9 & .1 \\ .3 & .7 \end{bmatrix}$ (B) $\begin{bmatrix} .2 & .8 \\ .4 & .6 \end{bmatrix}$

4. Determine the steady-state probabilities for each of the following transition matrices P:

   (A) $\begin{bmatrix} .8 & .1 & .1 \\ .3 & .2 & .5 \\ .6 & .3 & .1 \end{bmatrix}$ (B) $\begin{bmatrix} .1 & .2 & .7 \\ .2 & .3 & .5 \\ .5 & .4 & .1 \end{bmatrix}$

5. Find and interpret the steady-state probabilities for Exercise 1, Section 7–10.

6. Find and interpret the steady-state probabilities for Exercise 2, Section 7–10.

7. Find and interpret the steady-state probabilities for Exercise 3, Section 7–10.

8. Find and interpret the steady-state probabilities for Exercise 4, Section 7–10.

CASE E PROBABILITY DISTRIBUTIONS— NEXT DAY COURIER, INC.

Next Day Courier, Inc. claims that 40% of its customers have reduced their usage of its services from last year's level due to the recent increase in rates. In an effort to determine the validity of this claim, the marketing manager selects a random sample of 12 customers. The random variable x denotes the number of customers who have reduced their usage from last year's level.

EXERCISES

1. Determine the probability distribution of x.
2. Graph the probability distribution of x.
3. What is the probability that exactly 2 customers in the sample have reduced their usages from last year's levels?
4. What is the probability that from 2 to 5 customers in the sample have reduced their usages from last year's levels?
5. If the company's claim is true, how many customers in the sample should be expected to have reduced usages from last year's levels?
6. If exactly 1 customer in the sample has reduced usage from last year's level, what might we conclude?

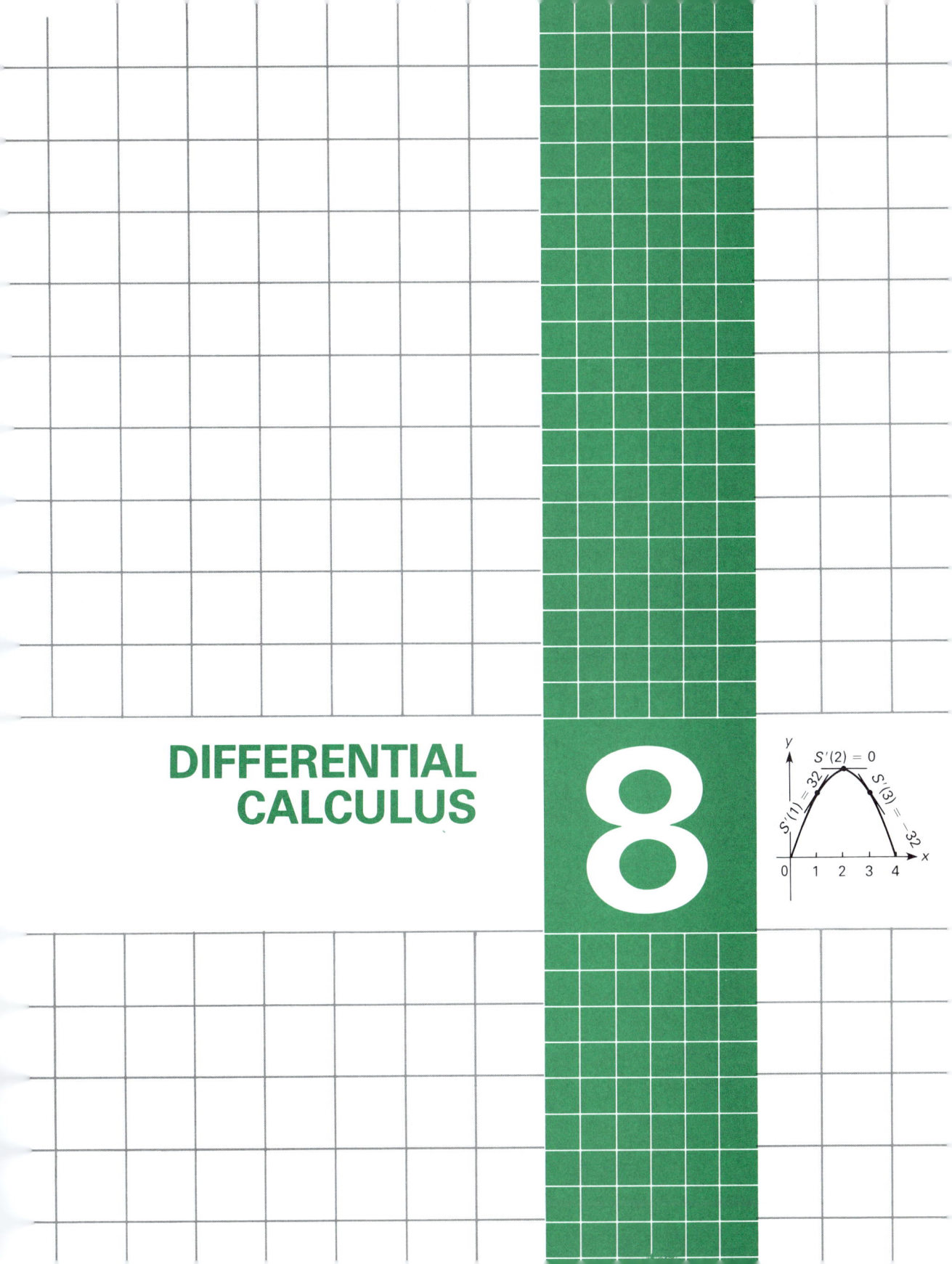

DIFFERENTIAL CALCULUS

8

CHAPTER EIGHT

In the remaining chapters of this text, we will study calculus. **Calculus** is the branch of mathematics that concerns itself with the *rate of change* of one quantity with respect to another quantity. Calculus is considered to have been invented by Isaac Newton and Gottfried Wilhelm von Leibnitz working independently at the close of the seventeenth century. Calculus is separated into two parts: **differential calculus** and **integral calculus**. Differential calculus is involved with a certain quantity called a **derivative**. In this chapter, we will learn what a derivative is, how to calculate a derivative, and how a derivative is used. Integral calculus will be considered in Chapter 10.

8-1

THE DERIVATIVE

Average Rate of Change

At the same instant that a test driver begins his journey around a track, a stopwatch is activated. The function defined by

$$y = f(x) = 10x^2$$

expresses the distance, y (in miles), traveled by the driver after x hours have elapsed. Thus, after 3 hours have elapsed, the driver has traveled

$$y = f(3) = 10(3)^2 = 90 \text{ miles}$$

After 5 hours have elapsed, the driver has traveled

$$y = f(5) = 10(5)^2 = 250 \text{ miles}$$

We now pose the following question:

What is the driver's average speed during the time interval between the end of the third hour and the end of the fifth hour?

Since the driver has traveled 90 miles during the first 3 hours and 250 miles during the first 5 hours, he has traveled

$$250 - 90 = 160 \text{ miles}$$

during the time interval from $x = 3$ to $x = 5$ (see Figure 8–1).

461 DIFFERENTIAL CALCULUS

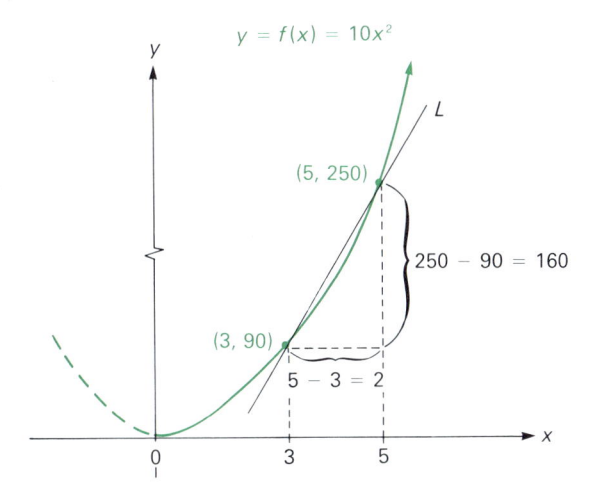

FIGURE 8–1

Dividing by the length of the time interval, we have

$$\frac{250 - 90}{5 - 3} = \frac{160}{2} = 80 \text{ miles per hour}$$

as the *average speed,* or **average rate of change** of distance with respect to time. Note that the average speed (or average rate of change) is the slope of the straight line L passing through the points (3, 90) and (5, 250) of the graph of $y = f(x) = 10x^2$ of Figure 8–1. Such a straight line is called a **secant line** (Figure 8.2).

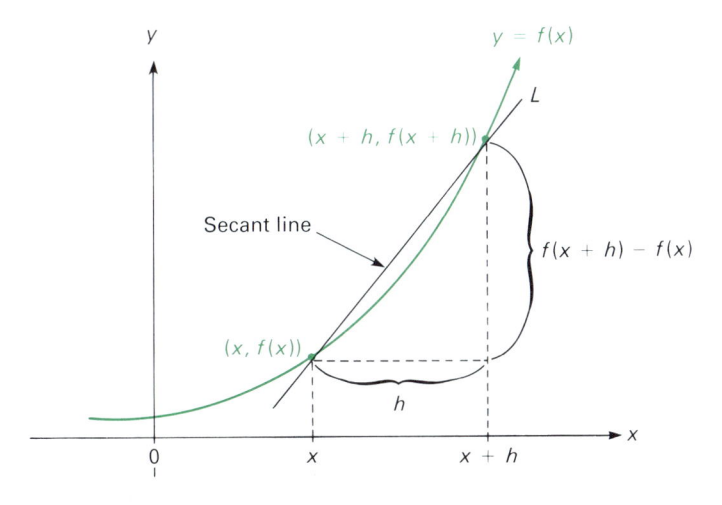

FIGURE 8–2

In general, the slope of a secant line passing through two points of the graph of a function is the average rate of change of that function over the respective interval. Observing Figure 8–2 on page 461, note that the secant line passes through the points $(x, f(x))$ and $(x + h, f(x + h))$. Thus, the average rate of change of the function $y = f(x)$ over the interval from x to $x + h$ is given by

$$\frac{f(x + h) - f(x)}{h}$$

Note that h is the horizontal distance between the two points $(x, f(x))$ and $(x + h, f(x + h))$.

We now show how the quantity $[f(x + h) - f(x)]/h$ can be calculated for a given function. As an example, we use the function defined by

$$y = f(x) = 10x^2$$

Since

$$f(x + h) = 10(x + h)^2 = 10x^2 + 20hx + 10h^2$$

then

$$\frac{f(x + h) - f(x)}{h} = \frac{(10x^2 + 20hx + 10h^2) - 10x^2}{h}$$

$$= \frac{20hx + 10h^2}{h} = 20x + 10h$$

Thus, the expression $20x + 10h$ is the average rate of change of the function $y = f(x) = 10x^2$ from x to $x + h$.

If $x = 3$ and $h = 2$, then

$$20x + 10h = 20(3) + 10(2)$$
$$= 80$$

is the average rate of change from $x = 3$ to $x + h = 3 + 2 = 5$. Note that this equals the result previously determined by using the coordinates of the points (3, 90) and (5, 250) (see Figure 8–3). If $x = 3$ and $h = 1$, then

$$20x + 10h = 20(3) + 10(1)$$
$$= 70$$

DIFFERENTIAL CALCULUS

is the average rate of change from $x = 3$ to $x + h = 3 + 1 = 4$. Observing Figure 8–3, note that this is the slope of the secant line M passing through the points (3, 90) and (4, 160).

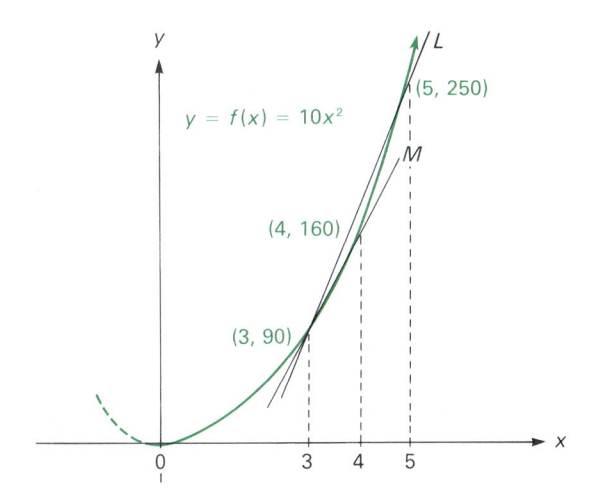

FIGURE 8–3

Instantaneous Rate of Change (The Derivative)

Continuing the preceding discussion and referring to Figure 8–3, we ask what the driver's actual speed is at $x = 3$. This result is called the **instantaneous speed,** or **instantaneous rate of change** of y with respect to x at $x = 3$. It is determined by calculating the average rate of change

$$\frac{f(x + h) - f(x)}{h} = 20x + 10h$$

and letting h get very small (i.e., equivalently, we write $h \to 0$ to indicate that h approaches 0; thus, the symbol $\to$ stands for the word *approaches*). Therefore, the instantaneous rate of change at $x = 3$ is

$$20x + 10h = 20(3) + 10(0)$$
$$= 60 \text{ miles per hour}$$

Graphically, this result is the slope of the straight line T tangent to the graph of the function $y = f(x) = 10x^2$ at $x = 3$ shown in Figure 8–4.

FIGURE 8-4

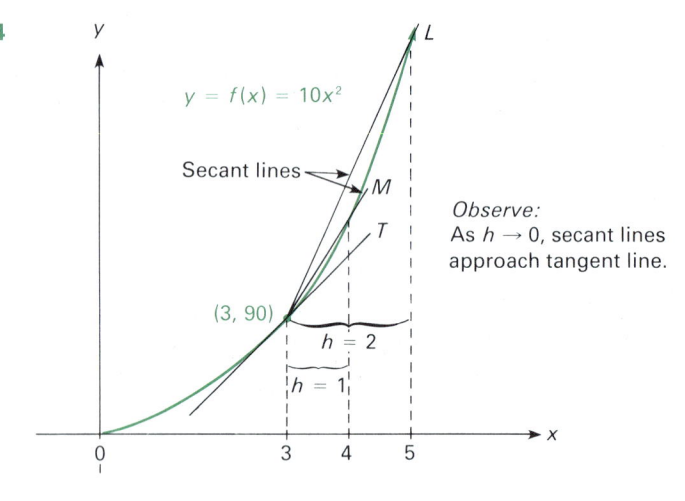

$y = f(x) = 10x^2$

Secant lines

Observe: As $h \to 0$, secant lines approach tangent line.

(3, 90)

$h = 2$
$h = 1$

In general, the instantaneous rate of change of a function $y = f(x)$ may be determined at any point x by the following procedure:

Step 1 Calculate $\dfrac{f(x+h) - f(x)}{h}$.

Step 2 Let h get very small (i.e., $h \to 0$).

The result is called the **derivative** of the function $f(x)$ and is denoted by $f'(x)$. Graphically, the derivative is the slope of the straight line tangent to the graph of the function at $(x, f(x))$ (see Figure 8-5).

FIGURE 8-5

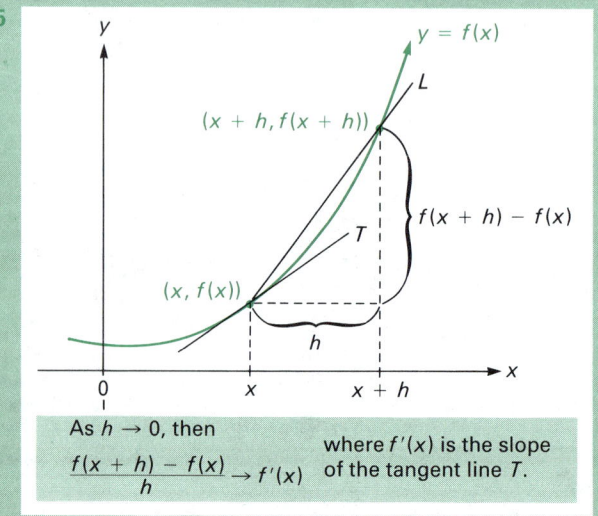

$y = f(x)$

$(x + h, f(x + h))$

$f(x + h) - f(x)$

$(x, f(x))$

h

As $h \to 0$, then

$\dfrac{f(x+h) - f(x)}{h} \to f'(x)$ where $f'(x)$ is the slope of the tangent line T.

DIFFERENTIAL CALCULUS

Thus, if $y = f(x) = 10x^2$, then its derivative, $f'(x)$, is calculated as follows:

Step 1 $\quad \dfrac{f(x+h) - f(x)}{h} = 20x + 10h$

Step 2 If $h \to 0$, then

$$f'(x) = 20x + 10(0) = 20x$$

Hence, the derivative at a point x is, in general,

$$f'(x) = 20x$$

At $x = 5$, the derivative (or instantaneous rate of change) is

$$f'(5) = 20(5) = 100$$

EXAMPLE 8–1 The Quality Hat Company manufactures hats. Its total sales revenue, y, is given by

$$y = f(x) = -3x^2 + 60x$$

where x is the number of hats sold.
(A) Calculate $f'(x)$.
(B) Calculate the instantaneous rate of change of sales revenue with respect to the number of hats sold at $x = 3$.
(C) Find the equation of the tangent line to the function $y = f(x) = -3x^2 + 60x$ at $x = 3$.

Solutions
(A) *Step 1* Since

$$\begin{aligned} f(x + h) &= -3(x+h)^2 + 60(x+h) \\ &= -3x^2 - 6hx - 3h^2 + 60x + 60h \end{aligned}$$

then $\dfrac{f(x+h) - f(x)}{h}$

$$= \dfrac{(-3x^2 - 6hx - 3h^2 + 60x + 60h) - (-3x^2 + 60x)}{h}$$

$$= \dfrac{-6hx - 3h^2 + 60h}{h}$$

$$= -6x - 3h + 60$$

Step 2 Letting $h \to 0$, we have

$$\begin{aligned} f'(x) &= -6x - 3(0) + 60 \\ &= -6x + 60 \end{aligned}$$

466 CHAPTER EIGHT

EXAMPLE 8-1 (continued)

(B) Since

$$f'(3) = -6(3) + 60 = 42$$

then, at a sales level of $x = 3$ hats, total sales revenue is increasing at a rate of $42 per hat.

(C) The slope of the tangent line at $x = 3$ is given by $f'(3) = 42$. Since $f(3) = -3(3)^2 + 60(3) = 153$, then the point of tangency is $(3, f(3)) = (3, 153)$. Thus, the tangent line passes through $(3, 153)$ and has a slope of 42. Using the point–slope form of the equation of a line, we have

$$y - y_1 = m(x - x_1)$$
$$y - 153 = 42(x - 3)$$
$$y = 42x + 27$$

The function and its tangent line appear in Figure 8-6.

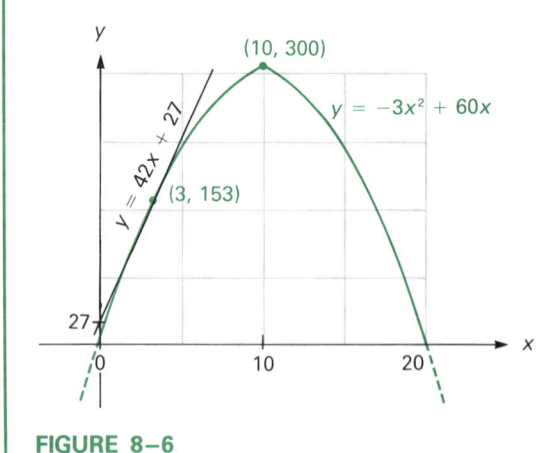

FIGURE 8-6

Notation As previously stated, the derivative of a function $y = f(x)$ is denoted by the symbol $f'(x)$. Alternative notations are

$$\frac{dy}{dx} \qquad y' \qquad \frac{d(f(x))}{dx} \qquad D_x Y$$

Thus, the derivative $f'(x) = 20x$ of the function $y = f(x) = 10x^2$ may also be expressed with any of the following notations:

$$\frac{dy}{dx} = 20x \qquad y' = 20x \qquad \frac{d}{dx}(10x^2) = 20x \qquad D_x Y = 20x$$

EXERCISES

1. For each of the following functions, find the average rate of change over the given interval. Also, graph the function and illustrate the graphical interpretation of the average rate of change.
   - (A) $f(x) = x^2 - 4x + 5$ from $x = 2$ to $x = 6$
   - (B) $f(x) = x^2 - 3x$ from $x = 1$ to $x = 3$
   - (C) $f(x) = 4x + 7$ from $x = 2$ to $x = 3$
   - (D) $f(x) = -2x^2 + 8$ from $x = 1$ to $x = 4$
   - (E) $f(x) = -3x^2 - 2x + 1$ from $x = 2$ to $x = 5$
   - (F) $f(x) = x^3 - 16x$ from $x = 0$ to $x = 2$

2. The Great Glove Company manufactures gloves. Its total sales revenue, R, is given by
   $$R = f(x) = x^2 - 6x + 9 \qquad x \geq 3$$
   where x is the number of pairs of gloves sold. Find the average rate of change of sales revenue with respect to number of pairs of gloves sold over the interval $4 \leq x \leq 6$. Show the graphical interpretation.

3. For each of the following functions, find the instantaneous rate of change of y with respect to x at the given point. Also, find the equation of the tangent line. Then, graph the function and its tangent line on the same axis system.
   - (A) $y = f(x) = x^2 - 4x + 5$ at $x = 2$
   - (B) $y = f(x) = x^2 - 3x$ at $x = 1$
   - (C) $y = f(x) = 4x + 7$ at $x = 2$
   - (D) $y = f(x) = -2x^2 + 8$ at $(1, 6)$
   - (E) $y = f(x) = x^3 - 16x$ at $x = 1$
   - (F) $y = f(x) = x^4 - 36x^2$ at $x = 3$

4. Referring to Exercise 2, find the instantaneous rate of change of sales revenue with respect to the number of pairs of gloves sold at $x = 7$. Include a graph illustrating the graphical interpretation.

5. If $f(x) = 3x^2 - 2x + 5$, find $f'(x)$.

6. If $y = x^2 - 8x$, find dy/dx.

7. If $y = x^3 - 16x$, find y'.

8. Find $\dfrac{d}{dx}(4x^2 - 5)$.

9. If $f(x) = x^2 - 3x + 4$, find each of the following:
   - (A) $f'(x)$ (B) $f'(1)$ (C) $f'(2)$

10. If $f(x) = -4x^2 + 6x$, find each of the following:
    - (A) $f'(x)$ (B) $f'(0)$ (C) $f'(2)$

8-2

LIMITS

In the previous section, we intuitively considered the derivative. We learned that the derivative $f'(x)$ of a function $f(x)$ is determined by calculating the quotient

$$\frac{f(x + h) - f(x)}{h}$$

and letting h approach 0 (i.e., $h \to 0$). The resulting quantity is called the derivative. In other words, as $h \to 0$, the difference quotient $[f(x + h) - f(x)]/h$ approaches $f'(x)$. Mathematically, we say that the derivative $f'(x)$ is the limit (if it exists) of $[f(x + h) - f(x)]/h$ as $h \to 0$. This is written in mathematical shorthand as

$$f'(x) = \lim_{h \to 0} \frac{f(x + h) - f(x)}{h}$$

The concept of limit is very important for the formal development of calculus. In this section, we will give a brief introduction to limits.

When we apply the concept of limit, we examine what happens to the y-values of a function $f(x)$ as x gets closer and closer to (but does not reach) some particular number, called a. If the y-values also get closer and closer to some number L, then the number L is said to be the *limit of the function as* x *approaches* a. Thus, we say that **L is the limit of $f(x)$ as x approaches a**. This is written in mathematical shorthand as

$$L = \lim_{x \to a} f(x)$$

Recall the symbol $\to$, which stands for the word *approaches*. If the y-values of the function do not get closer and closer to some number as x gets closer and closer to a, then the function has no limit as x approaches a. Figure 8–7 illustrates the graph of a function that has a limit as x approaches a particular number a, whereas Figure 8–8 illustrates the graph of a function that does not have a limit as x approaches a.

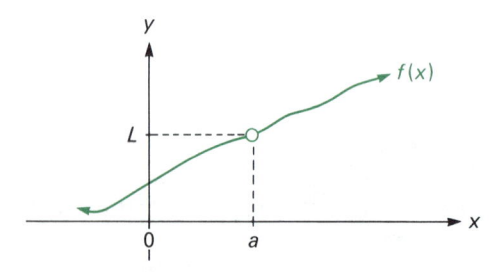

As the values of x get closer and closer to a, the y-values (values of $f(x)$) get closer and closer to L. Thus,

$$L = \lim_{x \to a} f(x)$$

FIGURE 8–7

DIFFERENTIAL CALCULUS

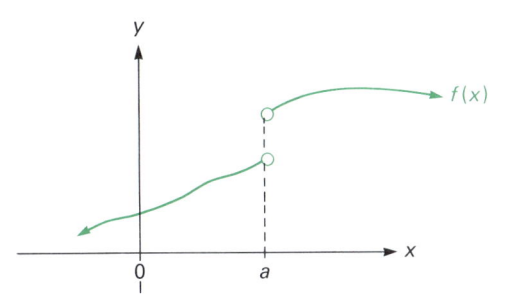

As the values of x get closer and closer to a, the y-values (values of $f(x)$) do not get closer and closer to a single number. Thus, $f(x)$ has no limit as $x \to a$.

FIGURE 8–8

To illustrate the concept of limit in a more analytical manner, we consider finding the limit of the function $f(x) = x^2$ as x approaches 3. Table 8–1 contains a listing of x and y values of $f(x) = x^2$ as x takes on values near 3. Observe that as x gets closer and closer to 3 from both sides, $f(x)$ gets closer to 9 from both sides. Thus, we say that 9 is the limit of $f(x)$ as x approaches 3. This is written in mathematical shorthand as

$$9 = \lim_{x \to 3} f(x)$$

TABLE 8–1

$y = f(x) = x^2$

x	y	x	y
2.9	8.41	3.1	9.61
2.99	8.9401	3.01	9.0601
2.999	8.994001	3.001	9.006001

Since $f(x) = x^2$, this expression may also be written as

$$9 = \lim_{x \to 3} x^2$$

We now give the following definition of limit.

> Let a and L be numbers and $f(x)$ a function. If, as x approaches the number a from either side, the values of $f(x)$ approach the single value of L, then L is said to be the **limit of $f(x)$ as x approaches a**. We write this in mathematical shorthand as
>
> $$L = \lim_{x \to a} f(x)$$

EXAMPLE 8–2 If $f(x) = x$, find $\lim_{x \to 4} f(x)$ (i.e., $\lim_{x \to 4} x$).
Solution We make a table of x and y values as x approaches 4 (see Table 8–2). Note that as x approaches 4 from either side, the values of $f(x)$ approach 4. Thus, $4 = \lim_{x \to 4} f(x)$ or, equivalently, $4 = \lim_{x \to 4} x$.

TABLE 8–2
$y = f(x) = x$

x	y	x	y
3.9	3.9	4.1	4.1
3.99	3.99	4.01	4.01
3.999	3.999	4.001	4.001

EXAMPLE 8–3 For each of the graphs of Figure 8–9, determine if $\lim_{x \to 7} h(x)$ exists. If the limit exists, state its numerical value.

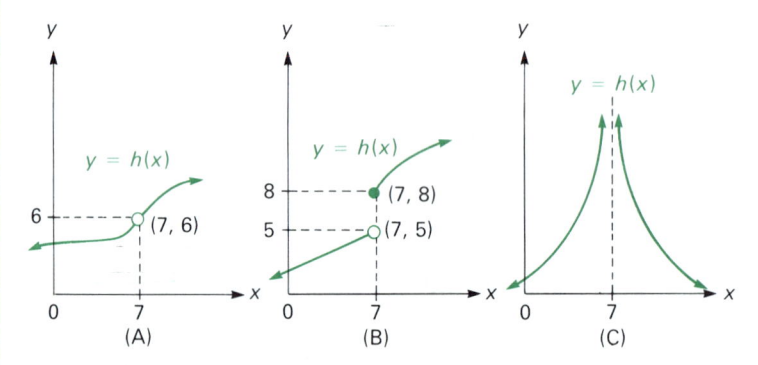

FIGURE 8–9

Solutions
(A) As x gets closer and closer to 7 from either side, the values of $h(x)$ get closer and closer to the single value 6. Thus, $\lim_{x \to 7} h(x) = 6$.
(B) As x gets closer and closer to 7 from either side, the values of $h(x)$ do not get closer and closer to a single value. Hence, the function $h(x)$ has no limit as $x \to 7$.
(C) As x gets closer and closer to 7, the values of $h(x)$ get larger and larger and do not approach a single number. Thus, $\lim_{x \to 7} h(x)$ does not exist.

DIFFERENTIAL CALCULUS

Limit Theorems

In the previous examples of this section, we determined the limits of functions by applying the definition of limit. Many times, limits of functions are determined in a much simpler manner by using **limit theorems**. The following limit theorems are stated without proof.

Limit Theorems

For each of the following, assume that $\lim_{x \to a} f(x)$ and $\lim_{x \to a} g(x)$ both exist. Also, let a, c, k, and r be constant real numbers, with $r > 0$.

Theorem 1 Limit of a Constant—

$$\lim_{x \to a} c = c$$

This theorem states that the limit of a constant function is the constant value. Thus, for example, we write

$$\lim_{x \to 2} 5 = 5 \qquad \lim_{x \to -1} 8 = 8 \qquad \lim_{x \to a} -4 = -4 \qquad \lim_{x \to 6} 9 = 9$$

Theorem 2 Limit of a Sum (or Difference)—

$$\lim_{x \to a}[f(x) \pm g(x)] = \lim_{x \to a} f(x) \pm \lim_{x \to a} g(x)$$

This theorem states that the limit of a sum (or difference) is the sum (or difference) of the individual limits provided that these limits exist. Thus, if $f(x) = x$ and $g(x) = 5$, then

$$\lim_{x \to 4}[f(x) + g(x)] = \lim_{x \to 4}(x + 5)$$
$$= \lim_{x \to 4} x + \lim_{x \to 4} 5$$
$$= 4 + 5$$
$$= 9$$

$$\lim_{x \to 4}[f(x) - g(x)] = \lim_{x \to 4}(x - 5)$$
$$= \lim_{x \to 4} x - \lim_{x \to 4} 5$$
$$= 4 - 5$$
$$= -1$$

Theorem 3 Limit of a Constant Times a Function—

$$\lim_{x \to a} kf(x) = k \lim_{x \to a} f(x)$$

This theorem states that the limit of a constant times a function is the constant times the limit of the function provided, of course, that the limit exists. Thus, if $f(x) = x$ and $k = 3$, then

$$\lim_{x \to 4} kf(x) = \lim_{x \to 4} 3x$$
$$= 3 \lim_{x \to 4} x$$
$$= 3(4)$$
$$= 12$$

Theorem 4 Limit of a Function to a Power —

$$\lim_{x \to a}[f(x)]^r = \left[\lim_{x \to a} f(x)\right]^r$$

This theorem states that the limit of a function raised to a power is the power of the limit provided, of course, that the limit exists. Thus, if $f(x) = x$ and $r = 3$, then

$$\lim_{x \to 4}[f(x)]^r = \lim_{x \to 4} x^3$$
$$= \left[\lim_{x \to 4} x\right]^3$$
$$= 4^3$$
$$= 64$$

Theorem 5 Limit of a Product —

$$\lim_{x \to a}[f(x)g(x)] = \left[\lim_{x \to a} f(x)\right]\left[\lim_{x \to a} g(x)\right]$$

This theorem states that the limit of a product is the product of the limits provided that these limits exist. Thus, if $f(x) = x^3$ and $g(x) = x + 5$, then

$$\lim_{x \to 4}[f(x)g(x)] = \lim_{x \to 4} x^3(x + 5)$$
$$= \left[\lim_{x \to 4} x^3\right]\left[\lim_{x \to 4}(x + 5)\right]$$
$$= 4^3(4 + 5)$$
$$= 64(9)$$
$$= 576$$

Note that we have previously determined that $\lim_{x \to 4} x^3 = 4^3 = 64$ and that $\lim_{x \to 4}(x + 5) = 4 + 5 = 9$.

473 DIFFERENTIAL CALCULUS

Theorem 6 Limit of a Quotient—

If $\lim_{x \to a} g(x) \neq 0$, then

$$\lim_{x \to a} \frac{f(x)}{g(x)} = \frac{\lim_{x \to a} f(x)}{\lim_{x \to a} g(x)}$$

This theorem states that the limit of a quotient is the quotient of the limits provided that the limits exist and that the limit of the denominator is not zero. Thus, if $f(x) = x^3$ and $g(x) = x + 5$, then

$$\lim_{x \to 4} \frac{f(x)}{g(x)} = \lim_{x \to 4} \frac{x^3}{x + 5}$$

$$= \frac{\lim_{x \to 4} x^3}{\lim_{x \to 4} (x + 5)}$$

$$= \frac{4^3}{(4 + 5)}$$

$$= \frac{64}{9}$$

Again, recall that we have previously determined that $\lim_{x \to 4} x^3 = 4^3 = 64$ and that $\lim_{x \to 4} (x + 5) = 9$.

EXAMPLE 8–4 Compute the following limits:
(A) $\lim_{x \to 5} x$ (B) $\lim_{x \to 5} x^4$ (C) $\lim_{x \to 5} 3x^4$

(D) $\lim_{x \to 5} (3x^4 - 275)$ (E) $\lim_{x \to 5} \sqrt{3x^4 - 275}$

(F) $\lim_{x \to 5} \frac{\sqrt{3x^4 - 275}}{x^3}$

Solutions
(A) In Example 8–2, we showed that $\lim_{x \to 4} x = 4$. In a similar manner, we could show that $\lim_{x \to 5} x = 5$.

(B) $\lim_{x \to 5} x^4 = \left(\lim_{x \to 5} x\right)^4$ By theorem 4

          $= 5^4$ By part (A)

          $= 625$

(C) $\lim_{x \to 5} 3x^4 = 3 \lim_{x \to 5} x^4$ By theorem 3

          $= 3(625)$ By part (B)

          $= 1875$

474 CHAPTER EIGHT

EXAMPLE 8-4 (continued)

(D) $\lim_{x \to 5}(3x^4 - 275) = \lim_{x \to 5} 3x^4 - \lim_{x \to 5} 275$ By theorem 2

Since $\lim_{x \to 5} 3x^4 = 1875$ by part (C) and $\lim_{x \to 5} 275 = 275$ by theorem 1, the result is

$$\lim_{x \to 5}(3x^4 - 275) = 1875 - 275$$
$$= 1600$$

(E) $\lim_{x \to 5} \sqrt{3x^4 - 275} = \lim_{x \to 5}(3x^4 - 275)^{1/2}$

$$= \left[\lim_{x \to 5}(3x^4 - 275)\right]^{1/2} \quad \text{By theorem 4}$$

Since $\lim_{x \to 5}(3x^4 - 275) = 1600$ from part (D), then we have

$$\left[\lim_{x \to 5}(3x^4 - 275)\right]^{1/2} = 1600^{1/2}$$
$$= 40$$

(F) $\lim_{x \to 5} \dfrac{\sqrt{3x^4 - 275}}{x^3} = \dfrac{\lim_{x \to 5} \sqrt{3x^4 - 275}}{\lim_{x \to 5} x^3}$ By theorem 6

In part (E), we determined that $\lim_{x \to 5} \sqrt{3x^4 - 275} = 40$. By theorem 4 and part (A), we determine that $\lim_{x \to 5} x^3 = (\lim_{x \to 5} x)^3 = 5^3 = 125$. Hence, we obtain

$$\dfrac{\lim_{x \to 5} \sqrt{3x^4 - 275}}{\lim_{x \to 5} x^3} = \dfrac{40}{125} = \dfrac{8}{25}$$

EXAMPLE 8-5 If

$$f(x) = \dfrac{x^2 - 36}{x - 6}$$

compute $\lim_{x \to 6} f(x)$.

Solution Note that $f(x)$ is not defined at $x = 6$ since $f(6) = (6^2 - 36)/6 - 6) = 0/0$, which is undefined. However, we can consider $\lim_{x \to 6} f(x)$ since the limit as x approaches 6 depends only on values of x near 6 without consideration of the value at $x = 6$. To evaluate $\lim_{x \to 6} f(x)$, note that

$$\dfrac{x^2 - 36}{x - 6} = \dfrac{\cancel{(x - 6)}(x + 6)}{\cancel{x - 6}}$$
$$= x + 6$$

Hence,
$$\lim_{x \to 6} \frac{x^2 - 36}{x - 6} = \lim_{x \to 6}(x + 6)$$
$$= 12$$

A graph of $f(x)$ appears in Figure 8–10. Observing Figure 8–10, note that the graph of $f(x)$ is the graph of the straight line $y = x + 6$ with the point (6, 12) excluded. The open circle at (6, 12) indicates that this point is excluded from the graph.

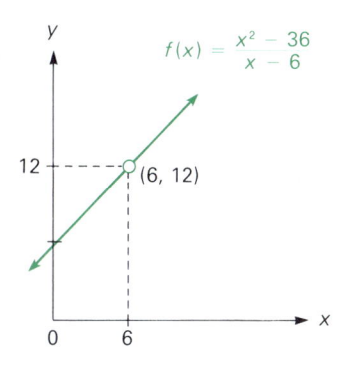

FIGURE 8–10

EXAMPLE 8–6 If $f(x) = x^3$, compute $f'(x)$.
Solution Since $f'(x) = \lim_{h \to 0} [f(x + h) - f(x)]/h$, we first calculate the quotient $[f(x + h) - f(x)]/h$ and then determine its limit as h approaches 0. Hence, we find

$$\frac{f(x + h) - f(x)}{h} = \frac{(x + h)^3 - x^3}{h}$$
$$= \frac{x^3 + 3x^2h + 3xh^2 + h^3 - x^3}{h}$$
$$= 3x^2 + 3xh + h^2$$

and then calculate

$$f'(x) = \lim_{h \to 0}(3x^2 + 3xh + h^2)$$
$$= 3x^2$$

Infinity and Limits

Consider the function

$$f(x) = \frac{x+3}{x-3}$$

The graph of this function appears in Figure 8–11. Observing Figure 8–11, note that as x takes on larger and larger positive numbers (i.e., as x increases without bound), the value of $f(x)$ approaches 1. (Recall that the horizontal line $y = 1$ is a horizontal asymptote in this case.) For such a situation, we say that 1 is the limit of $f(x)$ as x approaches infinity. The statement "x increases without bound" is equivalently expressed in mathematical terms as "x approaches infinity." Infinity is denoted by the symbol ∞. Thus, $x \to \infty$ means that x is increasing without bound (i.e., x exceeds any positive number we can think of). Similarly, $x \to -\infty$ means that x is decreasing without bound (i.e., x is getting more and more negative). Therefore, the fact that the function $f(x)$ of Figure 8–11 approaches 1 as x approaches ∞ is written as

$$\lim_{x \to \infty} f(x) = 1$$

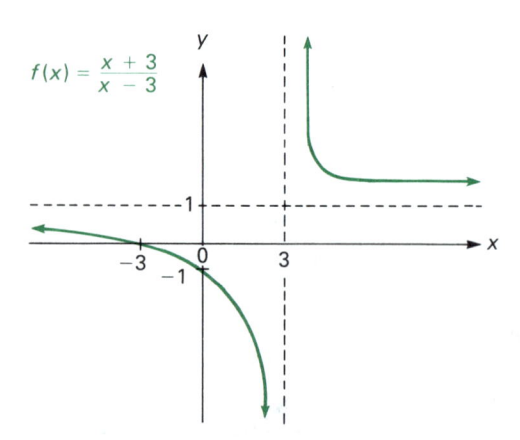

FIGURE 8–11

Observe also that as x becomes more and more negative (i.e., as x decreases without bound), the value of $f(x)$ approaches 1. Thus, we say that the limit of $f(x)$ as x approaches negative infinity is 1 and write

$$\lim_{x \to -\infty} f(x) = 1$$

EXAMPLE 8–7 Observe the graph of $f(x) = e^x$ of Figure 8–12. Determine

$$\lim_{x \to -\infty} e^x$$

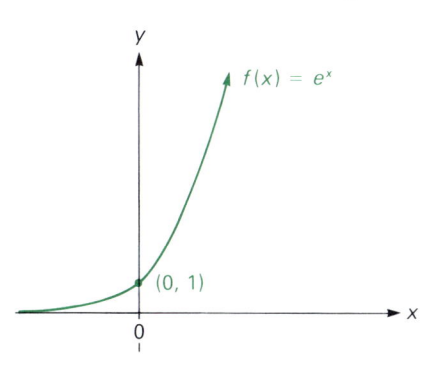

FIGURE 8–12

Solution Since e^x approaches 0 as x decreases without bound, we obtain

$$\lim_{x \to -\infty} e^x = 0$$

EXAMPLE 8–8 Determine

$$\lim_{x \to \infty} \frac{3}{x^2 + 5}$$

Solution As x increases without bound, so does $x^2 + 5$. Thus, the fraction $3/(x^2 + 5)$ approaches 0 as x approaches ∞. Hence, we have

$$\lim_{x \to \infty} \frac{3}{x^2 + 5} = 0$$

EXAMPLE 8–9 Determine

$$\lim_{x \to \infty} \frac{x^2 + 3x}{4x^2 + 9}$$

Solution Note that as x increases without bound, then so do both the numerator and the denominator. To determine the limit of their quotient, we factor out the highest powered

EXAMPLE 8-9 (continued)

term of the numerator and also the highest powered term of the denominator to obtain

$$\frac{x^2 + 3x}{4x^2 + 9} = \frac{x^2\left(1 + \frac{3}{x}\right)}{4x^2\left(1 + \frac{9}{4x^2}\right)}$$

$$= \frac{1 + \frac{3}{x}}{4\left(1 + \frac{9}{4x^2}\right)}$$

Now, as x increases without bound, $3/x$ and $9/4x^2$ both approach 0, and so the numerator, $1 + 3/x$, approaches 1 and the denominator, $4(1 + 9/4x^2)$, approaches 4. Thus, we calculate

$$\lim_{x \to \infty} \frac{x^2 + 3x}{4x^2 + 9} = \frac{1}{4}$$

EXERCISES

1. For each of the functions $f(x)$ of Figure 8-13, determine whether or not $\lim_{x \to 5} f(x)$ exists. If $\lim_{x \to 5} f(x)$ does exist, state its value.

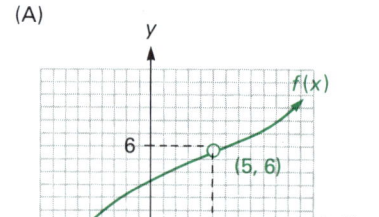

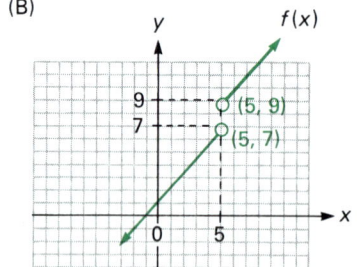

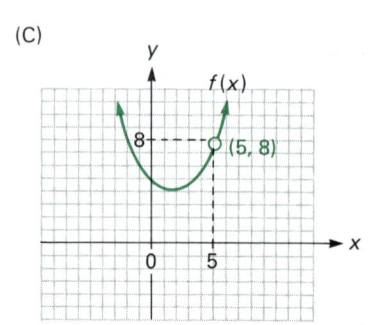

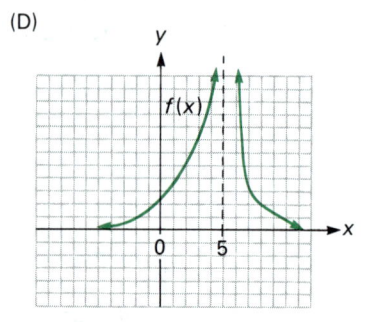

(A)

(B)

(C)

(D)

FIGURE 8-13

479 DIFFERENTIAL CALCULUS

2. For each of the functions $h(x)$ of Figure 8–14, determine whether or not $\lim_{x \to 7} h(x)$ exists. If $\lim_{x \to 7} h(x)$ does exist, state its value.

(A)

(B)

(C)

(D)

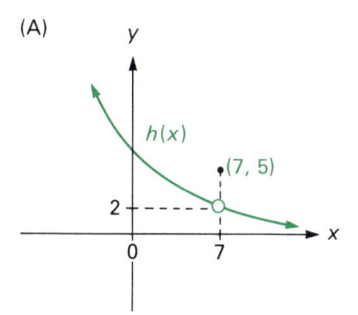

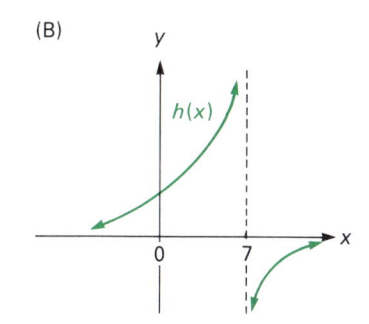

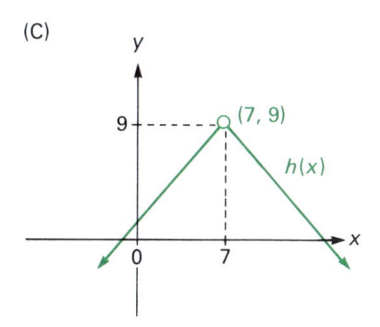

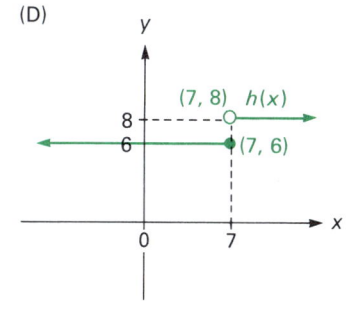

FIGURE 8–14

3. For each of the functions $f(x)$ of Figure 8–15, determine whether or not $\lim_{x \to \infty} f(x)$ exists. If $\lim_{x \to \infty} f(x)$ does exist, state its value.

(A)

(B)

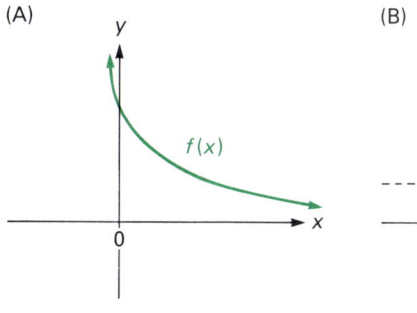

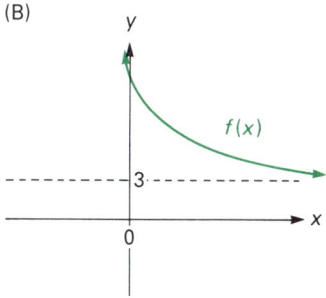

FIGURE 8–15

480 CHAPTER EIGHT

(C)

$f(x)$

(D)

$f(x)$

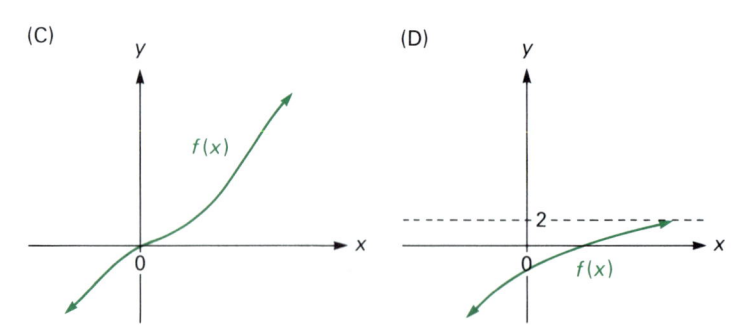

FIGURE 8–15 (continued)

4. For each of the functions $g(x)$ of Figure 8–16, determine whether or not $\lim_{x \to -\infty} g(x)$ exists. If $\lim_{x \to -\infty} g(x)$ does exist, state its value.

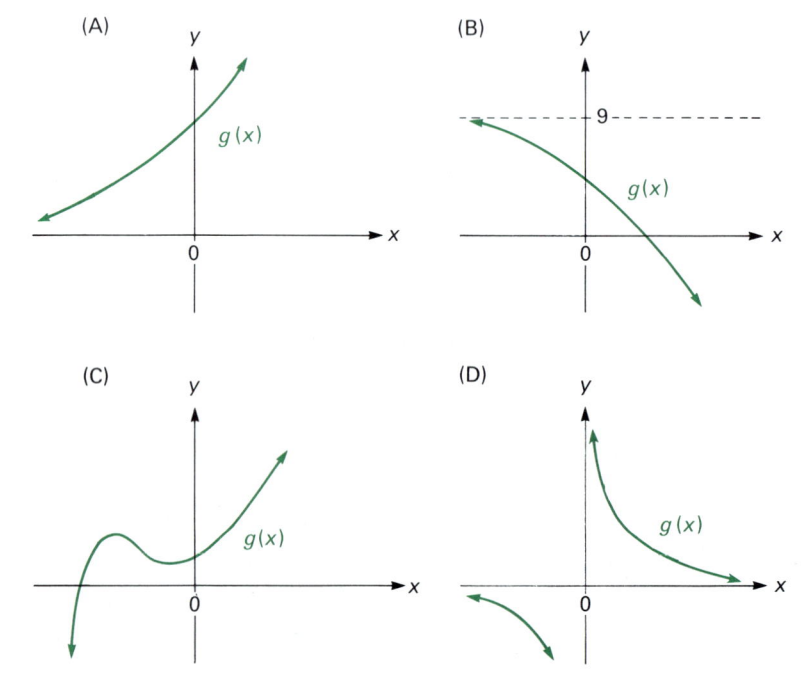

(A) $g(x)$

(B) $g(x)$

(C) $g(x)$

(D) $g(x)$

FIGURE 8–16

DIFFERENTIAL CALCULUS

5. Determine the value of each of the following limits. If the limit does not exist, then state so.

   (A) $\lim_{x \to 2}(4x + 7)$
   (B) $\lim_{x \to 1} \dfrac{2x^2 - 2x}{x - 1}$
   (C) $\lim_{x \to 5} \dfrac{x}{x - 5}$
   (D) $\lim_{x \to 2} \sqrt{3x^2 + 4}$
   (E) $\lim_{x \to 0} \dfrac{x^2 - 5x}{x}$
   (F) $\lim_{x \to 5} \dfrac{x^2 - 25}{x - 5}$
   (G) $\lim_{x \to 3} \dfrac{x^2 - x - 6}{x - 3}$
   (H) $\lim_{x \to 8} \dfrac{1}{(x - 8)^2}$
   (I) $\lim_{x \to \infty} \dfrac{9}{x^3}$
   (J) $\lim_{x \to \infty} \dfrac{5x^3 - 4x}{2x^3 - 3}$
   (K) $\lim_{x \to \infty} \dfrac{4}{3x - 7}$
   (L) $\lim_{x \to \infty} \dfrac{6x^4 - 8x^2}{3x^3 + 2x}$

6. Determine the value of each of the following limits. If the limit does not exist, then state so.

   (A) $\lim_{x \to 1}(8x - 5)$
   (B) $\lim_{x \to 3} \sqrt{5x^2 + 4}$
   (C) $\lim_{x \to -3} \dfrac{x}{x + 3}$
   (D) $\lim_{x \to 8} \dfrac{x^2 - 64}{x - 8}$
   (E) $\lim_{x \to 2} \dfrac{x^2 + 3x - 10}{x - 2}$
   (F) $\lim_{x \to 5} \dfrac{1}{x^2 - 25}$
   (G) $\lim_{x \to 0} \dfrac{x^3 - 4x^2 + 5x}{x}$
   (H) $\lim_{x \to \infty} \dfrac{2x + 3}{x + 7}$
   (I) $\lim_{x \to -\infty} \dfrac{1}{x^2}$
   (J) $\lim_{x \to -\infty} e^x$
   (K) $\lim_{x \to \infty} \dfrac{3x^2 - 2x}{5x^3 + x}$
   (L) $\lim_{x \to \infty} \dfrac{5x^3 - 7x}{2x^2 + 3}$

7. Using the limit formula for $f'(x)$

   $$f'(x) = \lim_{h \to 0} \dfrac{f(x + h) - f(x)}{h}$$

   compute $f'(x)$ for each of the following:
   (A) $f(x) = x^4$
   (B) $f(x) = x^5$
   (C) $f(x) = 3x^2 - 2x$
   (D) $f(x) = 2x + 6$
   (E) $f(x) = 5x^3 - 2x$
   (F) $f(x) = -2x^2 + 3x$

8. Using the limit formula for $g'(x)$

   $$g'(x) = \lim_{h \to 0} \dfrac{g(x + h) - g(x)}{h}$$

   compute the derivative $g'(x)$ for each of the following:
   (A) $g(x) = 3x + 7$
   (B) $g(x) = -4x^2$
   (C) $g(x) = 5x^2 - 3x$
   (D) $g(x) = 4x^3 - 5x$
   (E) $g(x) = x^3 - 2x^2$
   (F) $g(x) = 5x^4 + 8x$

8-3

DIFFERENTIABILITY AND CONTINUITY

Differentiability

Up to this point, we have defined the derivative of a function $f(x)$ to be

$$f'(x) = \lim_{h \to 0} \frac{f(x + h) - f(x)}{h}$$

If this limit does not exist at certain values of x, the function $f(x)$ does not have a derivative at those values of x. In general, if a function $f(x)$ has a derivative at $x = a$, then $f(x)$ is said to be **differentiable** at $x = a$.

We will approach the topic of differentiability from a graphical perspective. Recall that the derivative of a function $f(x)$ evaluated at $x = a$ is the slope of the tangent line to the function at $(a, f(a))$. Thus, a function $f(x)$ has a derivative at $x = a$ if its graph has a unique nonvertical tangent line at $x = a$.

As an example, consider the function

$$f(x) = x^{2/3}$$

Its graph appears in Figure 8–17. Observe that the graph of $f(x)$ has nonvertical tangent lines for all values of x except $x = 0$. At $x = 0$, the tangent line is the (vertical) y-axis. Notice how the graph of $f(x)$ comes to a sharp point at the origin. Let us see what happens to the derivative at $x = 0$. Since $f(x) = x^{2/3}$, then, as will be shown in Example 8–12, we determine that

$$f'(x) = \frac{2}{3} x^{-1/3}$$

$$= \frac{2}{3\sqrt[3]{x}}$$

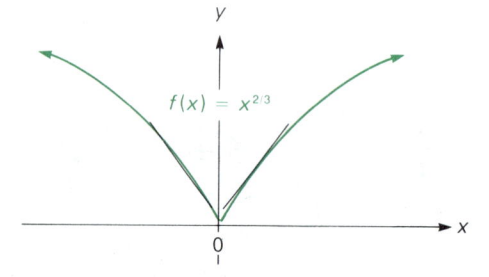

FIGURE 8–17

Hence, we have

$$f'(0) = \frac{2}{3\sqrt[3]{0}} \qquad \text{which is undefined.}$$

483 DIFFERENTIAL CALCULUS

Thus, the function $f(x)$ has no derivative at $x = 0$. Observe that $f(x)$ has a derivative at all other values of x since the ratio $2/3 \sqrt[3]{x}$ is defined for all values of x except $x = 0$. Figure 8–18 contains the graphs of other functions that are not differentiable at certain values of x.

$f(x)$ has no tangent line here.

$f(x)$ has a vertical tangent line here.

$f(x)$ has no tangent line here.

$f(x)$ has no tangent line here.

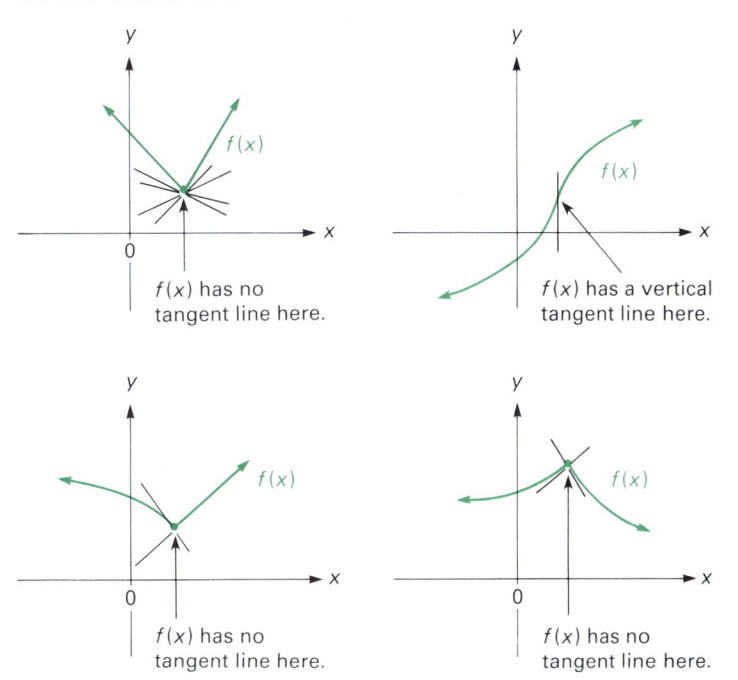

FIGURE 8–18

Of course, if a function is not defined at a value of x, then it is not differentiable there. Specifically, let us consider the function $f(x) = x^2$ for all values of x except $x = 2$. Thus, the graph of this function is the parabola $f(x) = x^2$ excluding the point $(2, 4)$ (see Figure 8–19). Since the point $(2, 4)$ does not belong to the function $f(x)$, as defined, the graph of $f(x)$ has a break at $(2, 4)$. Hence, there is no tangent line to $f(x)$ at $(2, 4)$ and $f(x)$ is not differentiable at $x = 2$.

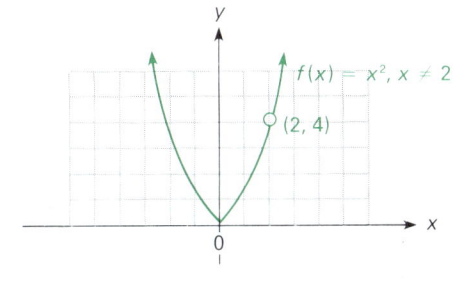

FIGURE 8–19

Continuity

The graph of Figure 8–19 brings up another topic which we will briefly discuss—the topic of **continuity**. A function $f(x)$ is continuous at all values of x if its graph has no breaks or gaps. Thus, the function of Figure 8–19 is not continuous at $x = 2$, but it is continuous at all other values of x.

Continuity can be defined in terms of limits. Specifically, a function $f(x)$ is continuous at $x = a$ if

$$\lim_{x \to a} f(x) = f(a)$$

Continuity and Differentiability

Observe that the rational function

$$f(x) = \frac{1}{x - 3}$$

whose graph appears in Figure 8–20, is continuous at all values of x except $x = 3$. Note that its derivative (which will be determined in Example 8–20),

$$f'(x) = \frac{-1}{(x - 3)^2}$$

is undefined at $x = 3$.

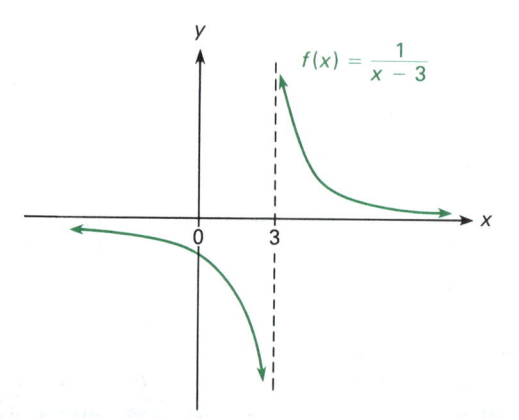

FIGURE 8–20

The functions of Figure 8–19 and 8–20 illustrate the fact that if a function $f(x)$ is not continuous at a value of x, then it is not differentiable there. Do not misinterpret this statement by concluding that if a function $f(x)$ is continuous at a value of x, then it is differentiable there. This is not true. If we observe the graph of Figure 8–17, we see a function, $f(x) = x^{2/3}$, which is continuous at $x = 0$ but not differentiable there. An inspection of the graphs of

485 DIFFERENTIAL CALCULUS

Figure 8–18 reveals functions continuous at certain values of x but not differentiable there.

We now state, without proof, the following theorem, which is usually proven in more formal calculus texts.

> If a function is differentiable at a value of x, then it is continuous there.

EXERCISES

1. Which of the functions of Figure 8–21 are differentiable at $x = a$?

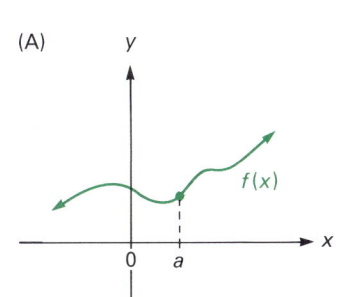

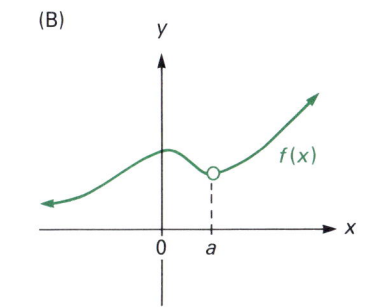

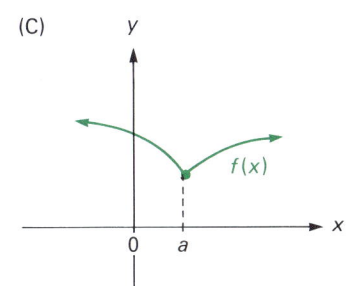

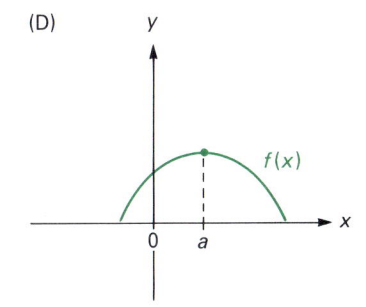

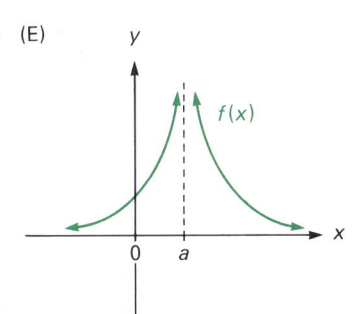

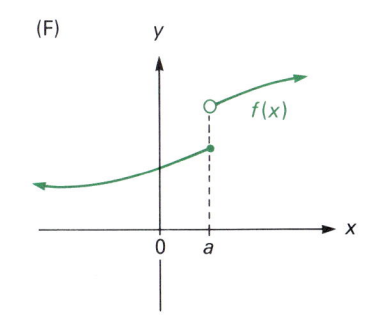

FIGURE 8–21

486 CHAPTER EIGHT

2. Which of the functions of Figure 8–21 are continuous at $x = a$?
3. Which of the functions of Figure 8–22 are continuous at $x = a$ but not differentiable at $x = a$?

(A) (B)

(C) (D)

(E) (F)

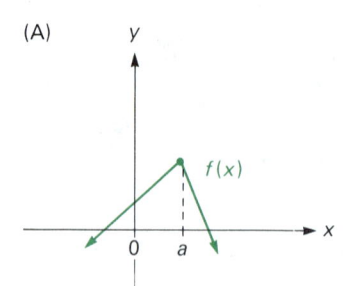

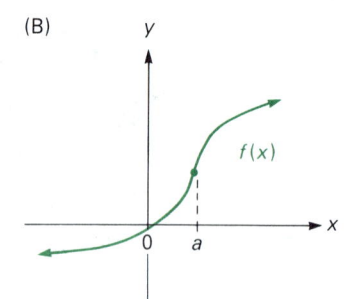

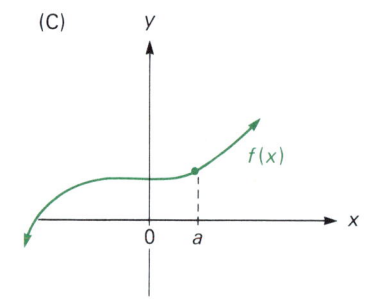

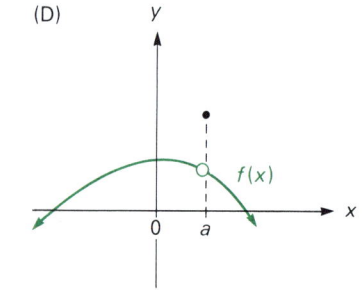

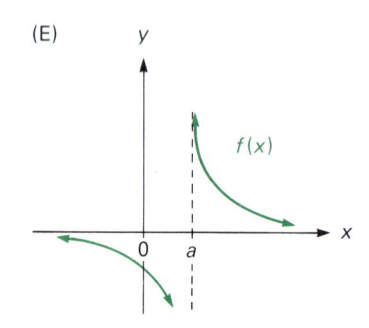

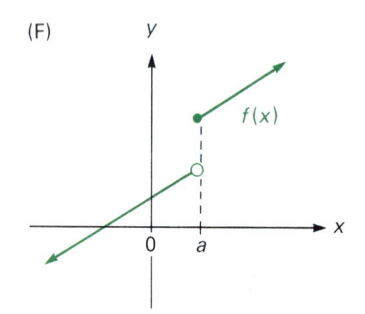

FIGURE 8–22

4. Given the function

$$f(x) = \frac{1}{(x-5)^2}$$

with

$$f'(x) = \frac{-2}{(x-5)^3}$$

(A) Graph $f(x)$.
(B) For which value(s) of x is $f(x)$ not continuous?
(C) For which value(s) of x is $f(x)$ not differentiable?

5. Given the function

$$f(x) = \frac{1}{(x-3)^2(x+5)}$$

with

$$f'(x) = \frac{-(3x+7)}{(x-3)^3(x+5)^2}$$

(A) Graph $f(x)$.
(B) For which value(s) of x is $f(x)$ not continuous?
(C) For which value(s) of x is $f(x)$ not differentiable?

6. Given the function

$$f(x) = x^{1/3}$$

with

$$f'(x) = \frac{1}{3\sqrt[3]{x^2}}$$

For which value(s) of x is $f(x)$ not differentiable?

8-4

RULES FOR FINDING DERIVATIVES

Up to this point, we have been calculating derivatives of functions $y = f(x)$ by first computing the quotient $[f(x+h) - f(x)]/h$ and then letting $h \to 0$. Since this is a tedious process, we will introduce some rules to expedite the calculation of derivatives.

The first such rule pertains to derivatives of functions of the form

$$f(x) = x^n$$

where n is a real number. It is called the **power rule** and is stated here. Its proof appears in the appendix at the end of this section.

Power Rule

If $f(x) = x^n$, where n is a real number, then

$$f'(x) = nx^{n-1}$$

Thus, if $f(x) = x^3$, then $f'(x) = 3x^2$. If $f(x) = x^6$, then $f'(x) = 6x^5$. We may, if we wish, verify each of these derivatives by calculating the quotient $[f(x + h) - f(x)]/h$ and letting $h \to 0$. However, since we now have the power rule, this method is no longer needed to find the derivatives of functions of the form $f(x) = x^n$.

EXAMPLE 8-10 If $y = f(x) = \sqrt{x}$, find
(A) dy/dx.
(B) $f'(16)$.
Solutions
(A) Since $y = f(x) = \sqrt{x} = x^{1/2}$, then, using the power rule, we have

$$\frac{dy}{dx} = \frac{1}{2}x^{\frac{1}{2}-1} = \frac{1}{2}x^{-\frac{1}{2}} = \frac{1}{2} \cdot \frac{1}{\sqrt{x}}$$

(B) Since $dy/dx = f'(x)$, then

$$f'(16) = \frac{1}{2} \cdot \frac{1}{\sqrt{16}} = \frac{1}{2} \cdot \frac{1}{4} = \frac{1}{8}$$

EXAMPLE 8-11 If $y = 1/x^2$, find dy/dx.
Solution Since $y = 1/x^2 = x^{-2}$, using the power rule, we have

$$\frac{dy}{dx} = -2x^{-2-1} = -2x^{-3} = \frac{-2}{x^3}$$

EXAMPLE 8-12 If $f(x) = x^{2/3}$, find $f'(x)$.
Solution By the power rule, we calculate

$$f'(x) = \frac{2}{3}x^{\frac{2}{3}-1} = \frac{2}{3}x^{-\frac{1}{3}} = \frac{2}{3\sqrt[3]{x}}$$

Using the power rule, we may easily calculate the derivative of a function of the form $y = x^n$, where n is a real number. However, additional rules are needed if we wish to determine the derivatives of such functions as

$$y = 3x^7$$
$$y = 6x^3 - 4x^2 + 8x - 5$$
$$y = (x^3 + 7)(x^2 - 3x + 5)$$

DIFFERENTIAL CALCULUS

One such rule, the **constant multiplier rule,** is stated here. Its proof appears in the appendix at the end of this section.

Constant Multiplier Rule

If $y = kf(x)$, where k is a constant and $f(x)$ is differentiable at x, then

$$\frac{dy}{dx} = kf'(x)$$

The constant multiplier rule states that if a function $f(x)$ is multiplied by a constant k, then the derivative of the new function, $kf(x)$, is k times the derivative of the original function.

Thus, if $y = 3x^7$, we have

$$\frac{dy}{dx} = 3(7x^6) = 21x^6$$

EXAMPLE 8–13 If $f(x) = 6/x^3$, find $f'(x)$.

Solution Since $f(x) = 6x^{-3}$, then by the constant multiplier rule and the power rule,

$$f'(x) = 6(-3x^{-3-1}) = -18x^{-4} = \frac{-18}{x^4}$$

Derivative of a Constant Function

If $f(x) = k$, where k is a constant, then $f'(x) = 0$ since the difference quotient

$$\frac{f(x+h) - f(x)}{h} = \frac{k - k}{h} = \frac{0}{h}$$

approaches 0 as $h \to 0$. Thus, we state the **constant function rule.**

Constant Function Rule

If $f(x) = k$, where k is a constant, then

$$f'(x) = 0$$

The constant function rule states that the derivative of a constant function is 0. Thus, if $f(x) = 5$, then $f'(x) = 0$. If $y = -7$, then $dy/dx = 0$.

Another useful rule for finding derivatives, the **sum** (or **difference**) **rule,** is stated here. Its proof appears in the appendix at the end of this section.

Sum (or Difference) Rule

If $y = f(x) \pm g(x)$, where $f(x)$ and $g(x)$ are differentiable functions at x, then

$$\frac{dy}{dx} = f'(x) \pm g'(x)$$

The sum (or difference) rule states that the derivative of a sum (or difference) of two functions is the sum (or difference) of their derivatives. It may be generalized to more than two functions.

Thus, the function

$$y = 6x^4 + 8x^2$$

is of the form

$$y = f(x) + g(x)$$

with $f(x) = 6x^4$ and $g(x) = 8x^2$. Since $f'(x) = 24x^3$ and $g'(x) = 16x$, then according to the sum rule,

$$\frac{dy}{dx} = f'(x) + g'(x)$$

$$= 24x^3 + 16x$$

EXAMPLE 8–14 If $y = x^3 - 4x^2 + 15x - 10$, find dy/dx.
Solution $dy/dx = 3x^2 - 8x + 15$

EXAMPLE 8–15 Find the equation of the tangent line to the parabola $f(x) = 3x^2 - 12x + 13$ at $x = 3$.
Solution The slope of the tangent line is given by $f'(3)$. Calculating $f'(x)$, we have

$$f'(x) = 6x - 12$$

DIFFERENTIAL CALCULUS

Hence,
$$f'(3) = 6(3) - 12$$
$$= 6$$

The point of tangency is (3, $f(3)$). Since
$$f(3) = 3(3)^2 - 12(3) + 13$$
$$= 4$$

the point of tangency is (3, 4). Using the point–slope form, we have
$$y - y_1 = m(x - x_1)$$
$$y - 4 = 6(x - 3)$$
$$y = 6x - 14$$

The parabola and its tangent line appear in Figure 8–23.

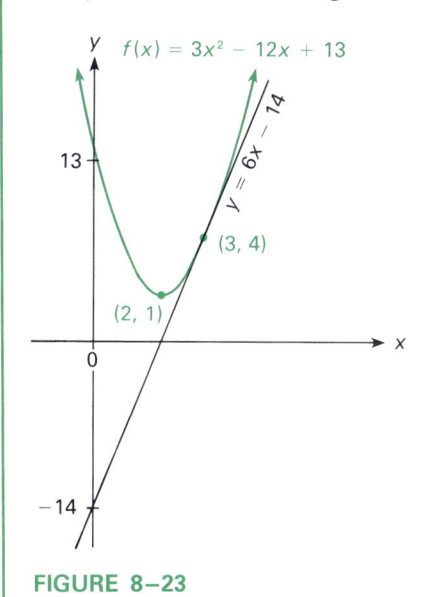

FIGURE 8–23

Often, we must find the derivative of a product of functions such as
$$y = (x^3 - 8x)(x^4 - 15)$$

The derivative, dy/dx, may be determined by the **product rule,** which is stated as follows.

> **Product Rule**
>
> If $y = f(x)s(x)$, where $f(x)$ and $s(x)$ are differentiable functions at x, then
>
> $$\frac{dy}{dx} = f(x)s'(x) + s(x)f'(x)$$

The product rule states that the derivative of the product $f(x)s(x)$ is $f(x)$ times the derivative of $s(x)$ plus $s(x)$ times the derivative of $f(x)$.

Thus, for the function

$$y = (x^3 - 8x)(x^4 - 15)$$

since $f(x) = x^3 - 8x$ and $s(x) = x^4 - 15$, we write

$$\frac{dy}{dx} = f(x)s'(x) + s(x)f'(x)$$
$$= (x^3 - 8x)(4x^3) + (x^4 - 15)(3x^2 - 8)$$

EXAMPLE 8–16 If $y = (x^5 - 6x^3 + 5)(x^{10} - 8x^2 + 5)$, find dy/dx.
Solution Let $f(x) = x^5 - 6x^3 + 5$ and $s(x) = x^{10} - 8x^2 + 5$. Then, by the product rule, we have

$$\frac{dy}{dx} = f(x)s'(x) + s(x)f'(x)$$
$$= (x^5 - 6x^3 + 5)(10x^9 - 16x) + (x^{10} - 8x^2 + 5)(5x^4 - 18x^2)$$

EXAMPLE 8–17 If $g(x) = (5x + 3)(2x - 1)$, find $g'(x)$
(A) By using the product rule.
(B) Without using the product rule.
Solutions
(A) Using the product rule, we have

$$g'(x) = (5x + 3)(2) + (2x - 1)(5)$$
$$= (10x + 6) + (10x - 5)$$
$$= 20x + 1$$

(B) Without using the product rule, we must multiply the binomial factors $5x + 3$ and $2x - 1$ of $g(x)$ to obtain

$$g(x) = 10x^2 + x - 3$$

Since the result is a polynomial, its derivative is

$$g'(x) = 20x + 1$$

Observe that this result agrees with the final answer for part (A).

To find the derivative of the quotient of functions such as

$$y = \frac{x^5 - 9x}{3x^2 - 8}$$

we must use the **quotient rule,** which is stated as follows.

Quotient Rule

If $y = n(x)/d(x)$, where $n(x)$ and $d(x)$ are differentiable functions at x and $d(x) \neq 0$, then

$$\frac{dy}{dx} = \frac{d(x)n'(x) - n(x)d'(x)}{[d(x)]^2}$$

The quotient rule states that the derivative of a quotient of two functions is the denominator times the derivative of the numerator minus the numerator times the derivative of the denominator, all divided by the denominator squared.

Thus, for the function

$$y = \frac{x^5 - 9x}{3x^2 - 8}$$

since $n(x) = x^5 - 9x$ and $d(x) = 3x^2 - 8$, we write

$$\frac{dy}{dx} = \frac{d(x)n'(x) - n(x)d'(x)}{[d(x)]^2}$$

$$= \frac{(3x^2 - 8)(5x^4 - 9) - (x^5 - 9x)(6x)}{(3x^2 - 8)^2}$$

EXAMPLE 8–18 If

$$y = \frac{x^8 - 3x}{2x^5 - 9}$$

find dy/dx.

Solution Since $n(x) = x^8 - 3x$ and $d(x) = 2x^5 - 9$, then, by the quotient rule, we write

$$\frac{dy}{dx} = \frac{d(x)n'(x) - n(x)d'(x)}{[d(x)]^2}$$

$$= \frac{(2x^5 - 9)(8x^7 - 3) - (x^8 - 3x)(10x^4)}{(2x^5 - 9)^2}$$

EXAMPLE 8–19 If

$$f(x) = \frac{6x^4 - 8x^2}{x}$$

find $f'(x)$
(A) By using the quotient rule.
(B) Without using the quotient rule.

Solutions

(A) Using the quotient rule, we have

$$f'(x) = \frac{(x)(24x^3 - 16x) - (6x^4 - 8x^2)(1)}{x^2}$$

$$= \frac{24x^4 - 16x^2 - 6x^4 + 8x^2}{x^2}$$

$$= \frac{18x^4 - 8x^2}{x^2}$$

$$= 18x^2 - 8$$

(B) Without using the quotient rule, we must divide $6x^4 - 8x^2$ by x to obtain

$$f(x) = 6x^3 - 8x$$

Hence, the derivative of the resulting polynomial is

$$f'(x) = 18x^2 - 8$$

Observe that this result agrees with the final answer for part (A).

DIFFERENTIAL CALCULUS

EXAMPLE 8–20 If
$$f(x) = \frac{1}{x-3}$$
find $f'(x)$.

Solution Using the quotient rule, we calculate
$$f'(x) = \frac{(x-3)(0) - (1)(1)}{(x-3)^2}$$
$$= \frac{-1}{(x-3)^2}$$

APPENDIX PROOFS OF DERIVATIVE RULES

Power Rule

If $f(x) = x^n$, where n is a real number, then
$$f'(x) = nx^{n-1}$$

We will prove the power rule for positive integers n only. Proofs for real numbers n are found in any standard calculus text. To prove the power rule, we must review the *binomial theorem* for expansion of binomials. The following formulas illustrate equivalent expressions for $(a + b)^n$ for positive integers n:

If $n = 2$, then $(a + b)^2 = a^2 + 2ab + b^2$
If $n = 3$, then $(a + b)^3 = a^3 + 3a^2b + 3ab^2 + b^3$
If $n = 4$, then $(a + b)^4 = a^4 + 4a^3b + 6a^2b^2 + 4ab^3 + b^4$
If $n = 5$, then $(a + b)^5 = a^5 + 5a^4b + 10a^3b^2 + 10a^2b^3 + 5ab^4 + b^5$

The **binomial theorem** provides the following general formula for $(a + b)^n$:

$$(a+b)^n = \binom{n}{0}a^n + \binom{n}{1}a^{n-1}b + \binom{n}{2}a^{n-2}b^2 + \ldots + \binom{n}{n-1}ab^{n-1} + \binom{n}{n}b^n$$

Note that the combinations $\binom{n}{0}, \binom{n}{1}, \binom{n}{2}, \ldots, \binom{n}{n-1},$ and $\binom{n}{n}$ are evaluated as follows:

$$\binom{n}{0} = \frac{n!}{0!n!} = 1$$

$$\binom{n}{1} = \frac{n!}{1!(n-1)!} = n$$

$$\binom{n}{2} = \frac{n!}{2!(n-2)!} = \frac{n(n-1)(n-2)!}{2!(n-2)!} = \frac{n(n-1)}{2}$$

$$\vdots$$

$$\binom{n}{n-1} = \frac{n!}{(n-1)!1!} = \frac{n(n-1)!}{(n-1)!1!} = n$$

$$\binom{n}{n} = \frac{n!}{n!0!} = 1$$

Thus, the binomial theorem may be rewritten as

$$(a + b)^n = a^n + na^{n-1}b + \frac{n(n-1)}{2} a^{n-2}b^2 + \ldots + nab^{n-1} + b^n$$

Returning to the power rule, we now use the binomial theorem to compute

$$f(x + h) = (x + h)^n = x^n + nx^{n-1}h + \frac{n(n-1)}{2} x^{n-2}h^2 + \ldots + nxh^{n-1} + h^n$$

For our purpose, it is only necessary to notice that

$$f(x + h) = (x + h)^n$$

$$= x^n + nx^{n-1}h + h^2 \text{ (a sum of products of powers of } x \text{ and } h\text{)}$$

Thus, the derivative of $f(x) = x^n$ is

$$f'(x) = \lim_{h \to 0} \frac{f(x+h) - f(x)}{h}$$

$$= \lim_{h \to 0} \frac{x^n + nx^{n-1}h + h^2 \text{ (a sum of products of powers of } x \text{ and } h\text{)} - x^n}{h}$$

$$= \lim_{h \to 0} \frac{nx^{n-1}h + h^2 \text{ (a sum of products of powers of } x \text{ and } h\text{)}}{h}$$

$$= \lim_{h \to 0} [nx^{n-1} + h \text{ (a sum of products of powers of } x \text{ and } h\text{)}]$$

$$= nx^{n-1}$$

Hence, $f'(x) = nx^{n-1}$.

Constant Multiplier Rule

> If $y = kf(x)$, where k is a constant and $f(x)$ is differentiable at x, then
>
> $$\frac{dy}{dx} = kf'(x)$$

If $y = kf(x)$, where k is a constant, then we have

$$\frac{dy}{dx} = \lim_{h \to 0} \frac{kf(x + h) - kf(x)}{h}$$

$$= \lim_{h \to 0} k \frac{f(x + h) - f(x)}{h}$$

By limit theorem 3 of Section 8–2, this result becomes

$$\frac{dy}{dx} = k \lim_{h \to 0} \frac{f(x + h) - f(x)}{h}$$

$$= kf'(x)$$

Sum (or Difference) Rule

> If $y = f(x) \pm g(x)$, where $f(x)$ and $g(x)$ are differentiable functions at x, then
>
> $$\frac{dy}{dx} = f'(x) \pm g'(x)$$

$$\frac{dy}{dx} = \lim_{h \to 0} \frac{[f(x + h) \pm g(x + h)] - [f(x) \pm g(x)]}{h}$$

$$= \lim_{h \to 0} \frac{[f(x + h) - f(x)] \pm [g(x + h) - g(x)]}{h}$$

$$= \lim_{h \to 0} \left[\frac{f(x + h) - f(x)}{h} \pm \frac{g(x + h) - g(x)}{h} \right]$$

By limit theorem 2 of Section 8–2, this result becomes

$$\frac{dy}{dx} = \lim_{h \to 0} \frac{f(x + h) - f(x)}{h} \pm \lim_{h \to 0} \frac{g(x + h) - g(x)}{h}$$

$$= f'(x) \pm g'(x)$$

Product Rule

> If $y = f(x)s(x)$, where $f(x)$ and $s(x)$ are differentiable functions at x, then
> $$\frac{dy}{dx} = f(x)s'(x) + s(x)f'(x)$$

$$\frac{dy}{dx} = \lim_{h \to 0} \frac{f(x + h)s(x + h) - f(x)s(x)}{h}$$

Adding and subtracting $f(x + h)s(x)$ to the numerator yields

$$\frac{dy}{dx} = \lim_{h \to 0} \frac{f(x + h)s(x + h) - f(x + h)s(x) + f(x + h)s(x) - f(x)s(x)}{h}$$

$$= \lim_{h \to 0} \left[f(x + h)\frac{s(x + h) - s(x)}{h} + s(x)\frac{f(x + h) - f(x)}{h} \right]$$

By limit theorem 2 of Section 8-2, this result becomes

$$\frac{dy}{dx} = \lim_{h \to 0} \left[f(x + h)\frac{s(x + h) - s(x)}{h} \right] + \lim_{h \to 0} \left[s(x)\frac{f(x + h) - f(x)}{h} \right]$$

By limit theorem 5 of Section 8-2, this becomes

$$\frac{dy}{dx} = \lim_{h \to 0} f(x + h) \lim_{h \to 0} \frac{s(x + h) - s(x)}{h} + \lim_{h \to 0} s(x) \lim_{h \to 0} \frac{f(x + h) - f(x)}{h}$$

Since $f(x)$ is differentiable at x, $f(x)$ is continuous at x, and so $\lim_{h \to 0} f(x + h) = f(x)$. Then, the preceding expression becomes

$$\frac{dy}{dx} = f(x)s'(x) + s(x)f'(x)$$

Quotient Rule

> If $y = n(x)/d(x)$, where $n(x)$ and $d(x)$ are differentiable functions at x and $d(x) \neq 0$, then
> $$\frac{dy}{dx} = \frac{d(x)n'(x) - n(x)d'(x)}{[d(x)]^2}$$

DIFFERENTIAL CALCULUS

By definition,

$$\frac{dy}{dx} = \lim_{h \to 0} \frac{\frac{n(x+h)}{d(x+h)} - \frac{n(x)}{d(x)}}{h}$$

If we multiply the first term of the numerator by $d(x)/d(x)$ and the second term by $d(x+h)/d(x+h)$, we have

$$\frac{dy}{dx} = \lim_{h \to 0} \frac{\frac{n(x+h)d(x) - d(x+h)n(x)}{d(x+h)d(x)}}{h}$$

$$= \lim_{h \to 0} \frac{n(x+h)d(x) - d(x+h)n(x)}{hd(x+h)d(x)}$$

Adding and subtracting $n(x)d(x)$ to the numerator yields

$$\frac{dy}{dx} = \lim_{h \to 0} \frac{n(x+h)d(x) - n(x)d(x) + n(x)d(x) - d(x+h)n(x)}{hd(x+h)d(x)}$$

$$= \lim_{h \to 0} \frac{d(x)\frac{n(x+h) - n(x)}{h} - n(x)\frac{d(x+h) - d(x)}{h}}{d(x+h)d(x)}$$

By limit theorems 2, 5, and 6 of Section 8–2, this expression becomes

$$\frac{dy}{dx} = \frac{\lim_{h \to 0} d(x) \lim_{h \to 0} \frac{n(x+h) - n(x)}{h} - \lim_{h \to 0} n(x) \lim_{h \to 0} \frac{d(x+h) - d(x)}{h}}{\lim_{h \to 0} d(x+h) \lim_{h \to 0} d(x)}$$

Since $d(x)$ is differentiable at x, it is continuous at x, and so $\lim_{h \to 0} d(x+h) = d(x)$. Hence, the preceding expression becomes

$$\frac{dy}{dx} = \frac{d(x)n'(x) - n(x)d'(x)}{[d(x)]^2}$$

EXERCISES

1. For each of the following, find dy/dx:
   (A) $y = x^3$
   (B) $y = x^{10}$
   (C) $y = x^{20}$
   (D) $y = x^{1/5}$
   (E) $y = 4x$
   (F) $y = x^{-3}$
   (G) $y = \frac{1}{x^6}$
   (H) $y = \frac{1}{x}$
   (I) $y = x^5$
   (J) $y = (3x)^7$
   (K) $y = \frac{1}{x^3}$
   (L) $y = \frac{1}{\sqrt{x^5}}$
   (M) $y = x^{-1/4}$
   (N) $y = -40$
   (O) $y = 6$

2. For each of the following, find $f'(x)$:
   - (A) $f(x) = x^5$
   - (B) $f(x) = -x^8$
   - (C) $f(x) = 4x^2$
   - (D) $f(x) = \dfrac{-3}{x^2}$
   - (E) $f(x) = \dfrac{4}{\sqrt{x}}$
   - (F) $f(x) = \dfrac{5}{\sqrt{x^3}}$
   - (G) $f(x) = \dfrac{-2}{(\sqrt[3]{x})^4}$
   - (H) $f(x) = \dfrac{1}{6}$

3. Find the derivative of each of the following:
   - (A) $y = 5x^3 - 8x^2 + 16x - 40$
   - (B) $f(x) = -2x^2 + 4x - 3$
   - (C) $f(x) = 5x^7 - 8x^3 + 7x^2 + \dfrac{3}{2}x + 11$
   - (D) $y = x^6 - x^5 + \dfrac{4}{x^2} - \dfrac{6}{5}x + 4$
   - (E) $y = x(x^3 - 4x^2 + 5)$
   - (F) $f(x) = x^3(x^2 - 3x + 7)$
   - (G) $f(x) = (x - 2)(x + 5)$
   - (H) $f(x) = x^2(x + 3)(x - 4)$
   - (I) $y = x^3 - 6x^2 + 5x - 3$
   - (J) $g(x) = -4x^6 + 8x^4 - 3x^2 + 2x - 50$
   - (K) $h(x) = 2x^5 - 8x^3 + 7x^2 - 3x - 91$

4. Find the derivative of each of the following:
   - (A) $y = (x^2 + 4x + 5)(x^3 - 2x^2 + 7)$
   - (B) $y = (4x^5 - 9x^3 + 7x)(x^7 - 8x^6 + 4)$
   - (C) $f(x) = (x^3 - 2x + 6)(x^2 - 3x + 2)$
   - (D) $g(x) = (5x^3 + 6)(x^2 - 8x + 4)$
   - (E) $y = (x^3 - x^2 - 2)(x^2 - x - 1)$

5. Find the derivative of each of the following:
   - (A) $y = \dfrac{x^2 - 4x + 5}{x^3 - 3x + 9}$
   - (B) $f(x) = \dfrac{x^3}{x + 5}$
   - (C) $g(x) = \dfrac{x^3 + 7}{x^4}$
   - (D) $y = \dfrac{x^3 - 10x^2 + 5x - 9}{x^5 - 7x^4 + 4x}$
   - (E) $y = \dfrac{x^3 + 9}{x^2 - 9}$

6. Find the derivative of each of the following:
   - (A) $y = (x^3 - 4x^2 + 5)(x^2 - 8x + 7)$
   - (B) $f(x) = \dfrac{x^2 - 3x}{x^3 + 16}$
   - (C) $y = \dfrac{(x^3 - 4x + 7)(x^2 - 2x)}{x^3 - 6x + 1}$

501 DIFFERENTIAL CALCULUS

(D) $f(x) = \dfrac{x^4 - 8x^2 + 6}{(x^3 + 5)(x^2 - 3x)}$

(E) $y = \dfrac{(x^3 - 6x^2 + 5)(x^2 - 4x)}{(x^5 - 8x^2)(x^3 + 1)}$

7. If $f(x) = \dfrac{-4}{x^3}$, find $f'(x)$ by using

   (A) The power rule.   (B) The quotient rule.

8. If $y = x^3(x^6 - 8x^2)$, find dy/dx
   (A) Without using the product rule.
   (B) By using the product rule.

9. If $f(x) = \dfrac{x - 1}{x + 3}$ find $f'(2)$.

10. If $f(x) = (x^2 + 2)(x^3 - 2x + 5)$, calculate $f'(1)$.

11. If $f(x) = (x + 1)(x + 2)(x - 5)$, find $f'(3)$.

12. Find the equation of the straight line tangent to $f(x) = (x^2 - 9)(x + 5)$ at $x = 1$. Graph $f(x)$ and its tangent line on the same axis system.

13. Find the equation of the straight line tangent to

$$f(x) = \dfrac{x^2}{x - 3}$$

at $x = 4$. Graph $f(x)$ and its tangent line on the same axis system.

14. Find the equation of the straight line tangent to

$$f(x) = \dfrac{(x^2 - 36)(x + 4)}{x^2 - 81} \text{ at } x = 1$$

Graph $f(x)$ and its tangent line on the same axis system.

15. At the same instant a ball is projected into the air, a stopwatch is activated. The function defined by

$$S(x) = -16x^2 + 64x$$

expresses the height above 0 (in feet), $S(x)$, of the ball after x seconds have elapsed.
   (A) At what speed (in feet per second) is the ball traveling after 1 second has elapsed (i.e., at $x = 1$)? Is the ball's height increasing or decreasing at this point?
   (B) At what speed is the ball traveling at $x = 2$? Is the ball's height increasing or decreasing at this point?
   (C) At what speed is the ball traveling at $x = 3$? Is the ball's height increasing or decreasing at this point?
   (D) Sketch $S(x)$ and illustrate the graphical interpretations of parts (A) through (C).

16. A company's annual profit, P, is related to time, x, by the equation

$$P(x) = 0.03x^2 + 5$$

where $P(x)$ is the profit (in millions of dollars) for the xth year the company has been operating.
(A) Find the rate of change of profit at the second year. Is profit increasing or decreasing at this point?
(B) Find the rate of change of profit at $x = 3$. Is profit increasing or decreasing at this point?
(C) If the rate of change of profit remains constant at and beyond the third year, find the equation relating P and x for $x \geq 3$. Calculate the profit for the seventh year.

8–5

CHAIN RULE

Different Variables

If $y = f(x)$, then its derivative is often denoted by dy/dx. If letters other than y and x are the dependent and independent variables, respectively, then the symbol dy/dx must be changed accordingly. Consider the function

$$z = t^3 - 8t^2 + 15t - 7$$

Since the dependent and independent variables are z and t, respectively, the derivative is denoted by dz/dt. Thus,

$$\frac{dz}{dt} = 3t^2 - 16t + 15$$

EXAMPLE 8–21 Find the derivative of the function $w = r^4 - 8r^2 + 5$.
Solution $dw/dr = 4r^3 - 16r$

EXAMPLE 8–22 If $y = u^3$, then find its derivative.
Solution $dy/du = 3u^2$

We are now ready to discuss another useful rule for finding derivatives. It is called the **chain rule** and is stated here without proof.

DIFFERENTIAL CALCULUS

> **Chain Rule**
>
> If $y = f(u)$ and $u = g(x)$, then
>
> $$\frac{dy}{dx} = \frac{dy}{du}\frac{du}{dx}$$

The chain rule states that if y is a function of u and u is a function of x, then the derivative of y with respect to x (dy/dx) equals the derivative of y with respect to u (dy/du) times the derivative of u with respect to x (du/dx).

As an example, consider the function

$$y = (6x^2 - 5)^{10}$$

If we let $u = 6x^2 - 5$, then the function may be expressed as

$$y = f(u) = u^{10}$$

Thus, the derivative, dy/dx is calculated by using the chain rule. Hence, we write

$$\frac{dy}{dx} = \frac{dy}{du}\frac{du}{dx}$$
$$= 10u^9(12x)$$

Replacing u by its equivalent expression $6x^2 - 5$, we have

$$\frac{dy}{dx} = 10(6x^2 - 5)^9(12x)$$

> **EXAMPLE 8–23** If $y = (x^3 - 16)^9$, find dy/dx.
> **Solution** Let $u = x^3 - 16$. Then, the function may be rewritten as $y = u^9$, where $u = x^3 - 16$. By the chain rule, we obtain
>
> $$\frac{dy}{dx} = \frac{dy}{du}\frac{du}{dx}$$
> $$= 9u^8(3x^2)$$
>
> Replacing u by its equivalent expression $x^3 - 16$, we have
>
> $$\frac{dy}{dx} = 9(x^3 - 16)^8(3x^2)$$

EXAMPLE 8–24 If $y = \sqrt{x^5 - 8}$, find dy/dx.
Solution Let $u = x^5 - 8$. Then, the function may be rewritten as $y = \sqrt{u} = u^{1/2}$, where $u = x^5 - 8$. By the chain rule, we obtain

$$\frac{dy}{dx} = \frac{dy}{du}\frac{du}{dx}$$

$$= \frac{1}{2}u^{-1/2}(5x^4)$$

Replacing u by its equivalent expression $x^5 - 8$, we have

$$\frac{dy}{dx} = \frac{1}{2}(x^5 - 8)^{-1/2}(5x^4)$$

$$= \frac{5x^4}{2\sqrt{x^5 - 8}}$$

Once mastered, the chain rule process is quickly performed without going through many of the intermediate steps. Specifically, since all of the preceding functions have been of the form

$$y = u^n$$

where u is a function of x, then the derivative may be written

$$\frac{dy}{dx} = \frac{dy}{du}\frac{du}{dx}$$

$$= nu^{n-1}\frac{du}{dx}$$

Thus, for such functions, we have the **special case chain rule**, stated as follows.

Special Case Chain Rule

If $y = u^n$, where u is a function of x, then

$$\frac{dy}{dx} = nu^{n-1}\frac{du}{dx}$$

Specifically, if

$$y = (x^6 - 8x)^{20}$$

DIFFERENTIAL CALCULUS

where $u = x^6 - 8x$ and $n = 20$, we find

$$\frac{dy}{dx} = nu^{n-1}\frac{du}{dx}$$
$$= 20(x^6 - 8x)^{19}(6x^5 - 8)$$

EXAMPLE 8–25 If $y = (x^5 - 9x^2)^{11}$, find dy/dx.
Solution Since $u = x^5 - 9x^2$ and $n = 11$, we obtain

$$\frac{dy}{dx} = nu^{n-1}\frac{du}{dx}$$
$$= 11(x^5 - 9x^2)^{10}(5x^4 - 18x)$$

EXAMPLE 8–26 If $y = (x^3 - 6)^5(2x + 9)^8$, find dy/dx.
Solution Since we have a product of two functions, we must first use the product rule. Hence, we calculate

$$\frac{dy}{dx} = (\text{first})\begin{pmatrix}\text{derivative}\\\text{of second}\end{pmatrix} + (\text{second})\begin{pmatrix}\text{derivative}\\\text{of first}\end{pmatrix}$$
$$= (x^3 - 6)^5 \cdot 8(2x + 9)^7 (2) + (2x + 9)^8 \cdot 5(x^3 - 6)^4 (3x^2)$$

Note that the chain rule was used to find the "derivative of second" and the "derivative of first." When simplified, the preceding result becomes

$$\frac{dy}{dx} = 16(x^3 - 6)^5 (2x + 9)^7 + 15x^2(2x + 9)^8 (x^3 - 6)^4$$
$$= (x^3 - 6)^4 (2x + 9)^7 (46x^3 + 135x^2 - 96)$$

EXAMPLE 8–27 If

$$y = \left(\frac{x^2 + 3}{x^5 + 4}\right)^{20}$$

find dy/dx.
Solution Since $u = (x^2 + 3)/(x^5 + 4)$ and $n = 20$, then we have

$$\frac{dy}{dx} = nu^{n-1}\frac{du}{dx}$$
$$= 20\left(\frac{x^2 + 3}{x^5 + 4}\right)^{19}\left[\frac{(x^5 + 4)(2x) - (x^2 + 3)(5x^4)}{(x^5 + 4)^2}\right]$$

EXAMPLE 8-27 (continued)

Note that du/dx is computed by the quotient rule. When simplified, this result becomes

$$\frac{dy}{dx} = 20\left(\frac{x^2+3}{x^5+4}\right)^{19}\left[\frac{(-3x^6-15x^4+8x)}{(x^5+4)^2}\right]$$

$$= \frac{20(x^2+3)^{19}(-3x^6-15x^4+8x)}{(x^5+4)^{21}}$$

EXERCISES

1. If $z = t^5 - 4t^3 + 7t - 8$, find dz/dt.
2. If $h = w^3 - 4w^2 + 8w - 9$, find dh/dw.
3. If $y = u^5$, find dy/du.
4. If $u = x^2$, find du/dx.
5. If $y = 3u^6 - 8u^5 + 4u - 8$, find dy/du.
6. If $u = 8x^2 - 3x + 5$, find du/dx.
7. If $f(u) = -7u^3 - 8u^2 + 6$, find $f'(u)$.
8. If $g(t) = 8t^2 - 4t + 5$, find $g'(t)$.
9. Find the derivative of each of the following:
   - (A) $y = (x^3 + 5)^{11}$
   - (B) $y = (x^6 - 4x + 9)^{1/2}$
   - (C) $y = x^3 - 6x$
   - (D) $f(x) = x^3 + x^2 + 9x$
   - (E) $f(x) = (x^4 - 9)^{15}$
   - (F) $y = \dfrac{1}{\sqrt{x^3 - 8x}}$
   - (G) $y = \dfrac{1}{\sqrt[3]{(x^6 - 8x)^2}}$
   - (H) $f(x) = (x^6 - 9x^2)^{-2/5}$
   - (I) $y = (x^3 - 4x^2 + 5)^{20}$
   - (J) $y = \dfrac{1}{(x^4 - 8x^3 + 5)^{10}}$
10. Find the derivative of each of the following:
    - (A) $y = (x^3 - 4x + 9)^{10}(x^4 - 8x^2 + 9)$
    - (B) $y = \dfrac{(x^2 - 3x)^5}{x^4 - 8x + 9}$
    - (C) $f(x) = \dfrac{(x^6 - 8x + 5)^7}{(x^5 - 4x)^{11}}$
    - (D) $y = \dfrac{x^2 - 4x}{(x^5 - 8x + 9)^8}$
    - (E) $f(x) = (x^6 - 4x + 5)(x^3 - 7x)^9$
    - (F) $y = (x^4 - 8x + 9)^3(x^2 - 3x^5)^{10}$

11. Find the equation of the straight line tangent to $f(x) = \dfrac{(x^2 - 25)^3}{(x - 4)^2}$ at $x = 3$. Graph $f(x)$ and its tangent line on the same axis system.

12. Find the equation of the straight line tangent to $f(x) = \dfrac{(x^2 - 4)^3}{(x - 1)^2}$ at $x = 3$. Graph $f(x)$ and its tangent line on the same axis system.

13. The total cost, C, of producing x units of some commodity is given by

$$C = x^3 + 9$$

The number of units produced, x, is a function of the number of days, t, subsequent to the start of a production run. The variable x is related to t by the equation

$$x = 5t + 4$$

(A) Find the formula for the rate of change of cost with respect to time.

(B) Evaluate the result of part (A) at $t = 6$ and interpret the result.

8-6

APPLICATIONS

Marginal Cost and Marginal Revenue

Suppose the total cost of producing x units of some commodity is given by the cost function

$$C(x) = -0.01x^2 + x + 175 \qquad (0 \leq x \leq 50)$$

Then, the total cost of producing 10 units is

$$C(10) = -0.01(10)^2 + 10 + 175$$
$$= \$184$$

Thus, at production level $x = 10$, the total cost is \$184.

If we want to determine the additional cost of producing one more unit at production level $x = 10$, we must calculate the total cost of producing $10 + 1 = 11$ units and subtract from this result the total cost of producing 10 units. Hence, we find

$$C(11) = -0.01(11)^2 + 11 + 175$$
$$= \$184.79$$

And the additional cost of producing one more unit is

$$C(11) - C(10) = 184.79 - 184$$
$$= \$0.79$$

Observing Figure 8-24, note that $C(11) - C(10)$ is the *vertical*

CHAPTER EIGHT

distance between the points (10, 184) and (11, 184.79) of the cost function.

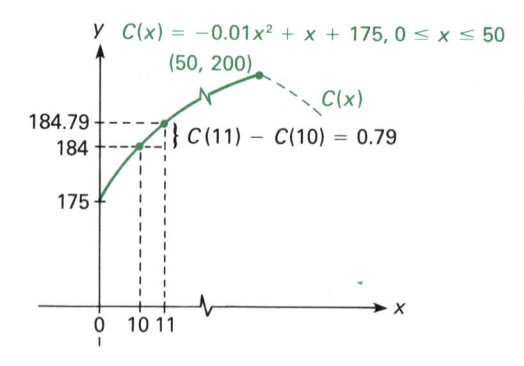

FIGURE 8–24

In general, for a cost function $C(x)$, the additional cost of producing *one more unit* at production level x is

$$C(x + 1) - C(x)$$

In practice, this quantity is usually approximated by the derivative $C'(x)$, which is called the **marginal cost**. Observing Figure 8–25, note that the marginal cost, $C'(x)$, is the slope of the tangent line T. However, since the horizontal distance between x and $x + 1$ is 1, then $C'(x)$ is the vertical distance indicated. Note that $C'(x)$ is an approximation to $C(x + 1) - C(x)$.

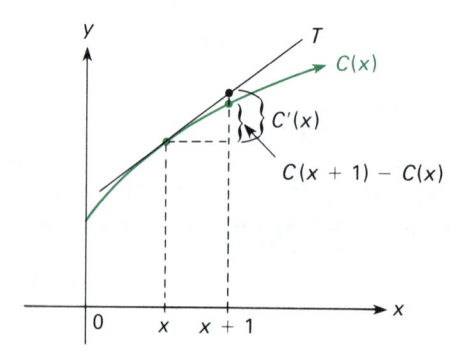

FIGURE 8–25

Thus, returning to the cost function

$$C(x) = -0.01x^2 + x + 175$$

DIFFERENTIAL CALCULUS

the marginal cost at production level $x = 10$ is $C'(10)$ and is determined by

$$C'(x) = -0.02x + 1$$
$$C'(10) = -0.02(10) + 1$$
$$= \$0.80$$

Thus, at production level $x = 10$, one more unit costs approximately $0.80.

Similarly, given a revenue function $R(x)$, its derivative, $R'(x)$, is called the **marginal revenue**. It approximates the quantity $R(x + 1) - R(x)$, which is the additional revenue derived from producing and selling one more unit.

EXAMPLE 8–28 Given the sales revenue function

$$R(x) = -x^2 + 10x \quad (0 \leq x \leq 10)$$

where $R(x)$ is the sales revenue derived from producing and selling x watches, find the marginal revenue at production level $x = 3$.

Solution The calculation is as follows:

$$R'(x) = -2x + 10$$
$$R'(3) = -2(3) + 10$$
$$= \$4$$

Thus, at production level $x = 3$, approximately $4 of additional sales revenue is derived from producing and selling one more watch.

Elasticity of Demand

Frequently, economists and managers wish to measure the responsiveness of consumers to changes in the price of a given commodity. This is accomplished by calculating a ratio, called elasticity of demand, for the commodity. Specifically, **elasticity of demand** is a ratio that compares the proportionate change in the quantity of a product demanded to the proportionate change in price. That is,

$$\frac{\text{Elasticity}}{\text{of demand}} = \frac{\% \text{ change in quantity demanded}}{\% \text{ change in price}}$$

To calculate elasticity of demand, we must be given a demand function that relates the commodity's demand with its unit price. If $D(p)$ is the number of units demanded at price p, then $D(p + h)$ is the number of units demanded at price $p + h$. Thus, when the unit price changes from p to $p + h$, the demand changes from $D(p)$ to $D(p + h)$. The percent change in demand is

$$\frac{D(p+h) - D(p)}{D(p)} \cdot 100$$

and the percent change in price is

$$\frac{(p+h) - p}{p} \cdot 100$$

Since elasticity of demand, E, is defined as

$$E = \frac{\% \text{ change in quantity demanded}}{\% \text{ change in price}}$$

then we write

$$E = \frac{\dfrac{D(p+h) - D(p)}{D(p)} \cdot 100}{\dfrac{(p+h) - p}{p} \cdot 100}$$

When simplified, this expression becomes

$$E = \frac{\dfrac{D(p+h) - D(p)}{h}}{\dfrac{D(p)}{p}}$$

If $h \to 0$, then the formula for elasticity of demand becomes

$$E = \frac{D'(p)}{\dfrac{D(p)}{p}}$$

EXAMPLE 8–29 The demand for E-Z Chew peanuts is given by

$$D(p) = -6p + 48 \qquad (0 \leq p \leq 7)$$

where p is the price of peanuts per pound.

(A) Find the equation for elasticity of demand, E.
(B) Find elasticity of demand at $p = 3$.
(C) Interpret the result of part (B).

Solutions

(A) Since $D'(p) = -6$, then

$$E = \frac{D'(p)}{\frac{D(p)}{p}}$$

$$= \frac{-6}{\frac{-6p + 48}{p}}$$

$$= \frac{-6p}{-6p + 48}$$

$$= \frac{p}{p - 8}$$

(B) If $p = 3$, then

$$E = \frac{3}{3 - 8} = \frac{3}{-5} = -0.6$$

(C) Since

$$E = \frac{\% \text{ change in quantity demanded}}{\% \text{ change in price}}$$

then $E = -0.6/1$ implies that a 1% increase in price is associated with a 0.6% decrease in demand.

Since

$$E = \frac{\% \text{ change in quantity demanded}}{\% \text{ change in price}}$$

we may identify three cases:

1. If $|E| = 1$, then the percent change in demand equals the percent change in price.
2. If $|E| > 1$, then the percent change in demand is greater than the percent change in price. In this case, demand is said to be **elastic**.
3. If $|E| < 1$, then the percent change in demand is less than the percent change in price. In this case, demand is said to be **inelastic**.

Newton's Method for Approximating the x-Intercepts of a Function

To find the x-intercepts of the graph of a function $y = f(x)$, we set $y = 0$ and solve the resulting equation for x. This process is quite simple for linear and quadratic equations. If the function is linear, i.e., of the form $y = mx + b$, we obtain $0 = mx + b$ with $x = -b/m$. If the function is quadratic, i.e., of the form $y = ax^2 + bx + c$, we can either try to factor $ax^2 + bx + c$ or use the quadratic formula

$$x = \frac{-b \pm \sqrt{b^2 - 4ac}}{2a}$$

If the function is of degree higher than 2, we attempt factorization. However, some polynomial functions are not factorable, and some are not easily factored. Thus, we need a method for *approximating* the x-intercepts of a function. Such a method exists—it is called **Newton's method.**

To understand Newton's method, consider the function $y = f(x)$ of Figure 8–26. Assume that we do not know the value of the x-intercept, r. We choose x_0 as an initial approximation of r. Then, we find the equation of the tangent line to the function at x_0. Since the slope of the tangent line is $f'(x_0)$ and the point of tangency is $(x_0, f(x_0))$, then using the point–slope form

$$y - y_1 = m(x - x_1)$$

the equation of the tangent line is shown to be

$$y - f(x_0) = f'(x_0)(x - x_0)$$

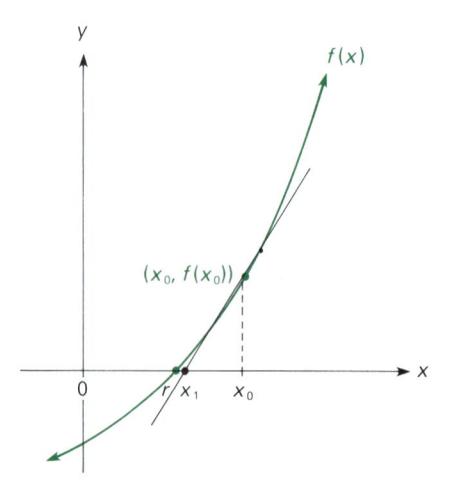

FIGURE 8–26

Observing Figure 8–26, note that the x-intercept of the tangent line is a better approximation of r. To find the x-intercept of the tangent line, we set $y = 0$ and solve for x. Hence, we write

$$0 - f(x_0) = f'(x_0)(x - x_0)$$

and obtain

$$x = x_0 - \frac{f(x_0)}{f'(x_0)}$$

If we denote the x-intercept of the tangent line by x_1, this equation becomes

$$x_1 = x_0 - \frac{f(x_0)}{f'(x_0)}$$

If we repeat this process using x_1 as an approximation of r, then a better approximation is

$$x_2 = x_1 - \frac{f(x_1)}{f'(x_1)}$$

Continuing this iterative process, if we have obtained x_n as an approximation of r, then a better approximation is

$$x_{n+1} = x_n - \frac{f(x_n)}{f'(x_n)}$$

It is essential that at each iteration, $f'(x_n) \neq 0$.

EXAMPLE 8–30 Use Newton's method to approximate a positive x-intercept of the function

$$f(x) = x^3 + x^2 - 6x + 1$$

Solution We first test various x-values to determine a sign change in $f(x)$. Since $f(1) = -3$ and $f(2) = +1$, then $f(x)$ crosses the x-axis at some value between $x = 1$ and $x = 2$ (see Figure 8–27). We may choose either $x_0 = 1$ or $x_0 = 2$ as our initial approximation. Choosing $x_0 = 2$, we must compute $f'(2)$. Hence, we calculate

$$f'(x) = 3x^2 + 2x - 6$$
$$f'(2) = 3(2)^2 + 2(2) - 6$$
$$= 10$$

EXAMPLE 8-30 (continued)

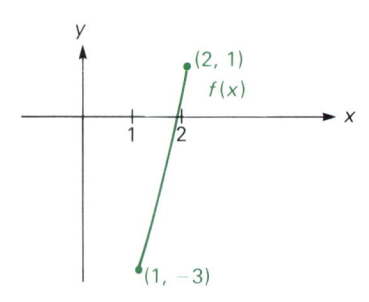

FIGURE 8-27

The next approximation, x_1, is calculated to be

$$x_1 = x_0 - \frac{f(x_0)}{f'(x_0)}$$

$$= 2 - \frac{1}{10}$$

$$= 1.9$$

We then calculate $f(x_1)$ and $f'(x_1)$. Thus, we obtain

$$f(x_1) = f(1.9) = (1.9)^3 + (1.9)^2 - 6(1.9) + 1 = 0.069$$
$$f'(x_1) = f'(1.9) = 3(1.9)^2 + 2(1.9) - 6 = 8.63$$

The next approximation, x_2, is calculated to be

$$x_2 = x_1 - \frac{f(x_1)}{f'(x_1)}$$

$$= 1.9 - \frac{0.069}{8.63}$$

$$\approx 1.8920$$

We then calculate $f(x_2)$ and $f'(x_2)$. Thus, we obtain

$$f(x_2) = f(1.8920)$$
$$= (1.8920)^3 + (1.8920)^2 - 6(1.8920) + 1$$
$$\approx 0.0004$$
$$f'(x_2) = f'(1.8920)$$
$$= 3(1.8920)^2 + 2(1.8920) - 6$$
$$\approx 8.5230$$

The next approximation, x_3, is

$$x_3 = x_2 - \frac{f(x_2)}{f'(x_2)}$$

$$= 1.8920 - \frac{0.0004}{8.5230}$$

$$\approx 1.8919$$

Since $f(1.8919) \approx -0.0004$ and $f(1.8920) \approx +0.0004$, we know that the x-intercept is between 1.8919 and 1.8920. Thus, the x-intercept is 1.892 to three decimal places. A graphical illustration appears in Figure 8–28.

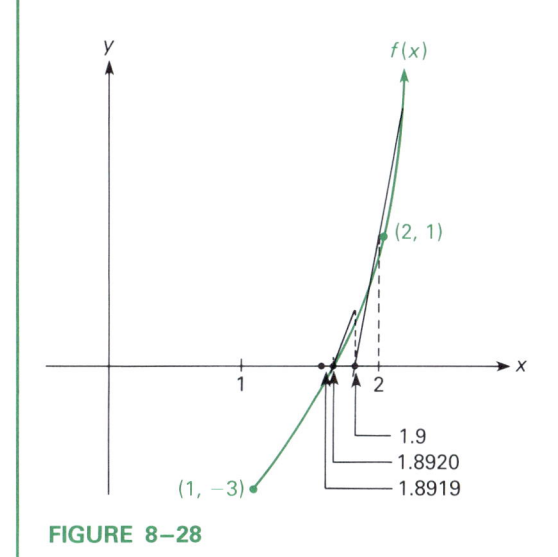

FIGURE 8–28

With Newton's method, the x_n's usually get closer and closer to the x-intercept very rapidly. However, in some instances, this may not be the case. As an example, consider the graph of Figure 8–29 on page 516. Note that if x_0 is the initial approximation of r, then the x_n's will go back and forth between x_0 and x_1. No matter how many iterations we perform using the formula

$$x_{n+1} = x_n - \frac{f(x_n)}{f'(x_n)}$$

we will obtain either x_0 or x_1 without getting any closer to r.

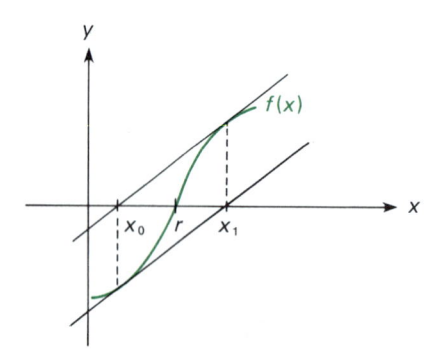

FIGURE 8–29

EXERCISES

1. The cost of producing x units of some commodity is given by the cost function defined by

$$C(x) = 5x^2 + 100 \quad (x \geq 0)$$

   (A) Find the equation for marginal cost.
   (B) Find the marginal cost at $x = 3$.
   (C) Interpret the answer to part (B).
   (D) Illustrate the graphical interpretation of the answer to part (B).

2. The revenue derived from selling x units of some item is given by the revenue function defined by

$$R(x) = -2x^2 + 60x \quad (0 \leq x \leq 30)$$

   (A) Find the equation for marginal revenue.
   (B) Find the marginal revenue at $x = 10$.
   (C) Interpret the answer to part (B).
   (D) Illustrate the graphical interpretation of the answer to part (B).

3. The demand for tricycles, x, is related to the price per tricycle, p, by the equation

$$p = -2x + 40 \quad (0 \leq x \leq 20)$$

   (A) Find the revenue function, $R(x)$.
   (B) Find the equation for marginal revenue.
   (C) Find the marginal revenue at $x = 5$. Is revenue increasing or decreasing at this point?
   (D) Find the marginal revenue at $x = 12$. Is revenue increasing or decreasing at this point?

4. The Ding Dong Company manufactures ornamental bells. The function

$$C(x) = x^2 - 100x + 2900 \quad (x \geq 0)$$

relates total daily production cost, C, with daily production, x, of bells.
(A) Find the equation for marginal cost.
(B) Find the marginal cost at $x = 40$. Is cost increasing or decreasing at this point?
(C) Find the marginal cost at $x = 50$. Is cost increasing or decreasing at this point?
(D) Find the marginal cost at $x = 70$. Is cost increasing or decreasing at this point?
(E) Graph $C(x)$ and illustrate the graphical interpretation of the answers to parts (B) through (D).

5. Consider the profit function defined by

$$P(x) = 4x - \frac{x^3}{1{,}000{,}000} \qquad (0 \le x \le 2000)$$

Marginal profit is defined in the same manner as marginal revenue and marginal cost.
(A) Find the equation for marginal profit.
(B) Find the marginal profit at $x = 100$. Is profit increasing or decreasing at this point?
(C) Find the marginal profit at $x = 1900$. Is profit increasing or decreasing at this point?
(D) Using the graph of $P(x)$, illustrate the graphical interpretation of the answers to parts (B) and (C).

6. Consider the revenue function defined by

$$R(x) = -\frac{1}{100}x^3 + 16x \qquad (0 \le x \le 40)$$

(A) Find the equation for marginal revenue.
(B) Find the marginal revenue at $x = 10$. Is revenue increasing or decreasing at this point?
(C) Find the marginal revenue at $x = 30$. Is revenue increasing or decreasing at this point?
(C) Using the graph of $R(x)$, illustrate the geometric interpretation of the answers to parts (B) and (C).

7. The demand for bracelets is given by

$$D(p) = -2p + 130 \qquad (0 \le p \le 65)$$

where p is the price per bracelet.
(A) Find the equation for elasticity of demand, E.
(B) Find elasticity of demand at $p = 50$.
(C) Interpret the result of part (B).

8. The demand for some commodity is given by

$$D(p) = -3p + 15 \qquad (0 \le p \le 5)$$

where p is the unit price.

(A) Find the equation for elasticity of demand, E.
(B) Find elasticity of demand at $p = 2$.
(C) Interpret the result of part (B).

9. Elasticity of demand for some commodity is given to be -1.5.
   (A) Interpret this result.
   (B) Is demand elastic or inelastic?

10. Elasticity of demand for some commodity is given to be -0.70.
    (A) Interpret this result.
    (B) Is demand elastic or inelastic?

11. Verify that the function defined by $f(x) = x^3 - 7x - 5$ has an x-intercept between $x = 2$ and $x = 3$. Use Newton's method to approximate this x-intercept.

12. Use Newton's method to approximate the negative x-intercept of the function defined by $f(x) = x^3 - 4x^2 - x + 5$.

13. Use Newton's method to approximate $\sqrt{5}$ by approximating the positive x-intercept of $f(x) = x^2 - 5$.

14. Use Newton's method to approximate $\sqrt[3]{2}$ by approximating the x-intercept of $f(x) = x^3 - 2$.

8–7

DERIVATIVES OF EXPONENTIAL AND LOGARITHMIC FUNCTIONS

In Chapter 3, we discussed exponential and logarithmic functions. In this section, we will cover the rules for finding their derivatives. Our first rule involves the exponential function $y = e^x$. Its proof appears in the appendix at the end of this section.

Rule 1

If $y = e^x$, then

$$\frac{dy}{dx} = e^x$$

This rule states that the derivative of e^x is itself e^x.

EXAMPLE 8–31 If $y = 8e^x$, find dy/dx.
Solution Using rule 1 and the constant multiplier rule, we have

$$\frac{dy}{dx} = 8e^x$$

> **EXAMPLE 8–32** If $y = e^u$, find dy/du.
> **Solution** Using rule 1, we write
> $$\frac{dy}{du} = e^u$$

We now consider exponential functions such as

$$y = e^{x^2-3x}$$

This function is of the form

$$y = e^u$$

where u is a function of x. The derivative of such a function is found by applying the chain rule to exponential functions. Recall that, according to the chain rule, if y is a function of u and u is a function of x, then

$$\frac{dy}{dx} = \frac{dy}{du}\frac{du}{dx}$$

Since $y = e^u$, then $dy/du = e^u$. Hence, we have

$$\frac{dy}{dx} = e^u \frac{du}{dx}$$

Thus, we have the following rule.

> **Rule 2**
>
> If $y = e^u$, where u is a function of x, then
> $$\frac{dy}{dx} = e^u \frac{du}{dx}$$

Returning to the function

$$y = e^{x^2-3x}$$

with $u = x^2 - 3x$, its derivative is

$$\frac{dy}{dx} = e^u \frac{du}{dx}$$
$$= e^{x^2-3x}(2x - 3)$$

EXAMPLE 8–33 If $y = e^{x^5-8x}$, find dy/dx.
Solution Since $u = x^5 - 8x$, then $du/dx = 5x^4 - 8$. Hence, we obtain

$$\frac{dy}{dx} = e^u \frac{du}{dx}$$
$$= e^{x^5-8x}(5x^4 - 8)$$

EXAMPLE 8–34 If $y = 7e^{5x}$, find dy/dx.
Solution Since $u = 5x$, then $du/dx = 5$. Using rule 2 and the constant multiplier rule, we have

$$\frac{dy}{dx} = 7e^{5x}(5)$$
$$= 35e^{5x}$$

EXAMPLE 8–35 If $f(t) = 800e^{-0.5t}$, find
(A) $f'(t)$.
(B) $f'(6)$.
Solutions
(A) $f'(t) = 800e^{-0.5t}(-0.5)$
$ = -400e^{-0.5t}$
(B) $f'(6) = -400e^{-0.5(6)}$
$ = -400e^{-3}$
$ = -400(0.049787)$
$ \approx -19.91$

EXAMPLE 8–36 If $1000 is invested for x years at 8% compounded continuously, then the total amount is given by $S(x) = 1000e^{0.08x}$
(A) Find $S'(x)$.
(B) Find $S'(2)$ and interpret.
(C) Find $S'(5)$ and interpret.
(D) If the total amount were to increase by only a constant amount for each year beyond the fifth, find the total amount at the end of the tenth year. Compare this result with what the total amount should be under normal conditions.

Solutions

(A) $S'(x) = 1000e^{0.08x}(0.08)$
$\quad\quad\quad = 80e^{0.08x}$

(B) $S'(2) = 80e^{0.08(2)}$
$\quad\quad\quad = 80e^{0.16}$
$\quad\quad\quad = 80(1.173511)$
$\quad\quad\quad \approx \$93.88$

Thus, at the end of the second year, the total amount is increasing by $93.88 per year.

(C) $S'(5) = 80e^{0.08(5)}$
$\quad\quad\quad = 80e^{0.40}$
$\quad\quad\quad = 80(1.491825)$
$\quad\quad\quad \approx \$119.35$

Thus, at the end of the fifth year, the total amount is increasing by $119.35 per year.

(D) If the total amount increases by a constant amount for each year beyond the fifth, then its graph is represented by the tangent line to $S(x)$ at $x = 5$ (see Figure 8–30). Since the slope of the tangent line is $S'(5) = 119.35$ and its point of tangency is $(5, S(5))$, or $(5, 1491.83)$, then according to the point–slope form

$$y - y_1 = m(x - x_1)$$

its equation is

$$y - 1491.83 = 119.35(x - 5)$$
$$y = 119.35x + 895.08$$

Thus, at the end of the tenth year ($x = 10$), the total amount is

$$y = 119.35(10) + 895.08$$
$$= \$2088.58$$

Note that under normal conditions of 8% compounded continuously, the total amount at the end of the tenth year is

$$S(10) = 1000e^{0.08(10)}$$
$$= 1000e^{0.80}$$
$$= 1000(2.225541)$$
$$\approx \$2225.54$$

EXAMPLE 8–36 (continued)

These amounts are graphically illustrated in Figure 8–30.

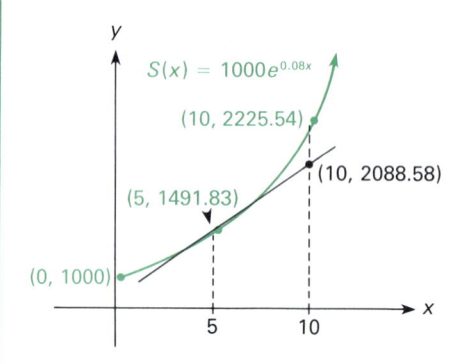

FIGURE 8–30

Up to this point, we have calculated the derivatives of exponential functions having base e. The next rule tells us how to calculate derivatives of exponential functions of the form $y = b^u$, where u is a function of x, $b > 0$, and $b \neq e$.

Rule 3

If $y = b^u$, where u is a function of x and $b > 0$, then

$$\frac{dy}{dx} = b^u \frac{du}{dx} \ln b$$

Note that if $b = e$, then rule 3 becomes rule 2. The proof of rule 3 appears in the appendix at the end of this section.

EXAMPLE 8–37 If $y = 5^{x^7 - 4x}$, find dy/dx.
Solution Note that $b = 5$, $u = x^7 - 4x$, and $du/dx = 7x^6 - 4$. Hence, we get

$$\frac{dy}{dx} = b^u \frac{du}{dx} \ln b$$
$$= 5^{x^7 - 4x}(7x^6 - 4)\ln 5$$
$$= 5^{x^7 - 4x}(7x^6 - 4)(1.609438)$$

We now consider derivatives of logarithmic functions.

> **Rule 4**
>
> If $y = \ln x$, then
> $$\frac{dy}{dx} = \frac{1}{x}$$

The proof of rule 4 appears in the appendix at the end of this section.

EXAMPLE 8–38 If $y = 5 \ln x$, find dy/dx.
Solution Using rule 4 and the constant multiplier rule, we have
$$\frac{dy}{dx} = 5 \cdot \frac{1}{x} = \frac{5}{x}$$

EXAMPLE 8–39 If $y = \ln u$, find dy/du.
Solution Using rule 4, we have
$$\frac{dy}{du} = \frac{1}{u}$$

By applying the chain rule, we can find the derivatives of functions of the form
$$y = \ln u$$
where u is a function of x. According to the chain rule, if y is a function of u and u is a function of x, then
$$\frac{dy}{dx} = \frac{dy}{du}\frac{du}{dx}$$
If $y = \ln u$, then $dy/du = 1/u$. Hence,
$$\frac{dy}{dx} = \frac{1}{u}\frac{du}{dx}$$
We now have the following rule.

Rule 5

If $y = \ln u$, where u is a function of x, then

$$\frac{dy}{dx} = \frac{1}{u}\frac{du}{dx}$$

EXAMPLE 8–40 If $y = \ln(x^3 - 5x^2 + 6)$, find dy/dx.
Solution Let $u = x^3 - 5x^2 + 6$. Then, $du/dx = 3x^2 - 10x$. Using rule 5, we have

$$\frac{dy}{dx} = \frac{1}{u}\frac{du}{dx}$$

$$= \frac{1}{x^3 - 5x^2 + 6}(3x^2 - 10x)$$

$$= \frac{3x^2 - 10x}{x^3 - 5x^2 + 6}$$

EXAMPLE 8–41 If $f(x) = e^{x^5-7}\ln(x^8 - 3x)$, find $f'(x)$.
Solution Using the product rule and rules 2 and 5, we have

$$f'(x) = e^{x^5-7} \cdot \frac{8x^7 - 3}{x^8 - 3x} + \ln(x^8 - 3x)e^{x^5-7} \cdot 5x^4$$

$$= \left[\frac{8x^7 - 3}{x^8 - 3x} + 5x^4\ln(x^8 - 3x)\right]e^{x^5-7}$$

Rules 4 and 5 for finding the derivatives of logarithmic functions have applied only to those functions involving natural logarithms. The following rule applies to logarithmic functions of the form $y = \log_b u$, where u is a function of x, $b > 0$, and $b \neq 1$.

Rule 6

If $y = \log_b u$, where u is a function of x, $b > 0$, and $b \neq 1$, then

$$\frac{dy}{dx} = \frac{1}{u \ln b}\frac{du}{dx}$$

DIFFERENTIAL CALCULUS

The proof of this rule appears in the appendix at the end of this section. Note that if $b = e$, then rule 6 becomes rule 5.

> **EXAMPLE 8–42** If $y = \log_{10}(x^6 - 9x^3)$, find dy/dx.
> **Solution** Since $b = 10$ and $u = x^6 - 9x^3$, then, using rule 6, we have
> $$\frac{dy}{du} = \frac{1}{u \ln b} \frac{du}{dx}$$
> $$= \frac{1}{(x^6 - 9x^3) \ln 10} (6x^5 - 27x^2)$$
> $$= \frac{6x^5 - 27x^2}{2.302585(x^6 - 9x^3)}$$

APPENDIX — PROOFS OF DERIVATIVE RULES

Rule 1

> If $y = e^x$, then
> $$\frac{dy}{dx} = e^x$$

If $y = f(x) = e^x$, then

$$\frac{dy}{dx} = f'(x) = \lim_{h \to 0} \frac{f(x + h) - f(x)}{h}$$
$$= \lim_{h \to 0} \frac{e^{x+h} - e^x}{h}$$
$$= \lim_{h \to 0} \frac{e^x e^h - e^x}{h}$$
$$= \lim_{h \to 0} e^x \cdot \frac{e^h - 1}{h}$$
$$= e^x \lim_{h \to 0} \frac{e^h - 1}{h}$$

It can be proven that

$$\lim_{h \to 0} \frac{e^h - 1}{h} = 1$$

We omit the proof here since it is long and difficult. Thus, the resulting expression for *dy/dx* becomes

$$\frac{dy}{dx} = e^x \cdot 1 = e^x$$

Rule 3

> If $y = b^u$, where *u* is a function of *x* and $b > 0$, then
>
> $$\frac{dy}{dx} = b^u \frac{du}{dx} \ln b$$

We make use of the fact stated in Section 3–3 that any positive number *b* can be expressed as $e^{\ln b}$. Hence,

$$y = b^u = (e^{\ln b})^u$$
$$= e^{u \ln b}$$

Using rule 2 of this section, we find

$$\frac{dy}{dx} = e^{u \ln b} \frac{du}{dx} \ln b$$

Replacing $e^{u \ln b}$ with b^u, this equation becomes

$$\frac{dy}{dx} = b^u \frac{du}{dx} \ln b$$

Rule 4

> If $y = \ln x$, then
>
> $$\frac{dy}{dx} = \frac{1}{x}$$

DIFFERENTIAL CALCULUS

When written in exponential form, the statement $y = \ln x$ is equivalent to

$$e^y = x$$

Taking the derivative of both sides with respect to x, we have

$$e^y \frac{dy}{dx} = 1$$

Solving for dy/dx, we obtain

$$\frac{dy}{dx} = \frac{1}{e^y}$$

Since $e^y = x$, this expression becomes

$$\frac{dy}{dx} = \frac{1}{x}$$

Rule 6

> If $y = \log_b u$, where u is a function of x, $b > 0$, and $b \neq 1$, then
>
> $$\frac{dy}{dx} = \frac{1}{u \ln b} \frac{du}{dx}$$

When written in exponential form, the statement $y = \log_b u$ becomes

$$b^y = u$$

Taking the derivative of both sides with respect to x, we obtain

$$b^y \frac{dy}{dx} \ln b = \frac{du}{dx}$$

Solving for dy/dx yields

$$\frac{dy}{dx} = \frac{1}{b^y \ln b} \frac{du}{dx}$$

Since $b^y = u$, this equation becomes

$$\frac{dy}{dx} = \frac{1}{u \ln b} \frac{du}{dx}$$

EXERCISES

1. Find the derivative of each of the following:
   - (A) $y = e^{4x}$
   - (B) $y = e^{-x}$
   - (C) $y = 4e^{-2x}$
   - (D) $y = -e^{-x}$
   - (E) $f(x) = -2e^{-0.1x}$
   - (F) $f(x) = e^x$

2. Find the derivative of each of the following:
   - (A) $y = e^{x^5 - 7x}$
   - (B) $f(x) = 6e^{x^3 - 2x}$
   - (C) $y = e^{2x-5}$
   - (D) $f(x) = -4e^{3x^2 + 4x}$

3. Find the derivative of each of the following:
   - (A) $y = 4^{x^2 - 3x}$
   - (B) $f(x) = 2^{3x^2 - 5x}$
   - (C) $y = 4^{-0.02x}$
   - (D) $f(x) = -2(3^{-x})$

4. Find the derivative of each of the following:
   - (A) $y = (x^5 - 4x^2)e^{x^4 - 7x}$
   - (B) $f(x) = (x^4 + 8x)e^{x^3 + 5x}$
   - (C) $y = \dfrac{e^x + e^{-x}}{x}$
   - (D) $y = \dfrac{e^x - e^{-x}}{x}$
   - (E) $f(x) = \dfrac{e^{x^3 - 4x}}{x^2 - 3x}$

5. Find the derivative of each of the following:
   - (A) $y = \ln 3x$
   - (B) $y = 4 \ln 6x$
   - (C) $f(x) = \ln(x^2 - 4x)$
   - (D) $f(x) = -2 \ln(x^2 + 1)$
   - (E) $y = \ln(x^2 - 4x + 1)$
   - (F) $y = \ln(-x^6 + 3x)$

6. Find the derivative of each of the following:
   - (A) $y = (x^2 + 4x)\ln(x^3 - 3x + 1)$
   - (B) $f(x) = (x^6 + 8x + 1)\ln(x^8 - 2x)$
   - (C) $y = \dfrac{x^5}{\ln x^3}$
   - (D) $f(x) = \dfrac{\ln(x^2 - 2x)}{x^4 - 3x + 1}$
   - (E) $y = \dfrac{\ln x^7}{x^9}$
   - (F) $y = [\ln(x^6 - 8x^3 - 4)]^{10}$
   - (G) $y = \ln \sqrt{x^3 + 7}$
   - (H) $y = \sqrt{\ln(x^3 + 7)}$

7. Find the derivative of each of the following:
   - (A) $y = \log_{10} 5x$
   - (B) $y = \log_6(3x - 4)$
   - (C) $y = \log_{10}(x^3 + 6x)$
   - (D) $f(x) = \log_3(2x^9 - 1)$

8. Find the derivative of each of the following:
   - (A) $y = \dfrac{e^x}{\ln x}$
   - (B) $f(x) = \dfrac{1000}{1 + 5e^{0.4x}}$
   - (C) $f(x) = \dfrac{800}{1 - 20e^{-0.3x}}$
   - (D) $y = e^x \ln x$

DIFFERENTIAL CALCULUS

(E) $y = \ln(x^3 - 2x)e^{x^4 - 8}$

(F) $y = \dfrac{x^3 e^x}{\ln x}$

(G) $y = \ln e^{3x}$

(H) $y = \ln e^{4x^2}$

9. If $10,000 is invested for x years at 10% compounded continuously, then the total amount is given by

$$S(x) = 10{,}000 e^{0.10x}$$

(A) Find $S'(x)$.
(B) Find $S'(3)$ and interpret.
(C) Find $S'(4)$ and interpret.
(D) Find the rate of change of the total amount at $x = 5$. Is the total amount increasing or decreasing at this point?
(E) If the total amount were to increase by only a constant amount for each year beyond the fifth, find the total amount at the end of the ninth year. Compare this with what the total amount should be under normal conditions.
(F) Graphically illustrate part (E).

10. Find the equation of the straight line tangent to $f(x) = e^x$ at $x = 0$. Graph $f(x)$ and its tangent line.

11. Find the equation of the straight line tangent to $f(x) = e^{-x}$ at $x = 1$. Graph $f(x)$ and its tangent line.

12. The number of employees, $P(x)$, of a certain company is related to annual sales, x (in millions of dollars), by the exponential function defined by

$$P(x) = 100{,}000 e^{0.06x}$$

(A) Find the rate of change of the number of employees at $x = 5$. Is the number of employees increasing or decreasing at this point?
(B) Find the rate of change of the number of employees at $x = 10$. Is the number of employees increasing or decreasing at this point?
(C) If the number of employees increases by a constant amount for each year beyond the tenth, find the number of employees at $x = 13$.
(D) Graphically illustrate part (C).

13. If a product is not advertised, then its monthly sales decay is in accordance with the equation

$$y(t) = 2000 e^{-0.60t}$$

where $y(t)$ represents monthly sales at time t (in months).
(A) Find $y'(t)$.
(B) Find the rate of change of sales at $t = 1/2$.
(C) How fast are sales decaying at $t = 2$?

14. The function defined by

$$N(x) = 50 - 50 e^{-0.3x}$$

is a learning curve where $N(x)$ represents the number of items produced by an assembly-line worker during the xth day after the training period. Is daily production increasing or decreasing at $t = 5$, and at what rate?

15. The temperature of a heated cup of coffee, y, is related to the time elapsed, t (in minutes), by

$$y = 150e^{-0.02t} + 65$$

Is the temperature increasing or decreasing at $t = 5$, and at what rate?

16. Find the equation of the straight line tangent to $y = \ln x$ at $x = 1$. Sketch $y = \ln x$ and its tangent line.

DIFFERENTIAL CALCULUS: OPTIMIZATION

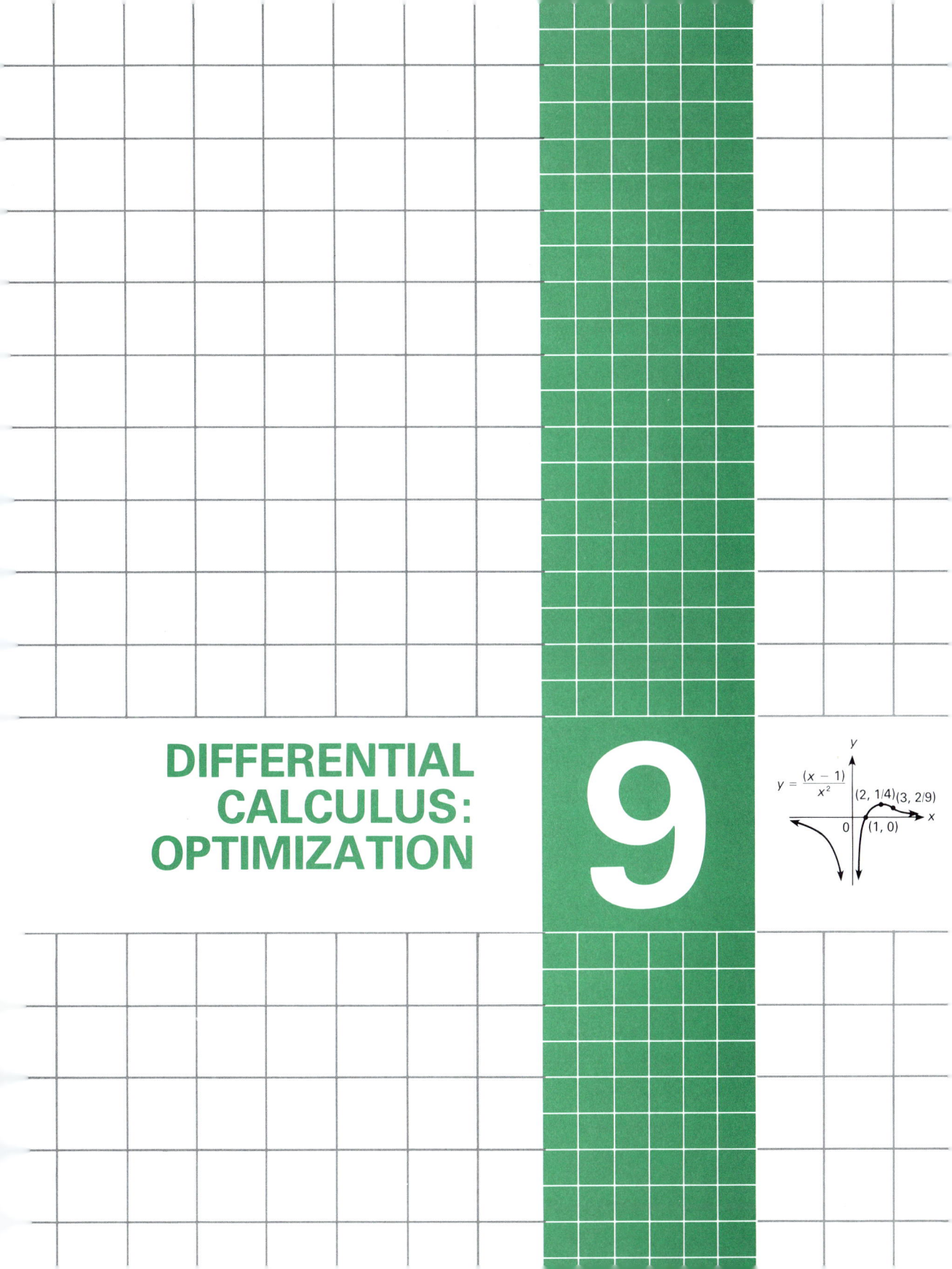

9

$y = \dfrac{(x-1)}{x^2}$

(2, 1/4) (3, 2/9)

(1, 0)

9-1

CRITICAL VALUES AND THE FIRST DERIVATIVE

Consider the graph of Figure 9-1. In this chapter, we will be concerned with identifying the x and y coordinates of points such as A and B. We first begin by identifying points of relative maxima and relative minima. Point $A(x_1, f(x_1))$ is an example of a **relative maximum** because it is *higher* than any of its neighboring points on the graph of $f(x)$. Mathematically, this is expressed by saying that $f(x_1) \geq f(x)$ for all values of x near x_1. Point $B(x_2, f(x_2))$ of Figure 9-1 is an example of a **relative minimum** because it is *lower* than any of its neighboring points on the graph of $f(x)$. Mathematically, this is expressed by saying that $f(x_2) \leq f(x)$ for all values of x near x_2. More precisely, a point $(x_1, f(x_1))$ is called a relative maximum of the function f if $f(x) \leq f(x_1)$ for all x-values in some open interval containing x_1 and in the domain of f (see Figure 9-2).* A point $(x_2, f(x_2))$ is called a relative minimum of the function f if $f(x) \geq f(x_2)$ for all x-values in some open interval containing x_2 and in the domain of f (see Figure 9-2).

FIGURE 9-1

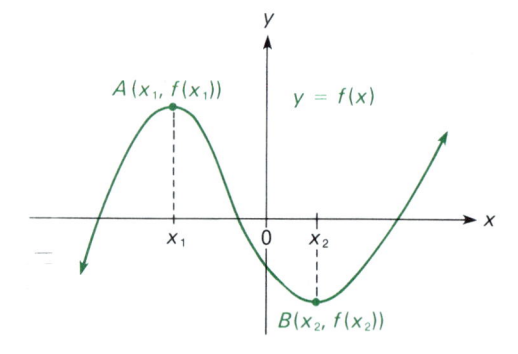

FIGURE 9-2

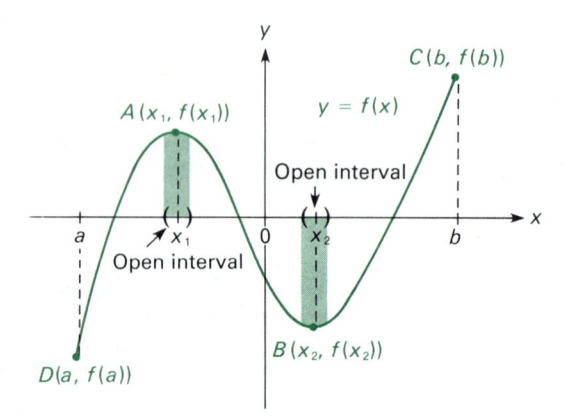

---

*An open interval does not contain its endpoints.

Points of relative maxima and relative minima should be distinguished from points of absolute maxima and absolute minima. Referring to Figure 9–2, point $A(x_1, f(x_1))$ is a relative maximum because it is higher than any of its neighboring points on the graph of $f(x)$. However, it is not the highest point on the graph of $f(x)$. Observe that point $C(b, f(b))$ is higher than point A. If we choose the domain of $f(x)$ to be the closed interval $a \leq x \leq b$, then point C is the *highest* point on the graph of $f(x)$.* Hence, it is called the **absolute maximum**. Similarly, observe that point $B(x_2, f(x_2))$ is a relative minimum because it is lower than any of its neighboring points on $f(x)$. However, it is not the lowest point on the function. That distinction is attributed to point $D(a, f(a))$. Thus, point $D(a, f(a))$ is called the **absolute minimum**. Note that if the domain of $f(x)$ is the set of all real numbers, then there is neither an absolute maximum nor an absolute minimum since the function $f(x)$ extends downward indefinitely to the left and upward indefinitely to the right. Therefore, for a continuous function defined on a closed interval, the absolute maximum occurs either at an endpoint of the interval or at a relative maximum. The absolute minimum occurs either at an endpoint of the interval or at a relative minimum.

In this section, we will be concerned mainly with determining relative maxima and relative minima of functions. In this regard, we consider the graph of the function illustrated in Figure 9–3. Observe that certain segments of the function are identified as either increasing or decreasing. Specifically, the function $f(x)$ is *increasing* as we move along its path from point D to point A and from point B to point C. Thus, the function is said to be increasing on each of the intervals $a < x < x_1$ and $x_2 < x < b$. Note that, within these intervals, any tangent line to the function has a *positive slope.* Since the slope of a tangent line is the derivative of the function at the point of tangency, we can conclude the following.

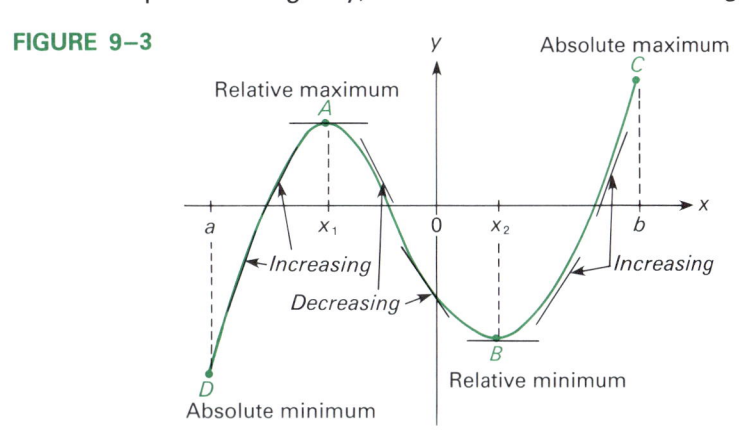

FIGURE 9–3

---

*A closed interval contains its endpoints.

> Given that $f(x)$ is differentiable on the interval $a < x < b$, then $f(x)$ is increasing on this interval, if $f'(x) > 0$ for all values of x within $a < x < b$.

Again referring to Figure 9–3, observe that the function is *decreasing* as we move along its path from point A to point B. The function is said to be decreasing on the interval $x_1 < x < x_2$. Any tangent line to the function within this interval has a *negative slope*. Thus, we conclude as follows.

> Given that $f(x)$ is differentiable on the interval $a < x < b$, then $f(x)$ is decreasing on this interval, if $f'(x) < 0$ for all values of x within $a < x < b$.

Once again observing Figure 9–3, note that at a point of relative maximum or relative minimum, the tangent line is *horizontal*. Hence, its *slope is 0*. Since the slope of a tangent line is the derivative of the function, then $f'(x) = 0$ for values of x that are either relative maxima or relative minima. We now state the following.

> If a function $f(x)$ has either a relative minimum or a relative maximum at $x = x_0$ and is differentiable at $x = x_0$, then
>
> $$f'(x_0) = 0$$

Thus, we can usually identify relative maxima and relative minima by finding the values of x for which $f'(x) = 0$. This assumes that $f(x)$ is differentiable at the points of relative maxima and minima. However, this is not always the case. An example of a function that is not differentiable at a relative minimum is $f(x) = x^{2/3}$. Observing its graph in Figure 9–4, note that $x = 0$ is a relative minimum since it is lower than any of its neighboring points. However, since the graph of $f(x)$ comes to a sharp point at $x = 0$, then $f(x)$ is not differentiable there. Hence, $f'(0)$ does not exist. Therefore, in order to identify points of relative maxima and relative minima of a continuous function $f(x)$, we should follow this procedure.

DIFFERENTIAL CALCULUS: OPTIMIZATION

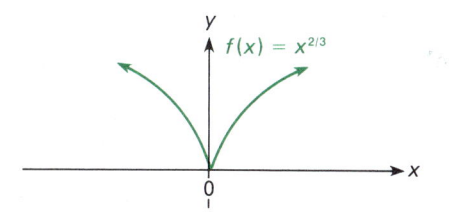

[Graph showing $f(x) = x^{2/3}$]

FIGURE 9-4

> Find $f'(x)$ and determine those values of x for which $f'(x) = 0$. Additionally, look for values of x at which $f'(x)$ does not exist.

Values of x at which either $f'(x) = 0$ or $f'(x)$ does not exist are called **critical values**.

EXAMPLE 9-1 Find the critical values of $f(x) = 2x^3 - 3x^2 - 36x + 7$.

Solution Here, we calculate

$$f'(x) = 6x^2 - 6x - 36$$

Since we seek values of x for which $f'(x) = 0$, we set $f'(x) = 0$ and solve for x to obtain

$$6x^2 - 6x - 36 = 0$$
$$6(x^2 - x - 6) = 0$$
$$6(x - 3)(x + 2) = 0$$

Thus, $x = 3$ and $x = -2$ are critical values. Since there are no values of x at which $f'(x)$ does not exist, these are the only critical values. Computing the corresponding y-coordinates, we have

$$f(3) = 2(3)^3 - 3(3)^2 - 36(3) + 7 = -74$$
$$f(-2) = 2(-2)^3 - 3(-2)^2 - 36(-2) + 7 = 51$$

Hence, the points $(3, -74)$ and $(-2, 51)$ are candidates for relative maximum and minimum.

EXAMPLE 9–2 Find the critical values of $f(x) = x^{2/3}$.
Solution Here, we calculate

$$f'(x) = \frac{2}{3} x^{-1/3}$$
$$= \frac{2}{3} \cdot \frac{1}{x^{1/3}}$$

Note that there are no values of x for which $f'(x) = 0$. However, $f'(x)$ does not exist at $x = 0$. Thus, $x = 0$ is the only critical value. Computing the corresponding y-coordinate, we have

$$f(0) = 0^{2/3} = 0$$

Therefore, (0, 0) is a candidate for a relative maximum or minimum.

First-Derivative Test

A critical value may or may not lead to a relative maximum or minimum. In other words, not every critical value yields a relative maximum or minimum. Thus, we need a test to determine whether a critical value is either a relative maximum or minimum or neither. One such test is called the **first-derivative test** since it is based on the first derivative, $f'(x)$, of a function $f(x)$.

We will explain the first derivative test by referring to the function of Figure 9–5. Observing the relative maximum $A(x_1, f(x_1))$, note that the tangent line to the left of x_1 has a positive slope whereas the tangent line to the right of x_1 has a negative slope. Thus, if x_1 is a critical value of $f(x)$, then x_1 is a relative maximum if $f'(x) > 0$ for all nearby values of x to the left of x_1 and $f'(x) < 0$ for all nearby values of x to the right of x_1. This is true even if $f'(x_1)$ does not exist (see Figure 9–6). For a relative minimum, the reverse holds. Observing the relative minimum $B(x_2, f(x_2))$ of Figure 9–5, note that the tangent line to the left of x_2 has a negative slope whereas the tangent line to the right of x_2 has a positive slope. Thus, if x_2 is a critical value of $f(x)$, then x_2 is a relative minimum if $f'(x) < 0$ for all nearby values of x to the left of x_2 and $f'(x) > 0$ for all nearby values of x to the right of x_2. This, also, is true even if $f'(x_2)$ does not exist, as illustrated in Figure 9–7. Thus, we now state the first-derivative test.

537 DIFFERENTIAL CALCULUS: OPTIMIZATION

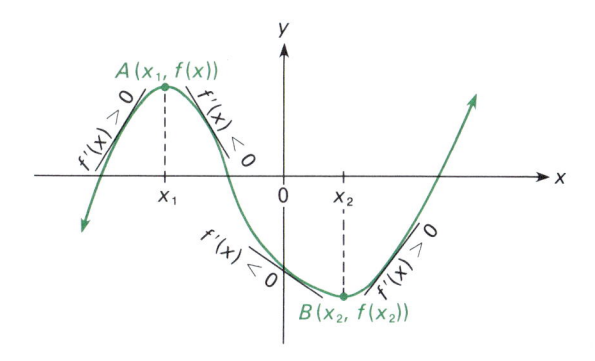

FIGURE 9–5

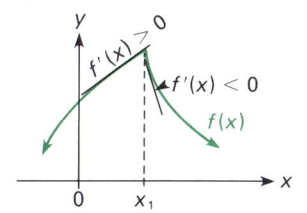

FIGURE 9–6

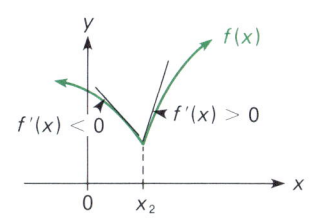

FIGURE 9–7

First-Derivative Test

If x_0 is a critical value of $f(x)$, then:

1. $(x_0, f(x_0))$ is a **relative maximum** if $f'(x) > 0$ for all nearby values of x to the left of x_0 and $f'(x) < 0$ for all nearby values of x to the right of x_0.
2. $(x_0, f(x_0))$ is a **relative minimum** if $f'(x) < 0$ for all nearby values of x to the left of x_0 and $f'(x) > 0$ for all nearby values of x to the right of x_0.

EXAMPLE 9–3 Find the relative maxima and minima of the function $f(x) = x^3 + 3x^2 - 72x + 9$.
Solution   First, we find

$$f'(x) = 3x^2 + 6x - 72$$

Setting $f'(x) = 0$ and solving for x yields

$$3x^2 + 6x - 72 = 0$$
$$3(x^2 + 2x - 24) = 0$$
$$3(x + 6)(x - 4) = 0$$

Thus, $x = -6$ and $x = 4$ are critical values. Since there are no values of x at which $f'(x)$ does not exist, these are the only critical values. We now apply the first-derivative test by analyzing the sign of $f'(x)$, as illustrated in Table 9–1.

TABLE 9–1

Sign of $x + 6$	$-----0+++++++++++++++++$ x $\qquad\qquad-6$
Sign of $x - 4$	$----------------0+++++++$ x $\qquad\qquad\qquad\qquad\quad 4$
Sign of $(x + 6)(x - 4)$	$+++++0-----------0+++++++$ x $\qquad-6\qquad\qquad\quad 4$

Studying Table 9–1, note that the sign of $f'(x) = 3(x + 6)(x - 4)$ is the same as the sign of $(x + 6)(x - 4)$ since the *positive* constant multiplier, 3, does not change the sign of $(x + 6)(x - 4)$. Since $f'(x) > 0$ for all nearby values of x to the left of $x = -6$ and $f'(x) < 0$ for all nearby values of x to the right of $x = -6$, then, by the first-derivative test, a relative maximum exists at $x = -6$. Also, since $f'(x) < 0$ for all nearby values of x to the left of $x = 4$ and $f'(x) > 0$ for all nearby values of x to the right of $x = 4$, then, by the first-derivative test, a relative minimum exists at $x = 4$.

Computing the y-coordinates corresponding to the relative maximum and minimum, we have

$$f(-6) = (-6)^3 + 3(-6)^2 - 72(-6) + 9 = 333$$
$$f(4) = 4^3 + 3(4)^2 - 72(4) + 9 = -167$$

Plotting the relative maximum and minimum and the y-intercept on the rectangular coordinate system gives us the graph of Figure 9–8. Observe that it is not difficult to deter-

539 DIFFERENTIAL CALCULUS: OPTIMIZATION

mine the general nature of the graph of $f(x)$, which appears in Figure 9–9 since $f(x)$ is increasing for those values of x at which $f'(x) > 0$ and decreasing for those values of x at which $f'(x) < 0$.

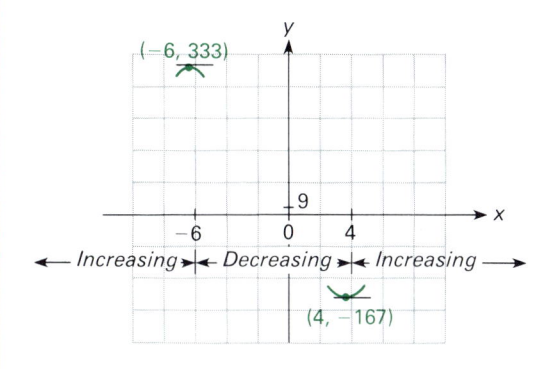

FIGURE 9–8

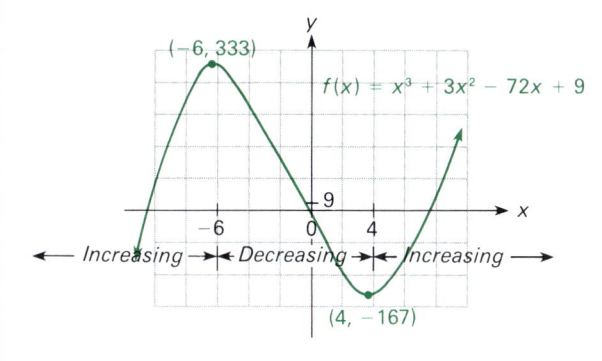

FIGURE 9–9

EXAMPLE 9–4 Find any relative maxima and minima of the function $f(x) = x^{2/3}$.

Solution In Example 9–2, we calculated

$$f'(x) = \frac{2}{3} \cdot \frac{1}{x^{1/3}}$$

which yielded a critical value of $x = 0$. Applying the first-derivative test, we analyze the sign of $f'(x)$. Note that the sign of $f'(x)$ is the same as the sign of $x^{1/3} = \sqrt[3]{x}$. If $x > 0$, then $\sqrt[3]{x} > 0$ and, hence, $f'(x) > 0$. If $x < 0$, then $\sqrt[3]{x} < 0$ and, hence, $f'(x) < 0$. Since $f'(x) < 0$ for all nearby values of x to the left of $x = 0$ and $f'(x) > 0$ for all nearby values of x to the

EXAMPLE 9–4 (continued)

right of $x = 0$, then, by the first-derivative test, a relative minimum exists at $x = 0$. Note that $f(0) = 0^{2/3} = 0$. Although the preceding information does not enable us to sketch a graph of $f(x)$, its graph is included in Figure 9–10.

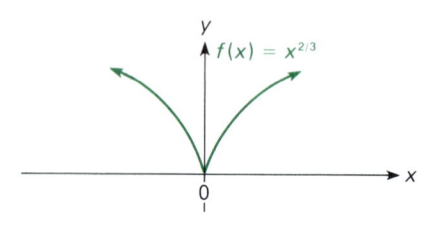

FIGURE 9–10

EXERCISES

1. Use the first-derivative test to determine any relative maxima and minima for each of the following:
   - (A) $y = x^2 + 2x - 15$
   - (B) $y = 2x^3 + 15x^2 - 36x + 1$
   - (C) $f(x) = 4x^3 - 72x^2 + 24x - 5$
   - (D) $f(x) = 3x^4 - 4x^3 + 10$

2. Find any relative maxima and minima of the function defined by $f(x) = x^{4/3}$. Use the first-derivative test.

3. Consider the function defined by
$$f(x) = x^3 + 6x^2 - 15x + 1 \quad (-10 \leq x \leq 11)$$
Use the first-derivative test to find any relative maxima and minima. Also determine the absolute maxima and minima and sketch $f(x)$.

4. Consider the function defined by
$$f(x) = -x^3 - 9x^2 + 21x + 4 \quad (-8 \leq x \leq 10)$$
Use the first-derivative test to determine any relative maxima and minima. Also, determine the absolute maxima and minima and sketch $f(x)$.

5. Use the first derivative to show that the x-coordinate of the vertex of a quadratic equation of the form $f(x) = ax^2 + bx + c$ is $-b/2a$.

6. Use the first derivative to show that the exponential function defined by $y = e^x$ is always increasing.

7. Use the first derivative to show that the exponential function defined by $y = e^{-x}$ is always decreasing.

8. Use the first derivative to show that the logarithmic function $y = \ln x$ is always increasing.

9-2

THE SECOND DERIVATIVE AND CURVE SKETCHING

In Section 9-1, we discussed the first-derivative test for determining whether a critical value is either a relative maximum or relative minimum or neither. There exists another test for making this determination. It is called the **second-derivative test**. In order to apply it, we must first consider the second derivative.

The Second Derivative

The derivative of a function $y = f(x)$, commonly denoted by $f'(x)$, dy/dx, y', or $D_x y$ is called a **first derivative**. The derivative of a first derivative is called the **second derivative** and is denoted by any of the following symbols:

$$f''(x) \qquad \frac{d^2 y}{dx^2} \qquad y'' \qquad D_x^2 y$$

Thus, if

$$f(x) = x^4 - 8x^3 + 7x^2 - 15$$

the first derivative is

$$f'(x) = 4x^3 - 24x^2 + 14x$$

and the second derivative is

$$f''(x) = 12x^2 - 48x + 14$$

EXAMPLE 9-5 If $f(x) = x^5 - 8x^4 + 7x^3 - 5$, find $f'(x)$, $f''(x)$, and $f''(1)$.

Solution $f'(x) = 5x^4 - 32x^3 + 21x^2$
$f''(x) = 20x^3 - 96x^2 + 42x$
$f''(1) = 20(1)^3 - 96(1)^2 + 42(1) = -34$

Graphical Interpretation of the Second Derivative

In order to discuss the graphical interpretation of the second derivative, we must define the terms *concave up* and *concave down* as they relate to functions. A function $f(x)$ is said to be **concave up** on an interval $a < x < b$ if each tangent line to the graph of $f(x)$ lies below the graph of $f(x)$ (see Figure 9-11). A function $f(x)$ is said to be **concave down** on an interval $a < x < b$ if each tangent line to the graph of $f(x)$ lies above its graph (see Figure 9-12.) In Chapter 2, we classified the graphs of quadratic functions (parabolas) as either opening up or opening down. In terms of concavity, a parabola that opens up is said to be concave up, and a parabola that opens down is said to be concave down (see Figure 9-13.)

542 CHAPTER NINE

FIGURE 9-11

FIGURE 9-12

FIGURE 9-13

FIGURE 9-14

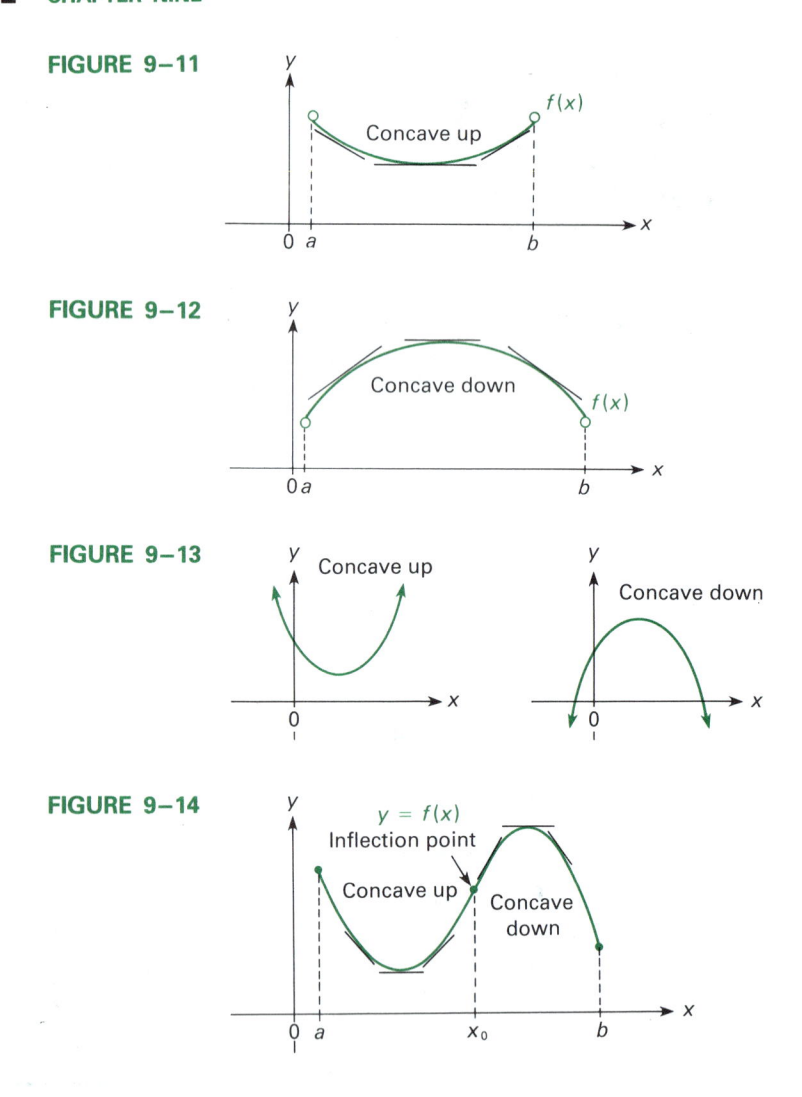

We now consider the graph of Figure 9-14. Observe that $f(x)$ is concave up on the interval $a < x < x_0$ and concave down on the interval $x_0 < x < b$. At $x = x_0$, the graph of $f(x)$ changes in concavity. Such a point, where a curve changes in concavity, is called an **inflection point**. Note that as we move along the concave-up portion of the curve from $(a, f(a))$ toward $(x_0, f(x_0))$, the slopes of the tangent lines are increasing. Thus, $f'(x)$ is increasing on the interval $a < x < x_0$. Since $f'(x)$ is increasing on the interval $a < x < x_0$, then its derivative, $f''(x) > 0$. Analogously, as we move along the concave-down portion of the curve from $(x_0, f(x_0))$ to $(b, f(b))$, the slopes of the tangent lines are decreasing. Thus, $f'(x)$ is decreasing on the interval $x_0 < x < b$. Since $f'(x)$ is de-

DIFFERENTIAL CALCULUS: OPTIMIZATION

creasing on the interval $x_0 < x < b$, then its derivative, $f''(x) < 0$. At the inflection point, x_0, the second derivative, $f''(x_0) = 0$. Therefore, as we move along a concave-up portion of the graph of a function, the first derivative is increasing, and thus the second derivative is positive. As we move along a concave-down portion, the first derivative is decreasing, and thus the second derivative is negative. At an inflection point, the first derivative is neither increasing nor decreasing, and thus the second derivative is often 0. However, this is not always the case (see Exercise 9 at the end of this section). A true test of the existence of an inflection point at $x = x_0$ is a change in sign of the second derivative at $x = x_0$. These relationships are illustrated by the graphs of $f(x)$, $f'(x)$, and $f''(x)$ of Figure 9–15. In summary, the second derivative indicates the concavity of a function $f(x)$ in accordance with the following rule.

> **Second-Derivative Rule**
> A function $f(x)$ is **concave up** at all values of x for which $f''(x) > 0$ and **concave down** at all values of x for which $f''(x) < 0$. If $f''(x)$ changes sign at $x = x_0$, then an inflection point exists at $x = x_0$.

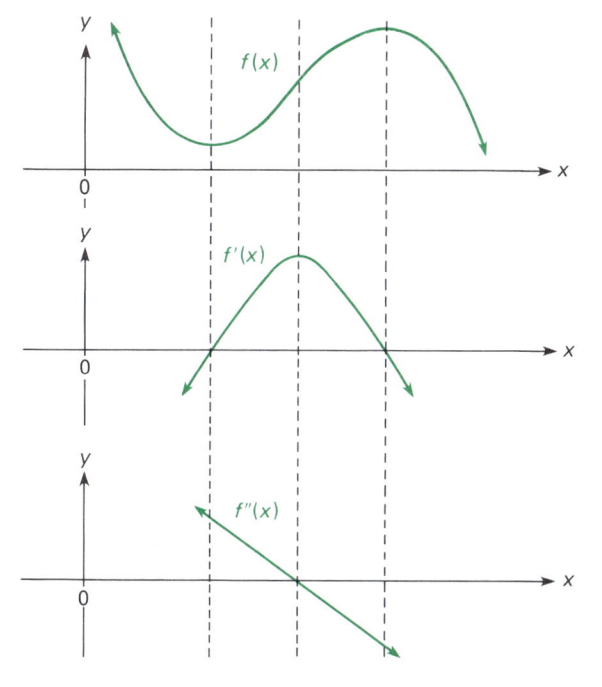

FIGURE 9–15

The second-derivative rule enables us to determine the concavity of a function $f(x)$ at a point $(x, f(x))$. It may also be used to determine whether a relative maximum or minimum or neither occurs at a value of x for which $f'(x) = 0$. This result is called the **second-derivative test** and is stated as follows.

Second-Derivative Test

If $f(x)$ is a function and x_0 is a number such that $f'(x_0) = 0$, then:

1. If $f''(x_0) < 0$, $f(x)$ has a relative maximum at $x = x_0$.
2. If $f''(x_0) > 0$, $f(x)$ has a relative minimum at $x = x_0$.
3. If $f''(x_0) = 0$, the test fails. In other words, the test gives no information regarding a relative maximum or relative minimum at $x = x_0$.

As stated, the second-derivative test fails if $f''(x_0) = 0$. In this instance, we might use the first-derivative test for determining the existence of relative maxima and minima. We might also use $f''(x)$ to check the concavity on both sides of x_0. If $f''(x)$ changes sign at $x = x_0$, then an inflection point exists at $x = x_0$.

EXAMPLE 9–6 Find all relative maxima and minima and inflection points of $f(x) = 2x^3 - 3x^2 - 36x + 7$ and sketch its graph.

Solution Here, we have

$$f'(x) = 6x^2 - 6x - 36$$
$$f''(x) = 12x - 6$$

Setting $f'(x) = 0$ and solving for x yields

$$6x^2 - 6x - 36 = 0$$
$$6(x - 3)(x + 2) = 0$$

Thus, $x = 3$ and $x = -2$ are critical values. Since there are no values of x at which $f'(x)$ does not exist, these are the only critical values. Applying the second-derivative test to the critical values, we have

$$f''(3) = 12(3) - 6 = 30$$
$$f''(-2) = 12(-2) - 6 = -30$$

Since $f''(3) > 0$, a relative minimum occurs at $x = 3$. Since $f''(-2) < 0$, a relative maximum occurs at $x = -2$. Computing the corresponding y-coordinates, we obtain

$$f(3) = 2(3)^3 - 3(3)^2 - 36(3) + 7 = -74$$
$$f(-2) = 2(-2)^3 - 3(-2)^2 - 36(-2) + 7 = 51$$

The points are plotted in Figure 9–16.

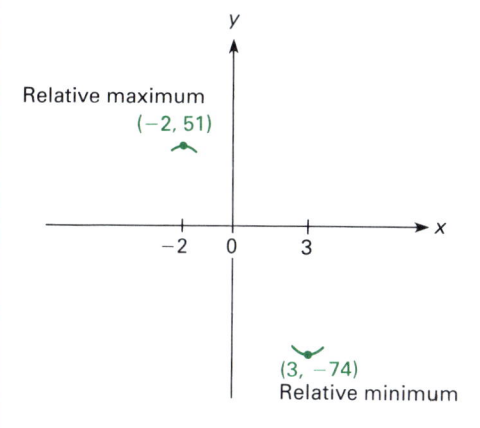

FIGURE 9–16

We now analyze the second derivative,

$$f''(x) = 12x - 6$$

for concavity and inflection points. Note that $f''(x) > 0$ when $x > 1/2$ and $f''(x) < 0$ when $x < 1/2$. Using the second-derivative rule, we conclude that:

1. $f(x)$ is concave up over the interval $x > 1/2$ since $f''(x) > 0$ there.
2. $f(x)$ is concave down over the interval $x < 1/2$ since $f''(x) < 0$ there.
3. $f(x)$ has an inflection point at $x = 1/2$. Its y-coordinate is

$$f\left(\frac{1}{2}\right) = 2\left(\frac{1}{2}\right)^3 - 3\left(\frac{1}{2}\right)^2 - 36\left(\frac{1}{2}\right) + 7 = -\frac{23}{2}$$

EXAMPLE 9–6 (continued)

Incorporating these conclusions into the results of Figure 9–16, we obtain the graph of Figure 9–17. Note also that the y-intercept is $y = f(0) = 7$.

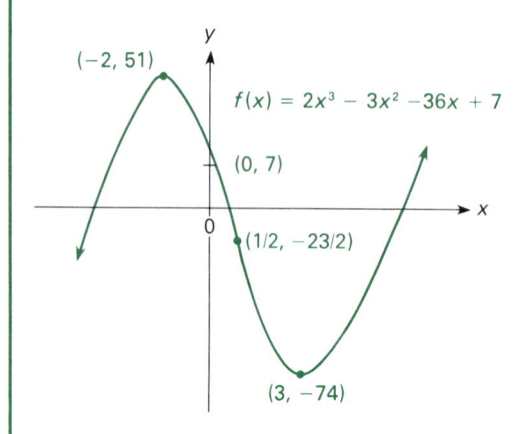

FIGURE 9–17

EXAMPLE 9–7 The ABC Container Company manufactures plastic bottles. Its cost function is defined by

$$C(x) = (x - 5)^3 + 1025$$

where x is the number of bottles produced (in millions) and $C(x)$ is the total cost (in thousands of dollars). Find all relative maxima and minima and inflection points of $C(x)$ and sketch its graph.

Solution Here, we have

$$C'(x) = 3(x - 5)^2$$
$$C''(x) = 6(x - 5)$$

Setting $C'(x) = 0$ and solving for x yields

$$3(x - 5)^2 = 0$$

Thus, $x = 5$ is a critical value. Since there are no values of x at which $C'(x)$ does not exist, this is the only critical value. Applying the second-derivative test to $x = 5$, we obtain

$$C''(5) = 6(5 - 5) = 0$$

Thus, the second-derivative test fails since $C''(5) = 0$. Hence, we have no indication as to whether a relative maximum or minimum or an inflection point occurs at $x = 5$. We test the concavity on both sides of $x = 5$ by analyzing the second derivative,

$$C''(x) = 6(x - 5)$$

and noting that $C''(x) > 0$ when $x > 5$ and $C''(x) < 0$ when $x < 5$. Using the second-derivative rule, we conclude that $C(x)$ is concave up for $x > 5$ and concave down for $x < 5$, which indicates that an inflection point occurs at $x = 5$. To sketch $C(x)$, we compute

$$C(5) = (5 - 5)^3 + 1025 = 1025$$

and plot the inflection point (5, 1025), as illustrated in Figure 9–18. Note also that the y-intercept is $C(0) = 900$. Observing the concavity, we sketch the function as shown in Figure 9–19. Our curve flattens at $x = 5$ since the tangent line there is horizontal because $f'(5) = 0$. Assuming $x \geq 0$, our curve is drawn as a solid line in the first quadrant.

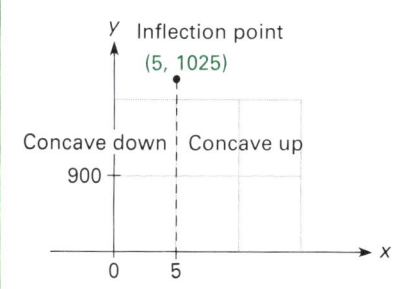

FIGURE 9–18

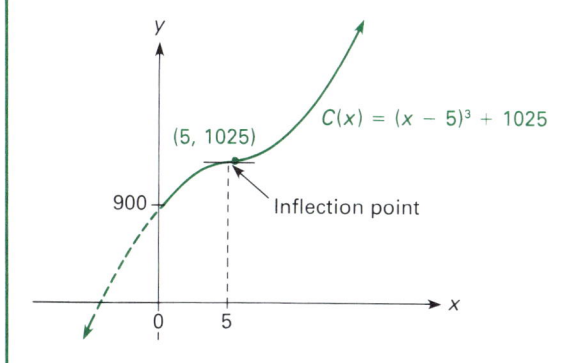

FIGURE 9–19

EXAMPLE 9-7 (continued)

The shape of $C(x)$ is typical of nonlinear cost functions of some large companies. The marginal cost, $C'(x)$, is decreasing at values of x to the left of the inflection point since the company is making efficient use of its fixed costs. However, at values of x to the right of the inflection point, the marginal cost, $C'(x)$, is increasing since the company has reached a point—the inflection point—at which the variable cost begins to escalate.

EXERCISES

1. For each of the following, find $f'(x)$ and $f''(x)$:
   - (A) $f(x) = x^3 - 4x^2 + 7x - 9$
   - (B) $f(x) = x^4 - 7x^3 + 8x - 3$
   - (C) $f(x) = \dfrac{x-1}{x+6}$
   - (D) $f(x) = xe^x$
   - (E) $f(x) = x^3 e^{x^5}$
   - (F) $f(x) = x^2 \cdot \ln(x+1)$
   - (G) $f(x) = (x+1)^4(x-2) + 6$

2. For each of the following, find dy/dx and d^2y/dx^2:
   - (A) $y = x^4 - 8x^3 + 6x - 9$
   - (B) $y = (x^3 + 1) \cdot \ln(x^6 - 5)$
   - (C) $y = \dfrac{x^4 + 6}{x^3 - 7}$
   - (D) $y = (x^2 - 3)e^{x^3 - 4x}$
   - (E) $y = (x-4)^3(x-7) + 8$

3. If $f(x) = x^3 - 4x^2 + 7x + 9$, find $f'(x)$, $f''(x)$, $f'(2)$, and $f''(4)$.

4. If $g(x) = \dfrac{x^2 - 1}{x - 5}$, find $g'(x)$, $g''(x)$, and $g''(2)$.

5. If $y(x) = (x+1)(x+3)^2$, find $y'(x)$, $y''(x)$, and $y''(2)$.

6. If $s(t) = (t^3 + 1)e^{t^2}$, find $s'(t)$, $s''(t)$, and $s''(0)$.

7. For each of the following, determine any relative maxima and minima and inflection points. Analyze for concavity and sketch a graph of the function.
   - (A) $y = x^2 - 2x + 4$
   - (B) $f(x) = x^3 + 6x^2 - 36x + 8$
   - (C) $y = x^3 - 12x^2 - 27x - 1$
   - (D) $f(x) = \dfrac{1}{3}x^3 - 4x + 8$
   - (E) $f(x) = 5x^4 - 20x^3 + 7$
   - (F) $f(x) = 5x^6 - 54x^5 + 60x^4 + 10$
   - (G) $y = x^3 - 3x + 2$
   - (H) $y = x^3 - 27x + 2$

549 DIFFERENTIAL CALCULUS: OPTIMIZATION

8. For each of the following, determine any relative maxima and minima and inflection points. Analyze for concavity and sketch a graph of the function.
   (A) $y = x^3$
   (B) $f(x) = x^6$
   (C) $f(x) = x^{4/3}$
   (D) $f(x) = e^{x^2}$
   (E) $y = e^{-x^2}$
   (F) $y = \dfrac{8}{x^2 + 1}$
   (G) $f(x) = \dfrac{x}{x^2 + 1}$
   (H) $y = \dfrac{x - 1}{x^2}$

9. Given the function defined by $f(x) = x^{1/3}$
   (A) For which values of x is $f(x)$ concave up?
   (B) For which values of x is $f(x)$ concave down?
   (C) At which values of x does $f(x)$ change in concavity? Is $f''(x)$ defined there?
   (D) Graph $f(x)$.

10. The profit from producing x units of some commodity is given by the function
    $$P(x) = -0.01x^3 + 1.20x^2 - 21x + 50{,}000 \qquad (10 \leq x \leq 80)$$
    (A) Find any relative maxima and minima and inflection points of $P(x)$ and sketch its graph.
    (B) How many units must be sold in order to yield the maximum profit?
    (C) What is the maximum profit? Is this an absolute maximum?

11. For each of the following functions, determine any relative maxima and minima and inflection points. Analyze for concavity and sketch the function.
    (A) $f(x) = (x - 5)^3$
    (B) $f(x) = (x - 6)^8$
    (C) $y = (x + 3)^{10} + 2$
    (D) $y = (x - 8)^5 + 2$
    (E) $y = (x + 2)(x - 5)^4 + 6$
    (F) $y = (x - 1)^3 (x + 6) + 1$

12. Sketch the graph of each of the following:
    (A) $f(x) = (x - 4)^3 (x + 2)^2$
    (B) $f(x) = x^3(x - 1)(x + 6)^4$

13. Sketch the graph of each of the following rational functions:
    (A) $f(x) = \dfrac{800}{x} + 2x$
    (B) $y = \dfrac{x - 1}{x^2}$
    (C) $f(x) = \dfrac{(x - 3)^2}{x + 1}$
    (D) $f(x) = \dfrac{1}{x} - 3$
    (E) $f(x) = \dfrac{x^2(x - 2)}{(x - 1)^3}$
    (F) $y = \dfrac{1}{(x - 3)^2}$
    (G) $f(x) = \dfrac{20}{x} + x + 9$
    (H) $y = \dfrac{6}{x} + x + 5$

14. The function defined by
    $$f(x) = \dfrac{1}{\sigma\sqrt{2\pi}} e^{-(x-\mu)^2/2\sigma^2} \qquad (-\infty < x < \infty)$$
    is called the **normal probability density function**. If μ, σ, and π are

550 CHAPTER NINE

constants, with $\sigma > 0$ and $\pi = 3.14\ldots$, then use calculus to sketch the graph of $f(x)$. Verify that a relative maximum occurs at $x = \mu$ and that inflection points occur at $x = \mu \pm \sigma$.

15. The function

$$f(x) = \begin{cases} 4xe^{-2x} & (x > 0) \\ 0 & (x \leq 0) \end{cases}$$

is an example of a **gamma distribution** of statistics. Use calculus to graph $f(x)$. Find its relative maximum and inflection point.

16. For each of the following, find the maximum and minimum values attained by the function on the given interval:
   (A) $f(x) = x^3 - 16x^2$ $(-5 \leq x \leq 6)$
   (B) $f(x) = -\frac{1}{100}x^3 + 16x$ $(-10 \leq x \leq 40)$
   (C) $f(x) = x^3 - 5x^2 + 7x - 3$ $(0 \leq x \leq 4)$
   (D) $f(x) = x^3 - 12x^2 + 45x - 50$ $(1 \leq x \leq 10)$

9-3

APPLICATIONS This section contains some applications of the relative maxima and minima concepts described in the previous sections.

EXAMPLE 9-8 *(Minimizing Average Cost)* The total cost of producing x thousand units of some product is given by

$$C(x) = x^2 - 6x + 16 \qquad (x > 0)$$

where $C(x)$ is in millions of dollars. How many units should be produced in order to minimize the average cost, $A(x)$?

Solution Since the average cost, $A(x)$, is $C(x)/x$, we write

$$A(x) = \frac{x^2 - 6x + 16}{x} \qquad (x > 0)$$

Using the quotient rule and simplifying, the first derivative is

$$A'(x) = \frac{(x - 4)(x + 4)}{x^2}$$

Note that $x = 4$, $x = -4$, and $x = 0$ are critical values. However, we exclude $x = -4$ and $x = 0$ since they are not in the

DIFFERENTIAL CALCULUS: OPTIMIZATION

domain, $x > 0$. Applying the second-derivative test to $x = 4$, we have

$$A''(x) = \frac{32}{x^3}$$

$$A''(4) = \frac{32}{4^3}$$

$$= \frac{1}{2}$$

Since $A''(4) > 0$, then, by the second-derivative test, a relative minimum occurs at $x = 4$. Also, since $A''(x) > 0$ for all x-values in the domain, $x > 0$, then, by the second-derivative rule, the average cost function is concave up for $x > 0$. This and the fact that $A'(4) = 0$ implies that the relative minimum at $x = 4$ is also an absolute minimum (see Figure 9–20). Thus, 4000 units (remember, $x = 4$ means 4000 units) should be produced in order to minimize the average cost.

$$A(x) = \frac{x^2 - 6x + 16}{x}, x > 0$$

(4, 2)

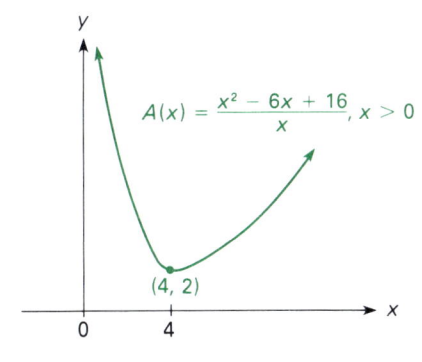

FIGURE 9–20

EXAMPLE 9–9 *(Minimizing Energy Costs)* The heating and cooling costs for an uninsulated office building are $850,000 per year. If x inches ($x \leq 9$) of insulation are added, the combined heating and cooling costs will be $2,550,000/(x + 3)$ dollars per year. Assume that it costs $200,000 for each inch (thickness) of insulation added.
(A) How many inches of insulation should be added in order to minimize total costs (insulated cost + combined heating and cooling costs) over a 5-year period?
(B) If the number of inches of insulation added must be a whole number, how many inches should be added in order to minimize the total cost?

EXAMPLE 9–9 (continued)

Solutions

(A) Total cost is given by

$$C(x) = 200{,}000x + 5\left(\frac{2{,}550{,}000}{x+3}\right) \quad (0 \le x \le 9)$$

or

$$C(x) = 200{,}000x + \frac{12{,}750{,}000}{x+3} \quad (0 \le x \le 9)$$

The first derivative is

$$C'(x) = 200{,}000 - \frac{12{,}750{,}000}{(x+3)^2}$$

A critical value is $x = -3$. However, we exclude $x = -3$ since it is not in the domain, $0 \le x \le 9$. We set $C'(x) = 0$ to obtain

$$200{,}000 - \frac{12{,}750{,}000}{(x+3)^2} = 0$$

$$(x+3)^2 = \frac{12{,}750{,}000}{200{,}000}$$

$$= 63.75$$

Taking the square root of both sides, we have

$$x + 3 = \sqrt{63.75} \quad \text{or} \quad x + 3 = -\sqrt{63.75}$$

Hence,

$$x = -3 + \sqrt{63.75} \quad \text{or} \quad x = -3 - \sqrt{63.75}$$

We exclude $x = -3 - \sqrt{63.75}$ since it is outside the domain, $0 \le x \le 9$. Thus, a workable critical value is

$$x = -3 + \sqrt{63.75}$$
$$= -3 + 7.98$$
$$= 4.98$$

Applying the second-derivative test to $x = 4.98$, we have

$$C''(x) = \frac{25{,}500{,}000}{(x+3)^3}$$

$$C''(4.98) = \frac{25{,}500{,}000}{(4.98+3)^3}$$

$$= 50{,}180.10$$

Since $C''(4.98) > 0$, then, by the second-derivative test, a relative minimum occurs at $x = 4.98$. Also, since $C''(x) > 0$ for all x-values in the domain, $0 \le x \le 9$, then, by the second-derivative rule, the cost function, C, is concave up for $0 \le x \le 9$. This and the fact that $C'(4.98) = 0$ implies that the relative minimum at $x = 4.98$ is also an absolute minimum (see Figure 9–21). Thus, 4.98 inches of insulation should be added in order to minimize total cost.

FIGURE 9–21

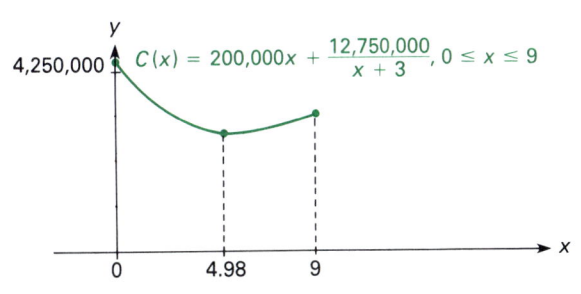

$$C(x) = 200{,}000x + \frac{12{,}750{,}000}{x+3}, \quad 0 \le x \le 9$$

(B) We must calculate the total cost at $x = 4$ and $x = 5$. Hence, we have

$$C(4) = 200{,}000(4) + \frac{12{,}750{,}000}{4+3} = \$2{,}621{,}428.60$$

$$C(5) = 200{,}000(5) + \frac{12{,}750{,}000}{5+3} = \$2{,}593{,}750.00$$

Thus, $x = 5$ inches of insulation should be added in order to yield a minimum cost of \$2,593,750.

EXAMPLE 9–10 *(Enclosure)* A farmer wishes to enclose a rectangular field of 2,000,000 square feet alongside a river. No fence is needed along the river (see Figure 9–22). Find the dimensions of the rectangular pasture that minimize the length of fence used.

FIGURE 9–22

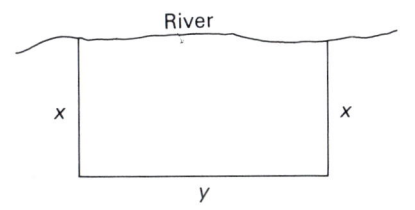

EXAMPLE 9–10 (continued)

Solution If L = length of fence, then from Figure 9–22, we determine that

$$L = 2x + y$$

Since the enclosed area is 2,000,000 square feet, we have

$$xy = 2,000,000$$

Solving for y, we obtain

$$y = \frac{2,000,000}{x}$$

Substituting this result for y in the equation $L = 2x + y$ yields

$$L(x) = 2x + \frac{2,000,000}{x} \qquad (x > 0)$$

We now seek the value of x that minimizes L. Hence, we find $L'(x)$, as follows:

$$L'(x) = 2 - \frac{2,000,000}{x^2}$$

Setting $L'(x) = 0$, we have

$$2 - \frac{2,000,000}{x^2} = 0$$

$$2x^2 = 2,000,000$$

$$x^2 = 1,000,000$$

$$x = \pm 1000$$

We disregard $x = -1000$ since x must be positive. Thus, we apply the second-derivative test to $x = 1000$ and get

$$L''(x) = \frac{4,000,000}{x^3}$$

$$L''(1000) = \frac{4,000,000}{1000^3}$$

$$= 0.004$$

Since $L''(1000) > 0$, a relative minimum occurs at $x = 1000$. Since $L''(x) > 0$ for all x-values in the domain, $x > 0$, then, by the second-derivative rule, the function $L(x)$ is concave up for $x > 0$. This and the fact that $L'(1000) = 0$ implies that the

DIFFERENTIAL CALCULUS: OPTIMIZATION

relative minimum at $x = 1000$ is also an absolute minimum. Thus, the minimum length of fence is

$$L(1000) = 2(1000) + \frac{2,000,000}{1000}$$

$$= 4000 \text{ feet}$$

Inventory Control A distributor of tires usually sells 100,000 tires per year. For each placement of an order for tires, it costs the distributor $20. Additionally, the cost of carrying one tire in inventory for a year is $4.

The graph of Figure 9–23 illustrates the depletion of the distributor's inventory of tires over time, t. Observe that at $t = 0$, the initial inventory is Q tires. At $t = 1$ and $t = 2$, the inventory has reached a level of 0 tires. Thus, at $t = 1$ and $t = 2$, an order is placed for Q tires, and the inventory level is immediately replenished.

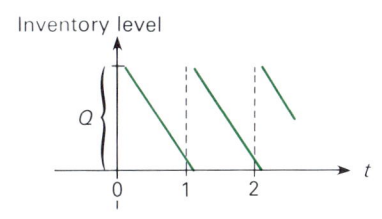

FIGURE 9–23

The distributor must find an answer to the following question:

Each time an order is placed for tires, how many tires should be ordered so that the total annual inventory cost is minimized?

The total annual inventory cost, C, is calculated by the following formula:

$$C = \text{ordering cost} + \text{carrying cost}$$

The **ordering cost** is calculated by multiplying the cost of placing an order by the number of orders placed in a year. Since 100,000 tires are usually ordered annually in quantities of Q tires per order, the number of orders per year is $100,000/Q$. Thus, since the cost of placing each order is $20, we have

$$\text{ordering cost} = 20\left(\frac{100,000}{Q}\right)$$

The **carrying cost** is calculated by multiplying the cost of carrying one unit in inventory for a year by the average number of units in inventory. Since the inventory level varies from 0 to Q, the average inventory is $(0 + Q)/2 = Q/2$. Thus, since the cost of carrying one tire in inventory for a year is $4, we obtain

$$\text{carrying cost} = 4\left(\frac{Q}{2}\right)$$

Therefore, the total inventory cost is calculated as

$$C(Q) = 20\left(\frac{100{,}000}{Q}\right) + 4\left(\frac{Q}{2}\right)$$

Simplifying, we obtain

$$C(Q) = \frac{2{,}000{,}000}{Q} + 2Q \qquad (Q > 0)$$

Since we must find the value of Q that minimizes $C(Q)$, we calculate $C'(Q)$, as follows:

$$C'(Q) = \frac{-2{,}000{,}000}{Q^2} + 2$$

A critical value is $Q = 0$. However, we exclude $Q = 0$ since it is not in the domain, $Q > 0$. We set $C'(Q) = 0$ to obtain

$$0 = \frac{-2{,}000{,}000}{Q^2} + 2$$

Solving for Q^2 yields

$$Q^2 = 1{,}000{,}000$$
$$Q = \pm 1000$$

We disregard the critical value $Q = -1000$ since it is outside the domain, $Q > 0$. Applying the second-derivative test to $Q = 1000$, we have

$$C''(Q) = \frac{4{,}000{,}000}{Q^3}$$

$$C''(1000) = \frac{4{,}000{,}000}{1000^3}$$

$$= \frac{4}{1000}$$

Since $C''(1000) > 0$, a relative minimum occurs at $Q = 1000$. Since $C''(Q) > 0$ for all values of Q in the domain, $Q > 0$, the cost function, C, is concave up for $Q > 0$. This and the fact that $C'(1000) = 0$ implies that the relative minimum at $Q = 1000$ is also an absolute minimum.

In general, if D is the annual demand for a given inventory product, K is the cost of placing an order, H is the cost of carrying or holding one unit of inventory for a year, and Q is the order size, then the total annual inventory cost, $C(Q)$, is given by

$$C(Q) = K\frac{D}{Q} + H\frac{Q}{2}$$

where D/Q is the number of orders placed per year at a cost of K dollars per order, and $Q/2$ is the average number of units in inventory at a carrying or holding cost of H dollars per unit.

EXERCISES

1. The total cost of producing x million units of some product is given by

$$C(x) = x^2 - 8x + 36 \quad (x > 0)$$

where $C(x)$ is in millions. How many units should be produced in order to minimize the average cost, $A(x)$?

2. The heating and cooling costs for a building with no insulation are $670,000 per year. If x inches ($x \leq 10$) of insulation are added, the heating and cooling costs will be $2,680,000/(x + 4)$ dollars per year. Assume that it costs $175,000 for each inch of insulation added.
   (A) How many inches of insulation should be added in order to minimize total cost over a 6-year period?
   (B) Is the answer to part (A) an absolute minimum?
   (C) If the number of inches of insulation added must be a whole number, how many inches should be added in order to minimize the total cost?
   (D) Using the results of part (C), what are the savings over a 6-year period?

3. A company manufactures open boxes from square pieces of tin 18 inches on each side. The process involves cutting equal squares from the corners of each piece of tin and folding up the flaps to form sides. What size square should be cut from each corner in order to maximize the volume of the box?

4. A company manufactures the open trash bin illustrated in Figure 9–24 on the next page. Each bin must be 6 feet high and have a volume of 294 cubic feet. Find the dimensions x and y of the base that minimize the total surface area of a bin. (*Note:* The total surface area is the sum of the areas of the sides and bottom.)

FIGURE 9-24

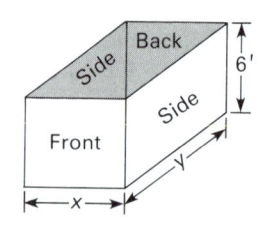

5. A company manufactures the open trash bin illustrated in Figure 9-25. Each bin must be 6 feet high and have a volume of 192 cubic feet. If the material for the front and back costs $5 per square foot, the material for the sides costs $10 per square foot, and the material for the bottom costs $20 per square foot, find the dimensions x and y of the base that minimize the total cost.

FIGURE 9-25

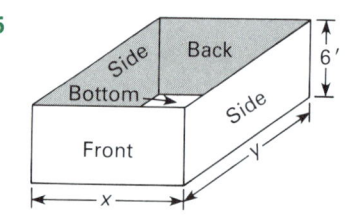

6. A zoo manager wishes to enclose an area of 20,000 square feet into three cages of equal size, as illustrated in Figure 9-26. Find the dimensions x and y that minimize the length of fence used.

FIGURE 9-26

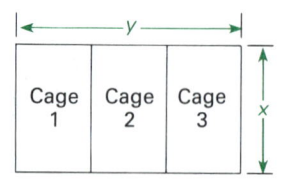

7. A city recreation department is planning to enclose a rectangular area of 125,000 square feet for a playground (see Figure 9-27). If the fence along the sides costs $10 per foot and the fence along the front and back costs $20 per foot, find the dimensions of the playground that minimize the fence cost.

FIGURE 9-27

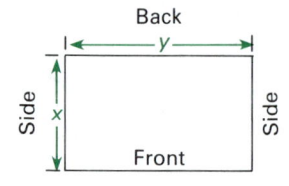

DIFFERENTIAL CALCULUS: OPTIMIZATION

8. It costs a distributor of mopeds $100 to place an order for Q mopeds. Also, it costs $2 per year to keep one moped in inventory. The distributor usually sells 250,000 mopeds annually.
   (A) Find the formula for total inventory cost as a function of Q.
   (B) What order size Q minimizes the total inventory cost?

9. A bicycle manufacturer finds that it costs $1500 to set up a production run for Q bicycles. Also, it costs $3 per year to keep a bicycle in inventory. The manufacturer expects to sell 4,900,000 bicycles this year.
   (A) Find the formula for total inventory cost as a function of Q.
   (B) What production lot size Q minimizes the total inventory cost?
   (C) What is the minimum total inventory cost?

10. A park manager has 400 feet of fence to enclose a rectangular area alongside a river. No fence is needed along the river (see Figure 9–28). Use calculus to determine the dimensions x and y that maximize the enclosed area.

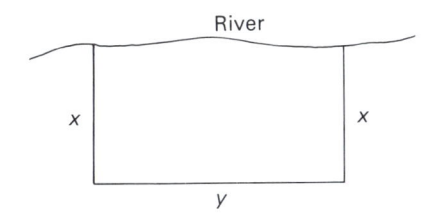

FIGURE 9–28

11. A ball is projected vertically into the air. The function $S(t) = -16t^2 + 192t$ gives the height, S (in feet), above the ground of the ball at time t (in seconds). When does the ball reach maximum height, and what is the maximum height? Use calculus.

12. If an apple grower harvests her crop now, she will pick on the average 120 pounds per tree. She will get $0.48 per pound for the apples. From past experience, the grower knows that for each additional week she waits, the yield per tree will increase by about 10 pounds while the price will decrease by about $0.03 per pound. Use calculus to determine how many weeks the grower should wait in order to maximize sales revenue. What is the maximum sales revenue?

13. A park manager has 2000 feet of fence with which to enclose a rectangular area. Use calculus to determine the dimensions of the rectangle that yield the maximum area.

14. Using calculus, show that the maximum area enclosed by a rectangle will always result in a square.

15. Find two numbers whose sum is 24 and whose product is as large as possible.

16. Find two numbers whose difference is 30 and whose product is as small as possible.

17. In biology, the number

$$R = a \ln x - bx \quad (x > 0)$$

where a and b are positive constants, is called the **Reynolds number**. It is related to the turbulence of blood flow in an aorta of radius x. For what value of x is R maximum?

18. If $C(x)$ is the total cost of producing x units of some item, then the average cost per unit is $A(x) = C(x)/x$, $x > 0$. If C is concave up and $C''(x)$ exists for $x > 0$, show that average cost is minimized when average cost equals marginal cost.

9-4

PARTIAL DERIVATIVES

Multivariate Functions

A function such as

$$y = f(x) = x^3 + \frac{4}{x} + 17$$

has dependent variable y and independent variable x. Since there is only one independent variable, $f(x)$ is called a **function of one variable**. If a function has two independent variables, say x and y, it is called a **function of two variables** and is usually denoted by $f(x, y)$. If a function has three independent variables, say x, y, and w, it is called a **function of three variables** and may be denoted by $f(x, y, w)$. In general, functions of more than one variable are called **multivariate functions**.

As a specific example, we consider a company producing metal tanks. It has been determined that the daily production cost, z, is dependent upon the daily number of tanks produced, x, and the daily number of man-hours used, y. These quantities are related by the multivariate function

$$z = f(x, y) = x^2 - 8x + y^2 - 12y + 1500$$

If, during a given day, the company produces $x = 2$ tanks and uses $y = 5$ man-hours of labor, then the daily production cost is

$$z = f(2, 5) = 2^2 - 8(2) + 5^2 - 12(5) + 1500 = \$1453$$

The graph of $z = f(x, y) = x^2 - 8x + y^2 - 12y + 1500$ appears in Figure 9-29. Observe that the graph of $f(x, y)$ is three-dimensional. Also, since $f(2, 5) = 1453$, notice that the point $(2, 5, 1453)$ lies on the surface of the function. Although we will not be concerned with graphing multivariate functions, Figure 9-29 has been included to provide an intuitive understanding of the graphs of functions of two variables.

561 DIFFERENTIAL CALCULUS: OPTIMIZATION

$$z = f(x,y) = x^2 - 8x + y^2 - 12y + 1500$$

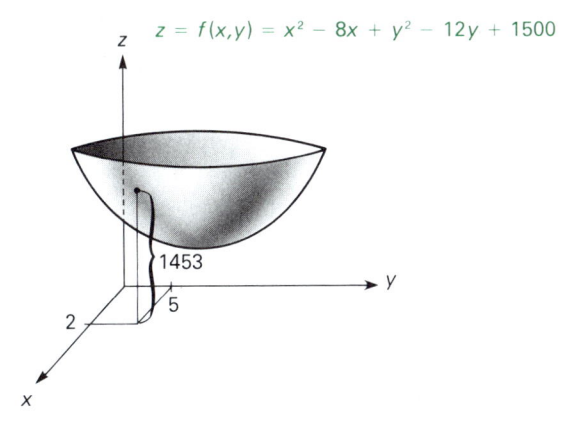

FIGURE 9–29

Partial Derivatives

Returning to the cost function

$$z = f(x, y) = x^2 - 8x + y^2 - 12y + 1500$$

we now consider finding the *instantaneous rate of change of z with respect to* x (i.e., the instantaneous rate of change of cost with respect to the number of tanks produced). This is expressed by finding the derivative of z with respect to x. However, since $z = f(x, y)$ has two independent variables, the derivative of z with respect to one of the independent variables is called a **partial derivative**, or simply **partial**. Specifically, the derivative of z with respect to x is called the **partial derivative of z with respect to x** and is denoted by any of the following:

$$\frac{\partial z}{\partial x} \qquad f_x(x, y) \qquad f_x$$

The partial derivative of z with respect to x is found by treating x as a variable and the remaining independent variables (in this example, y) as constants and applying the differentiation rules of this chapter. Thus, if

$$z = f(x, y) = x^2 - 8x + y^2 - 12y + 1500$$

then

$$f_x(x, y) = 2x - 8$$

Similarly, the *instantaneous rate of change of z with respect to* y is expressed by the **partial derivative of z with respect to y** and is denoted by any of the following:

$$\frac{\partial z}{\partial y} \qquad f_y(x, y) \qquad f_y$$

The partial derivative of z with respect to y is found by treating y as a variable and the remaining independent variables (in this example, x) as constants. Hence, if

$$z = f(x, y) = x^2 - 8x + y^2 - 12y + 1500$$

then

$$f_y(x, y) = 2y - 12$$

EXAMPLE 9–11 The revenue, z, derived from selling x calculators and y adding machines is given by the function

$$z = f(x, y) = -x^2 + 8x - 2y^2 + 6y + 2xy + 50$$

(A) At a sales level of $x = 4$ calculators and $y = 3$ adding machines, find the marginal revenue resulting from the sale of one additional calculator.

(B) At a sales level of $x = 4$ calculators and $y = 3$ adding machines, find the marginal revenue resulting from the sale of one additional adding machine.

Solutions

(A) Since the marginal revenue resulting from the sale of one additional calculator is the instantaneous rate of change of z with respect to x, we calculate f_x by treating x as a variable and y as a constant. We thus obtain

$$f_x(x, y) = -2x + 8 + 2y$$
$$f_x(4, 3) = -2(4) + 8 + 2(3) = 6$$

Therefore, at $x = 4$ and $y = 3$, sales revenue is increasing at the rate of $6 per calculator sold. Hence, the marginal revenue is $6.

(B) Since the marginal revenue resulting from the sale of one additional adding machine is the instantaneous rate of change of z with respect to y, we calculate f_y by treating y as a variable and x as a constant. Here, we get

$$f_y(x, y) = -4y + 6 + 2x$$
$$f_y(4, 3) = -4(3) + 6 + 2(4) = 2$$

Thus, at $x = 4$ and $y = 3$, sales revenue is increasing at the rate of $2 per calculator. Hence, the marginal revenue is $2.

563 DIFFERENTIAL CALCULUS: OPTIMIZATION

Second Partial Derivatives   The partial derivative of a partial derivative is called a **second partial derivative,** or simply **second partial.** Specifically, if $z = f(x, y)$, then there are four second partials:

1. The partial derivative of $\partial z/\partial x$ with respect to x, or

$$\frac{\partial}{\partial x}\left(\frac{\partial z}{\partial x}\right)$$

   This is denoted by any of the following:

$$\frac{\partial^2 z}{\partial x^2} \qquad f_{xx}(x, y) \qquad f_{xx}$$

2. The partial derivative of $\partial z/\partial y$ with respect to y, or

$$\frac{\partial}{\partial y}\left(\frac{\partial z}{\partial y}\right)$$

   This is denoted by any of the following:

$$\frac{\partial^2 z}{\partial y^2} \qquad f_{yy}(x, y) \qquad f_{yy}$$

3. The partial derivative of $\partial z/\partial y$ with respect to x, or

$$\frac{\partial}{\partial x}\left(\frac{\partial z}{\partial y}\right)$$

   This is denoted by any of the following:

$$\frac{\partial^2 z}{\partial x \partial y} \qquad f_{yx}(x, y) \qquad f_{yx}$$

4. The partial derivative of $\partial z/\partial x$ with respect to y, or

$$\frac{\partial}{\partial y}\left(\frac{\partial z}{\partial x}\right)$$

   This is denoted by any of the following:

$$\frac{\partial^2 z}{\partial y \partial x} \qquad f_{xy}(x, y) \qquad f_{xy}$$

Therefore, if

$$z = f(x, y) = x^5 - y^4 + 3x^2 y^6 + 18$$

then

$$f_x = 5x^4 + 6xy^6$$

The derivative of f_x with respect to x is f_{xx}. Thus, treating x as a variable and y as a constant, we have

$$f_{xx} = 20x^3 + 6y^6$$

The derivative of f_x with respect to y is f_{xy}. Returning to f_x, we treat y as a variable and x as a constant to obtain

$$f_{xy} = 36xy^5$$

Returning to $z = f(x, y) = x^5 - y^4 + 3x^2y^6 + 18$, we calculate

$$f_y = -4y^3 + 18x^2y^5$$

The derivative of f_y with respect to y is f_{yy}. Treating x as a constant and y as a variable, we have

$$f_{yy} = -12y^2 + 90x^2y^4$$

The derivative of f_y with respect to x is f_{yx}. Returning to f_y, we treat x as a variable and y as a constant to obtain

$$f_{yx} = 36xy^5$$

Observe that $f_{xy} = f_{yx}$. This is true if f_x, f_y, f_{xy}, and f_{yx} are all continuous. In this text, $f_{xy} = f_{yx}$ for all functions $z = f(x, y)$.

EXAMPLE 9–12 If $z = f(x, y) = 3x^2 + 4y^5 - 8x^3y^6 + 15$, calculate each of the following:
(A) f_x (B) f_{xx} (C) $f_{xx}(1, 0)$ (D) f_{xy}
(E) f_y (F) f_{yy} (G) f_{yx} (H) $f_{yx}(-1, 1)$

Solutions
(A) $f_x = 6x - 24x^2y^6$
(B) $f_{xx} = 6 - 48xy^6$
(C) $f_{xx}(1, 0) = 6 - 48(1)(0)^6 = 6$
(D) $f_{xy} = -144x^2y^5$
(E) $f_y = 20y^4 - 48x^3y^5$
(F) $f_{yy} = 80y^3 - 240x^3y^4$
(G) $f_{yx} = -144x^2y^5$
(H) $f_{yx}(-1, 1) = -144(-1)^2(1)^5 = -144$

EXERCISES

1. If $f(x, y) = 3x^2 - 4y + 6xy + y^2 + 5$, find each of the following:
   (A) $f(0, 1)$ (B) $f(2, 5)$ (C) $f(1, 0)$
2. If $h(s, t) = 4s^3 - 2s^2t + 3t^2 + 8t$, find each of the following:
   (A) $h(1, 2)$ (B) $h(0, 2)$ (C) $h(1, 0)$

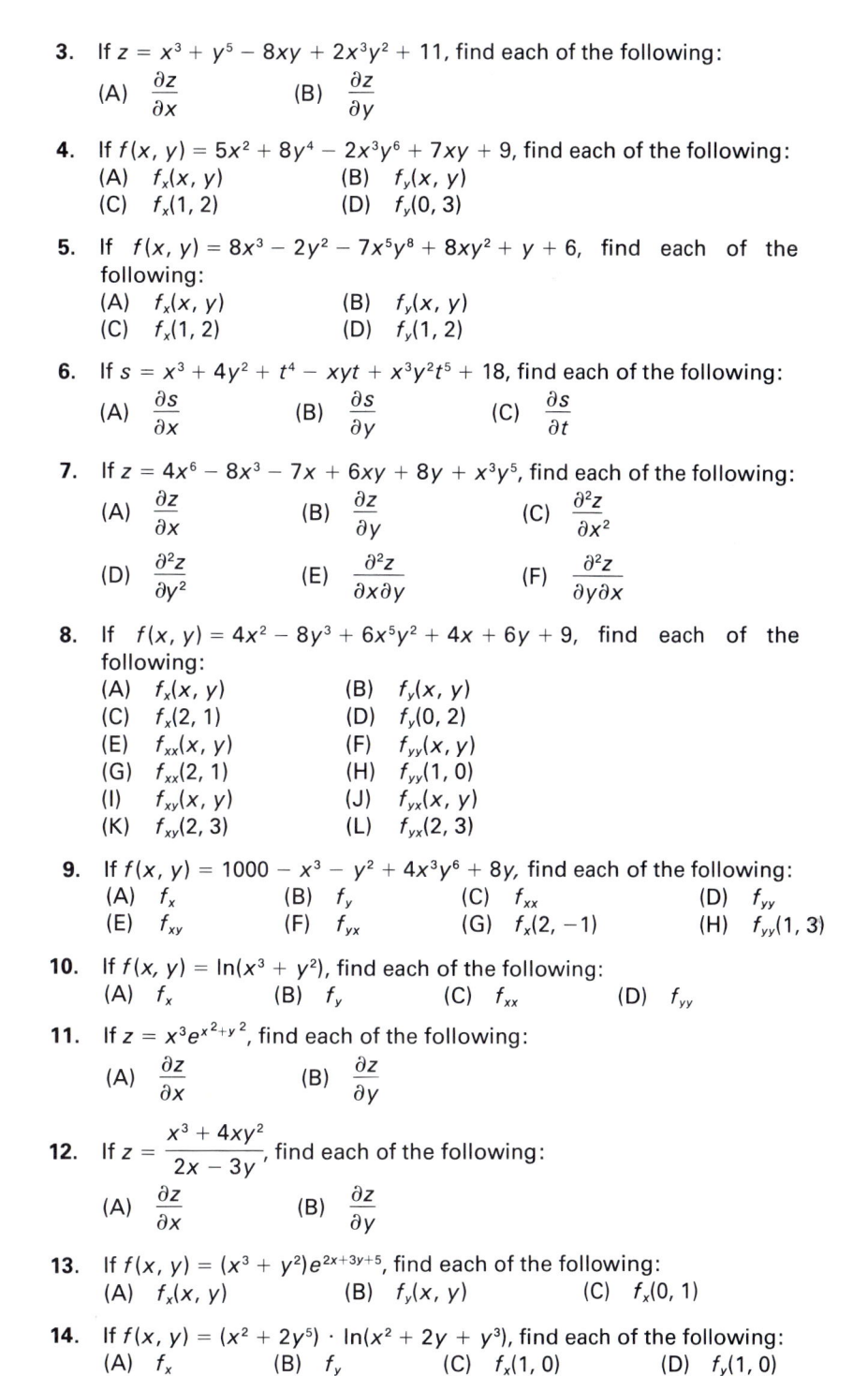

3. If $z = x^3 + y^5 - 8xy + 2x^3y^2 + 11$, find each of the following:
   (A) $\dfrac{\partial z}{\partial x}$
   (B) $\dfrac{\partial z}{\partial y}$

4. If $f(x, y) = 5x^2 + 8y^4 - 2x^3y^6 + 7xy + 9$, find each of the following:
   (A) $f_x(x, y)$
   (B) $f_y(x, y)$
   (C) $f_x(1, 2)$
   (D) $f_y(0, 3)$

5. If $f(x, y) = 8x^3 - 2y^2 - 7x^5y^8 + 8xy^2 + y + 6$, find each of the following:
   (A) $f_x(x, y)$
   (B) $f_y(x, y)$
   (C) $f_x(1, 2)$
   (D) $f_y(1, 2)$

6. If $s = x^3 + 4y^2 + t^4 - xyt + x^3y^2t^5 + 18$, find each of the following:
   (A) $\dfrac{\partial s}{\partial x}$
   (B) $\dfrac{\partial s}{\partial y}$
   (C) $\dfrac{\partial s}{\partial t}$

7. If $z = 4x^6 - 8x^3 - 7x + 6xy + 8y + x^3y^5$, find each of the following:
   (A) $\dfrac{\partial z}{\partial x}$
   (B) $\dfrac{\partial z}{\partial y}$
   (C) $\dfrac{\partial^2 z}{\partial x^2}$
   (D) $\dfrac{\partial^2 z}{\partial y^2}$
   (E) $\dfrac{\partial^2 z}{\partial x \partial y}$
   (F) $\dfrac{\partial^2 z}{\partial y \partial x}$

8. If $f(x, y) = 4x^2 - 8y^3 + 6x^5y^2 + 4x + 6y + 9$, find each of the following:
   (A) $f_x(x, y)$
   (B) $f_y(x, y)$
   (C) $f_x(2, 1)$
   (D) $f_y(0, 2)$
   (E) $f_{xx}(x, y)$
   (F) $f_{yy}(x, y)$
   (G) $f_{xx}(2, 1)$
   (H) $f_{yy}(1, 0)$
   (I) $f_{xy}(x, y)$
   (J) $f_{yx}(x, y)$
   (K) $f_{xy}(2, 3)$
   (L) $f_{yx}(2, 3)$

9. If $f(x, y) = 1000 - x^3 - y^2 + 4x^3y^6 + 8y$, find each of the following:
   (A) f_x
   (B) f_y
   (C) f_{xx}
   (D) f_{yy}
   (E) f_{xy}
   (F) f_{yx}
   (G) $f_x(2, -1)$
   (H) $f_{yy}(1, 3)$

10. If $f(x, y) = \ln(x^3 + y^2)$, find each of the following:
    (A) f_x
    (B) f_y
    (C) f_{xx}
    (D) f_{yy}

11. If $z = x^3 e^{x^2 + y^2}$, find each of the following:
    (A) $\dfrac{\partial z}{\partial x}$
    (B) $\dfrac{\partial z}{\partial y}$

12. If $z = \dfrac{x^3 + 4xy^2}{2x - 3y}$, find each of the following:
    (A) $\dfrac{\partial z}{\partial x}$
    (B) $\dfrac{\partial z}{\partial y}$

13. If $f(x, y) = (x^3 + y^2)e^{2x+3y+5}$, find each of the following:
    (A) $f_x(x, y)$
    (B) $f_y(x, y)$
    (C) $f_x(0, 1)$

14. If $f(x, y) = (x^2 + 2y^5) \cdot \ln(x^2 + 2y + y^3)$, find each of the following:
    (A) f_x
    (B) f_y
    (C) $f_x(1, 0)$
    (D) $f_y(1, 0)$

566 CHAPTER NINE

15. The cost of producing x washers and y dryers is given by

$$C(x, y) = 40x + 200y + 10xy + 300$$

Presently, 50 washers and 90 dryers are being produced. Find the marginal cost of producing
    (A) One more washer.   (B) One more dryer.

16. The revenue derived from selling x toasters and y broilers is given by

$$R(x, y) = 2x^2 + y^2 + 4x + 5y + 1000$$

At present, the retailer is selling 30 toasters and 50 broilers. Which of these two product lines should be expanded in order to yield the greater increase in revenue?

17. The annual profit of a certain hotel is given by

$$P(x, y) = 100x^2 + 4y^2 + 2x + 5y + 100,000$$

where x is the number of rooms available for rent and y is the monthly advertising expenditures. Presently, the hotel has 90 rooms available and is spending $1000 per month on advertising.
   (A) If an additional room is constructed in an unfinished area, how will this affect annual profits?
   (B) If an additional dollar is spent on monthly advertising expenditures, how will this affect annual profit?

18. Two brands of ice cream, Farmer's Delight and Mellow Creme, are competing for the same market. The demands (in cases) for Farmer's Delight and Mellow Creme are represented by D_f and D_m, respectively. If x is the price of one case of Farmer's Delight and y is the price of one case of Mellow Creme, then

$$D_f = 5000 - 50x + 25y - 2xy$$
$$D_m = 6000 + 30x - 20y - xy$$

In economics, these two products are said to be **competitive** at those values of x and y for which

$$\frac{\partial D_f}{\partial y} > 0 \quad \text{and} \quad \frac{\partial D_m}{\partial x} > 0$$

Find those prices at which these two products are competitive.

9–5

MAXIMA AND MINIMA

Relative maxima and minima of multivariate functions are defined in a manner similar to that used for single variable functions. The graphs of Figure 9–30 illustrate relative maximum and minimum points of multivariate functions. Notice that the point labeled "Relative minimum" is lower than any of its neighboring points and that the point labeled "Relative maximum" is higher than any of its neighboring points.

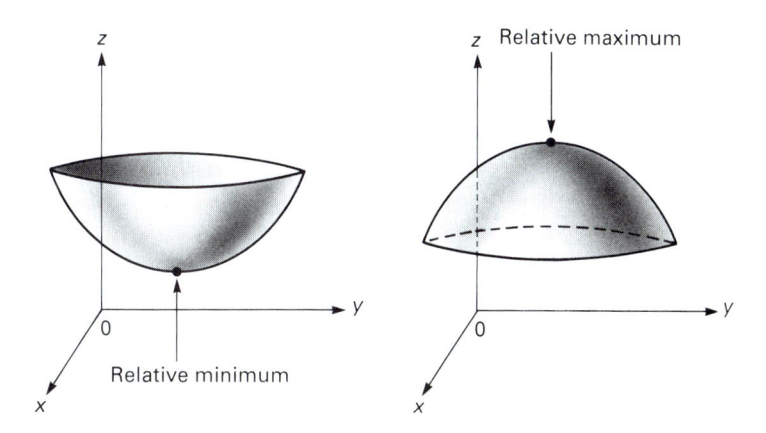

FIGURE 9–30

To determine the relative maxima and minima of multivariate functions, we first search for critical points. A point (x_1, y_1) is called a **critical point** of the multivariate function $z = f(x, y)$ if

$$f_x(x_1, y_1) = 0 \quad \text{and} \quad f_y(x_1, y_1) = 0$$

Consider the multivariate function

$$z = f(x, y) = x^2 - 8x + y^2 - 12y + 1500$$

of Section 9–4. To find its critical point(s), we must calculate f_x and f_y, set them equal to 0, and solve for x and y. Since

$$f_x = 2x - 8 \qquad f_y = 2y - 12$$

then

$$0 = 2x - 8 \qquad 0 = 2y - 12$$
$$x = 4 \qquad y = 6$$

Thus, the point $(x_1, y_1) = (4, 6)$ is a critical point of $z = f(x, y)$.

A critical point may or may not be either a relative maximum or a relative minimum. Recall, this is also the case with single-variable functions. Just as we have a second-derivative test for single-variable functions, we also have a similar test to identify the relative maxima and minima of multivariate functions. This test is stated as follows.

> **Second-Derivative Test for Multivariate Functions**
>
> Let $z = f(x, y)$ be a multivariate function such that f_x, f_y, f_{xx}, and f_{yy} are continuous. Let (x_1, y_1) be a critical point of $z = f(x, y)$. Also, let
>
> $$A = f_{xx}(x_1, y_1)$$
> $$B = f_{yy}(x_1, y_1)$$
> $$C = f_{xy}(x_1, y_1)$$
>
> Then:
>
> 1. $f(x_1, y_1)$ is a relative maximum if $AB - C^2 > 0$ and $A < 0$.
> 2. $f(x_1, y_1)$ is a relative minimum if $AB - C^2 > 0$ and $A > 0$.
> 3. $f(x_1, y_1)$ is a **saddle point** if $AB - C^2 < 0$. (A saddle point is illustrated in the graph of Figure 9–31.)
>
> Note that the critical point labeled "Saddle point" is neither a relative maximum nor a relative minimum.
>
> **FIGURE 9–31**
>
> 4. The test fails and no information is given about the point (x_1, y_1) if $AB - C^2 = 0$.

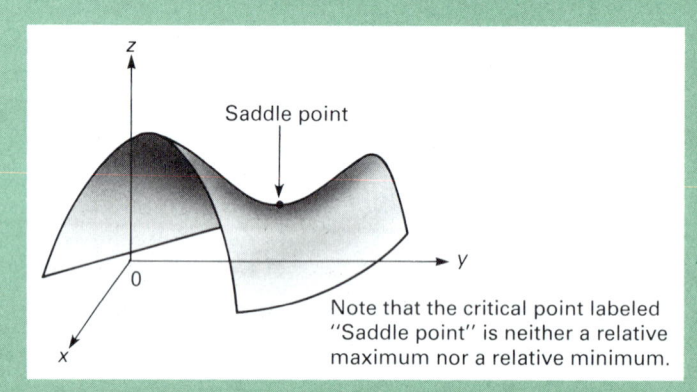

Returning to our function

$$z = f(x, y) = x^2 - 8x + y^2 - 12y + 1500$$

and its first-order partials

$$f_x = 2x - 8 \quad \text{and} \quad f_y = 2y - 12$$

we now calculate

$$f_{xx} = 2 \quad f_{yy} = 2 \quad f_{xy} = 0$$

DIFFERENTIAL CALCULUS: OPTIMIZATION

Since the critical point is $(x_1, y_1) = (4, 6)$, we have

$$A = f_{xx}(4, 6) = 2$$
$$B = f_{yy}(4, 6) = 2$$
$$C = f_{xy}(4, 6) = 0$$
$$AB - C^2 = 2(2) - 0^2 = 4$$

Since $AB - C^2 > 0$ and $A > 0$, then, according to the second-derivative test, a relative minimum occurs at $(4, 6)$. Thus,

$$z = f(4, 6) = (4)^2 - 8(4) + 6^2 - 12(6) + 1500 = \$1448$$

is a relative minimum. Since z is the daily production cost, x is the number of tanks produced, and y is the man-hours of labor used daily, a minimum daily production cost of $z = \$1448$ occurs when $x = 4$ tanks are produced and $y = 6$ man-hours of labor are used.

EXAMPLE 9–13 In Example 9–11, we considered the function

$$z = f(x, y) = -x^2 + 8x - 2y^2 + 6y + 2xy + 50$$

where z is the sales revenue from selling x calculators and y adding machines.

(A) How many calculators and adding machines should be sold in order to maximize sales revenue?
(B) What is the maximum sales revenue?

Solutions
(A) We first calculate f_x and f_y, as follows:

$$f_x = -2x + 8 + 2y$$
$$f_y = -4y + 6 + 2x$$

Setting f_x and f_y equal to 0, we have

$$0 = -2x + 8 + 2y$$
$$0 = -4y + 6 + 2x$$

Solving this linear system for x and y, we obtain $x = 11$ and $y = 7$. Thus, a critical point is $(11, 7)$. We now calculate

$$f_{xx} = -2 \qquad f_{yy} = -4 \qquad f_{xy} = 2$$

EXAMPLE 9–13 (continued)

Since the critical point is $(x_1, y_1) = (11, 7)$, then

$$A = f_{xx}(11, 7) = -2$$
$$B = f_{yy}(11, 7) = -4$$
$$C = f_{xy}(11, 7) = 2$$
$$AB - C^2 = -2(-4) - 2^2 = 4$$

Since $AB - C^2 > 0$ and $A < 0$, then, according to the second-derivative test, a relative maximum occurs at $(11, 7)$. Thus, in order to maximize revenue, $x = 11$ calculators and $y = 7$ adding machines must be sold.

(B) The maximum sales revenue is

$$z = f(11, 7) = -(11)^2 + 8(11) - 2(7)^2 + 6(7) + 2(11)(7) + 50$$
$$= \$115$$

EXERCISES

1. Find any relative maxima or relative minima of each of the following:
   (A) $z = x^2 + 2y^2 - 8x - 20y + 18$
   (B) $f(x, y) = 2x^2 + y^2 - 28x - 20y + 80$
   (C) $f(x, y) = 9x - 50y + x^2 + 5y^2 + 100$
   (D) $z = 40x + 160y - 2x^2 - 4y^2 + 1000$
   (E) $f(x, y) = 1000 + 80x + 100y - 2x^2 - y^2$

2. Find any relative maxima or relative minima of each of the following:
   (A) $f(x, y) = 4x^2 + 2y^2 + 3xy - 70x - 55y + 1000$
   (B) $z = 3x^2 + 4y^2 + 2xy - 30x - 32y + 50$
   (C) $f(x, y) = 4x^2 + 5y^2 + 5xy - 73x - 80y + 6$
   (D) $z = 200 - 2x^2 - 6y^2 + 2xy + 32x + 28y$
   (E) $f(x, y) = 80 + 3xy + 42x - 16y - 3x^2 - 2y^2$

3. Show that the function defined by

   $$f(x, y) = 500 + x^2 - 2y^2 - 18x + 16y$$

   has neither a relative maximum nor a relative minimum. Additionally, show that $f(x, y)$ has a saddle point. Find the saddle point.

4. Show that the function defined by

   $$z = x^2 + 2y^2 + 3xy - 40x - 55y + 100$$

   has a saddle point. Find the saddle point. Does the function have any relative maxima or minima?

571 DIFFERENTIAL CALCULUS: OPTIMIZATION

5. The profit of a company is given by
$$P(x, y) = 1{,}000{,}000 + 1600x + 2000y - 4x^2 - 2y^2$$
where x is the unit labor cost and y is the unit raw material cost.
   - (A) Find the unit labor cost and unit raw material cost that maximizes profit.
   - (B) Find the maximum profit.

6. The manager of Freddy's Frogurt Stand has determined that the cost of producing x gallons of strawberry frogurt and y gallons of blueberry frogurt is given by
$$C(x, y) = 2x^2 + 3y^2 + 2xy - 800x - 1400y + 185{,}000$$
   - (A) How many gallons of each flavor should be produced in order to minimize cost?
   - (B) Find the minimum cost.

7. A large bottling company manufactures two competing brands of soda: Crystal Club and Mineral Club. The demands (in cases) for Crystal Club and Mineral Club are represented by D_c and D_m, respectively. If x is the price for one case of Crystal Club and y is the price for one case of Mineral Club, then
$$D_c = 200 - 20x + y$$
$$D_m = 300 - 15y + 2x$$
   - (A) Find the sales revenue, $R(x, y)$, from both brands. (*Hint:* Solve for $R(x, y) = xD_c + yD_m$.)
   - (B) How should each brand be priced in order to maximize sales revenue, $R(x, y)$?
   - (C) Find the maximum sales revenue.
   - (D) Find the demand for each brand at the optimal prices x and y.

8. The weekly output of a firm is given by
$$z(x, y) = 1000x + 1600y + 2xy - 5x^2 - 2y^2$$
where x is the number of hours of labor and y is the number of units of raw material used weekly.
   - (A) How many hours of labor and how many units of raw material should be used weekly in order to maximize output?
   - (B) Find the maximum output.

CASE F BACK-ORDER INVENTORY MODEL— BFI, INC

A distributor of plumbing supplies, BFI, Inc., allows back orders to be taken when demand exceeds the available supply of inventory. This results in the imposition of a stockout cost (or penalty cost) upon the average number of back orders. If

D = annual demand (in units) for a given inventory product
K = cost of placing an order
H = annual carrying cost per unit
B = annual stockout cost (or back-order cost) per unit
Q = number of units ordered per order
S = maximum number of back orders allowed

then the total annual inventory cost in such situations is given by

$$C(Q, S) = K\frac{D}{Q} + H\frac{(Q - S)^2}{2Q} + B\frac{S^2}{2Q}$$

Figure 1 illustrates a graph of inventory level versus time for such an inventory model.

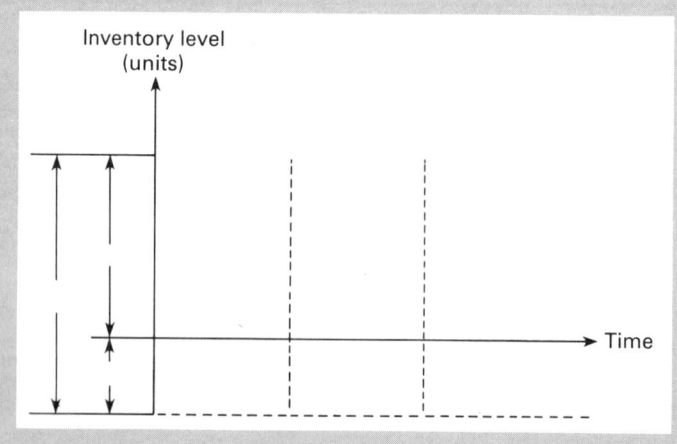

FIGURE 1

One of BFI's products, the BFI307, has an annual demand of 1200, an ordering cost of $5.00 per order, an annual carrying cost of $1.15 per unit, and an annual stockout cost of $2.40 per unit.

EXERCISES
1. Determine the equation defining the total annual inventory cost.
2. Determine the values of Q and S that minimize the total annual inventory cost.
3. What is the minimum total annual inventory cost?
4. How many back orders are allowed?
5. Every time an order is placed for the BFI307, how many are ordered? After back orders have been set aside, how many are available for sale?

INTEGRATION

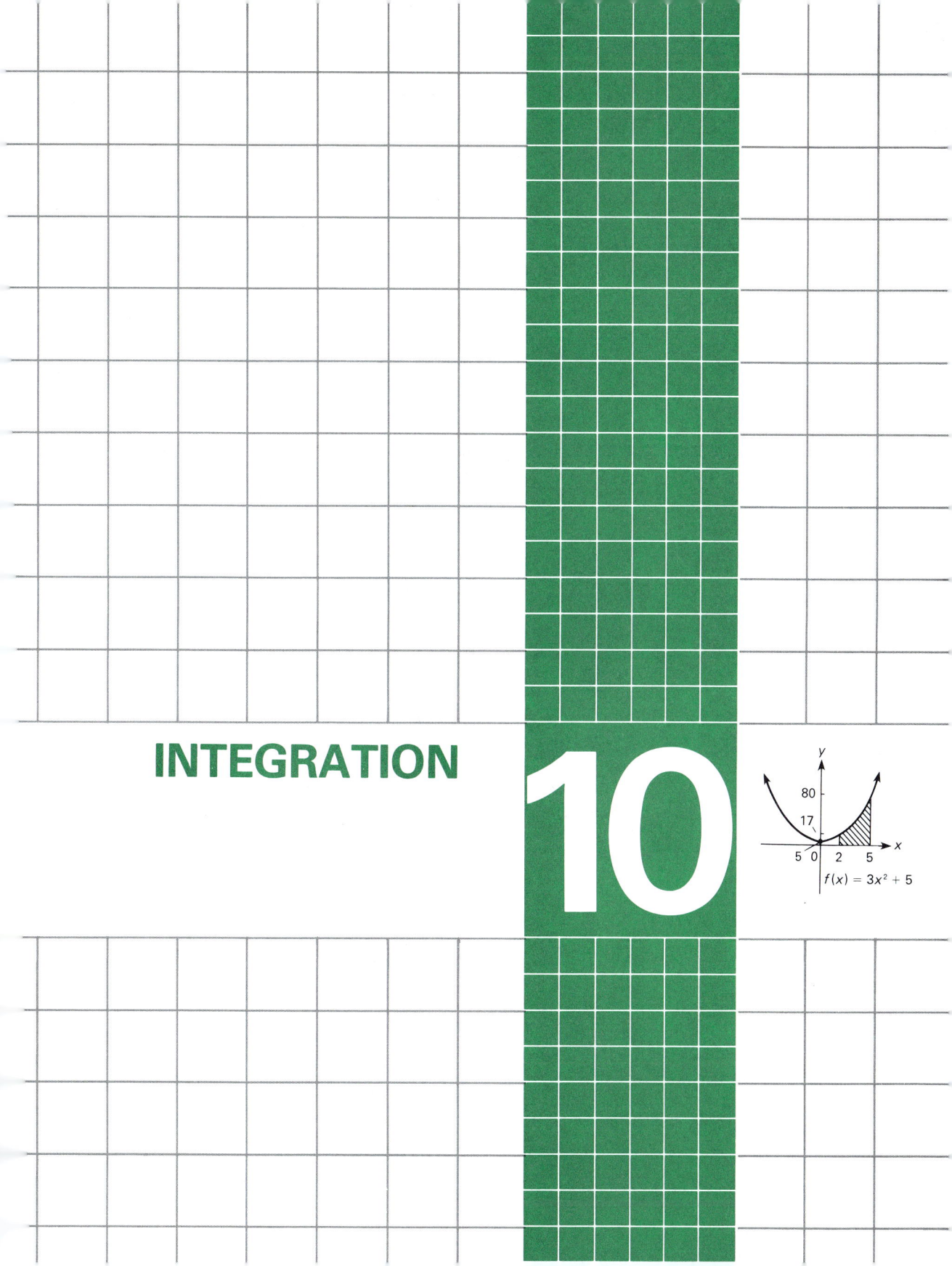

10

$f(x) = 3x^2 + 5$

10–1

ANTI-DIFFERENTIATION

In Chapters 8 and 9, we discussed the calculations of derivatives $F'(x)$ of functions $F(x)$. The operation of calculating the derivative of a function is called *differentiation*. The result, $F'(x)$, is called the *derivative* of the function $F(x)$. In this chapter, we will perform an operation that is the *reverse* of differentiation. It is called **anti-differentiation**. We will encounter problems in which we are given the derivative $F'(x)$ and must determine the **antiderivative $F(x)$**.

EXAMPLE 10–1 Find an antiderivative of $f(x) = x^2$.

Solution We seek a function $F(x)$ whose derivative is x^2. One such function is

$$F(x) = \frac{1}{3} x^3$$

since $F'(x) = x^2$. Another such function is

$$F(x) = \frac{1}{3} x^3 + 5$$

since $F'(x) = x^2$. In summary, the antiderivative of $f(x)$ is any function of the form

$$F(x) = \frac{1}{3} x^3 + c$$

where c is an arbitrary constant. The antiderivative $F(x)$ represents a family of curves, as is illustrated in Figure 10–1.

FIGURE 10–1

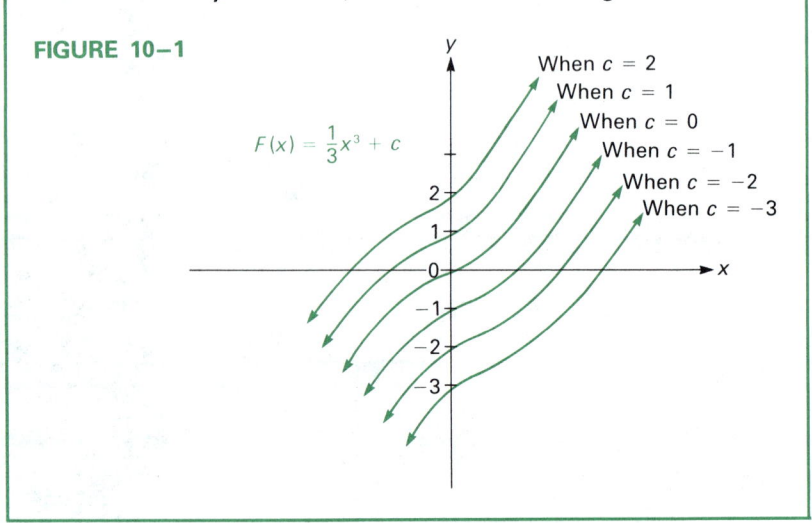

$F(x) = \frac{1}{3}x^3 + c$

When $c = 2$
When $c = 1$
When $c = 0$
When $c = -1$
When $c = -2$
When $c = -3$

EXAMPLE 10-2 Find the antiderivatives of $f(x) = x^3$.
Solution We seek all functions $F(x)$ such that $F'(x) = x^3$. One such function is

$$F(x) = \frac{1}{4} x^4$$

since $F'(x) = x^3$. In fact, any function of the form

$$F(x) = \frac{1}{4} x^4 + c$$

where c is an arbitrary constant, is an antiderivative of $f(x) = x^3$ since $F'(x) = x^3$.

Note that in Examples 10-1 and 10-2, we have determined antiderivatives of functions of the form

$$f(x) = x^n \quad (n \neq -1)$$

We now state a general rule for finding the antiderivatives of such functions.

Rule 10-1

Antiderivatives of $x^n = \dfrac{1}{n+1} x^{n+1} + c$, where c is an arbitrary constant and $n \neq -1$.

Note that if $F(x) = [1/(n+1)]x^{n+1} + c$, then $F'(x) = x^n$.

Indefinite Integrals

The antiderivatives of a function $f(x)$ are denoted by

$$\int f(x)\, dx$$

The symbol $\int$ is an integral sign, and the entire expression $\int f(x)\, dx$ is called the **indefinite integral** of $f(x)$. Thus, rule 10-1 may be restated as follows.

Rule 10-1 restated

$$\int x^n \, dx = \frac{1}{n+1} x^{n+1} + c \quad (n \neq -1)$$

EXAMPLE 10-3 If $f(x) = \sqrt{x}$, find $\int f(x) \, dx$.
Solution Since $f(x) = \sqrt{x} = x^{1/2}$, then, using rule 10-1 with $n = 1/2$, we have

$$\int x^{1/2} \, dx = \frac{1}{\frac{1}{2} + 1} x^{\frac{1}{2}+1} + c$$

$$= \frac{2}{3} x^{3/2} + c$$

Note that the derivative of $\frac{2}{3} x^{3/2} + c$ is $x^{1/2}$.

EXAMPLE 10-4 Determine $\int (1/x^2) \, dx$.
Solution Since $1/x^2 = x^{-2}$, then, using rule 10-1 with $n = -2$, we have

$$\int x^{-2} \, dx = \frac{1}{-2+1} x^{-2+1} + c$$

$$= -x^{-1} + c$$

$$= -\frac{1}{x} + c$$

Observe that the derivative of $-x^{-1} + c$ is x^{-2}.

Up to this point, we have used rule 10-1 to determine the antiderivatives, or integrals, of functions of the form $f(x) = x^n$, where $n \neq -1$. We now consider functions of the form

$$f(x) = kx^n \quad (n \neq 1)$$

where k is a constant. Rule 10-2 provides a useful property of integrals which enables us to integrate such functions.

Rule 10-2

$$\int kf(x)\,dx = k\int f(x)\,dx$$

Rule 10-2 states that the integral of a constant k times a function $f(x)$ is equal to the constant k times the integral of the function $f(x)$. Specifically, we have

$$\int 5x^3\,dx = 5\int x^3\,dx$$
$$= 5\left(\frac{1}{4}x^4 + c\right)$$
$$= \frac{5}{4}x^4 + 5c$$

Since c is an arbitrary constant, then $5c$ is an arbitrary constant and may be written as c. Hence, the expression becomes

$$\int 5x^3\,dx = \frac{5}{4}x^4 + c$$

EXAMPLE 10-5 Find $\int 8x\,dx$.
Solution Using rules 10-2 and 10-1, we have

$$\int 8x\,dx = 8\int x\,dx$$
$$= 8\left(\frac{1}{2}x^2 + c\right)$$
$$= \frac{8}{2}x^2 + 8c$$
$$= 4x^2 + 8c$$

Rewriting the arbitrary constant $8c$ as c, we obtain

$$\int 8x\,dx = 4x^2 + c$$

> **EXAMPLE 10–6** Find $\int 9\,dx$.
> **Solution** Since $9 = 9x^0$, we have
>
> $$\int 9x^0\,dx = 9\int x^0\,dx$$
> $$= 9(x + c)$$
> $$= 9x + 9c$$
>
> Rewriting the arbitrary constant $9c$ as c, we obtain
>
> $$\int 9\,dx = 9x + c$$

In general, if k is a constant, then

$$\int k\,dx = kx + c$$

where c is an arbitrary constant. Note that the derivative of $kx + c$ is k.

Integral of a Sum (or Difference) Sometimes, we must determine the integral of a sum (or difference), $f(x) \pm g(x)$, of two functions $f(x)$ and $g(x)$. Rule 10–3 provides a useful property of integrals which enables us to integrate such functions.

> **Rule 10–3**
>
> $$\int (f(x) \pm g(x))\,dx = \int f(x)\,dx \pm \int g(x)\,dx$$

Rule 10–3 states that the integral of a sum (or difference) of two functions is the sum (or difference) of their integrals. For example,

$$\int (5x^7 + 7x)\,dx = \int 5x^7\,dx + \int 7x\,dx$$
$$= \frac{5}{8}x^8 + \frac{7}{2}x^2 + c$$

Rule 10–3 may be extended to sums (or differences) that involve three or more functions. Hence,

581 INTEGRATION

$$\int \left(9x^6 - \frac{1}{2}x^2 + 4\right) dx = \int 9x^6 \, dx - \int \frac{1}{2}x^2 \, dx + \int 4 \, dx$$

$$= \frac{9}{7}x^7 - \frac{1}{6}x^3 + 4x + c$$

EXAMPLE 10–7 Determine $\int (3x^2 - x^{-5} - 7) \, dx$.

Solution

$$\int (3x^2 - x^{-5} - 7) \, dx = \int 3x^2 \, dx - \int x^{-5} \, dx - \int 7 \, dx$$

$$= \frac{3}{3}x^3 - \frac{1}{-4}x^{-4} - 7x + c$$

$$= x^3 + \frac{1}{4}x^{-4} - 7x + c$$

$$= x^3 + \frac{1}{4x^4} - 7x + c$$

Sometimes, the independent variable is denoted by a letter other than x. Such a case is illustrated in Example 10–8.

EXAMPLE 10–8 Find $\int (u^5 - 8u + 3) \, du$.

Solution

$$\int (u^5 - 8u + 3) \, du = \int u^5 \, du - \int 8u \, du + \int 3 \, du$$

$$= \frac{1}{6}u^6 - \frac{8}{2}u^2 + 3u + c$$

$$= \frac{1}{6}u^6 - 4u^2 + 3u + c$$

EXAMPLE 10–9 A firm produces picture frames. At a production level of x frames, the marginal cost is

$$C'(x) = 4x + 5$$

Find the cost equation, $C(x)$, if the fixed cost is $500.

EXAMPLE 10-9 (continued)

Solution Here, we calculate

$$C(x) = \int C'(x)\, dx$$
$$= \int (4x + 5)\, dx$$
$$= 2x^2 + 5x + c$$

Since the fixed cost is $500, then $C(0) = 500$. Hence,

$$500 = 2(0^2) + 5(0) + c$$
$$= c$$

Thus, we obtain

$$C(x) = 2x^2 + 5x + 500$$

EXAMPLE 10-10 *(Capital Formation)* The value, V, of an investment fund changes over time, t (in years), at the rate

$$V'(t) = 18{,}000t^2$$

The initial (i.e., at $t = 0$) amount in the fund is $750,000.
(A) Find the equation defining V as a function of t.
(B) Find the value of the investment fund at the end of the fourth year (i.e., at $t = 4$).

Solutions
(A) Here,

$$V(t) = \int 18{,}000t^2\, dt$$
$$= 6000t^3 + c$$

Since $V(0) = 750{,}000$, then

$$750{,}000 = 6000(0^3) + c$$

Hence, $c = 750{,}000$, and

$$V(t) = 6000t^3 + 750{,}000$$

583 INTEGRATION

(B) The value of the fund at the end of the fourth year is

$$V(4) = 6000(4^3) + 750{,}000$$
$$= \$1{,}134{,}000$$

EXERCISES

1. For each of the following, verify that $F(x)$ is an antiderivative of $f(x)$:
   (A) $f(x) = 7x^6$; $F(x) = x^7$
   (B) $f(x) = 7x^6$; $F(x) = x^7 - 5$
   (C) $f(x) = x^2 - 8x + 5$; $F(x) = \frac{1}{3}x^3 - 4x^2 + 5x + 1$
   (D) $f(x) = \frac{1}{\sqrt[5]{x}}$; $F(x) = \frac{5}{4}\sqrt[5]{x^4} + 9$

2. Find each of the following integrals:
   (A) $\displaystyle\int x^{12}\, dx$ (B) $\displaystyle\int 5\, dx$
   (C) $\displaystyle\int \frac{dx}{x^3}$ (D) $\displaystyle\int \sqrt[4]{x}\, dx$
   (E) $\displaystyle\int \frac{1}{x^9}\, dx$ (F) $\displaystyle\int x^{-1/6}\, dx$
   (G) $\displaystyle\int \frac{1}{\sqrt[7]{x}}\, dx$ (H) $\displaystyle\int \frac{-4}{\sqrt[7]{x^2}}\, dx$
   (I) $\displaystyle\int \frac{-3}{\sqrt{x}}$ (J) $\displaystyle\int \frac{8}{x^{1/3}}\, dx$

3. Find each of the following integrals:
   (A) $\displaystyle\int (3x^2 - 8x + 5)\, dx$ (B) $\displaystyle\int (4x^3 - 16)\, dx$
   (C) $\displaystyle\int (2x^7 - 6x^4 - 1)\, dx$ (D) $\displaystyle\int (x^5 - 7x)\, dx$
   (E) $\displaystyle\int \left(5x^4 - \sqrt{x} + \frac{7}{\sqrt{x}}\right) dx$ (F) $\displaystyle\int \left(x^3 - \frac{4}{x^2} + 6\right) dx$

4. Find each of the following integrals:
   (A) $\displaystyle\int (u^4 - 6u^2 + 5)\, du$ (B) $\displaystyle\int (t^3 - 2t^2)\, dt$
   (C) $\displaystyle\int (v^2 - 1)\, dv$ (D) $\displaystyle\int (y^6 - 5y^4 + 1)\, dy$
   (E) $\displaystyle\int \left(t^6 - \frac{2}{t^2} + \frac{6}{\sqrt{t}}\right) dt$ (F) $\displaystyle\int \left(4u^3 - \frac{8}{\sqrt{u}} + \frac{5}{u^2}\right) du$

5. The Safe Ride Company produces tires. At a production level of x tires, the marginal cost is

$$C'(x) = 8x + 2$$

Find the cost function, $C(x)$, if the fixed cost is $1000.

6. Given the marginal cost function

$$C'(x) = 6x^2 + 4x - 5$$

find the cost function, $C(x)$, if the fixed cost is $800.

7. Given the marginal cost function

$$C'(x) = 6x + 1$$

find the cost function, $C(x)$, if the total cost of producing 2 units is $900.

8. The value, V, of a mutual fund changes over time, t (in years), at the rate

$$V'(t) = 24{,}000t^2$$

The initial amount (i.e., at $t = 0$) in the fund is $1,000,000.
   (A) Find the equation defining V as a function of t.
   (B) Find the value of the mutual fund at the end of the fifth year (i.e., at $t = 5$).

9. The marginal propensity to save, $S'(x)$, is a function of a nation's income, x (in billions of dollars), as defined by the equation

$$S'(x) = 0.5 - 0.12x^{-1/2} \qquad (x > 0)$$

If $S = 0$ when $x = 81$, then
   (A) Find the equation defining total savings, $S(x)$.
   (B) Find the total savings at a national income of $144 billion.

10–2

THE DEFINITE INTEGRAL AND AREA UNDER A CURVE

Definite Integrals

In Section 10–1, we discussed indefinite integrals of functions $f(x)$. We will now consider the mechanics of computing a *definite integral*. A definite integral results in a numerical value. A definite integral of a function $f(x)$ is evaluated within a definite range of x-values. The endpoints of this range of x-values are called *limits of integration*. We now illustrate the concept of a definite integral by the following example.

Consider a firm producing some commodity. At a production level of x units, the marginal cost is

$$C'(x) = 6x + 8$$

The antiderivative of the marginal cost is

585 INTEGRATION

$$C(x) = \int (6x + 8) \, dx$$
$$= 3x^2 + 8x + c$$

If we are told that the fixed cost is $600, then we determine the arbitrary constant to be $c = 600$. Hence, the cost function is

$$C(x) = 3x^2 + 8x + 600$$

Suppose we wish to determine the *total change* in cost if production rises from $x = 10$ to $x = 15$. This total change in cost is determined by evaluating

$$C(15) - C(10)$$

Since

$$C(15) = 3(15^2) + 8(15) + 600 = \$1395$$
$$C(10) = 3(10^2) + 8(10) + 600 = \$980$$

the total change in cost is

$$C(15) - C(10) = \$1395 - \$980$$
$$= \$415$$

In general, if a and b are numbers and $F(x)$ is a function, then the quantity

$$F(b) - F(a)$$

is the **total change** of $F(x)$ over the interval from $x = a$ to $x = b$. The quantity $F(b) - F(a)$ is often abbreviated by the symbol

$$F(x) \Big|_a^b$$

We now define a **definite integral**.

Definite Integral

Let a and b be numbers and $f(x)$ a continuous function with an antiderivative $F(x)$. Then, the **definite integral** of $f(x)$ from $x = a$ to $x = b$ is denoted by

$$\int_a^b f(x) \, dx = F(x) \Big|_a^b = F(b) - F(a)$$

The numbers a and b are called **limits of integration**. The definite integral $\int_a^b f(x) \, dx$ is the total change of the antiderivative $F(x)$ over the interval from $x = a$ to $x = b$.

It is assumed that $f(x)$ is continuous over the interval $a \leq x \leq b$.

EXAMPLE 10–11 Find $\int_1^2 x^2 \, dx$.
Solution Here, we write

$$\int_1^2 x^2 \, dx = F(x) \Big|_1^2 = F(2) - F(1)$$

where

$$F(x) = \frac{1}{3}x^3 + c$$

Hence, we have

$$F(2) - F(1) = \left[\frac{1}{3}(2^3) + c\right] - \left[\frac{1}{3}(1^3) + c\right]$$

$$= \left(\frac{8}{3} + c\right) - \left(\frac{1}{3} + c\right)$$

$$= \frac{7}{3}$$

Notice that the definite integral does not depend upon the choice of the arbitrary constant, c. Thus, we will choose $c = 0$ when computing definite integrals.

EXAMPLE 10–12 Find $\int_{-1}^3 (8x^3 - 4x + 5) \, dx$.
Solution

$$\int_{-1}^3 (8x^3 - 4x + 5) \, dx = \underbrace{2x^4 - 2x^2 + 5x}_{F(x)} \Big|_{-1}^3$$

$$= F(3) - F(-1)$$
$$= [2(3^4) - 2(3^2) + 5(3)] -$$
$$\quad [2(-1^4) - 2(-1^2) + 5(-1)]$$
$$= (162 - 18 + 15) - (2 - 2 - 5)$$
$$= 164$$

Area under a Curve   Suppose we had to compute the area bounded by the curve $f(x) = x^2$, the x-axis, and the vertical lines $x = 0$ and $x = 1$ (see Figure 10–2). We could obtain an approximation by arbitrarily

587 INTEGRATION

dividing the interval $0 \leq x \leq 1$ into, say, four subintervals and then cover the shaded area with the four rectangles as shown in Figure 10-3.

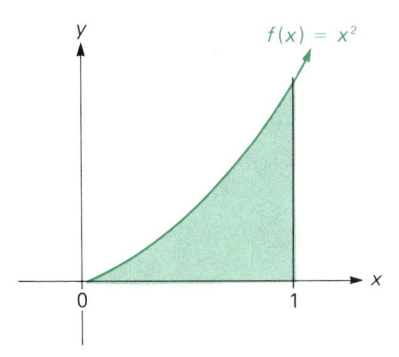

FIGURE 10-2

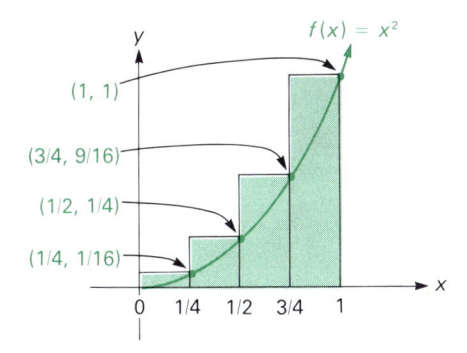

FIGURE 10-3

Observe that each rectangle has a width of 1/4 unit. Also, note that the height of each rectangle is given by the y-coordinate of the upper right-hand corner point (of the rectangle) on the function,

$$f(x) = x^2$$

Thus, the height of the first rectangle is

$$f\left(\frac{1}{4}\right) = \left(\frac{1}{4}\right)^2 = \frac{1}{16}$$

the height of the second rectangle is

$$f\left(\frac{1}{2}\right) = \left(\frac{1}{2}\right)^2 = \frac{1}{4}$$

the height of the third rectangle is

$$f\left(\frac{3}{4}\right) = \left(\frac{3}{4}\right)^2 = \frac{9}{16}$$

and the height of the fourth rectangle is

$$f(1) = (1)^2 = 1$$

Since the area of each rectangle is the width times the height, the sum of the areas of the rectangles is

$$A_1 = \frac{1}{4} \cdot f\left(\frac{1}{4}\right) + \frac{1}{4} \cdot f\left(\frac{1}{2}\right) + \frac{1}{4} \cdot f\left(\frac{3}{4}\right) + \frac{1}{4} \cdot f(1)$$

$$= \frac{1}{4} \cdot \frac{1}{16} + \frac{1}{4} \cdot \frac{1}{4} + \frac{1}{4} \cdot \frac{9}{16} + \frac{1}{4} \cdot 1$$

$$= \frac{15}{32} \text{ square unit}$$

Of course, this approximation is greater than the actual area. Figure 10–4 illustrates the use of rectangles to obtain an approximation less than the actual area. Observing Figure 10–4, note that the sum of the areas of the rectangles is

$$A_2 = \frac{1}{4} \cdot f\left(\frac{1}{4}\right) + \frac{1}{4} \cdot f\left(\frac{1}{2}\right) + \frac{1}{4} \cdot f\left(\frac{3}{4}\right)$$

$$= \frac{1}{4} \cdot \frac{1}{16} + \frac{1}{4} \cdot \frac{1}{4} + \frac{1}{4} \cdot \frac{9}{16}$$

$$= \frac{7}{32} \text{ square unit}$$

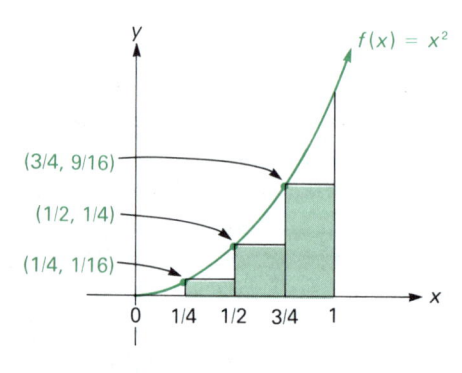

FIGURE 10–4

The actual area, A, lies somewhere between A_1 and A_2. Hence,

$$\frac{7}{32} < A < \frac{15}{32}$$

589 INTEGRATION

Riemann Sum A more accurate approximation of this area A of Figure 10–2 is obtained by dividing the interval $0 \le x \le 1$ into a greater number of subintervals and summing the areas of the respective rectangles. In Figure 10–5, we divide the interval $0 \le x \le 1$ into n subintervals each of length $1/n$. Again, the height of each rectangle is given by the y-coordinate of the upper right-hand corner point (of the rectangle) on the function,

$$f(x) = x^2$$

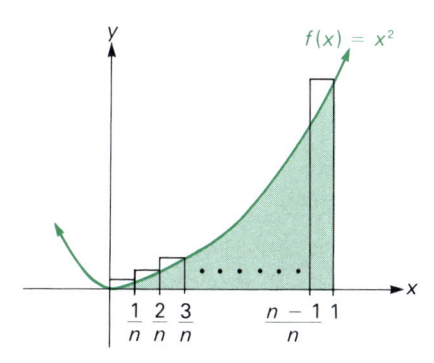

FIGURE 10–5

Thus, the height of the first rectangle is

$$f\left(\frac{1}{n}\right) = \left(\frac{1}{n}\right)^2 = \frac{1^2}{n^2}$$

the height of the second rectangle is

$$f\left(\frac{2}{n}\right) = \left(\frac{2}{n}\right)^2 = \frac{2^2}{n^2}$$

and the height of the nth rectangle is

$$f\left(\frac{n}{n}\right) = \left(\frac{n}{n}\right)^2 = \frac{n^2}{n^2}$$

Since the area of each rectangle is the width times the height, the sum of the areas of the respective rectangles is

$$S = \frac{1}{n} \cdot f\left(\frac{1}{n}\right) + \frac{1}{n} \cdot f\left(\frac{2}{n}\right) + \ldots + \frac{1}{n} \cdot f\left(\frac{n}{n}\right)$$

$$= \frac{1}{n} \cdot \frac{1^2}{n^2} + \frac{1}{n} \cdot \frac{2^2}{n^2} + \ldots + \frac{1}{n} \cdot \frac{n^2}{n^2}$$

$$= \frac{1}{n^3}(1^2 + 2^2 + \ldots + n^2)$$

It can be verified that

$$1^2 + 2^2 + \ldots + n^2 = \frac{n(n+1)(2n+1)}{6}$$

Substituting this result into the preceding equation for S yields

$$S = \frac{1}{n^3}\left[\frac{n(n+1)(2n+1)}{6}\right]$$

$$= \frac{(n+1)(2n+1)}{6n^2}$$

$$= \frac{2n^2 + 3n + 1}{6n^2}$$

$$= \frac{2n^2}{6n^2} + \frac{3n}{6n^2} + \frac{1}{6n^2}$$

$$= \frac{1}{3} + \frac{1}{2n} + \frac{1}{6n^2}$$

The sum S is called a **Riemann sum**. The Riemann sum S approximates the shaded area of Figure 10-5. The area is obtained by letting the number of rectangles increase without bound (i.e., let $n \to \infty$). Hence, as $n \to \infty$, $1/2n \to 0$, $1/6n^2 \to 0$, and $S \to 1/3$. Thus, the area, A, equals 1/3 square unit.

In summary, if $f(x)$ is a nonnegative continuous function over the interval $a \leq x \leq b$, then the area, A, between $f(x)$ and the x-axis from $x = a$ to $x = b$ may be approximated by the sum of areas of rectangles as illustrated in Figure 10-6.

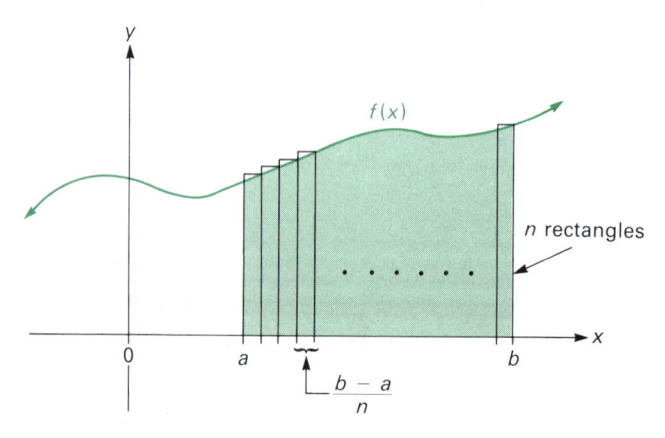

FIGURE 10-6

Each rectangle is constructed as follows:

1. The interval $a \leq x \leq b$ is divided into n subintervals each of width $dx = (b - a)/n$.

591 INTEGRATION

2. The height of the ith rectangle is $f(x_i)$, where x_i is a point in the ith subinterval (see Figure 10–7).

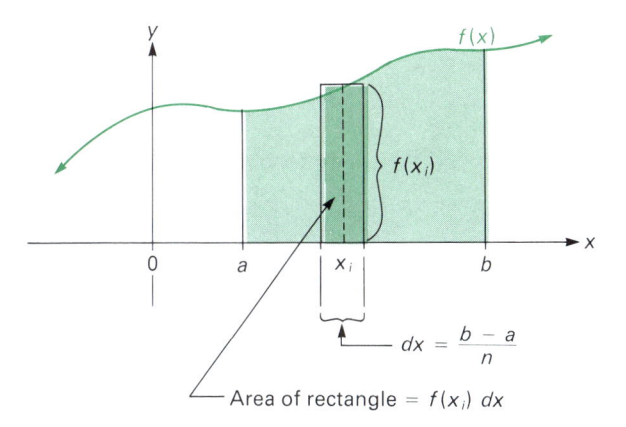

$$dx = \frac{b-a}{n}$$

Area of rectangle = $f(x_i)\, dx$

FIGURE 10–7

Thus, the area between $f(x)$ and the x-axis from $x = a$ to $x = b$ is approximated by the Riemann sum

$$S = f(x_1)\, dx + f(x_2)\, dx + \ldots + f(x_n)\, dx$$

As $n \to \infty$, the Riemann sum approaches the area, A.

Fundamental Theorem of Calculus

We now state a result that relates area under a curve to the antiderivative. This result is called the **fundamental theorem of calculus**. An informal argument for the theorem appears in the appendix at the end of this section.

> **Fundamental Theorem of Calculus**
>
> Let $f(x)$ be defined, nonnegative, and continuous over the interval $a \le x \le b$. Let $F(x)$ be an antiderivative of $f(x)$. Then, the limit of a Riemann sum equals
>
> $$\int_a^b f(x)\, dx = F(b) - F(a)$$

The fundamental theorem of calculus is the link between differentiation and integration. It allows us to compute the area under the graph of a continuous nonnegative function $f(x)$ over an interval $a \le x \le b$ by $\int_a^b f(x)\, dx = F(b) - F(a)$, where $F(x)$ is an antiderivative of $f(x)$ (see Figure 10–8 on next page).

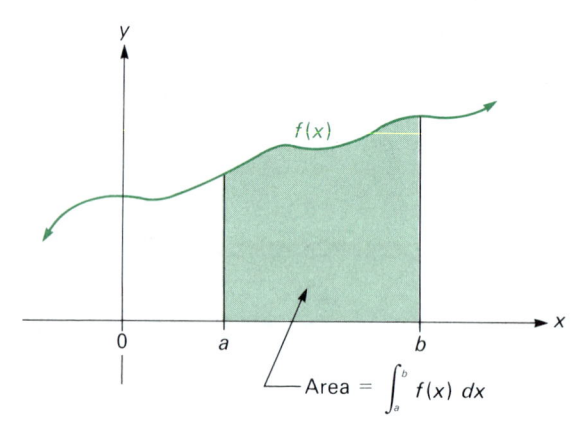

FIGURE 10–8

Area = $\int_a^b f(x)\, dx$

Thus, the area between $f(x) = x^2$ and the x-axis from $x = 0$ to $x = 1$ is given by

$$\int_0^1 x^2\, dx = \underbrace{\frac{1}{3}x^3}_{F(x)}\Big|_0^1$$

$$= F(1) - F(0)$$

$$= \frac{1}{3} - 0$$

$$= \frac{1}{3} \text{ square unit}$$

This result is illustrated in Figure 10–9.

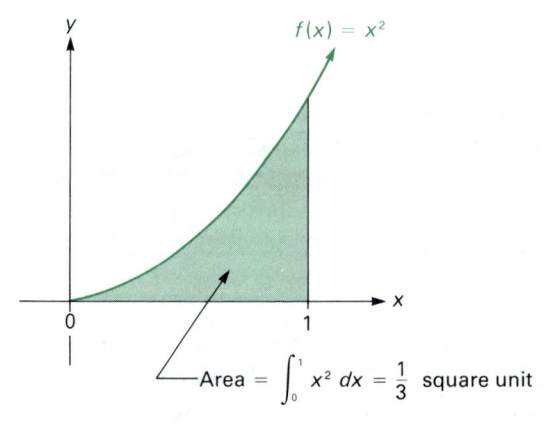

Area = $\int_0^1 x^2\, dx = \frac{1}{3}$ square unit

FIGURE 10–9

EXAMPLE 10-13 Find the area between the x-axis and the curve $f(x) = -x^2 + 25$ from $x = 1$ to $x = 4$.

Solution A sketch of $f(x)$ and the desired area appears in Figure 10-10. The shaded area is computed by the definite integral

$$\int_1^4 (-x^2 + 25)\, dx = \underbrace{-\frac{1}{3}x^3 + 25x}_{F(x)} \bigg|_1^4$$

$$= F(4) - F(1)$$

$$= \left[-\frac{1}{3}(4^3) + 25(4)\right] - \left[-\frac{1}{3}(1^3) + 25(1)\right]$$

$$= \frac{236}{3} - \frac{74}{3}$$

$$= \frac{162}{3}$$

$$= 54 \text{ square units}$$

FIGURE 10-10

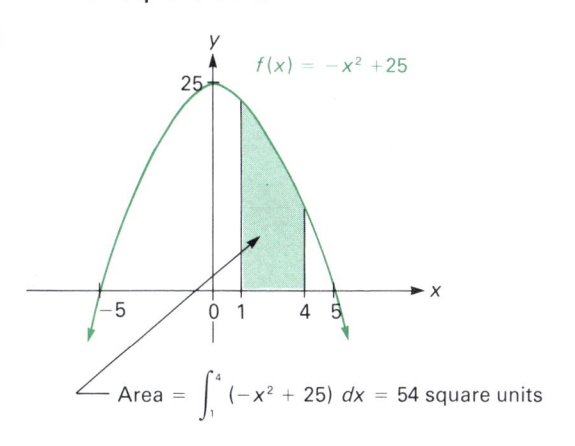

Area = $\int_1^4 (-x^2 + 25)\, dx = 54$ square units

Note that the fundamental theorem requires the function $f(x)$ to be nonnegative over the interval $a \leq x \leq b$. If $f(x)$ is *negative* over the interval $a \leq x \leq b$, the definite integral $\int_a^b f(x)\, dx$ results in a value that is the negative of the area between $f(x)$ and the x-axis from $x = a$ to $x = b$. In such a case, the area is the absolute value of the definite integral $\int_a^b f(x)\, dx$.

EXAMPLE 10-14 Find the area between the x-axis and the curve $f(x) = x^2 - 9$ from $x = 1$ to $x = 3$.

594 CHAPTER TEN

EXAMPLE 10–14 (continued)

Solution A sketch of $f(x)$ and the indicated area appears in Figure 10–11. Observe that $f(x) \leq 0$ over the interval $1 \leq x \leq 3$, and thus the shaded area appears below the x-axis. To determine the area, A, we begin by computing the definite integral, as follows:

$$\int_1^3 (x^2 - 9)\, dx = \underbrace{\frac{1}{3}x^3 - 9x}_{F(x)} \Big|_1^3$$

$$= F(3) - F(1)$$

$$= \left[\frac{1}{3}(3^3) - 9(3)\right] - \left[\frac{1}{3}(1^3) - 9(1)\right]$$

$$= -18 - \left(-8\frac{2}{3}\right)$$

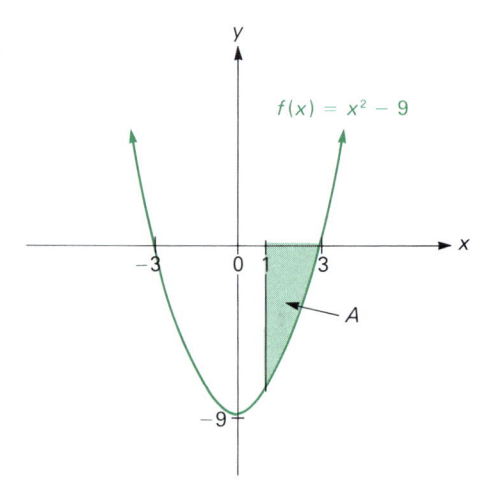

$$= -9\frac{1}{3}$$

FIGURE 10–11

The definite integral results in a negative number because the area is located below the x-axis. Thus,

$$A = \left|-9\frac{1}{3}\right| = 9\frac{1}{3} \text{ square units}$$

EXAMPLE 10–15
Find the area between the x-axis and the curve $f(x) = x^2 - 9$ from $x = 1$ to $x = 4$.

INTEGRATION

Solution A sketch of $f(x)$ and the indicated area appears in Figure 10–12. Observe that $f(x) \leq 0$ over the interval $1 \leq x \leq 3$ and $f(x) \geq 0$ over the interval $3 \leq x \leq 4$. Thus, part of the area appears below the x-axis and part appears above the x-axis. Each part must be determined separately. Since A_1, the area below, was determined in Example 10–14, we now compute A_2, as follows:

$$A_2 = \int_3^4 (x^2 - 9)\, dx = \underbrace{\frac{1}{3}x^3 - 9x}_{F(x)}\bigg|_3^4$$

$$= F(4) - F(3)$$

$$= \left[\frac{1}{3}(4^3) - 9(4)\right] - \left[\frac{1}{3}(3^3) - 9(3)\right]$$

$$= \left(\frac{64}{3} - 36\right) - (-18)$$

$$= 3\frac{1}{3} \text{ square units}$$

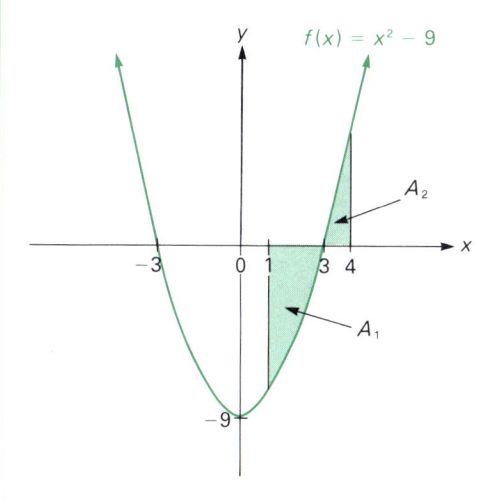

FIGURE 10–12

Thus, the total area is

$$A = A_1 + A_2$$

$$= 9\frac{1}{3} + 3\frac{1}{3}$$

$$= 12\frac{2}{3} \text{ square units}$$

APPENDIX   Fundamental Theorem of Calculus

We now present an informal argument for the fundamental theorem of calculus. Let $A(x)$ be a function that represents the area between $f(x)$ and the x-axis (see Figure 10–13) from a to x, where x is any number within the interval $a \leq x \leq b$. Note that $A(a) = 0$ and $A(b) - A(a)$ represents the area over the entire interval $a \leq x \leq b$. If h is a small positive number, then $A(x + h) - A(x)$ is the shaded region of Figure 10–14. This shaded region is approximately a rectangle with height $f(x)$, width h, and area $h \cdot f(x)$. Hence, we may write

$$A(x + h) - A(x) \approx h \cdot f(x)$$

where the approximation gets better as h approaches 0. Dividing this expression by h, we obtain

$$\frac{A(x + h) - A(x)}{h} \approx f(x)$$

FIGURE 10–13

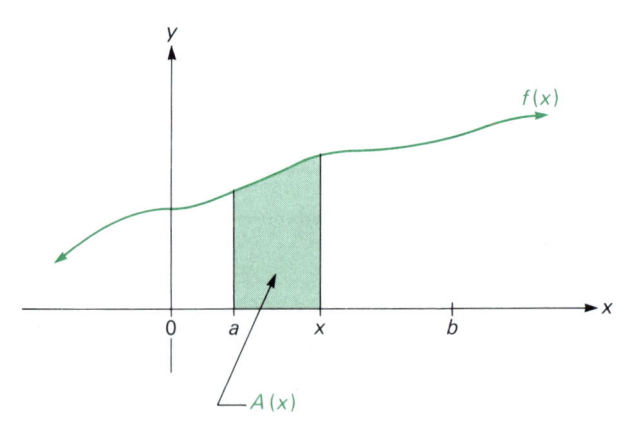

FIGURE 10–14

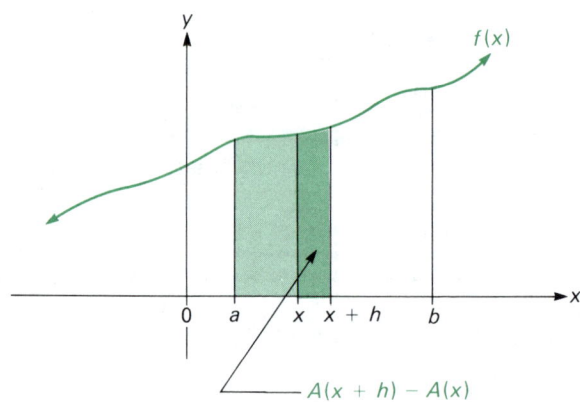

INTEGRATION

Since the approximation becomes better as h approaches 0, the quotient $[A(x + h) - A(x)]/h$ approaches $f(x)$. From the definition of a derivative, we know that the quotient $[A(x + h) - A(x)]/h$ approaches $A'(x) = f(x)$. Since x is any number within the interval $a \le x \le b$, then $A(x)$ is an antiderivative of $f(x)$. Thus, if $F(x)$ is an antiderivative of $f(x)$, we have

$$A(x) = \int f(x)\, dx = F(x) + c$$

Since $A(a) = 0$, then $A(a) = F(a) + c = 0$ and

$$c = -F(a)$$

Substituting this into $A(x) = F(x) + c$ gives us

$$A(x) = F(x) - F(a)$$

Hence,

$$A(b) = F(b) - F(a)$$

$$= \int_a^b f(x)\, dx$$

EXERCISES

1. Evaluate each of the following:

   (A) $\displaystyle\int_1^3 4x^3\, dx$ (B) $\displaystyle\int_2^5 \frac{-6}{x^2}\, dx$

   (C) $\displaystyle\int_0^1 (4x + 1)\, dx$ (D) $\displaystyle\int_{-2}^4 (8x^3 - 6x^2 + 2)\, dx$

   (E) $\displaystyle\int_1^4 \frac{4}{x^3}\, dx$ (F) $\displaystyle\int_1^{27} 5\sqrt[3]{x}\, dx$

2. Using the formula

$$1 + 2 + \ldots + n = \frac{n(n + 1)}{2}$$

in conjunction with a Riemann sum as developed in this section, find the area bounded by the x-axis and the graph of $f(x) = x$ from $x = 0$ to $x = 1$. Check your answer by using $F(1) - F(0)$.

3. Using the formula

$$1^3 + 2^3 + \ldots + n^3 = \left[\frac{n(n + 1)}{2}\right]^2$$

in conjunction with a Riemann sum as developed in this section, find the area bounded by the x-axis and the graph of $f(x) = x^3$ over the interval $0 \le x \le 1$. Check your answer by using $F(1) - F(0)$.

4. Find the area between the curve $f(x) = 3x^2$ and the x-axis from $x = 0$ to $x = 1$ by using
   (A) The definite integral.
   (B) The Riemann sum as developed in this section.

5. Find both lower and upper approximations of the area bounded by the x-axis and the curve $f(x) = -x^2 + 4$ from $x = 0$ to $x = 1$ by dividing the interval $0 \le x \le 1$ into
   (A) 4 subintervals.
   (B) 10 subintervals.

6. Find both lower and upper approximations of the area bounded by the x-axis and the curve $f(x) = 2x^2$ from $x = 1$ to $x = 2$ by dividing the interval $1 \le x \le 2$ into
   (A) 5 subintervals.
   (B) 10 subintervals.

7. Use the definite integral to determine the area bounded by the x-axis and the curve $f(x) = -x^2 + 4$ from $x = 0$ to $x = 1$. Compare your answer with those of Exercise 5.

8. Use the definite integral to determine the area bounded by the x-axis and the curve $f(x) = 2x^2$ from $x = 1$ to $x = 2$. Compare your answer with those of Exercise 6.

9. Use the definite integral to find the area bounded by the x-axis and the curve $f(x) = 3x^2 + 5$ from $x = 2$ to $x = 5$. Graph $f(x)$ and shade the desired area.

10. Find the area between the x-axis and the curve $y = x^3$ from $x = 0$ to $x = 4$. Graph $y = x^3$ and shade the desired area.

11. Find the area between the x-axis and the curve $f(x) = -3x^2 + 24x$ from $x = 0$ to $x = 8$. Graph $f(x)$ and shade the desired area.

12. Find the area between the x-axis and the curve $f(x) = x^3 - 4x$ from $x = -2$ to $x = 0$. Graph $f(x)$ and shade the desired area.

13. Find the area between the x-axis and the curve $f(x) = x^3 - 4x$ from $x = 0$ to $x = 4$. Graph $f(x)$ and shade the desired area.

14. Find the area between the x-axis and the curve $f(x) = 3x^2 - 27$ from $x = 0$ to $x = 4$. Graph $f(x)$ and shade the desired area.

15. Find the area between the x-axis and the curve $y = x^3$ from $x = -1$ to $x = 2$. Graph $y = x^3$ and shade the desired area.

16. Find the area between the x-axis and $f(x) = -6x + 12$ from $x = 0$ to $x = 3$. Graph $f(x)$ and shade the desired area.

17. Find the area between the x-axis and the curve $f(x) = 5x^2 - 10x$ from $x = 0$ to $x = 4$. Graph $f(x)$ and shade the desired area.

18. Find the area between the x-axis and the curve $f(x) = -x^2 + 16$ from $x = 0$ to $x = 5$. Graph $f(x)$ and shade the desired area.

19. Find the area between the x-axis and the curve $f(x) = x^4 - 25x^2$ from $x = 0$ to $x = 6$. Graph $f(x)$ and shade the desired area.

20. Find the area bounded by the x-axis and the curve $f(x) = 1/x^2$ from $x = 1$ to $x = 3$. Graph $f(x)$ and shade the desired area.

10-3

AREA BETWEEN TWO CURVES

Sometimes, we must determine the area between the graphs of two continuous functions $f(x)$ and $g(x)$ over an interval $a \leq x \leq b$ as illustrated in Figure 10-15. Such an area may be approximated by n rectangles each of width $dx = (b - a)/n$ as shown in Figure 10-16. Since $f(x) \geq g(x)$ over the interval $a \leq x \leq b$, the height of one typical rectangle, say, the ith rectangle, is given by $f(x_i) - g(x_i)$, as illustrated in Figure 10-17. Thus, the area of the ith rectangle is

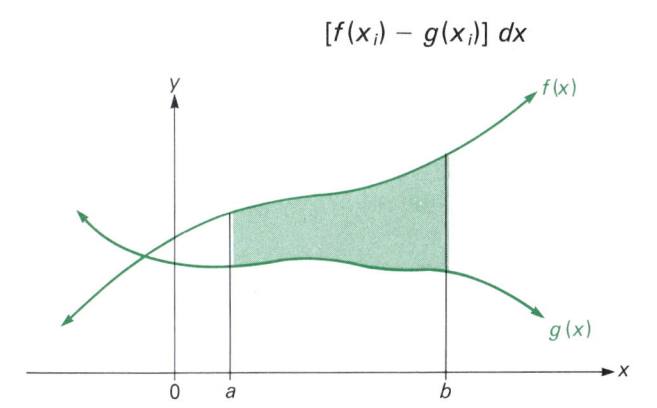

$$[f(x_i) - g(x_i)] \, dx$$

FIGURE 10-15

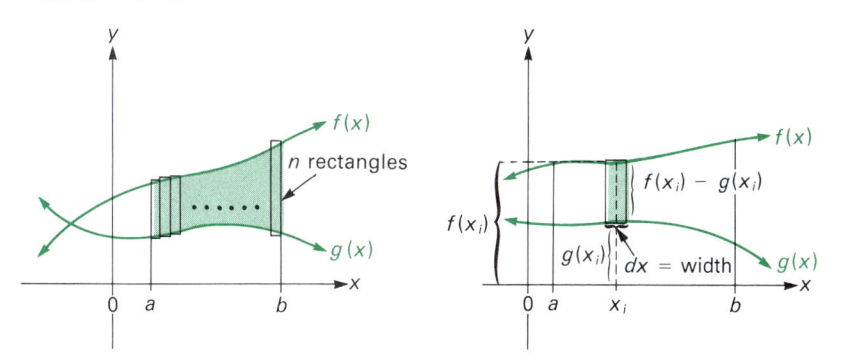

FIGURE 10-16

FIGURE 10-17

Hence, the area from $x = a$ to $x = b$ is approximated by the Riemann sum

$$S = [f(x_1) - g(x_1)] \, dx + [f(x_2) - g(x_2)] \, dx + \ldots + [f(x_n) - g(x_n)] \, dx$$

If the number of rectangles increases without bound (i.e., $n \to \infty$), then the Riemann sum approaches the definite integral

$$\int_a^b [f(x) - g(x)] \, dx$$

(See Figure 10–18.)

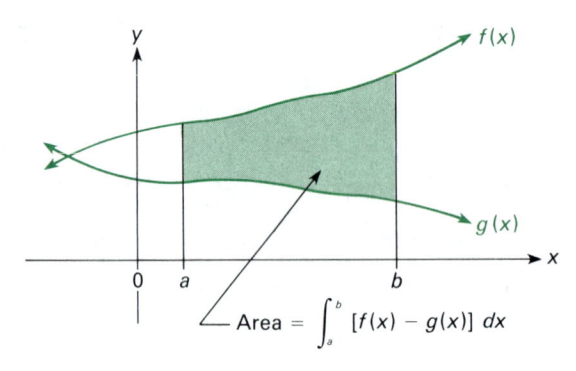

FIGURE 10–18

Area = $\int_a^b [f(x) - g(x)]\, dx$

EXAMPLE 10–16 Find the area between $f(x) = x + 6$ and $g(x) = x^2$ from $x = 1$ to $x = 2$.

Solution A sketch of both functions appears in Figure 10–19. We must determine the x-coordinate of the intersection points of both functions. Setting

$$f(x) = g(x)$$

we have

$$x + 6 = x^2$$

Solving for x yields

$$0 = x^2 - x - 6$$
$$0 = (x - 3)(x + 2)$$

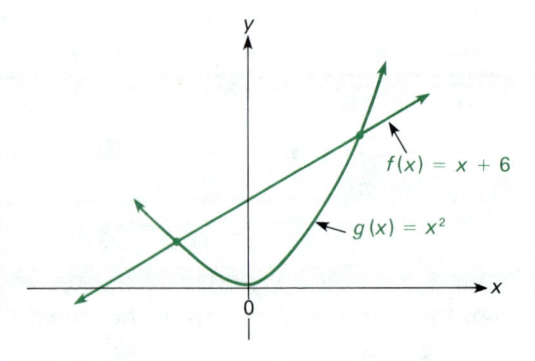

$$x - 3 = 0 \qquad x + 2 = 0$$
$$x = 3 \qquad x = -2$$

FIGURE 10–19

601 INTEGRATION

The area between $x = 1$ and $x = 2$ is now defined and shaded in Figure 10–20. This area is determined by the definite integral

$$\int_1^2 [f(x) - g(x)]\, dx$$

where

$$f(x) - g(x) = x + 6 - x^2$$
$$= -x^2 + x + 6$$

Thus, we have

$$\int_1^2 [f(x) - g(x)]\, dx = \int_1^2 (-x^2 + x + 6)\, dx$$

$$= \underbrace{-\frac{1}{3}x^3 + \frac{1}{2}x^2 + 6x}_{F(x)} \Big|_1^2$$

$$= F(2) - F(1)$$

$$= \frac{34}{3} - \frac{37}{6}$$

$$= \frac{31}{6} \text{ square units}$$

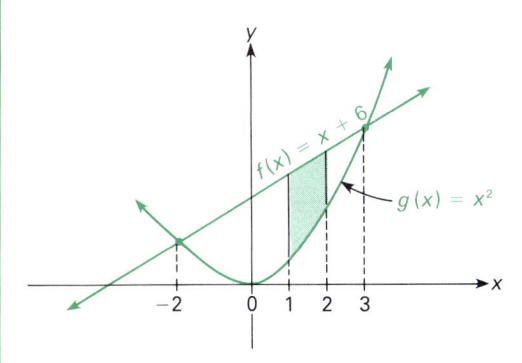

FIGURE 10–20

EXAMPLE 10–17 Find the area completely bounded by the graphs of $f(x) = x + 6$ and $g(x) = x^2$ of Example 10–16.

EXAMPLE 10–17 (continued)

Solution We seek the shaded area of Figure 10–21. This area is determined by the definite integral

$$\int_{-2}^{3} [f(x) - g(x)]\, dx = \int_{-2}^{3} (-x^2 + x + 6)\, dx$$

$$= \underbrace{-\frac{1}{3}x^3 + \frac{1}{2}x^2 + 6x}_{F(x)} \Big|_{-2}^{3}$$

$$= F(3) - F(-2)$$

$$= \frac{27}{2} - \left(-\frac{22}{3}\right)$$

$$= \frac{125}{6} \text{ square units}$$

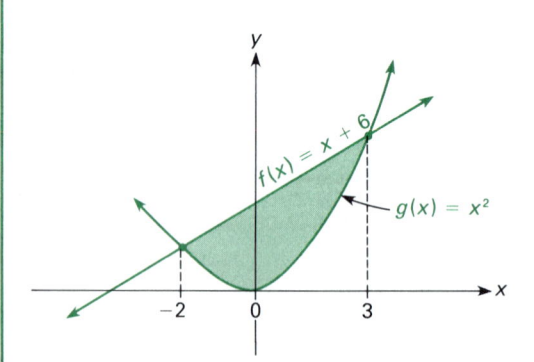

FIGURE 10–21

EXAMPLE 10–18 Find the area between the functions $f(x) = x + 6$ and $g(x) = x^2$ of Examples 10–16 and 10–17 from $x = -2$ to $x = 4$.

Solution We seek the shaded area of Figure 10–22. Observing Figure 10–22, note that for the area between $x = 3$ and $x = 4$, $g(x) \geq f(x)$. Thus, we must determine area A_2 separately by the definite integral

$$\int_{3}^{4} [g(x) - f(x)]\, dx$$

FIGURE 10–22

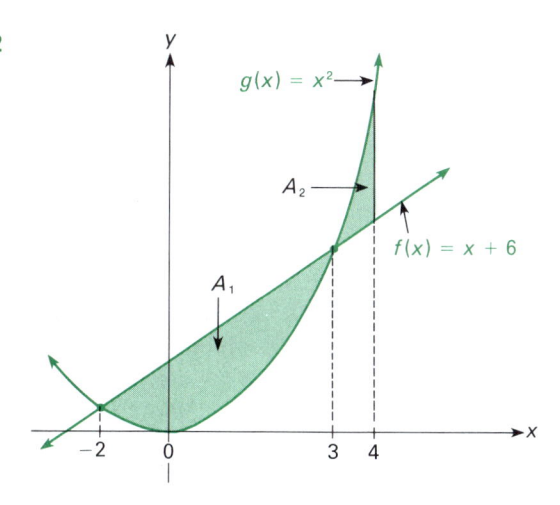

Since

$$g(x) - f(x) = x^2 - (x + 6)$$
$$= x^2 - x - 6$$

then we have

$$\int_3^4 [g(x) - f(x)]\,dx = \int_3^4 (x^2 - x - 6)\,dx$$
$$= \underbrace{\frac{1}{3}x^3 - \frac{1}{2}x^2 - 6x}_{F(x)}\Big|_3^4$$
$$= F(4) - F(3)$$
$$= -\frac{32}{3} - \left(-\frac{27}{2}\right)$$
$$= -\frac{64}{6} + \frac{81}{6}$$
$$= \frac{17}{6} \text{ square units}$$

Since area A_1 was determined in Example 10–16, the total area, A, is

$$A = A_1 + A_2$$
$$= \frac{125}{6} + \frac{17}{6}$$
$$= \frac{142}{6} \text{ square units}$$

EXERCISES

1. Find the area between $f(x) = 2x + 20$ and $g(x) = x^2 + 5$ from $x = 1$ to $x = 3$. Graph $f(x)$ and $g(x)$ and shade the desired area.
2. Find the area completely bounded by $f(x) = 2x + 20$ and $g(x) = x^2 + 5$. Graph $f(x)$ and $g(x)$ and shade the desired area.
3. Find the area between $f(x) = 2x + 20$ and $g(x) = x^2 + 5$ from $x = 0$ to $x = 7$. Graph $f(x)$ and $g(x)$ and shade the desired area.
4. Find the area completely enclosed by $f(x) = x^2$ and $g(x) = x^3$. Graph both functions and shade the desired area.
5. Find the area between $f(x) = x$ and $g(x) = x^3$ from $x = 0$ to $x = 1$. Graph both functions and shade the desired area.
6. Find the area completely enclosed by $f(x) = x$ and $g(x) = x^3$. Graph both functions and shade the desired area.
7. Find the area between $f(x) = 2x^2 + 4$ and $g(x) = x^2 + 3$ from $x = 0$ to $x = 3$. Graph both functions and shade the desired area.
8. Find the area between $f(x) = 3x^2 + 5$ and $g(x) = x^2 + 5$ from $x = 0$ to $x = 2$. Graph both $f(x)$ and $g(x)$ and shade the desired area.

10–4

APPLICATION: CONSUMERS' AND PRODUCERS' SURPLUSES

Suppose a product has a supply and demand function as illustrated in Figure 10–23. Observe that if the unit price, x, exceeds the equilibrium price, p, then consumers would have to pay more for each unit. The shaded area under the demand function from $x = p$ to $x = b$, denoted by CS, represents the total amount of money saved by consumers as a result of the market being at equilibrium. This area CS is called the **consumers' surplus**. It is determined by the definite integral

$$CS = \int_p^b D(x)\, dx$$

Note that the limit of the sum of such increments of area, $D(x)\, dx$, of Figure 10–23 represents *demand × unit price*, or *revenue gained by consumers* who are willing to pay more than the equilibrium price. Similarly, the shaded area in Figure 10–23 under the supply function from $x = a$ to $x = p$, denoted by PS, is the **producers' surplus**. It is determined by the definite integral

INTEGRATION

$$PS = \int_a^p S(x)\, dx$$

FIGURE 10-23

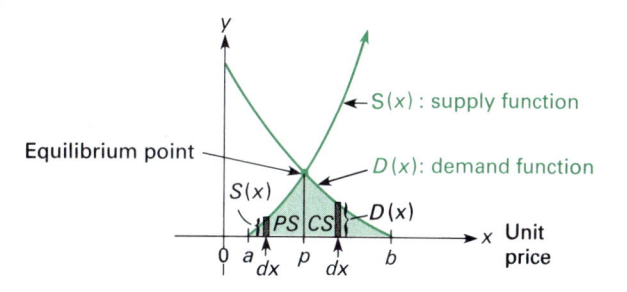

Note that the limit of the sum of such increments of area, $S(x)\, dx$, of Figure 10-23 represents *supply × unit price*, or *revenue gained by producers* who are willing to supply the product at prices lower than the equilibrium price.

EXAMPLE 10-19 Given the demand function defined by

$$D(x) = -\frac{3}{2}x + 27$$

and supply function defined by

$$S(x) = \frac{1}{2}x^2$$

find the consumers' and producers' surpluses.
Solution The graphs of $D(x)$ and $S(x)$ appear in Figure 10-24. (See Exercise 1 at the end of this section.) The consumers' surplus is

$$CS = \int_6^{18} D(x)\, dx$$

$$= \int_6^{18} \left(-\frac{3}{2}x + 27\right) dx$$

$$= \underbrace{-\frac{3}{4}x^2 + 27x}_{F(x)} \Big|_6^{18}$$

$$= F(18) - F(6)$$

$$= 243 - 135$$

$$= \$108$$

EXAMPLE 10–19 (continued)

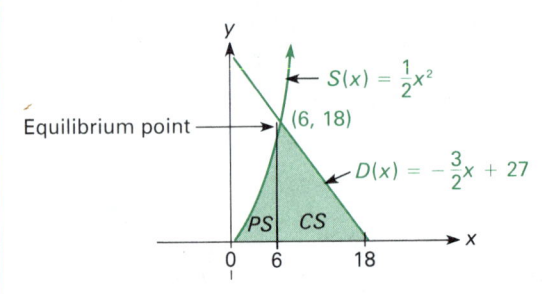

FIGURE 10–24

The producers' surplus is

$$PS = \int_0^6 S(x)\, dx$$

$$= \int_0^6 \frac{1}{2}x^2\, dx$$

$$= \underbrace{\frac{1}{6}x^3}_{F(x)} \Big|_0^6$$

$$= F(6) - F(0)$$

$$= 36 - 0$$

$$= \$36$$

EXERCISES

1. Verify that the supply and demand functions of Example 10–19 are graphed as shown in Figure 10–24. Also, verify the equilibrium point.

2. Given the demand equation $D(x) = -6x + 27$ and supply equation $S(x) = x^2$,
   (A) Graph both functions.
   (B) Find the equilibrium point.
   (C) Find the consumers' surplus.
   (D) Find the producers' surplus.

3. Given the demand equation $D(x) = -10x + 500$ and supply equation $S(x) = 15x - 250$,
   (A) Graph both functions.
   (B) Find the equilibrium point.
   (C) Find the consumers' surplus.
   (D) Find the producers' surplus.

607 INTEGRATION

4. Given the demand equation $D(x) = -3x + 45$ and the supply equation $S(x) = 4x - 18$,
   (A) Graph both functions.
   (B) Find the equilibrium point.
   (C) Find the consumers' surplus.
   (D) Find the producers' surplus.

10–5

INTEGRATION BY SUBSTITUTION

Up to this point, we have been integrating functions consisting of sums of terms of the form

$$kx^n$$

where k and n are constants and $n \neq -1$. In this section, we will learn a technique that will enable us to integrate a greater variety of functions. The technique will involve the **substitution** principle. In order to apply the substitution principle, we must learn to find differentials.

Differentials If $u = f(x)$ defines a differentiable function of x, then we define the quantity

$$du = f'(x)\, dx$$

where du is termed the **differential of u** and dx is termed the **differential of x**. (Differentials have a more profound meaning, which will not be explored in this text.) Thus, if

$$u = f(x) = x^3$$

then

$$du = f'(x)\, dx$$
$$= 3x^2\, dx$$

EXAMPLE 10–20 If $u = x^6$, find the differential du.
Solution Since $u = f(x) = x^6$, then $f'(x) = 6x^5$. Hence,

$$du = f'(x)\, dx$$
$$= 6x^5\, dx$$

EXAMPLE 10–21 If $u = x^4 - 8x^2 + 16$, find du.
Solution Since $u = f(x) = x^4 - 8x^2 + 16$, then $f'(x) = 4x^3 - 16x$. Hence,

$$du = f'(x)\, dx$$
$$= (4x^3 - 16x)\, dx$$

In rule 10–1, Section 10–1, we saw that

$$\int x^n\, dx = \frac{1}{n+1} x^{n+1} + c \qquad (n \neq -1)$$

This rule may be restated in terms of the variable u. If x is replaced by u and dx by du, then we have rule 10–4.

Rule 10–4

$$\int u^n\, du = \frac{1}{n+1} u^{n+1} + c \qquad (n \neq -1)$$

Rule 10–4 will be used in conjunction with the substitution principle to integrate the functions of this section.
We now consider

$$\int (x^3 - 4)^7\, 3x^2\, dx$$

In order to determine this integral, we will restate the function in a simpler form by substituting the variable u in place of $x^3 - 4$. Hence, we let

$$u = x^3 - 4$$

Then we have

$$du = 3x^2\, dx$$

Thus, the integral

$$\int \underbrace{(x^3 - 4)^7}_{u}\, \underbrace{3x^2\, dx}_{du}$$

becomes

$$\int u^7\, du$$

INTEGRATION

This result is integrated by using rule 10–4. Hence, we obtain

$$\int u^7 \, du = \frac{1}{8} u^8 + c$$

Since the solution

$$\frac{1}{8} u^8 + c$$

is written in terms of u, we must rewrite it in terms of x. Since

$$u = x^3 - 4$$

replacing u with $x^3 - 4$ in the solution yields

$$\frac{1}{8}(x^3 - 4)^8 + c$$

Hence, we have

$$\int (x^3 - 4)^7 \, 3x^2 \, dx = \frac{1}{8}(x^3 - 4)^8 + c$$

EXAMPLE 10–22 Find $\int (x^5 - 6)^9 \, 5x^4 \, dx$.
Solution Let $u = x^5 - 6$. Then

$$du = 5x^4 \, dx$$

Hence,

$$\int (x^5 - 6)^9 \, 5x^4 \, dx = \int u^9 \, du$$

$$= \frac{1}{10} u^{10} + c$$

$$= \frac{1}{10}(x^5 - 6)^{10} + c$$

We now consider a problem in which du does not appear in the function to be integrated. Consider the integral

$$\int (x^3 - 7)^5 \, x^2 \, dx$$

CHAPTER TEN

If $u = x^3 - 7$, then

$$du = 3x^2 \, dx$$

In order to rewrite the integral in terms of u, the $x^2 \, dx$ term must equal du.

$$\int (x^3 - 7)^5 \, x^2 \, dx = \int u^5 \underline{\quad ?\quad}$$

Since $du = 3x^2 \, dx$, we must multiply $x^2 \, dx$ by 3 and compensate by multiplying the integral by 1/3 to obtain

$$\frac{1}{3} \int \underbrace{(x^3 - 7)}_{u}{}^5 \, \underbrace{3x^2 \, dx}_{du}$$

Our multiplying $x^2 \, dx$ by 3 and the integral by 1/3 is allowed by rule 10–2 of Section 10–1, which states that the integral of a constant times a function is equal to the constant times the integral of the function. It must be noted that this works only for a *constant multiplier*. It does not apply to variables. Thus, in our example,

$$\frac{1}{3} \int \underbrace{(x^3 - 7)}_{u}{}^5 \, \underbrace{3x^2 \, dx}_{du} = \frac{1}{3} \int u^5 \, du$$

$$= \frac{1}{3} \cdot \frac{1}{6} u^6 + c$$

$$= \frac{1}{18}(x^3 - 7)^6 + c$$

EXAMPLE 10–23 Find

$$\int \frac{x^4 \, dx}{\sqrt{x^5 - 9}}$$

Solution We write

$$\int \frac{x^4 \, dx}{\sqrt{x^5 - 9}} = \int (x^5 - 9)^{-1/2} \, x^4 \, dx$$

Letting $u = x^5 - 9$ gives us

$$du = 5x^4 \, dx$$

Hence, we obtain

$$\int (x^5 - 9)^{-1/2} x^4 \, dx = \frac{1}{5} \int \underbrace{(x^5 - 9)^{-1/2}}_{u} \underbrace{5x^4 \, dx}_{du}$$

$$= \frac{1}{5} \int u^{-1/2} \, du$$

$$= \frac{1}{5} \cdot \frac{1}{1/2} u^{1/2} + c$$

$$= \frac{2}{5}(x^5 - 9)^{1/2} + c$$

EXAMPLE 10-24 Find $\int (2x^3 - 5)^{11} 5x^2 \, dx$.
Solution Letting $u = 2x^3 - 5$ gives us

$$du = 6x^2 \, dx$$

Hence, we obtain

$$\int (2x^3 - 5)^{11} 5x^2 \, dx = \frac{5}{6} \int \underbrace{(2x^3 - 5)^{11}}_{u} \cdot \underbrace{\frac{6}{5} \cdot 5x^2 \, dx}_{du}$$

$$= \frac{5}{6} \int u^{11} \, du$$

$$= \frac{5}{6} \cdot \frac{1}{12} u^{12} + c$$

$$= \frac{5}{72}(2x^3 - 5)^{12} + c$$

EXAMPLE 10-25 Find the area between the x-axis and the curve $f(x) = (x - 2)^3$ from $x = 2$ to $x = 4$.
Solution A graph of $f(x)$ and the desired area appear in Figure 10-25. The shaded area is determined by the definite integral

$$\int_2^4 (x - 2)^3 \, dx$$

EXAMPLE 10–25 (continued)

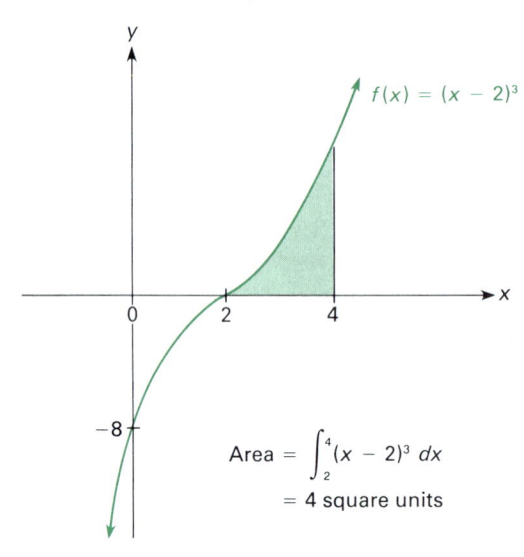

FIGURE 10–25

$$\text{Area} = \int_2^4 (x - 2)^3 \, dx = 4 \text{ square units}$$

Using the substitution principle, we let $u = x - 2$. Then

$$du = dx$$

and we write

$$\int_2^4 \underbrace{(x - 2)^3}_{u} \underbrace{dx}_{du} = \underbrace{\frac{1}{4}(x - 2)^4}_{F(x)} \Big|_2^4$$

$$= F(4) - F(2)$$
$$= 4 - 0$$
$$= 4 \text{ square units}$$

EXAMPLE 10–26 If the rate of flow of revenue into a firm is given by

$$R'(t) = \frac{10{,}000{,}000}{\sqrt{5 + t}}$$

where t is time (in years), determine the total revenue flowing in during the time interval $4 \leq t \leq 20$.

INTEGRATION

Solution We seek $\int_4^{20} R'(t)\, dt = R(20) - R(4)$. Hence,

$$\int_4^{20} R'(t)\, dt = \int_4^{20} \frac{10{,}000{,}000}{\sqrt{5+t}}\, dt$$

$$= \int_4^{20} 10{,}000{,}000(5+t)^{-1/2}\, dt$$

$$= \underbrace{10{,}000{,}000 \cdot \frac{2}{1} \cdot (5+t)^{1/2}}_{R(t)} \Big|_4^{20}$$

$$= R(20) - R(4)$$

$$= 100{,}000{,}000 - 60{,}000{,}000$$

$$= \$40{,}000{,}000$$

Thus, the total revenue flowing into the firm during the time interval $4 \le t \le 20$ is \$40,000,000.

EXERCISES

1. Determine each of the following:

(A) $\displaystyle\int (x^3 - 7)^{10}\, 3x^2\, dx$

(B) $\displaystyle\int (x^2 - 3)^4\, 2x\, dx$

(C) $\displaystyle\int (x^3 - 4x)^5\, (3x^2 - 4)\, dx$

(D) $\displaystyle\int (5x + 6)^{12}\, 5\, dx$

2. Determine each of the following:

(A) $\displaystyle\int (x^4 - 8)^9\, x^3\, dx$

(B) $\displaystyle\int (2x^5 - 7)^6\, x^4\, dx$

(C) $\displaystyle\int (x - 3)^9\, dx$

(D) $\displaystyle\int (x^3 - 4)^{1/2}\, x^2\, dx$

(E) $\displaystyle\int (x^4 + 6)^5\, 5x^3\, dx$

(F) $\displaystyle\int (x^4 - 8x + 5)^7\, (2x^3 - 4)\, dx$

(G) $\displaystyle\int (x^6 + 9)^{10}\, 2x^5\, dx$

(H) $\displaystyle\int (4x^3 - 1)^{11}\, 5x^2\, dx$

3. Determine each of the following:

(A) $\displaystyle\int \frac{x^2}{(x^3 - 5)^{10}}\, dx$

(B) $\displaystyle\int (\sqrt{4x^2 + 5})x\, dx$

(C) $\displaystyle\int \frac{x^3}{\sqrt{x^4 - 6}}\, dx$

(D) $\displaystyle\int (\sqrt[3]{x^3 - 9})^4 x^2\, dx$

(E) $\displaystyle\int \frac{dx}{(3x - 5)^4}$

(F) $\displaystyle\int \frac{5x^2\, dx}{\sqrt[5]{(x^3 - 9)^2}}$

CHAPTER TEN

4. Evaluate each of the following:

   (A) $\int_{2}^{4} (\sqrt{x^2 + 9})x\, dx$

   (B) $\int_{1}^{2} (x^4 - 3)^3\, x^3\, dx$

   (C) $\int_{2}^{5} \frac{x\, dx}{(x^2 - 5)^2}$

   (D) $\int_{0}^{1} (x^2 + 2x)^3 (x + 1)\, dx$

5. Find the area between the x-axis and the curve $f(x) = (x - 5)^3$ from $x = 5$ to $x = 7$. Sketch $f(x)$ and shade the desired area.

6. Find the area between the x-axis and the curve $f(x) = (x + 4)^3$ from $x = -4$ to $x = 0$. Sketch $f(x)$ and shade the desired area.

7. Find the area between the x-axis and the curve $y = x(x^2 - 16)^3$ from $x = 0$ to $x = 5$. Sketch the function and shade the desired area.

8. Find the area between the x-axis and the function $f(x) = 1/(x - 5)^3$ from $x = 6$ to $x = 8$. Sketch $f(x)$ and shade the desired area.

9. Consider the integral

$$\int (x^3 - 5)^{10}\, x\, dx$$

   Why can this integral not be determined by the techniques of this section?

10. If the rate of flow of revenue into a mutual fund is given by

$$R'(t) = 50t + 20t\sqrt{1 + t^2}$$

    where t is given in months, find the total revenue obtained during the first 2 years.

11. An oil tanker is leaking crude oil at the rate of $B'(t)$ barrels per hour, where t is the number of hours since the leak began. If

$$B'(t) = 200t + 50$$

    how many barrels of oil will have leaked into the water during the first 2 days?

12. Executive Realty Corporation has recently completed a new development called Hidden Valley Estates. The population of Hidden Valley Estates has been growing at the rate of $P'(t)$ individuals per month, where t is the number of months elapsed since December 31 of the past year. If

$$P'(t) = 100 + 18t$$

    find the total increase in population during the first three-quarters of the present year.

10-6

INTEGRALS INVOLVING EXPONENTIAL AND LOGARITHMIC FUNCTIONS

In Chapter 8, we learned the following rule for differentiating exponential functions:

> If $y = e^u$, where u is a function of x, then
> $$\frac{dy}{dx} = e^u \frac{du}{dx}$$

Since integration is the reverse operation of differentiation, this rule implies rule 10–5.

> **Rule 10–5**
> $$\int e^u \, du = e^u + c$$

Rule 10–5 may be used in conjunction with the substitution principle to integrate exponential functions with base e. Consider,

$$\int e^{x^8-5} x^7 \, dx$$

If we let $u = x^8 - 5$, then

$$du = 8x^7 \, dx$$

Thus, we obtain

$$\int e^{x^8-5} x^7 \, dx = \frac{1}{8} \int \underbrace{e^{x^8-5}}_{u} \underbrace{8x^7 \, dx}_{du}$$

$$= \frac{1}{8} \int e^u \, du$$

$$= \frac{1}{8} e^u + c$$

$$= \frac{1}{8} e^{x^8-5} + c$$

EXAMPLE 10–27 Find $\int e^{x^3-6} 2x^2 \, dx$
Solution Let $u = x^3 - 6$. Then

$$du = 3x^2 \, dx$$

Thus, we have

$$\int e^{x^3-6} 2x^2 \, dx = \frac{2}{3} \int \overbrace{e^{x^3-6}}^{u} \cdot \underbrace{\frac{3}{2} \cdot 2x^2 \, dx}_{du}$$

$$= \frac{2}{3} \int e^u \, du$$

$$= \frac{2}{3} e^u + c$$

$$= \frac{2}{3} e^{x^3-6} + c$$

EXAMPLE 10–28 Find the area bounded by the x-axis and the curve $y = e^{3x}$ from $x = 1$ to $x = 2$.
Solution The graph of $y = e^{3x}$ and the desired area appear in Figure 10–26. The shaded area is determined by the definite integral

$$\int_1^2 e^{3x} \, dx$$

Thus, we get

$$\int_1^2 e^{3x} \, dx = \underbrace{\frac{1}{3} e^{3x}}_{F(x)} \Big|_1^2$$

$$= F(2) - F(1)$$

$$= \frac{1}{3} e^6 - \frac{1}{3} e^3$$

$$= \frac{1}{3}(e^6 - e^3)$$

$$= \frac{1}{3}(403.428793 - 20.085537)$$

$$\approx 127.78 \text{ square units}$$

617 INTEGRATION

FIGURE 10-26

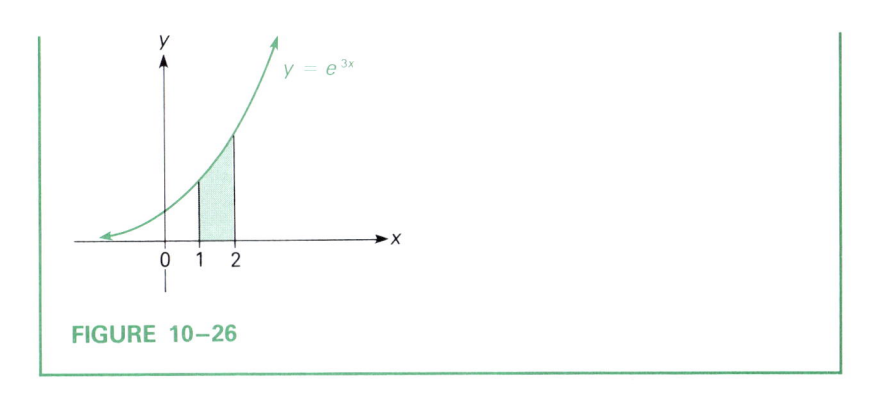

In Chapter 8, we also learned the following rule for differentiating logarithmic functions.

> If $y = \ln u$, where u is a function of x, then
> $$\frac{du}{dx} = \frac{1}{u}\frac{du}{dx}$$

Since integration is the reverse operation of differentiation, this rule implies rule 10-6.

Rule 10-6

$$\int \frac{du}{u} = \ln u + c$$

Rule 10-6 may be used in conjunction with the substitution principle to integrate functions that may be expressed in the form du/u. Consider, for instance,

$$\int \frac{x \, dx}{x^2 - 1}$$

If we let $u = x^2 - 1$, then

$$du = 2x \, dx$$

Hence, we have

$$\int \frac{x\,dx}{x^2-1} = \frac{1}{2}\int \underbrace{\frac{\overbrace{2x\,dx}^{du}}{x^2-1}}_{u}$$

$$= \frac{1}{2}\int \frac{du}{u}$$

$$= \frac{1}{2}\ln u + c$$

$$= \frac{1}{2}\ln(x^2-1) + c$$

EXAMPLE 10-29 Find

$$\int \frac{x^6\,dx}{x^7+15}$$

Solution Let $u = x^7 + 15$. Then

$$du = 7x^6\,dx$$

Hence, we find

$$\int \frac{x^6\,dx}{x^7+15} = \frac{1}{7}\int \underbrace{\frac{\overbrace{7x^6\,dx}^{du}}{x^7+15}}_{u}$$

$$= \frac{1}{7}\int \frac{du}{u}$$

$$= \frac{1}{7}\ln u + c$$

$$= \frac{1}{7}\ln(x^7+15) + c$$

EXAMPLE 10-30 Find the area bounded by the x-axis and the curve $f(x) = 1/x$ from $x = 1$ to $x = 3$.

Solution The graph of $f(x)$ and the desired area appear in Figure 10-27. The shaded area is determined by the definite integral

$$\int_1^3 \frac{1}{x}\,dx$$

INTEGRATION

Thus, we obtain

$$\int_1^3 \frac{1}{x}\,dx = \int_1^3 \frac{dx}{x}$$

$$= \underbrace{\ln x}_{F(x)} \Big|_1^3$$

$$= F(3) - F(1)$$

$$= \ln 3 - \ln 1$$

$$= 1.098612 - 0$$

$$\approx 1.10 \text{ square units}$$

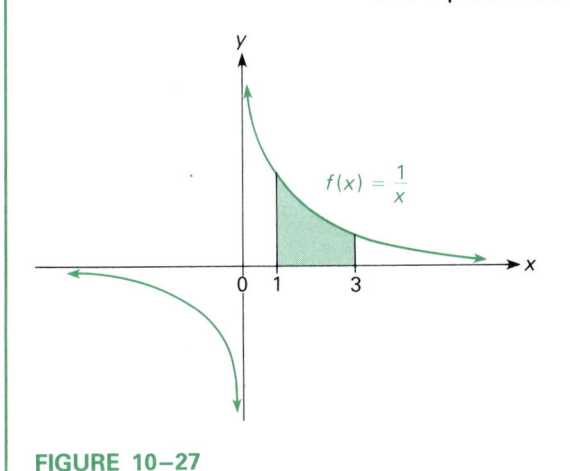

FIGURE 10–27

EXERCISES

1. Determine each of the following:

   (A) $\displaystyle\int e^x\,dx$

   (B) $\displaystyle\int e^{-x}\,dx$

   (C) $\displaystyle\int e^{4x}\,dx$

   (D) $\displaystyle\int_0^1 e^{-3x}\,dx$

   (E) $\displaystyle\int e^{x/2}\,dx$

   (F) $\displaystyle\int 6e^{-0.3x}\,dx$

2. Determine each of the following:

   (A) $\displaystyle\int e^{x^3-5}x^2\,dx$

   (B) $\displaystyle\int \frac{x\,dx}{e^{x^2}}$

   (C) $\displaystyle\int x^3 e^{x^4+6}\,dx$

   (D) $\displaystyle\int_0^1 xe^{x^2+1}\,dx$

(E) $\int (x+1)e^{x^2+2x}\, dx$ (F) $\int e^{5x-2}\, dx$

3. Determine each of the following:

   (A) $\int \dfrac{5\, dx}{x}$ (B) $\int_0^1 \dfrac{dx}{x+1}$

   (C) $\int \dfrac{x}{x^2+1}\, dx$ (D) $\int \dfrac{x^3\, dx}{x^4-1}$

   (E) $\int \dfrac{6x^2\, dx}{x^3+4}$ (F) $\int \dfrac{5x\, dx}{x^2-4}$

   (G) $\int \dfrac{-x\, dx}{5x^2-6}$ (H) $\int \dfrac{-4x^2}{x^3+5}\, dx$

4. Find the area bounded by the x-axis and $f(x) = e^{-2x}$ from $x = 0$ to $x = 1$. Sketch $f(x)$ and shade the desired area.

5. Find the area bounded by the x-axis and the curve $f(x) = 1/(x-2)$ from $x = 3$ to $x = 6$. Sketch $f(x)$ and shade the desired area.

6. Find the area bounded by the x-axis and $f(x) = x/(x^2 - 9)$ from $x = -2$ to $x = 0$. Sketch $f(x)$ and shade the desired area.

7. If the rate of flow of revenue into a firm is given by

$$R'(t) = \dfrac{10{,}000{,}000}{t+5}$$

where t is measured in years, find the total revenue obtained during the interval $0 \le t \le 4$.

8. If the rate of flow of revenue into a firm is given by

$$R'(t) = 100t + 10e^{-t}$$

where t is measured in years, find the total revenue obtained during the interval $1 \le t \le 10$.

9. After a certain cancer-inducing drug is injected into a mouse, cancer cells increase at the rate of $D'(t)$ cells per day, where t is the number of days following the drug injection. If

$$D'(t) = 100e^{5t}$$

find the total increase in cancer cells during the first 2 days following the drug injection.

10–7

USING TABLES OF INTEGRALS In Sections 10–5 and 10–6, we integrated functions by using the substitution principle in conjunction with the formulas

$$\int u^n\, du = \dfrac{1}{n+1} u^{n+1} + c \qquad (n \ne -1)$$

INTEGRATION

$$\int e^u \, du = e^u + c$$

$$\int \frac{du}{u} = \ln u + c$$

These formulas allow us to integrate a variety of functions. However, the functions are limited to those that can be expressed in one of these forms. There are many functions that must be integrated by other methods.

Table 10 of the Appendix at the end of the text lists additional integral forms. Such a table is called a **table of integrals.** A more complete table of integrals appears in *CRC Standard Mathematical Tables.** Table 10–1 is an abridged version of a table of integrals. We now illustrate how a table of integrals is used.

TABLE 10–1 Table of Integrals (Abridged)

1. $\displaystyle\int \frac{du}{\sqrt{a^2 + u^2}} = \ln(u + \sqrt{a^2 + u^2}) + c$

2. $\displaystyle\int \frac{du}{a^2 - u^2} = \frac{1}{2a} \ln \frac{a + u}{a - u} + c$

3. $\displaystyle\int u^n e^u \, du = u^n e^u - n \int u^{n-1} e^u \, du$

4. $\displaystyle\int \frac{du}{u^2(a + bu)} = -\frac{1}{au} + \frac{b}{a^2} \ln \frac{a + bu}{u} + c$

EXAMPLE 10–31 Find

$$\int \frac{dx}{\sqrt{36 + x^2}}$$

Solution Scanning the integral forms of Table 10–1, note that this integral is of the form

$$\int \frac{du}{\sqrt{a^2 + u^2}} = \ln(u + \sqrt{a^2 + u^2}) + c$$

**CRC Standard Mathematical Tables,* 19th ed. (Cleveland, Ohio: The Chemical Rubber Co., 1971).

EXAMPLE 10-31 (continued)

with $a^2 = 36$ and $u = x$. Substituting 36 for a^2 and x for u into this form, we have

$$\int \frac{dx}{\sqrt{36 + x^2}} = \ln(x + \sqrt{36 + x^2}) + c$$

EXAMPLE 10-32 Find

$$\int \frac{dx}{x^2(3 - 5x)}$$

Solution Scanning the integral forms of Table 10-1, we see that this integral is of the form

$$\int \frac{du}{u^2(a + bu)} = -\frac{1}{au} + \frac{b}{a^2} \ln\left(\frac{a + bu}{u}\right) + c$$

with $a = 3$, $b = -5$, and $u = x$. Substituting 3 for a, -5 for b, and x for u into this form, we obtain

$$\int \frac{dx}{x^2(3 - 5x)} = -\frac{1}{3x} + \frac{-5}{3^2} \ln\left(\frac{3 - 5x}{x}\right) + c$$

$$= -\frac{1}{3x} - \frac{5}{9} \ln\left(\frac{3 - 5x}{x}\right) + c$$

EXAMPLE 10-33 Find $\int x^2 e^x \, dx$.

Solution Scanning the integral forms of Table 10-1, we determine this integral to be of the form

$$\int u^n e^u \, du = u^n e^u - n \int u^{n-1} e^u \, du$$

with $n = 2$ and $u = x$. Substituting 2 for n and x for u into this form yields

$$\int x^2 e^x \, dx = x^2 e^x - 2 \int x e^x \, dx$$

We must use this form again to evaluate $\int x e^x \, dx$. This time, $n = 1$. Hence,

INTEGRATION

$$\int xe^x \, dx = xe^x - \int e^x \, dx$$
$$= xe^x - e^x + c_1$$

And,

$$\int x^2 e^x \, dx = x^2 e^x - 2\int xe^x \, dx$$
$$= x^2 e^x - 2(xe^x - e^x + c_1)$$
$$= x^2 e^x - 2xe^x + 2e^x - 2c_1$$
$$= x^2 e^x - 2xe^x + 2e^x + c$$

where c is an arbitrary constant.

EXERCISES

1. Using Table 10–1, find each of the following:

   (A) $\displaystyle\int \frac{dx}{\sqrt{81+x^2}}$
   (B) $\displaystyle\int \frac{dx}{\sqrt{x^2+64}}$
   (C) $\displaystyle\int_0^8 \frac{dx}{\sqrt{x^2+36}}$
   (D) $\displaystyle\int \frac{dx}{4-x^2}$
   (E) $\displaystyle\int \frac{-7\,dx}{81-x^2}$
   (F) $\displaystyle\int \frac{dx}{x^2(5+3x)}$
   (G) $\displaystyle\int \frac{3\,dx}{x^2(4-7x)}$
   (H) $\displaystyle\int x^2 e^{5x}\,dx$
   (I) $\displaystyle\int_0^1 x^3 e^x\,dx$
   (J) $\displaystyle\int_0^2 \frac{dx}{\sqrt{x^2+25}}$

2. Using Table 10 of the Appendix at the end of the text, find each of the following:

   (A) $\displaystyle\int \ln 5x\,dx$
   (B) $\displaystyle\int \ln(3x-1)\,dx$
   (C) $\displaystyle\int \frac{dx}{x\sqrt{5-2x}}$
   (D) $\displaystyle\int \frac{dx}{x\sqrt{2+6x}}$
   (E) $\displaystyle\int \frac{dx}{\sqrt{x^2-100}}$
   (F) $\displaystyle\int \frac{dx}{x^2-64}$
   (G) $\displaystyle\int \frac{dx}{x\sqrt{25-x^2}}$
   (H) $\displaystyle\int \frac{dx}{x\sqrt{x^2+4}}$
   (I) $\displaystyle\int (36-x^2)^{-3/2}\,dx$
   (J) $\displaystyle\int_0^7 \frac{dx}{(x^2+49)^{3/2}}$

10–8

APPLICATION: CONTINUOUS CASH FLOWS

In Chapter 4, we learned that if an amount of money, P, is compounded continuously at a nominal rate, r, then its future value after t years is given by

$$S = Pe^{rt}$$

In many business situations (e.g., revenues received by a store or a large manufacturing operation or toll roads and bridges), the flow of revenue is most accurately approximated by a *continuous* flow of cash into a fund. An analogy may be drawn between a liquid flowing into a container and money flowing into a fund. In this section, we let $P(t)$ (where t = time in years) represent the rate of flow of a continuous stream of money into a fund during some time interval $a \leq t \leq b$. If t_1 and t_2 are points in the interval $a \leq t \leq b$, then $P(t_1)$ is the rate of flow of money into the fund at time t_1, and $P(t_2)$ is the rate of flow of money into the fund at time t_2.

Let us assume that money flows continuously into a fund at a rate of $P(t)$ dollars a year (where t = time in years) during the time interval $a \leq t \leq b$. If this continuous flow is compounded continuously at a nominal rate, r, during the time interval $a \leq t \leq b$, then:

1. The **future value** of this money flow (i.e., its total accumulated amount including interest during the time interval $a \leq t \leq b$) is given by

$$\int_a^b P(t)e^{rt}\, dt$$

2. The **present value** of this money flow at $t = a$ is given by

$$\int_a^b P(t)e^{-rt}\, dt$$

EXAMPLE 10–34 Revenues flow continuously into a manufacturing operation at the rate of $P(t)$ dollars per year (t = time in years), where $P(t) = 8000t$ during the time interval $0 \leq t \leq 4$. The money flow is compounded continuously at 10%.

(A) Find the future value of this money flow.
(B) Find the present value of this money flow.

Solutions
(A) The future value is given by

$$\int_0^4 P(t)e^{rt}\, dt = \int_0^4 8000te^{0.10t}\, dt$$

This integral may be evaluated by using formula 7 of our table of integrals (Appendix Table 10). This form is

$$\int ue^u\, du = e^u(u - 1) + c$$

We must evaluate $\int 8000te^{0.10t}\, dt$ by using this form with $u = 0.10t$ and $du = 0.10\, dt$. Thus, we must rewrite $\int 8000te^{0.10t}\, dt$ as

$$\frac{8000}{0.10^2} \int \underbrace{0.10t}_{u}\ \underbrace{e^{0.10t}}_{e^u}\ \underbrace{(0.10)\, dt}_{du}$$

Since $\int ue^u\, du = e^u(u - 1) + c$, we have

$$\frac{8000}{0.10^2} \int \underbrace{0.10t}_{u}\ \underbrace{e^{0.10t}}_{e^u}\ \underbrace{(0.10)\, dt}_{du} = \frac{8000}{0.10^2} e^{0.10t}(0.10t - 1) + c$$

Hence,

$$\int_0^4 8000te^{0.10t}\, dt = \underbrace{800{,}000e^{0.10t}(0.10t - 1)}_{F(t)}\ \Big|_0^4$$

$$= F(4) - F(0)$$
$$= -716{,}076 - (-800{,}000)$$
$$= \$83{,}924$$

Thus, the future value (or total accumulated amount) is $83,924.

(B) The present value is given by

$$\int_0^4 P(t)e^{-rt}\, dt = \int_0^4 8000te^{-0.10t}\, dt$$

The integral $\int_0^4 8000te^{-0.10t}\, dt$ may be evaluated by using the integral form of part (A) with $u = -0.10t$ and $du = -0.10\, dt$ to yield

EXAMPLE 10-34 (continued)

$$\int_0^4 8000te^{-0.10t}\, dt = \underbrace{800{,}000e^{-0.10t}(-0.10t - 1)}_{F(t)}\Big|_0^4$$

$$= F(4) - F(0)$$
$$= -750{,}758.4 - (-800{,}000)$$
$$= \$49{,}241.60$$

Thus, the present value is $49,241.60.

EXERCISES

1. Sales revenue flows continuously into a supermarket at the rate of $P(t)$ dollars per year (t = time in years), where $P(t) = 10{,}000{,}000t$ during the time interval $0 \le t \le 5$. If this money flow is compounded continuously at 12%,
   (A) Find its future value.
   (B) Find its present value.

2. Money flows continuously into a fund at a constant rate of $P(t) = \$10{,}000$ per year (t = time in years) during the time interval $0 \le t \le 7$. If this money flow is compounded continuously at 10%,
   (A) Find its future value.
   (B) Find its present value.

3. A toll booth yields a continuous revenue flow of $P(t)$ dollars per year (t = time in years), where $P(t) = 9{,}500{,}000t$ during the time interval $2 \le t \le 5$. If this money flow is compounded continuously at 8%,
   (A) Find its future value.
   (B) Find its present value.

4. Revenue flows continuously into a retail store at the rate of $P(t)$ dollars per year (t = time in years), where $P(t) = 10{,}000{,}000t^2$ during the time interval $0 \le t \le 6$. If this money flow is compounded continuously at 10%,
   (A) Find its future value.
   (B) Find its present value.

10-9

IMPROPER INTEGRALS

Suppose we must determine the area between the x-axis and the curve $f(x) = e^x$ over the interval $-\infty < x \le 0$ (see Figure 10-28). The area is determined by the definite integral

$$\int_{-\infty}^{0} e^x\, dx$$

INTEGRATION

Observing Figure 10–28, we note that the shaded region is not completely bounded as $x \to -\infty$. Since the shaded region is unbounded over the interval $-\infty < x \leq 0$, the corresponding definite integral $\int_{-\infty}^{0} e^x \, dx$ is called an **improper integral**.

We now illustrate a procedure for evaluating an improper integral using $\int_{-\infty}^{0} e^x \, dx$ as an example. Since the shaded region of Figure 10–28 is unbounded as $x \to -\infty$, we replace the lower limit of integration, $-\infty$, with h and evaluate the result as $h \to -\infty$. Hence, we have

$$\int_{h}^{0} e^x \, dx = e^x \Big|_{h}^{0}$$
$$= e^0 - e^h$$
$$= 1 - e^h$$

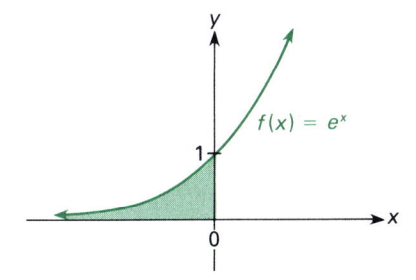

FIGURE 10–28

We now evaluate the resulting expression,

$$1 - e^h$$

as $h \to -\infty$. Since $e^h \to 0$ as $h \to -\infty$, then

$$\int_{-\infty}^{0} e^x \, dx \to 1 - 0 = 1$$

Since this improper integral approaches a *finite number* as $h \to -\infty$, it is said to be **convergent**. If this were not the case, the improper integral would be said to be **divergent**.

EXAMPLE 10–35 Find the area between the x-axis and the curve $f(x) = 1/(x - 1)$ over the interval $1 \leq x \leq 2$ (see Figure 10–29).

EXAMPLE 10–35 (continued)

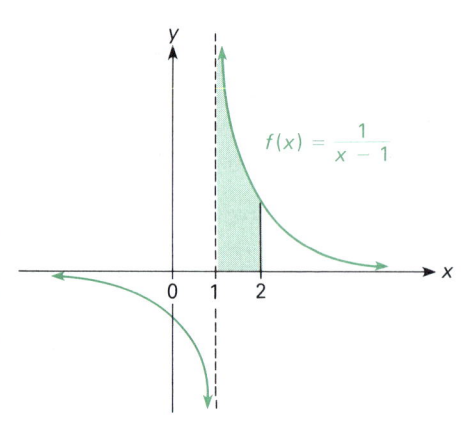

FIGURE 10–29

Solution The area of the shaded region of Figure 10–29 is represented by the definite integral

$$\int_1^2 \frac{1}{x-1}\,dx$$

Since the shaded region is unbounded as $x \to 1$, we replace the limit of integration 1 with h and evaluate the result as $h \to 1$. Hence, we find

$$\int_h^2 \frac{1}{x-1}\,dx = \ln(x-1)\Big|_h^2$$
$$= \ln 1 - \ln(h-1)$$
$$= 0 - \ln(h-1)$$
$$= -\ln(h-1)$$

Evaluating the resulting expression,

$$-\ln(h-1)$$

as $h \to 1$, we have

$$-\ln(h-1) \to \infty$$

Thus, $\int_1^2 1/(x-1)\,dx$ does not approach a finite number and is therefore divergent. Therefore, the area of the shaded region of Figure 10–29 is **indeterminate**.

INTEGRATION

Present Value of a Perpetual Flow

In certain situations (e.g., interest from a perpetual bond or income from an indestructible asset such as land), a continuous stream of money flows forever (i.e., perpetually). If a continuous stream of money flows perpetually into a fund at a rate of $P(t)$ dollars per year (t = time in years) during the time interval $0 \leq t < \infty$, and if this money flow is compounded continuously at a nominal rate, r, then its present value at $t = 0$ is given by the improper integral

$$\int_0^\infty P(t) e^{-rt}\, dt$$

EXAMPLE 10–36 Income from a piece of land flows continuously into a trust fund at a constant rate of $P(t) =$ \$600,000 per year (t = time in years) and is compounded continuously at 10%. If this flow continues forever (i.e., $0 \leq t < \infty$), find its present value.

Solution The present value at $t = 0$ is given by

$$\int_0^\infty 600{,}000 e^{-0.10t}\, dt = \$6{,}000{,}000$$

Thus, the present value is \$6,000,000.

EXERCISES

1. Determine the area between the x-axis and the curve $y = e^{-x}$ over the interval $0 \leq x < \infty$. Sketch the function and shade the desired area.
2. Find the area bounded by the x-axis and the curve $f(x) = 1/(x - 3)^2$ over the interval $10 \leq x < \infty$. Sketch the function and shade the desired area.
3. Find the area bounded by the x-axis and the curve $f(x) = 1/(x - 3)^2$ over the interval $3 \leq x \leq 5$. Sketch the function and shade the desired area.
4. Find the area bounded by the x-axis and the curve $f(x) = 1/(x - 1)^3$ over the interval $2 \leq x < \infty$. Sketch the function and shade the desired area.
5. Find the area bounded by the x-axis and the curve $f(x) = 1/(x - 1)^3$ over the interval $1 \leq x \leq 4$. Sketch the function and shade the desired area.
6. Evaluate whichever of the following improper integrals are convergent:

    (A) $\displaystyle\int_1^\infty \frac{dx}{x^2}$ (B) $\displaystyle\int_0^1 \frac{dx}{x^4}$

(C) $\displaystyle\int_0^1 \frac{dx}{x}$ (D) $\displaystyle\int_0^\infty e^{-2x}\,dx$

(E) $\displaystyle\int_9^\infty \frac{dx}{x+1}$ (F) $\displaystyle\int_0^\infty xe^{-x}\,dx$

7. A perpetual bond yields a continuous flow of interest at a constant rate of $P(t) = \$40,000$ per year (t = time in years). If this perpetual flow is compounded continuously at 9%, find its present value.

8. A piece of land yields a continuous flow of revenue at a constant rate of $800,000 per year. If this perpetual flow is compounded continuously at 12%, find its present value.

10–10

CONTINUOUS PROBABILITY DISTRIBUTIONS

Discrete versus Continuous Distributions

In Chapter 7, we discussed probability distributions. In general, there are two categories of probability distributions: discrete distributions and continuous distributions. If a random variable x takes on values from a discrete set of possibilities, such as that illustrated in Figure 10–30, then it is called a **discrete random variable** and its respective probability distribution is called a **discrete distribution**. If a random variable x can take on any value from a continuous range of values, such as $0 \leq x \leq 6$ (see Figure 10–31), then it is called a **continuous random variable** and its respective probability distribution is called a **continuous probability distribution**. The probability distributions of Chapter 7, including the binomial, are discrete distributions. We have not yet encountered continuous probability distributions. In this section, we will consider a few.

Discrete values of x

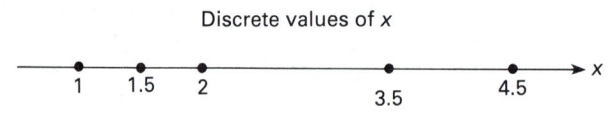

FIGURE 10–30

Continuous range of values of x

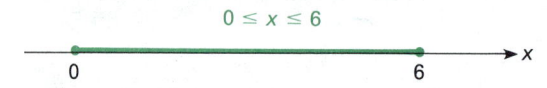

$0 \leq x \leq 6$

FIGURE 10–31

Uniform Distributions

Consider for a moment a box of numbers whose distribution is given by the discrete probability distribution of Table 10–2. A graph of this probability distribution appears in Figure 10–32. If a chance experiment consists of selecting one number from the box,

631 INTEGRATION

the probabilities of selecting either a 1, 3, 5, or 7 are equal. Thus, we say the numbers are **uniformly distributed.**

TABLE 10–2

x	P(x)
1	.25
3	.25
5	.25
7	.25
	1.00

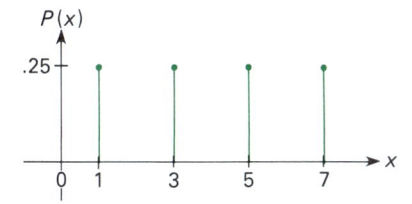

FIGURE 10–32

Now, consider a uniformly distributed set of numbers consisting of the interval $1 \leq x \leq 7$. Since the numbers are uniformly distributed, the graph of the probability distribution must take on a *constant value* throughout the interval $1 \leq x \leq 7$ (see Figure 10–33).

FIGURE 10–33

With a continuous probability distribution, the area between its graph and the x-axis measures probability. Thus, if a chance experiment consists of selecting a number from this uniformly distributed set, the probability that the selected number lies between 2 and 4, i.e.,

$$P(2 < x < 4)$$

is represented by the shaded area of Figure 10–34. (Note that for a continuous probability distribution, $P(a < x < b) = P(a \leq x \leq b)$.

632 CHAPTER TEN

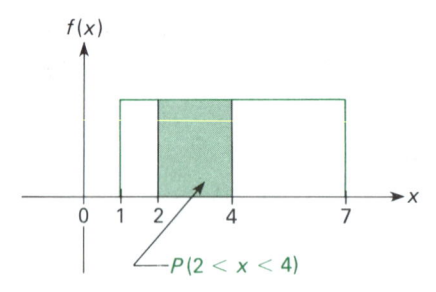

FIGURE 10–34

Thus, $P(2 < x < 4) = P(2 \leq x \leq 4)$. Since the random variable x represents all values within the interval $1 \leq x \leq 7$, then $P(1 \leq x \leq 7) = 1$. Therefore, the total area between the graph of the probability distribution and the x-axis over the interval $1 \leq x \leq 7$ equals 1, and the height of the graph is 1/6 (see Figure 10–35). Hence,

$$P(2 < x < 4) = (\text{width})(\text{height})$$
$$= (4 - 2)\left(\frac{1}{6}\right)$$
$$= \frac{1}{3}$$

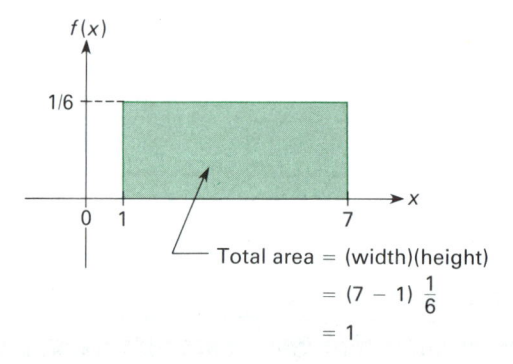

FIGURE 10–35

A continuous probability distribution is usually described by a function called its **probability density function**. The probability density function for the preceding uniform distribution is given by

$$f(x) = \begin{cases} \dfrac{1}{6} & \text{if } 1 \leq x \leq 7 \\ 0 & \text{otherwise} \end{cases}$$

INTEGRATION

Thus, the probability $P(2 < x < 4)$ can also be determined by the definite integral

$$\int_2^4 f(x)\, dx = \int_2^4 \frac{1}{6}\, dx$$

$$= \frac{1}{6}x \Big|_2^4$$

$$= \frac{1}{6}(4) - \frac{1}{6}(2)$$

$$= \frac{1}{3}$$

(See Figure 10–36.)

FIGURE 10–36

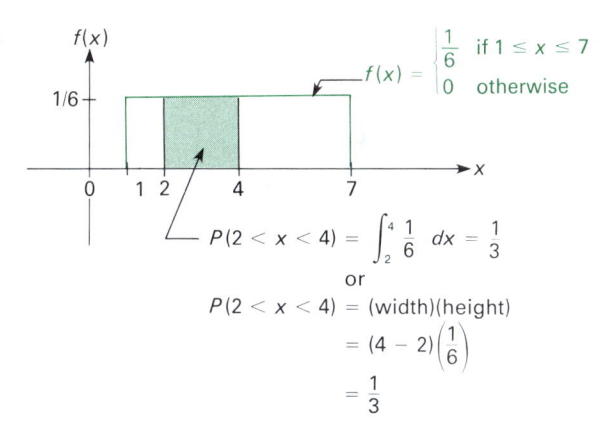

$f(x) = \begin{cases} \frac{1}{6} & \text{if } 1 \le x \le 7 \\ 0 & \text{otherwise} \end{cases}$

$P(2 < x < 4) = \int_2^4 \frac{1}{6}\, dx = \frac{1}{3}$

or

$P(2 < x < 4) =$ (width)(height)

$= (4 - 2)\left(\frac{1}{6}\right)$

$= \frac{1}{3}$

In general, if a random variable x is uniformly distributed over the interval $a \le x \le b$, its probability density function is given by

$$f(x) = \begin{cases} \dfrac{1}{b-a} & \text{for } a \le x \le b \\ 0 & \text{otherwise} \end{cases}$$

The graph of $f(x)$ appears in Figure 10–37. Observe that the total area equals 1.

FIGURE 10–37

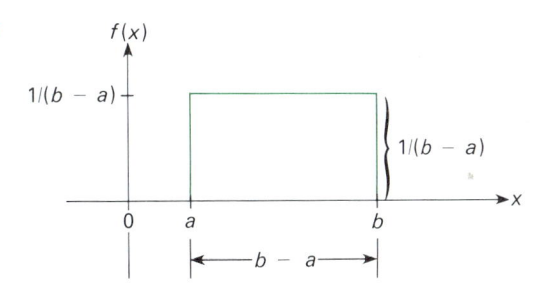

EXAMPLE 10-37 The time a customer waits at a department store's counter never exceeds 5 minutes. However, upon arriving at the counter, it is just as likely that one will wait 0 minutes as it is that one will wait 1.5 minutes, etc. In other words, the waiting time, x, is uniformly distributed over the interval $0 \leq x \leq 5$. Find the probability that an arriving customer will wait at most 4 minutes.

Solution The probability we seek is represented by the shaded area of Figure 10-38. Since the shaded area is rectangular, the probability may be determined by the product $4(1/5)$. Hence,

$$P(0 \leq x \leq 4) = 4\left(\frac{1}{5}\right)$$
$$= .8$$

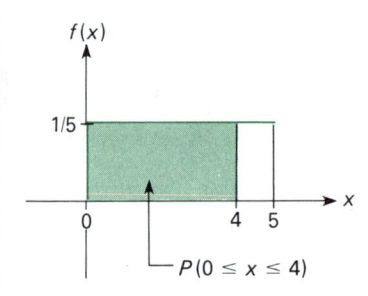

FIGURE 10-38

Using the definite integral to find the shaded area yields the same result. Therefore,

$$\int_0^4 \frac{1}{5}\, dx = \frac{1}{5}x \Big|_0^4$$
$$= \frac{1}{5}(4) - \frac{1}{5}(0)$$
$$= .8$$

Exponential Distributions If a random variable x is distributed in accordance with the density function

$$f(x) = \begin{cases} ke^{-kx} & \text{if } x \geq 0 \\ 0 & \text{otherwise} \end{cases}$$

635 INTEGRATION

where the constant $k > 0$, then x is said to be **exponentially distributed**. A graph of the exponential density function appears in Figure 10–39.

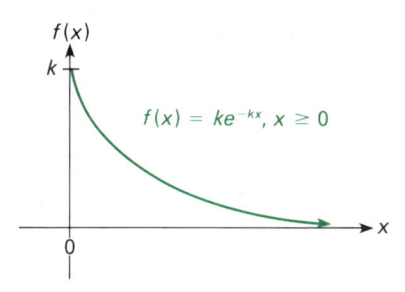

FIGURE 10–39

EXAMPLE 10–38 Show that the total area between the exponential density function and the x-axis is 1.
Solution We must show that the improper integral

$$\int_0^\infty ke^{-kx}\, dx$$

with constant $k > 0$, converges to 1. Hence,

$$\int_0^h ke^{-kx}\, dx = -e^{-kx} \Big|_0^h$$

$$= -e^{-kh} - (-e^{-k(0)})$$
$$= -e^{-kh} - (-1)$$
$$= -e^{-kh} + 1$$

Since $-e^{-kh} \to 0$ as $h \to \infty$, then

$$\int_0^\infty ke^{-kx}\, dx \to 0 + 1 = 1$$

EXAMPLE 10–39 The interarrival time x (i.e., the time between two successive arrivals) of customers at a particular bank is exponentially distributed, with $k = 10$. Find the probability that the interarrival time is between 0.1 and 0.5 minute.

EXAMPLE 10–39 (continued)

Solution We seek the shaded area of Figure 10–40. Thus, we have

$$P(0.1 < x < 0.5) = \int_{0.1}^{0.5} 10e^{-10x}\, dx$$

$$= -e^{-10x} \Big|_{0.1}^{0.5}$$

$$= -e^{-10(0.5)} - (-e^{-10(0.1)})$$

$$= -e^{-5} + e^{-1}$$

$$= -.006738 + .367879$$

$$\approx .3611$$

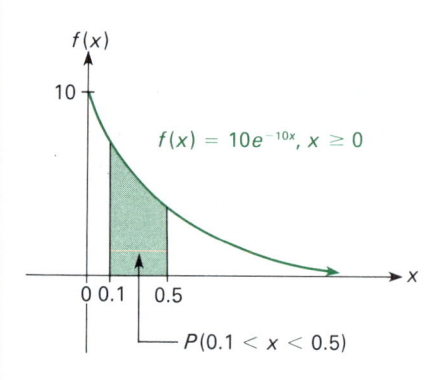

FIGURE 10–40

EXERCISES

1. A customer appears at a certain restaurant every 20 minutes. If a customer arrives at this restaurant, find the probability that he or she must wait
   (A) At least 12 minutes.
   (B) Less than 5 minutes.
   (C) Between 5 and 10 minutes.

2. The length of life (in years) of a certain brand of battery is a random variable x distributed in accordance with the probability density function

$$f(x) = \begin{cases} -6x^2 + 6x & \text{if } 0 \leq x \leq 1 \\ 0 & \text{otherwise} \end{cases}$$

(A) Sketch $f(x)$.
(B) Find the probability that a battery lasts longer than 3/4 year.
(C) Find $P(0 \leq x \leq 0.2)$.
(D) If the batteries are unconditionally guaranteed for 2 months, what percentage will be returned?

3. A random variable x is uniformly distributed over the interval $2 \leq x \leq 7$.
   (A) Determine its probability density function $f(x)$.
   (B) Find $P(2 \leq x \leq 3)$.
   (C) Find $P(4 \leq x \leq 6)$.

4. A random variable x is distributed in accordance with a probability density function of the form $f(x) = kx$ over the interval $0 \leq x \leq 4$.
   (A) Determine k.
   (B) Sketch $f(x)$.
   (C) Calculate $P(0 \leq x \leq 2)$.
   (D) Calculate $P(3 \leq x \leq 4)$.
   (E) Calculate $P(1 \leq x \leq 2)$.

5. The interarrival times (in minutes) of incoming calls at a certain switchboard are exponentially distributed, with $k = 5$. Find the probability that the time between two successive arrivals is
   (A) Less than 0.1 minute.
   (B) Between 0.2 and 0.6 minute.
   (C) Longer than 0.4 minute.

6. At a certain self-service gas station, the interarrival times (in minutes) are exponentially distributed, with $k = 0.5$. Find the probability that the interarrival time is
   (A) Less than 0.6 minute.
   (B) Between 0.2 and 0.8 minute.
   (C) Longer than 1 minute.

7. A random variable x is exponentially distributed, with $k = 2$. Find each of the following:
   (A) $P(0.5 \leq x \leq 1)$
   (B) $P(0 \leq x \leq 1.5)$
   (C) $P(x \geq 3)$

8. *(Reliability)* Let the random variable x represent the length of life of some mechanical component with probability density function $f(x)$. If t is a given length of time, then

$$P(x > t) = \int_{t}^{\infty} f(x)\, dx$$

Since this integral represents the probability that the component's lifetime exceeds t units of time (see Figure 10–41), it is called the **reliability function** and is denoted by $R(t)$. Hence,

$$R(t) = P(x > t) = \int_{t}^{\infty} f(x)\, dx$$

$$P(x > t) = \int_t^\infty f(x)\, dx$$

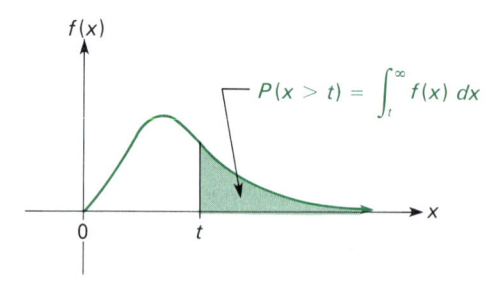

FIGURE 10–41

(A) Find $R(t)$ for an exponential density function with $k = 0.8$.
(B) Find $R(1)$ and interpret.
(C) Find $R(5)$ and interpret.

9. The length of life (in years) of circuitry in a certain brand of calculator is exponentially distributed, with $k = 0.5$.
   (A) Find the reliability function, $R(t)$.
   (B) Find $R(1)$ and interpret.
   (C) Find $R(4)$ and interpret.

10. Show that, in general, an exponential density function has the reliability function

$$R(t) = e^{-kt}$$

APPENDIX

TABLE 1 Common Logarithms

x	0	1	2	3	4	5	6	7	8	9
1.0	0.0000	0.0043	0.0086	0.0128	0.0170	0.0212	0.0253	0.0294	0.0334	0.0374
1.1	0.0414	0.0453	0.0492	0.0531	0.0569	0.0607	0.0645	0.0682	0.0719	0.0755
1.2	0.0792	0.0828	0.0864	0.0899	0.0934	0.0969	0.1004	0.1038	0.1072	0.1106
1.3	0.1139	0.1173	0.1206	0.1239	0.1271	0.1303	0.1335	0.1367	0.1399	0.1430
1.4	0.1461	0.1492	0.1523	0.1553	0.1584	0.1614	0.1644	0.1673	0.1703	0.1732
1.5	0.1761	0.1790	0.1818	0.1847	0.1875	0.1903	0.1931	0.1959	0.1987	0.2014
1.6	0.2041	0.2068	0.2095	0.2122	0.2148	0.2175	0.2201	0.2227	0.2253	0.2279
1.7	0.2304	0.2330	0.2355	0.2380	0.2405	0.2430	0.2455	0.2480	0.2504	0.2529
1.8	0.2553	0.2577	0.2601	0.2625	0.2648	0.2672	0.2695	0.2718	0.2742	0.2765
1.9	0.2788	0.2810	0.2833	0.2856	0.2878	0.2900	0.2923	0.2945	0.2967	0.2989
2.0	0.3010	0.3032	0.3054	0.3075	0.3096	0.3118	0.3139	0.3160	0.3181	0.3201
2.1	0.3222	0.3243	0.3263	0.3284	0.3304	0.3324	0.3345	0.3365	0.3385	0.3404
2.2	0.3424	0.3444	0.3464	0.3483	0.3502	0.3522	0.3541	0.3560	0.3579	0.3598
2.3	0.3617	0.3636	0.3655	0.3674	0.3692	0.3711	0.3729	0.3747	0.3766	0.3784
2.4	0.3802	0.3820	0.3838	0.3856	0.3874	0.3892	0.3909	0.3927	0.3945	0.3962
2.5	0.3979	0.3997	0.4014	0.4031	0.4048	0.4065	0.4082	0.4099	0.4116	0.4133
2.6	0.4150	0.4166	0.4183	0.4200	0.4216	0.4232	0.4249	0.4265	0.4281	0.4298
2.7	0.4314	0.4330	0.4346	0.4362	0.4378	0.4393	0.4409	0.4425	0.4440	0.4456
2.8	0.4472	0.4487	0.4502	0.4518	0.4533	0.4548	0.4564	0.4579	0.4594	0.4609
2.9	0.4624	0.4639	0.4654	0.4669	0.4683	0.4698	0.4713	0.4728	0.4742	0.4757
3.0	0.4771	0.4786	0.4800	0.4814	0.4829	0.4843	0.4857	0.4871	0.4886	0.4900
3.1	0.4914	0.4928	0.4942	0.4955	0.4969	0.4983	0.4997	0.5011	0.5024	0.5038
3.2	0.5052	0.5065	0.5079	0.5092	0.5105	0.5119	0.5132	0.5145	0.5159	0.5172
3.3	0.5185	0.5198	0.5211	0.5224	0.5237	0.5250	0.5263	0.5276	0.5289	0.5302
3.4	0.5315	0.5328	0.5340	0.5353	0.5366	0.5378	0.5391	0.5403	0.5416	0.5428
3.5	0.5441	0.5453	0.5465	0.5478	0.5490	0.5502	0.5515	0.5527	0.5539	0.5551
3.6	0.5563	0.5575	0.5587	0.5599	0.5611	0.5623	0.5635	0.5647	0.5658	0.5670
3.7	0.5682	0.5694	0.5705	0.5717	0.5729	0.5740	0.5752	0.5763	0.5775	0.5786

TABLE 1 Common Logarithms (continued)

x	0	1	2	3	4	5	6	7	8	9
3.8	0.5798	0.5809	0.5821	0.5832	0.5843	0.5855	0.5866	0.5877	0.5888	0.5899
3.9	0.5911	0.5922	0.5933	0.5944	0.5955	0.5966	0.5977	0.5988	0.5999	0.6010
4.0	0.6021	0.6031	0.6042	0.6053	0.6064	0.6075	0.6085	0.6096	0.6107	0.6117
4.1	0.6128	0.6138	0.6149	0.6160	0.6170	0.6180	0.6191	0.6201	0.6212	0.6222
4.2	0.6232	0.6243	0.6253	0.6263	0.6274	0.6284	0.6294	0.6304	0.6314	0.6325
4.3	0.6335	0.6345	0.6355	0.6365	0.6375	0.6385	0.6395	0.6405	0.6415	0.6425
4.4	0.6435	0.6444	0.6454	0.6464	0.6474	0.6484	0.6493	0.6503	0.6513	0.6522
4.5	0.6532	0.6542	0.6551	0.6561	0.6571	0.6580	0.6590	0.6599	0.6609	0.6618
4.6	0.6628	0.6637	0.6646	0.6656	0.6665	0.6675	0.6684	0.6693	0.6702	0.6712
4.7	0.6721	0.6730	0.6739	0.6749	0.6758	0.6767	0.6776	0.6785	0.6794	0.6803
4.8	0.6812	0.6821	0.6830	0.6839	0.6848	0.6857	0.6866	0.6875	0.6884	0.6893
4.9	0.6902	0.6911	0.6920	0.6928	0.6937	0.6946	0.6955	0.6964	0.6972	0.6981
5.0	0.6990	0.6998	0.7007	0.7016	0.7024	0.7033	0.7042	0.7050	0.7059	0.7067
5.1	0.7076	0.7084	0.7093	0.7101	0.7110	0.7118	0.7126	0.7135	0.7143	0.7152
5.2	0.7160	0.7168	0.7177	0.7185	0.7193	0.7202	0.7210	0.7218	0.7226	0.7235
5.3	0.7243	0.7251	0.7259	0.7267	0.7275	0.7284	0.7292	0.7300	0.7308	0.7316
5.4	0.7324	0.7332	0.7340	0.7348	0.7356	0.7364	0.7372	0.7380	0.7388	0.7396
5.5	0.7404	0.7412	0.7419	0.7427	0.7435	0.7443	0.7451	0.7459	0.7466	0.7474
5.6	0.7482	0.7490	0.7497	0.7505	0.7513	0.7520	0.7528	0.7536	0.7543	0.7551
5.7	0.7559	0.7566	0.7574	0.7582	0.7589	0.7597	0.7604	0.7612	0.7619	0.7627
5.8	0.7634	0.7642	0.7649	0.7657	0.7664	0.7672	0.7679	0.7686	0.7694	0.7701
5.9	0.7709	0.7716	0.7723	0.7731	0.7738	0.7745	0.7752	0.7760	0.7767	0.7774
6.0	0.7782	0.7789	0.7796	0.7803	0.7810	0.7818	0.7825	0.7832	0.7839	0.7846
6.1	0.7853	0.7860	0.7868	0.7875	0.7882	0.7889	0.7896	0.7903	0.7910	0.7917
6.2	0.7924	0.7931	0.7938	0.7945	0.7952	0.7959	0.7966	0.7973	0.7980	0.7987
6.3	0.7993	0.8000	0.8007	0.8014	0.8021	0.8028	0.8035	0.8041	0.8048	0.8055
6.4	0.8062	0.8069	0.8075	0.8082	0.8089	0.8096	0.8102	0.8109	0.8116	0.8122
6.5	0.8129	0.8136	0.8142	0.8149	0.8156	0.8162	0.8169	0.8176	0.8182	0.8189
6.6	0.8195	0.8202	0.8209	0.8215	0.8222	0.8228	0.8235	0.8241	0.8248	0.8254
6.7	0.8261	0.8267	0.8274	0.8280	0.8287	0.8293	0.8299	0.8306	0.8312	0.8319
6.8	0.8325	0.8331	0.8338	0.8344	0.8351	0.8357	0.8363	0.8370	0.8376	0.8382
6.9	0.8388	0.8395	0.8401	0.8407	0.8414	0.8420	0.8426	0.8432	0.8439	0.8445
7.0	0.8451	0.8457	0.8463	0.8470	0.8476	0.8482	0.8488	0.8494	0.8500	0.8506
7.1	0.8513	0.8519	0.8525	0.8531	0.8537	0.8543	0.8549	0.8555	0.8561	0.8567
7.2	0.8573	0.8579	0.8585	0.8591	0.8597	0.8603	0.8609	0.8615	0.8621	0.8627
7.3	0.8633	0.8639	0.8645	0.8651	0.8657	0.8663	0.8669	0.8675	0.8681	0.8686
7.4	0.8692	0.8698	0.8704	0.8710	0.8716	0.8722	0.8727	0.8733	0.8739	0.8745
7.5	0.8751	0.8756	0.8762	0.8768	0.8774	0.8779	0.8785	0.8791	0.8797	0.8802
7.6	0.8808	0.8814	0.8820	0.8825	0.8831	0.8837	0.8842	0.8848	0.8854	0.8859
7.7	0.8865	0.8871	0.8876	0.8882	0.8887	0.8893	0.8899	0.8904	0.8910	0.8915
7.8	0.8921	0.8927	0.8932	0.8938	0.8943	0.8949	0.8954	0.8960	0.8965	0.8971
7.9	0.8976	0.8982	0.8987	0.8993	0.8998	0.9004	0.9009	0.9015	0.9020	0.9025
8.0	0.9031	0.9036	0.9042	0.9047	0.9053	0.9058	0.9063	0.9069	0.9074	0.9079
8.1	0.9085	0.9090	0.9096	0.9101	0.9106	0.9112	0.9117	0.9122	0.9128	0.9133
8.2	0.9138	0.9143	0.9149	0.9154	0.9159	0.9165	0.9170	0.9175	0.9180	0.9186

TABLE 1 Common Logarithms (continued)

x	0	1	2	3	4	5	6	7	8	9
8.3	0.9191	0.9196	0.9201	0.9206	0.9212	0.9217	0.9222	0.9227	0.9232	0.9238
8.4	0.9243	0.9248	0.9253	0.9258	0.9263	0.9269	0.9274	0.9279	0.9284	0.9289
8.5	0.9294	0.9299	0.9304	0.9309	0.9315	0.9320	0.9325	0.9330	0.9335	0.9340
8.6	0.9345	0.9350	0.9355	0.9360	0.9365	0.9370	0.9375	0.9380	0.9385	0.9390
8.7	0.9395	0.9400	0.9405	0.9410	0.9415	0.9420	0.9425	0.9430	0.9435	0.9440
8.8	0.9445	0.9450	0.9455	0.9460	0.9465	0.9469	0.9474	0.9479	0.9484	0.9489
8.9	0.9494	0.9499	0.9504	0.9509	0.9513	0.9518	0.9523	0.9528	0.9533	0.9538
9.0	0.9542	0.9547	0.9552	0.9557	0.9562	0.9566	0.9571	0.9576	0.9581	0.9586
9.1	0.9590	0.9595	0.9600	0.9605	0.9609	0.9614	0.9619	0.9624	0.9628	0.9633
9.2	0.9638	0.9643	0.9647	0.9652	0.9657	0.9661	0.9666	0.9671	0.9675	0.9680
9.3	0.9685	0.9689	0.9694	0.9699	0.9703	0.9708	0.9713	0.9717	0.9722	0.9727
9.4	0.9731	0.9736	0.9741	0.9745	0.9750	0.9754	0.9759	0.9764	0.9768	0.9773
9.5	0.9777	0.9782	0.9786	0.9791	0.9795	0.9800	0.9805	0.9809	0.9814	0.9818
9.6	0.9823	0.9827	0.9832	0.9836	0.9841	0.9845	0.9850	0.9854	0.9859	0.9863
9.7	0.9868	0.9872	0.9877	0.9881	0.9886	0.9890	0.9894	0.9899	0.9903	0.9908
9.8	0.9912	0.9917	0.9921	0.9926	0.9930	0.9934	0.9939	0.9943	0.9948	0.9952
9.9	0.9956	0.9961	0.9965	0.9969	0.9974	0.9978	0.9983	0.9987	0.9991	0.9996

TABLE 2 Natural Logarithms (continued)

x	ln x	x	ln x	x	ln x
0.1	2.302585	2.2	0.788457	4.3	1.458615
0.2	1.609438	2.3	0.832909	4.4	1.481605
0.3	1.203973	2.4	0.875469	4.5	1.504077
0.4	0.916291	2.5	0.916291	4.6	1.526056
0.5	0.693147	2.6	0.955511	4.7	1.547563
0.6	0.510826	2.7	0.993252	4.8	1.568616
0.7	0.356675	2.8	1.029619	4.9	1.589235
0.8	0.223144	2.9	1.064711	5.0	1.609438
0.9	0.105361	3.0	1.098612	5.1	1.629241
1.0	0.000000	3.1	1.131402	5.2	1.648659
1.1	0.095310	3.2	1.163151	5.3	1.667707
1.2	0.182322	3.3	1.193922	5.4	1.686399
1.3	0.262364	3.4	1.223775	5.5	1.704748
1.4	0.336472	3.5	1.252763	5.6	1.722767
1.5	0.405465	3.6	1.280934	5.7	1.740466
1.6	0.470004	3.7	1.308333	5.8	1.757858
1.7	0.530628	3.8	1.335001	5.9	1.774952
1.8	0.587787	3.9	1.360977	6.0	1.791759
1.9	0.641854	4.0	1.386294	6.1	1.808289
2.0	0.693147	4.1	1.410987	6.2	1.824549
2.1	0.741937	4.2	1.435085	6.3	1.840550

TABLE 2 Natural Logarithms (continued)

x	ln x	x	ln x	x	ln x
6.4	1.856298	8.5	2.140066	16.0	2.772589
6.5	1.871802	8.6	2.151762	17.0	2.833213
6.6	1.887070	8.7	2.163323	18.0	2.890372
6.7	1.902108	8.8	2.174752	19.0	2.944439
6.8	1.916923	8.9	2.186051	20.0	2.995732
6.9	1.931521	9.0	2.197225	25.0	3.218876
7.0	1.945910	9.1	2.208274	30.0	3.401197
7.1	1.960095	9.2	2.219203	35.0	3.555348
7.2	1.974081	9.3	2.230014	40.0	3.688879
7.3	1.987874	9.4	2.240710	45.0	3.806662
7.4	2.001480	9.5	2.251292	50.0	3.912023
7.5	2.014903	9.6	2.261763	55.0	4.007333
7.6	2.028148	9.7	2.272126	60.0	4.094345
7.7	2.041220	9.8	2.282382	65.0	4.174387
7.8	2.054124	9.9	2.292535	70.0	4.248495
7.9	2.066863	10.0	2.302585	75.0	4.317488
8.0	2.079442	11.0	2.397895	80.0	4.382027
8.1	2.091864	12.0	2.484907	85.0	4.442651
8.2	2.104134	13.0	2.564949	90.0	4.499810
8.3	2.116256	14.0	2.639057	95.0	4.553877
8.4	2.128232	15.0	2.708050	100.0	4.605170

TABLE 3 Exponential Functions

x	e^x	e^{-x}	x	e^x	e^{-x}
0.00	1.000000	1.000000	0.15	1.161834	0.860708
0.01	1.010050	0.990050	0.16	1.173511	0.852144
0.02	1.020201	0.980199	0.17	1.185305	0.843665
0.03	1.030455	0.970446	0.18	1.197217	0.835270
0.04	1.040811	0.960789	0.19	1.209250	0.826959
0.05	1.051271	0.951229	0.20	1.221403	0.818731
0.06	1.061837	0.941765	0.30	1.349859	0.740818
0.07	1.072508	0.932394	0.40	1.491825	0.670320
0.08	1.083287	0.923116	0.50	1.648721	0.606531
0.09	1.094174	0.913931	0.60	1.822119	0.548812
0.10	1.105171	0.904837	0.70	2.013753	0.496585
0.11	1.116278	0.895834	0.80	2.225541	0.449329
0.12	1.127497	0.886920	0.90	2.459603	0.406570
0.13	1.138828	0.878095	1.00	2.718282	0.367879
0.14	1.150274	0.869358	1.10	3.004166	0.332871

TABLE 3 Exponential Functions (continued)

x	e^x	e^{-x}	x	e^x	e^{-x}
1.20	3.320117	0.301194	2.90	18.174145	0.055023
1.30	3.669297	0.272532	3.00	20.085537	0.049787
1.40	4.055200	0.246597	3.50	33.115452	0.030197
1.50	4.481689	0.223130	4.00	54.598150	0.018316
1.60	4.953032	0.201897	4.50	90.017131	0.011109
1.70	5.473947	0.182684	5.00	148.413159	0.006738
1.80	6.049647	0.165299	5.50	244.691932	0.004087
1.90	6.685894	0.149569	6.00	403.428793	0.002479
2.00	7.389056	0.135335	6.50	665.141633	0.001503
2.10	8.166170	0.122456	7.00	1096.633158	0.000912
2.20	9.025013	0.110803	7.50	1808.042414	0.000553
2.30	9.974182	0.100259	8.00	2980.957987	0.000335
2.40	11.023176	0.090718	8.50	4914.768840	0.000203
2.50	12.182494	0.082085	9.00	8103.083928	0.000123
2.60	13.463738	0.074274	9.50	13359.726830	0.000075
2.70	14.879732	0.067206	10.00	22026.465795	0.000045
2.80	16.444647	0.060810			

TABLE 4 Compound Amount $(1 + i)^n$

n	$\frac{1}{2}$%	$\frac{3}{4}$%	1%	$1\frac{1}{4}$%	$1\frac{1}{2}$%	$1\frac{3}{4}$%	2%
1	1.005000	1.007500	1.010000	1.012500	1.015000	1.017500	1.020000
2	1.010025	1.015056	1.020100	1.025156	1.030225	1.035306	1.040400
3	1.015075	1.022669	1.030301	1.037971	1.045678	1.053424	1.061208
4	1.020151	1.030339	1.040604	1.050945	1.061364	1.071859	1.082432
5	1.025251	1.038067	1.051010	1.064082	1.077284	1.090617	1.104081
6	1.030378	1.045852	1.061520	1.077383	1.093443	1.109702	1.126162
7	1.035529	1.053696	1.072135	1.090850	1.109845	1.129122	1.148686
8	1.040707	1.061599	1.082857	1.104486	1.126493	1.148882	1.171659
9	1.045911	1.069561	1.093685	1.118292	1.143390	1.168987	1.195093
10	1.051140	1.077583	1.104622	1.132271	1.160541	1.189444	1.218994
11	1.056396	1.085664	1.115668	1.146424	1.177949	1.210260	1.243374
12	1.061678	1.093807	1.126825	1.160755	1.195618	1.231439	1.268242
13	1.066986	1.102010	1.138093	1.175264	1.213552	1.252990	1.293607
14	1.072321	1.110276	1.149474	1.189955	1.231756	1.274917	1.319479
15	1.077683	1.118603	1.160969	1.204829	1.250232	1.297228	1.345868
16	1.083071	1.126992	1.172579	1.219890	1.268986	1.319929	1.372786
17	1.088487	1.135445	1.184304	1.235138	1.288020	1.343028	1.400241
18	1.093929	1.143960	1.196147	1.250577	1.307341	1.366531	1.428246
19	1.099399	1.152540	1.208109	1.266210	1.326951	1.390445	1.456811
20	1.104896	1.161184	1.220190	1.282037	1.346855	1.414778	1.485947

TABLE 4 Compound Amount $(1 + i)^n$ (continued)

n	$\frac{1}{2}$%	$\frac{3}{4}$%	1%	$1\frac{1}{4}$%	$1\frac{1}{2}$%	$1\frac{3}{4}$%	2%
21	1.110420	1.169893	1.232392	1.298063	1.367058	1.439537	1.515666
22	1.115972	1.178667	1.244716	1.314288	1.387564	1.464729	1.545980
23	1.121552	1.187507	1.257163	1.330717	1.408377	1.490361	1.576899
24	1.127160	1.196414	1.269735	1.347351	1.429503	1.516443	1.608437
25	1.132796	1.205387	1.282432	1.364193	1.450945	1.542981	1.640606
26	1.138460	1.214427	1.295256	1.381245	1.472710	1.569983	1.673418
27	1.144152	1.223535	1.308209	1.398511	1.494800	1.597457	1.706886
28	1.149873	1.232712	1.321291	1.415992	1.517222	1.625413	1.741024
29	1.155622	1.241957	1.334504	1.433692	1.539981	1.653858	1.775845
30	1.161400	1.251272	1.347849	1.451613	1.563080	1.682800	1.811362
31	1.167207	1.260656	1.361327	1.469759	1.586526	1.712249	1.847589
32	1.173043	1.270111	1.374941	1.488131	1.610324	1.742213	1.884541
33	1.178908	1.279637	1.388690	1.506732	1.634479	1.772702	1.922231
34	1.184803	1.289234	1.402577	1.525566	1.658996	1.803725	1.960676
35	1.190727	1.298904	1.416603	1.544636	1.683881	1.835290	1.999890
36	1.196681	1.308645	1.430769	1.563944	1.709140	1.867407	2.039887
37	1.202664	1.318460	1.445076	1.583493	1.734777	1.900087	2.080685
38	1.208677	1.328349	1.459527	1.603287	1.760798	1.933338	2.122299
39	1.214721	1.338311	1.474123	1.623328	1.787210	1.967172	2.164745
40	1.220794	1.348349	1.488864	1.643619	1.814018	2.001597	2.208040
41	1.226898	1.358461	1.503752	1.664165	1.841229	2.036625	2.252200
42	1.233033	1.368650	1.518790	1.684967	1.868847	2.072266	2.297244
43	1.239198	1.378915	1.533978	1.706029	1.896880	2.108531	2.343189
44	1.245394	1.389256	1.549318	1.727354	1.925333	2.145430	2.390053
45	1.251621	1.399676	1.564811	1.748946	1.954213	2.182975	2.437854
46	1.257879	1.410173	1.580459	1.770808	1.983526	2.221177	2.486611
47	1.264168	1.420750	1.596263	1.792943	2.013279	2.260048	2.536344
48	1.270489	1.431405	1.612226	1.815355	2.043478	2.299599	2.587070
49	1.276842	1.442141	1.628348	1.838047	2.074130	2.339842	2.638812
50	1.283226	1.452957	1.644632	1.861022	2.105242	2.380789	2.691588
51	1.289642	1.463854	1.661078	1.884285	2.136821	2.422453	2.745420
52	1.296090	1.474833	1.677689	1.907839	2.168873	2.464846	2.800328
53	1.302571	1.485894	1.694466	1.931687	2.201406	2.507980	2.856335
54	1.309083	1.497038	1.711410	1.955833	2.234428	2.551870	2.913461
55	1.315629	1.508266	1.728525	1.980281	2.267944	2.596528	2.971731
56	1.322207	1.519578	1.745810	2.005034	2.301963	2.641967	3.031165
57	1.328818	1.530975	1.763268	2.030097	2.336493	2.688202	3.091789
58	1.335462	1.542457	1.780901	2.055473	2.371540	2.735245	3.153624
59	1.342139	1.554026	1.798710	2.081167	2.407113	2.783112	3.216697
60	1.348850	1.565681	1.816697	2.107181	2.443220	2.831816	3.281031
61	1.355594	1.577424	1.834864	2.133521	2.479868	2.881373	3.346651
62	1.362372	1.589254	1.853212	2.160190	2.517066	2.931797	3.413584
63	1.369184	1.601174	1.871744	2.187193	2.554822	2.983104	3.481856
64	1.376030	1.613183	1.890462	2.214532	2.593144	3.035308	3.551493
65	1.382910	1.625281	1.909366	2.242214	2.632042	3.088426	3.622523

TABLE 4 Compound Amount $(1 + i)^n$ (continued)

n	$\frac{1}{2}$%	$\frac{3}{4}$%	1%	$1\frac{1}{4}$%	$1\frac{1}{2}$%	$1\frac{3}{4}$%	2%
66	1.389825	1.637471	1.928460	2.270242	2.671522	3.142473	3.694974
67	1.396774	1.649752	1.947745	2.298620	2.711595	3.197466	3.768873
68	1.403758	1.662125	1.967222	2.327353	2.752269	3.253422	3.844251
69	1.410777	1.674591	1.986894	2.356444	2.793553	3.310357	3.921136
70	1.417831	1.687151	2.006763	2.385900	2.835456	3.368288	3.999558
71	1.424920	1.699804	2.026831	2.415724	2.877988	3.427233	4.079549
72	1.432044	1.712553	2.047099	2.445920	2.921158	3.487210	4.161140
73	1.439204	1.725397	2.067570	2.476494	2.964975	3.548236	4.244363
74	1.446401	1.738337	2.088246	2.507450	3.009450	3.610330	4.329250
75	1.453633	1.751375	2.109128	2.538794	3.054592	3.673511	4.415835
76	1.460901	1.764510	2.130220	2.570529	3.100411	3.737797	4.504152
77	1.468205	1.777744	2.151522	2.602660	3.146917	3.803209	4.594235
78	1.475546	1.791077	2.173037	2.635193	3.194120	3.869765	4.686120
79	1.482924	1.804510	2.194768	2.668133	3.242032	3.937486	4.779842
80	1.490339	1.818044	2.216715	2.701485	3.290663	4.006392	4.875439
81	1.497790	1.831679	2.238882	2.735254	3.340023	4.076504	4.972948
82	1.505279	1.845417	2.261271	2.769444	3.390123	4.147843	5.072407
83	1.512806	1.859258	2.283884	2.804062	3.440975	4.220430	5.173855
84	1.520370	1.873202	2.306723	2.839113	3.492590	4.294287	5.277332
85	1.527971	1.887251	2.329790	2.874602	3.544978	4.369437	5.382879
86	1.535611	1.901405	2.353088	2.910534	3.598153	4.445903	5.490536
87	1.543289	1.915666	2.376619	2.946916	3.652125	4.523706	5.600347
88	1.551006	1.930033	2.400385	2.983753	3.706907	4.602871	5.712354
89	1.558761	1.944509	2.424389	3.021049	3.762511	4.683421	5.826601
90	1.566555	1.959092	2.448633	3.058813	3.818949	4.765381	5.943133
91	1.574387	1.973786	2.473119	3.097048	3.876233	4.848775	6.061996
92	1.582259	1.988589	2.497850	3.135761	3.934376	4.933629	6.183236
93	1.590171	2.003503	2.522829	3.174958	3.993392	5.019967	6.306900
94	1.598122	2.018530	2.548057	3.214645	4.053293	5.107816	6.433038
95	1.606112	2.033669	2.573538	3.254828	4.114092	5.197203	6.561699
96	1.614143	2.048921	2.599273	3.295513	4.175804	5.288154	6.692933
97	1.622213	2.064288	2.625266	3.336707	4.238441	5.380697	6.826792
98	1.630324	2.079770	2.651518	3.378416	4.302017	5.474859	6.963328
99	1.638476	2.095369	2.678033	3.420646	4.366547	5.570669	7.102594
100	1.646668	2.111084	2.704814	3.463404	4.432046	5.668156	7.244646

TABLE 4 Compound Amount $(1 + i)^n$

n	3%	4%	5%	6%	7%	8%	9%
1	1.030000	1.040000	1.050000	1.060000	1.070000	1.080000	1.090000
2	1.060900	1.081600	1.102500	1.123600	1.144900	1.166400	1.188100
3	1.092727	1.124864	1.157625	1.191016	1.225043	1.259712	1.295029
4	1.125509	1.169859	1.215506	1.262477	1.310796	1.360489	1.411582

TABLE 4 Compound Amount $(1 + i)^n$ (continued)

n	3%	4%	5%	6%	7%	8%	9%
5	1.159274	1.216653	1.276282	1.338226	1.402552	1.469328	1.538624
6	1.194052	1.265319	1.340096	1.418519	1.500730	1.586874	1.677100
7	1.229874	1.315932	1.407100	1.503630	1.605781	1.713824	1.828039
8	1.266770	1.368569	1.477455	1.593848	1.718186	1.850930	1.992563
9	1.304773	1.423312	1.551328	1.689479	1.838459	1.999005	2.171893
10	1.343916	1.480244	1.628895	1.790848	1.967151	2.158925	2.367364
11	1.384234	1.539454	1.710339	1.898299	2.104852	2.331639	2.580426
12	1.425761	1.601032	1.795856	2.012196	2.252192	2.518170	2.812665
13	1.468534	1.665074	1.885649	2.132928	2.409845	2.719624	3.065805
14	1.512590	1.731676	1.979932	2.260904	2.578534	2.937194	3.341727
15	1.557967	1.800944	2.078928	2.396558	2.759032	3.172169	3.642482
16	1.604706	1.872981	2.182875	2.540352	2.952164	3.425943	3.970306
17	1.652848	1.947900	2.292018	2.692773	3.158815	3.700018	4.327633
18	1.702433	2.025817	2.406619	2.854339	3.379932	3.996019	4.717120
19	1.753506	2.106849	2.526950	3.025600	3.616528	4.315701	5.141661
20	1.806111	2.191123	2.653298	3.207135	3.869684	4.660957	5.604411
21	1.860295	2.278768	2.785963	3.399564	4.140562	5.033834	6.108808
22	1.916103	2.369919	2.925261	3.603537	4.430402	5.436540	6.658600
23	1.973587	2.464716	3.071524	3.819750	4.740530	5.871464	7.257874
24	2.032794	2.563304	3.225100	4.048935	5.072367	6.341181	7.911083
25	2.093778	2.665836	3.386355	4.291871	5.427433	6.848475	8.623081
26	2.156591	2.772470	3.555673	4.549383	5.807353	7.396353	9.399158
27	2.221289	2.883369	3.733456	4.822346	6.213868	7.988061	10.245082
28	2.287928	2.998703	3.920129	5.111687	6.648838	8.627106	11.167140
29	2.356566	3.118651	4.116136	5.418388	7.114257	9.317275	12.172182
30	2.427262	3.243398	4.321942	5.743491	7.612255	10.062657	13.267678
31	2.500080	3.373133	4.538039	6.088101	8.145113	10.867669	14.461770
32	2.575083	3.508059	4.764941	6.453387	8.715271	11.737083	15.763329
33	2.652335	3.648381	5.003189	6.840590	9.325340	12.676050	17.182028
34	2.731905	3.794316	5.253348	7.251025	9.978114	13.690134	18.728411
35	2.813862	3.946089	5.516015	7.686087	10.676581	14.785344	20.413968
36	2.898278	4.103933	5.791816	8.147252	11.423942	15.968172	22.251225
37	2.985227	4.268090	6.081407	8.636087	12.223618	17.245626	24.253835
38	3.074783	4.438813	6.385477	9.154252	13.079271	18.625276	26.436680
39	3.167027	4.616366	6.704751	9.703507	13.994820	20.115298	28.815982
40	3.262038	4.801021	7.039989	10.285718	14.974458	21.724521	31.409420
41	3.359899	4.993061	7.391988	10.902861	16.022670	23.462483	34.236268
42	3.460696	5.192784	7.761588	11.557033	17.144257	25.339482	37.317532
43	3.564517	5.400495	8.149667	12.250455	18.344355	27.366640	40.676110
44	3.671452	5.616515	8.557150	12.985482	19.628460	29.555972	44.336960
45	3.781596	5.841176	8.985008	13.764611	21.002452	31.920449	48.327286
46	3.895044	6.074823	9.434258	14.590487	22.472623	34.474085	52.676742
47	4.011895	6.317816	9.905971	15.465917	24.045707	37.232012	57.417649
48	4.132252	6.570528	10.401270	16.393872	25.728907	40.210573	62.585237
49	4.256219	6.833349	10.921333	17.377504	27.529930	43.427419	68.217908

TABLE 4 Compound Amount $(1 + i)^n$ (continued)

n	3%	4%	5%	6%	7%	8%	9%
50	4.383906	7.106683	11.467400	18.420154	29.457025	46.901613	74.357520
51	4.515423	7.390951	12.040770	19.525364	31.519017	50.653742	81.049697
52	4.650886	7.686589	12.642808	20.696885	33.725348	54.706041	88.344170
53	4.790412	7.994052	13.274949	21.938698	36.086122	59.082524	96.295145
54	4.934125	8.313814	13.938696	23.255020	38.612151	63.809126	104.961708
55	5.082149	8.646367	14.635631	24.650322	41.315001	68.913856	114.408262
56	5.234613	8.992222	15.367412	26.129341	44.207052	74.426965	124.705005
57	5.391651	9.351910	16.135783	27.697101	47.301545	80.381122	135.928456
58	5.553401	9.725987	16.942572	29.358927	50.612653	86.811612	148.162017
59	5.720003	10.115026	17.789701	31.120463	54.155539	93.756540	161.496598
60	5.891603	10.519627	18.679186	32.987691	57.946427	101.257064	176.031292
61	6.068351	10.940413	19.613145	34.966952	62.002677	109.357629	191.874108
62	6.250402	11.378029	20.593802	37.064969	66.342864	118.106239	209.142778
63	6.437914	11.833150	21.623493	39.288868	70.986865	127.554738	227.965628
64	6.631051	12.306476	22.704667	41.646200	75.955945	137.759117	248.482535
65	6.829983	12.798735	23.839901	44.144972	81.272861	148.779847	270.845963
66	7.034882	13.310685	25.031896	46.793670	86.961962	160.682234	295.222099
67	7.245929	13.843112	26.283490	49.601290	93.049299	173.536813	321.792088
68	7.463307	14.396836	27.597665	52.577368	99.562750	187.419758	350.753376
69	7.687206	14.972710	28.977548	55.732010	106.532142	202.413339	382.321180
70	7.917822	15.571618	30.426426	59.075930	113.989392	218.606406	416.730086
71	8.155357	16.194483	31.947747	62.620486	121.968650	236.094918	454.235794
72	8.400017	16.842262	33.545134	66.377715	130.506455	254.982512	495.117015
73	8.652018	17.515953	35.222391	70.360378	139.641907	275.381113	539.677547
74	8.911578	18.216591	36.983510	74.582001	149.416840	297.411602	588.248526
75	9.178926	18.945255	38.832686	79.056921	159.876019	321.204530	641.190893
76	9.454293	19.703065	40.774320	83.800336	171.067341	346.900892	698.898074
77	9.737922	20.491187	42.813036	88.828356	183.042055	374.652964	761.798900
78	10.030060	21.310835	44.953688	94.158058	195.854998	404.625201	830.360801
79	10.330962	22.163268	47.201372	99.807541	209.564848	436.995217	905.093274
80	10.640891	23.049799	49.561441	105.795993	224.234388	471.954834	986.551668
81	10.960117	23.971791	52.039513	112.143753	239.930795	509.711221	1075.341318
82	11.288921	24.930663	54.641489	118.872378	256.725950	550.488119	1172.122037
83	11.627588	25.927889	57.373563	126.004721	274.696767	594.527168	1277.613020
84	11.976416	26.965005	60.242241	133.565004	293.925541	642.089342	1392.598192
85	12.335709	28.043605	63.254353	141.578904	314.500328	693.456489	1517.932029
86	12.705780	29.165349	66.417071	150.073639	336.515351	748.933008	1654.545912
87	13.086953	30.331963	69.737925	159.078057	360.071426	808.847649	1803.455044
88	13.479562	31.545242	73.224821	168.622741	385.276426	873.555461	1965.765998
89	13.883949	32.807051	76.886062	178.740105	412.245776	943.439897	2142.684938
90	14.300467	34.119333	80.730365	189.464511	441.102980	1018.915089	2335.526582
91	14.729481	35.484107	84.766883	200.832382	471.980188	1100.428296	2545.723975
92	15.171366	36.903471	89.005227	212.882325	505.018802	1188.462560	2774.839132
93	15.626507	38.379610	93.455489	225.655264	540.370118	1283.539565	3024.574654
94	16.095302	39.914794	98.128263	239.194580	578.196026	1386.222730	3296.786373

APPENDIX

TABLE 4 Compound Amount $(1 + i)^n$ (continued)

n	3%	4%	5%	6%	7%	8%	9%
95	16.578161	41.511386	103.034676	253.546255	618.669748	1497.120549	3593.497147
96	17.075506	43.171841	108.186410	268.759030	661.976630	1616.890192	3916.911890
97	17.587771	44.898715	113.595731	284.884572	708.314994	1746.241408	4269.433960
98	18.115404	46.694664	119.275517	301.977646	757.897044	1885.940720	4653.683016
99	18.658866	48.562450	125.239293	320.096305	810.949837	2036.815978	5072.514488
100	19.218632	50.504948	131.501258	339.302084	867.716326	2199.761256	5529.040792

TABLE 4 Compound Amount $(1 + i)^n$

n	10%	11%	12%	13%	14%	15%	16%
1	1.100000	1.110000	1.120000	1.130000	1.140000	1.150000	1.16000
2	1.210000	1.232100	1.254400	1.276900	1.299600	1.322500	1.34560
3	1.331000	1.367631	1.404928	1.442897	1.481544	1.520875	1.56090
4	1.464100	1.518070	1.573519	1.630474	1.688960	1.749006	1.81064
5	1.610510	1.685058	1.762342	1.842435	1.925415	2.011357	2.10034
6	1.771561	1.870415	1.973823	2.081952	2.194973	2.313061	2.43640
7	1.948717	2.076160	2.210681	2.352605	2.502269	2.660020	2.82622
8	2.143589	2.304538	2.475963	2.658444	2.852586	3.059023	3.27841
9	2.357948	2.558037	2.773079	3.004042	3.251949	3.517876	3.80296
10	2.593742	2.839421	3.105848	3.394567	3.707221	4.045558	4.41144
11	2.853117	3.151757	3.478550	3.835861	4.226232	4.652391	5.11726
12	3.138428	3.498451	3.895976	4.334523	4.817905	5.350250	5.93603
13	3.452271	3.883280	4.363493	4.898011	5.492411	6.152788	6.88579
14	3.797498	4.310441	4.887112	5.534753	6.261349	7.075706	7.98752
15	4.177248	4.784589	5.473566	6.254270	7.137938	8.137062	9.26552
16	4.594973	5.310894	6.130394	7.067326	8.137249	9.357621	10.74800
17	5.054470	5.895093	6.866041	7.986078	9.276464	10.761264	12.46768
18	5.559917	6.543553	7.689966	9.024268	10.575169	12.375454	14.46251
19	6.115909	7.263344	8.612762	10.197423	12.055693	14.231772	16.77652
20	6.727500	8.062312	9.646293	11.523088	13.743490	16.366537	19.46076
21	7.400250	8.949166	10.803848	13.021089	15.667578	18.821518	22.57448
22	8.140275	9.933574	12.100310	14.713831	17.861039	21.644746	26.18640
23	8.954302	11.026267	13.552347	16.626629	20.361585	24.891458	30.37622
24	9.849733	12.239157	15.178629	18.788091	23.212207	28.625176	35.23642
25	10.834706	13.585464	17.000064	21.230542	26.461916	32.918953	40.87424
26	11.918177	15.079865	19.040072	23.990513	30.166584	37.856796	47.41412
27	13.109994	16.738650	21.324881	27.109279	34.389906	43.535315	55.00038
28	14.420994	18.579901	23.883866	30.633486	39.204493	50.065612	63.80044
29	15.863093	20.623691	26.749930	34.615839	44.693122	57.575454	74.00851
30	17.449402	22.892297	29.959922	39.115898	50.950159	66.211772	85.84988
31	19.194342	25.410449	33.555113	44.200965	58.083181	76.143538	99.58586
32	21.113777	28.205599	37.581726	49.947090	66.214826	87.565068	115.51959
33	23.225154	31.308214	42.091533	56.440212	75.484902	100.699829	134.00273

TABLE 4 Compound Amount $(1 + i)^n$ (continued)

n	10%	11%	12%	13%	14%	15%	16%
34	25.547670	34.752118	47.142517	63.777439	86.052788	115.804803	155.44317
35	28.102437	38.574851	52.799620	72.068506	98.100178	133.175523	180.31407
36	30.912681	42.818085	59.135574	81.437412	111.834203	153.151852	209.16432
37	34.003949	47.528074	66.231843	92.024276	127.490992	176.124630	242.63062
38	37.404343	52.756162	74.179664	103.987432	145.339731	202.543324	281.45151
39	41.144778	58.559340	83.081224	117.505798	165.687293	232.924823	326.48376
40	45.259256	65.000867	93.050970	132.781552	188.883514	267.863546	378.72116
41	49.785181	72.150963	104.217087	150.043153	215.327206	308.043078	439.31654
42	54.763699	80.087569	116.723137	169.548763	245.473015	354.249540	509.60719
43	60.240069	88.897201	130.729914	191.590103	279.839237	407.386971	591.14434
44	66.264076	98.675893	146.417503	216.496816	319.016730	468.495017	685.72744
45	72.890484	109.530242	163.987604	244.641402	363.679072	538.769269	795.44383
46	80.179532	121.578568	183.666116	276.444784	414.594142	619.584659	922.71484
47	88.197485	134.952211	205.706050	312.382606	472.637322	712.522358	1070.34921
48	97.017234	149.796954	230.390776	352.992345	538.806547	819.400712	1241.60509
49	106.718957	166.274619	258.037669	398.881350	614.239464	942.310819	1440.26190
50	117.390853	184.564827	289.002190	450.735925	700.232988	1083.657442	1670.70380

TABLE 5 Present Value $(1 + i)^{-n}$

n	$\frac{1}{2}$%	$\frac{3}{4}$%	1%	$1\frac{1}{4}$%	$1\frac{1}{2}$%	$1\frac{3}{4}$%	2%
1	0.995025	0.992556	0.990099	0.987654	0.985222	0.982801	0.980392
2	0.990075	0.985167	0.980296	0.975461	0.970662	0.965898	0.961169
3	0.985149	0.977833	0.970590	0.963418	0.956317	0.949285	0.942322
4	0.980248	0.970554	0.960980	0.951524	0.942184	0.932959	0.923845
5	0.975371	0.963329	0.951466	0.939777	0.928260	0.916913	0.905731
6	0.970518	0.956158	0.942045	0.928175	0.914542	0.901143	0.887971
7	0.965690	0.949040	0.932718	0.916716	0.901027	0.885644	0.870560
8	0.960885	0.941975	0.923483	0.905398	0.887711	0.870412	0.853490
9	0.956105	0.934963	0.914340	0.894221	0.874592	0.855441	0.836755
10	0.951348	0.928003	0.905287	0.883181	0.861667	0.840729	0.820348
11	0.946615	0.921095	0.896324	0.872277	0.848933	0.826269	0.804263
12	0.941905	0.914238	0.887449	0.861509	0.836387	0.812058	0.788493
13	0.937219	0.907432	0.878663	0.850873	0.824027	0.798091	0.773033
14	0.932556	0.900677	0.869963	0.840368	0.811849	0.784365	0.757875
15	0.927917	0.893973	0.861349	0.829993	0.799852	0.770875	0.743015
16	0.923300	0.887318	0.852821	0.819746	0.788031	0.757616	0.728446
17	0.918707	0.880712	0.844377	0.809626	0.776385	0.744586	0.714163
18	0.914136	0.874156	0.836017	0.799631	0.764912	0.731780	0.700159
19	0.909588	0.867649	0.827740	0.789759	0.753607	0.719194	0.686431
20	0.905063	0.861190	0.819544	0.780009	0.742470	0.706825	0.672971
21	0.900560	0.854779	0.811430	0.770379	0.731498	0.694668	0.659776
22	0.896080	0.848416	0.803396	0.760868	0.720688	0.682720	0.646839

TABLE 5 Present Value $(1 + i)^{-n}$ (continued)

n	$\frac{1}{2}$%	$\frac{3}{4}$%	1%	$1\frac{1}{4}$%	$1\frac{1}{2}$%	$1\frac{3}{4}$%	2%
23	0.891622	0.842100	0.795442	0.751475	0.710037	0.670978	0.634156
24	0.887186	0.835831	0.787566	0.742197	0.699544	0.659438	0.621721
25	0.882772	0.829609	0.779768	0.733034	0.689206	0.648096	0.609531
26	0.878380	0.823434	0.772048	0.723984	0.679021	0.636950	0.597579
27	0.874010	0.817304	0.764404	0.715046	0.668986	0.625995	0.585862
28	0.869662	0.811220	0.756836	0.706219	0.659099	0.615228	0.574375
29	0.865335	0.805181	0.749342	0.697500	0.649359	0.604647	0.563112
30	0.861030	0.799187	0.741923	0.688889	0.639762	0.594248	0.552071
31	0.856746	0.793238	0.734577	0.680384	0.630308	0.584027	0.541246
32	0.852484	0.787333	0.727304	0.671984	0.620993	0.573982	0.530633
33	0.848242	0.781472	0.720103	0.663688	0.611816	0.564111	0.520229
34	0.844022	0.775654	0.712973	0.655494	0.602774	0.554408	0.510028
35	0.839823	0.769880	0.705914	0.647402	0.593866	0.544873	0.500028
36	0.835645	0.764149	0.698925	0.639409	0.585090	0.535502	0.490223
37	0.831487	0.758461	0.692005	0.631515	0.576443	0.526292	0.480611
38	0.827351	0.752814	0.685153	0.623719	0.567924	0.517240	0.471187
39	0.823235	0.747210	0.678370	0.616019	0.559531	0.508344	0.461948
40	0.819139	0.741648	0.671653	0.608413	0.551262	0.499601	0.452890
41	0.815064	0.736127	0.665003	0.600902	0.543116	0.491008	0.444010
42	0.811009	0.730647	0.658419	0.593484	0.535089	0.482563	0.435304
43	0.806974	0.725208	0.651900	0.586157	0.527182	0.474264	0.426769
44	0.802959	0.719810	0.645445	0.578920	0.519391	0.466107	0.418401
45	0.798964	0.714451	0.639055	0.571773	0.511715	0.458090	0.410197
46	0.794989	0.709133	0.632728	0.564714	0.504153	0.450212	0.402154
47	0.791034	0.703854	0.626463	0.557742	0.496702	0.442469	0.394268
48	0.787098	0.698614	0.620260	0.550856	0.489362	0.434858	0.386538
49	0.783182	0.693414	0.614119	0.544056	0.482130	0.427379	0.378958
50	0.779286	0.688252	0.608039	0.537339	0.475005	0.420029	0.371528
51	0.775409	0.683128	0.602019	0.530705	0.467985	0.412805	0.364243
52	0.771551	0.678043	0.596058	0.524153	0.461069	0.405705	0.357101
53	0.767713	0.672995	0.590156	0.517682	0.454255	0.398727	0.350099
54	0.763893	0.667986	0.584313	0.511291	0.447542	0.391869	0.343234
55	0.760093	0.663013	0.578528	0.504979	0.440928	0.385130	0.336504
56	0.756311	0.658077	0.572800	0.498745	0.434412	0.378506	0.329906
57	0.752548	0.653178	0.567129	0.492587	0.427992	0.371996	0.323437
58	0.748804	0.648316	0.561514	0.486506	0.421667	0.365598	0.317095
59	0.745079	0.643490	0.555954	0.480500	0.415435	0.359310	0.310878
60	0.741372	0.638700	0.550450	0.474568	0.409296	0.353130	0.304782
61	0.737684	0.633945	0.545000	0.468709	0.403247	0.347057	0.298806
62	0.734014	0.629226	0.539604	0.462922	0.397288	0.341088	0.292947
63	0.730362	0.624542	0.534261	0.457207	0.391417	0.335221	0.287203
64	0.726728	0.619893	0.528971	0.451563	0.385632	0.329456	0.281572
65	0.723113	0.615278	0.523734	0.445988	0.379933	0.323790	0.276051
66	0.719515	0.610698	0.518548	0.440482	0.374318	0.318221	0.270638
67	0.715935	0.606152	0.513414	0.435044	0.368787	0.312748	0.265331

TABLE 5 Present Value $(1 + i)^{-n}$ (continued)

n	$\frac{1}{2}$%	$\frac{3}{4}$%	1%	$1\frac{1}{4}$%	$1\frac{1}{2}$%	$1\frac{3}{4}$%	2%
68	0.712374	0.601639	0.508331	0.429673	0.363337	0.307369	0.260129
69	0.708829	0.597161	0.503298	0.424368	0.357967	0.302082	0.255028
70	0.705303	0.592715	0.498315	0.419129	0.352677	0.296887	0.250028
71	0.701794	0.588303	0.493381	0.413955	0.347465	0.291781	0.245125
72	0.698302	0.583924	0.488496	0.408844	0.342330	0.286762	0.240319
73	0.694828	0.579577	0.483659	0.403797	0.337271	0.281830	0.235607
74	0.691371	0.575262	0.478871	0.398811	0.332287	0.276983	0.230987
75	0.687932	0.570980	0.474129	0.393888	0.327376	0.272219	0.226458
76	0.684509	0.566730	0.469435	0.389025	0.322538	0.267537	0.222017
77	0.681104	0.562511	0.464787	0.384222	0.317771	0.262936	0.217664
78	0.677715	0.558323	0.460185	0.379479	0.313075	0.258414	0.213396
79	0.674343	0.554167	0.455629	0.374794	0.308448	0.253969	0.209212
80	0.670988	0.550042	0.451118	0.370167	0.303890	0.249601	0.205110
81	0.667650	0.545947	0.446651	0.365597	0.299399	0.245308	0.201088
82	0.664329	0.541883	0.442229	0.361083	0.294975	0.241089	0.197145
83	0.661023	0.537849	0.437851	0.356625	0.290615	0.236943	0.193279
84	0.657735	0.533845	0.433515	0.352223	0.286321	0.232868	0.189490
85	0.654462	0.529871	0.429223	0.347874	0.282089	0.228862	0.185774
86	0.651206	0.525927	0.424974	0.343580	0.277920	0.224926	0.182132
87	0.647967	0.522012	0.420766	0.339338	0.273813	0.221058	0.178560
88	0.644743	0.518126	0.416600	0.335148	0.269767	0.217256	0.175059
89	0.641535	0.514269	0.412475	0.331011	0.265780	0.213519	0.171627
90	0.638344	0.510440	0.408391	0.326924	0.261852	0.209847	0.168261
91	0.635168	0.506641	0.404348	0.322888	0.257982	0.206238	0.164962
92	0.632008	0.502869	0.400344	0.318902	0.254170	0.202691	0.161728
93	0.628863	0.499126	0.396380	0.314965	0.250414	0.199204	0.158556
94	0.625735	0.495410	0.392456	0.311076	0.246713	0.195778	0.155448
95	0.622622	0.491722	0.388570	0.307236	0.243067	0.192411	0.152400
96	0.619524	0.488062	0.384723	0.303443	0.239475	0.189102	0.149411
97	0.616442	0.484428	0.380914	0.299697	0.235936	0.185850	0.146482
98	0.613375	0.480822	0.377142	0.295997	0.232449	0.182653	0.143609
99	0.610323	0.477243	0.373408	0.292342	0.229014	0.179512	0.140794
100	0.607287	0.473690	0.369711	0.288733	0.225629	0.176424	0.138033

TABLE 5 Present Value $(1 + i)^{-n}$

n	3%	4%	5%	6%	7%	8%	9%
1	0.970874	0.961538	0.952381	0.943396	0.934579	0.925926	0.917431
2	0.942596	0.924556	0.907029	0.889996	0.873439	0.857339	0.841680
3	0.915142	0.888996	0.863838	0.839619	0.816298	0.793832	0.772183
4	0.888487	0.854804	0.822702	0.792094	0.762895	0.735030	0.708425
5	0.862609	0.821927	0.783526	0.747258	0.712986	0.680583	0.649931
6	0.837484	0.790315	0.746215	0.704961	0.666342	0.630170	0.596267

TABLE 5 Present Value $(1 + i)^{-n}$ (continued)

n	3%	4%	5%	6%	7%	8%	9%
7	0.813092	0.759918	0.710681	0.665057	0.622750	0.583490	0.547034
8	0.789409	0.730690	0.676839	0.627412	0.582009	0.540269	0.501866
9	0.766417	0.702587	0.644609	0.591898	0.543934	0.500249	0.460428
10	0.744094	0.675564	0.613913	0.558395	0.508349	0.463193	0.422411
11	0.722421	0.649581	0.584679	0.526788	0.475093	0.428883	0.387533
12	0.701380	0.624597	0.556837	0.496969	0.444012	0.397114	0.355535
13	0.680951	0.600574	0.530321	0.468839	0.414964	0.367698	0.326179
14	0.661118	0.577475	0.505068	0.442301	0.387817	0.340461	0.299246
15	0.641862	0.555265	0.481017	0.417265	0.362446	0.315242	0.274538
16	0.623167	0.533908	0.458112	0.393646	0.338735	0.291890	0.251870
17	0.605016	0.513373	0.436297	0.371364	0.316574	0.270269	0.231073
18	0.587395	0.493628	0.415521	0.350344	0.295864	0.250249	0.211994
19	0.570286	0.474642	0.395734	0.330513	0.276508	0.231712	0.194490
20	0.553676	0.456387	0.376889	0.311805	0.258419	0.214548	0.178431
21	0.537549	0.438834	0.358942	0.294155	0.241513	0.198656	0.163698
22	0.521893	0.421955	0.341850	0.277505	0.225713	0.183941	0.150182
23	0.506692	0.405726	0.325571	0.261797	0.210947	0.170315	0.137781
24	0.491934	0.390121	0.310068	0.246979	0.197147	0.157699	0.126405
25	0.477606	0.375117	0.295303	0.232999	0.184249	0.146018	0.115968
26	0.463695	0.360689	0.281241	0.219810	0.172195	0.135202	0.106393
27	0.450189	0.346817	0.267848	0.207368	0.160930	0.125187	0.097608
28	0.437077	0.333477	0.255094	0.195630	0.150402	0.115914	0.089548
29	0.424346	0.320651	0.242946	0.184557	0.140563	0.107328	0.082155
30	0.411987	0.308319	0.231377	0.174110	0.131367	0.099377	0.075371
31	0.399987	0.296460	0.220359	0.164255	0.122773	0.092016	0.069148
32	0.388337	0.285058	0.209866	0.154957	0.114741	0.085200	0.063438
33	0.377026	0.274094	0.199873	0.146186	0.107235	0.078889	0.058200
34	0.366045	0.263552	0.190355	0.137912	0.100219	0.073045	0.053395
35	0.355383	0.253415	0.181290	0.130105	0.093663	0.067635	0.048986
36	0.345032	0.243669	0.172657	0.122741	0.087535	0.062625	0.044941
37	0.334983	0.234297	0.164436	0.115793	0.081809	0.057986	0.041231
38	0.325226	0.225285	0.156605	0.109239	0.076457	0.053690	0.037826
39	0.315754	0.216621	0.149148	0.103056	0.071455	0.049713	0.034703
40	0.306557	0.208289	0.142046	0.097222	0.066780	0.046031	0.031838
41	0.297628	0.200278	0.135282	0.091719	0.062412	0.042621	0.029209
42	0.288959	0.192575	0.128840	0.086527	0.058329	0.039464	0.026797
43	0.280543	0.185168	0.122704	0.081630	0.054513	0.036541	0.024584
44	0.272372	0.178046	0.116861	0.077009	0.050946	0.033834	0.022555
45	0.264439	0.171198	0.111297	0.072650	0.047613	0.031328	0.020692
46	0.256737	0.164614	0.105997	0.068538	0.044499	0.029007	0.018984
47	0.249259	0.158283	0.100949	0.064658	0.041587	0.026859	0.017416
48	0.241999	0.152195	0.096142	0.060998	0.038867	0.024869	0.015978
49	0.234950	0.146341	0.091564	0.057546	0.036324	0.023027	0.014659
50	0.228107	0.140713	0.087204	0.054288	0.033948	0.021321	0.013449
51	0.221463	0.135301	0.083051	0.051215	0.031727	0.019742	0.012338

TABLE 5 Present Value $(1 + i)^{-n}$ (continued)

n	3%	4%	5%	6%	7%	8%	9%
52	0.215013	0.130097	0.079096	0.048316	0.029651	0.018280	0.011319
53	0.208750	0.125093	0.075330	0.045582	0.027711	0.016925	0.010385
54	0.202670	0.120282	0.071743	0.043001	0.025899	0.015672	0.009527
55	0.196767	0.115656	0.068326	0.040567	0.024204	0.014511	0.008741
56	0.191036	0.111207	0.065073	0.038271	0.022621	0.013436	0.008019
57	0.185472	0.106930	0.061974	0.036105	0.021141	0.012441	0.007357
58	0.180070	0.102817	0.059023	0.034061	0.019758	0.011519	0.006749
59	0.174825	0.098863	0.056212	0.032133	0.018465	0.010666	0.006192
60	0.169733	0.095060	0.053536	0.030314	0.017257	0.009876	0.005681
61	0.164789	0.091404	0.050986	0.028598	0.016128	0.009144	0.005212
62	0.159990	0.087889	0.048558	0.026980	0.015073	0.008467	0.004781
63	0.155330	0.084508	0.046246	0.025453	0.014087	0.007840	0.004387
64	0.150806	0.081258	0.044044	0.024012	0.013166	0.007259	0.004024
65	0.146413	0.078133	0.041946	0.022653	0.012304	0.006721	0.003692
66	0.142149	0.075128	0.039949	0.021370	0.011499	0.006223	0.003387
67	0.138009	0.072238	0.038047	0.020161	0.010747	0.005762	0.003108
68	0.133989	0.069460	0.036235	0.019020	0.010044	0.005336	0.002851
69	0.130086	0.066788	0.034509	0.017943	0.009387	0.004940	0.002616
70	0.126297	0.064219	0.032866	0.016927	0.008773	0.004574	0.002400
71	0.122619	0.061749	0.031301	0.015969	0.008199	0.004236	0.002201
72	0.119047	0.059374	0.029811	0.015065	0.007662	0.003922	0.002020
73	0.115580	0.057091	0.028391	0.014213	0.007161	0.003631	0.001853
74	0.112214	0.054895	0.027039	0.013408	0.006693	0.003362	0.001700
75	0.108945	0.052784	0.025752	0.012649	0.006255	0.003113	0.001560
76	0.105772	0.050754	0.024525	0.011933	0.005846	0.002883	0.001431
77	0.102691	0.048801	0.023357	0.011258	0.005463	0.002669	0.001313
78	0.099700	0.046924	0.022245	0.010620	0.005106	0.002471	0.001204
79	0.096796	0.045120	0.021186	0.010019	0.004772	0.002288	0.001105
80	0.093977	0.043384	0.020177	0.009452	0.004460	0.002119	0.001014
81	0.091240	0.041716	0.019216	0.008917	0.004168	0.001962	0.000930
82	0.088582	0.040111	0.018301	0.008412	0.003895	0.001817	0.000853
83	0.086002	0.038569	0.017430	0.007936	0.003640	0.001682	0.000783
84	0.083497	0.037085	0.016600	0.007487	0.003402	0.001557	0.000718
85	0.081065	0.035659	0.015809	0.007063	0.003180	0.001442	0.000659
86	0.078704	0.034287	0.015056	0.006663	0.002972	0.001335	0.000604
87	0.076412	0.032969	0.014339	0.006286	0.002777	0.001236	0.000554
88	0.074186	0.031701	0.013657	0.005930	0.002596	0.001145	0.000509
89	0.072026	0.030481	0.013006	0.005595	0.002426	0.001060	0.000467
90	0.069928	0.029309	0.012387	0.005278	0.002267	0.000981	0.000428
91	0.067891	0.028182	0.011797	0.004979	0.002119	0.000909	0.000393
92	0.065914	0.027098	0.011235	0.004697	0.001980	0.000841	0.000360
93	0.063994	0.026056	0.010700	0.004432	0.001851	0.000779	0.000331
94	0.062130	0.025053	0.010191	0.004181	0.001730	0.000721	0.000303
95	0.060320	0.024090	0.009705	0.003944	0.001616	0.000668	0.000278
96	0.058563	0.023163	0.009243	0.003721	0.001511	0.000618	0.000255

TABLE 5 Present Value $(1 + i)^{-n}$ (continued)

n	3%	4%	5%	6%	7%	8%	9%
97	0.056858	0.022272	0.008803	0.003510	0.001412	0.000573	0.000234
98	0.055202	0.021416	0.008384	0.003312	0.001319	0.000530	0.000215
99	0.053594	0.020592	0.007985	0.003124	0.001233	0.000491	0.000197
100	0.052033	0.019800	0.007604	0.002947	0.001152	0.000455	0.000181

TABLE 5 Present Value $(1 + i)^{-n}$

n	10%	11%	12%	13%	14%	15%	16%
1	0.909091	0.900901	0.892857	0.884956	0.877193	0.869565	0.862069
2	0.826446	0.811622	0.797194	0.783147	0.769468	0.756144	0.743163
3	0.751315	0.731191	0.711780	0.693050	0.674972	0.657516	0.640658
4	0.683013	0.658731	0.635518	0.613319	0.592080	0.571753	0.552291
5	0.620921	0.593451	0.567427	0.542760	0.519369	0.497177	0.476113
6	0.564474	0.534641	0.506631	0.480319	0.455587	0.432328	0.410442
7	0.513158	0.481658	0.452349	0.425061	0.399637	0.375937	0.353830
8	0.466507	0.433926	0.403883	0.376160	0.350559	0.326902	0.305025
9	0.424098	0.390925	0.360610	0.332885	0.307508	0.284262	0.262953
10	0.385543	0.352184	0.321973	0.294588	0.269744	0.247185	0.226684
11	0.350494	0.317283	0.287476	0.260698	0.236617	0.214943	0.195417
12	0.318631	0.285841	0.256675	0.230706	0.207559	0.186907	0.168463
13	0.289664	0.257514	0.229174	0.204165	0.182069	0.162528	0.145227
14	0.263331	0.231995	0.204620	0.180677	0.159710	0.141329	0.125195
15	0.239392	0.209004	0.182696	0.159891	0.140096	0.122894	0.107927
16	0.217629	0.188292	0.163122	0.141496	0.122892	0.106865	0.093041
17	0.197845	0.169633	0.145644	0.125218	0.107800	0.092926	0.080207
18	0.179859	0.152822	0.130040	0.110812	0.094561	0.080805	0.069144
19	0.163508	0.137678	0.116107	0.098064	0.082948	0.070265	0.059607
20	0.148644	0.124034	0.103667	0.086782	0.072762	0.061100	0.051385
21	0.135131	0.111742	0.092560	0.076798	0.063826	0.053131	0.044298
22	0.122846	0.100669	0.082643	0.067963	0.055988	0.046201	0.038188
23	0.111678	0.090693	0.073788	0.060144	0.049112	0.040174	0.032920
24	0.101526	0.081705	0.065882	0.053225	0.043081	0.034934	0.028380
25	0.092296	0.073608	0.058823	0.047102	0.037790	0.030378	0.024465
26	0.083905	0.066314	0.052521	0.041683	0.033149	0.026415	0.021091
27	0.076278	0.059742	0.046894	0.036888	0.029078	0.022970	0.018182
28	0.069343	0.053822	0.041869	0.032644	0.025507	0.019974	0.015674
29	0.063039	0.048488	0.037383	0.028889	0.022375	0.017369	0.013512
30	0.057309	0.043683	0.033378	0.025565	0.019627	0.015103	0.011648
31	0.052099	0.039354	0.029802	0.022624	0.017217	0.013133	0.010042
32	0.047362	0.035454	0.026609	0.020021	0.015102	0.011420	0.008657
33	0.043057	0.031940	0.023758	0.017718	0.013248	0.009931	0.007463
34	0.039143	0.028775	0.021212	0.015680	0.011621	0.008635	0.006433
35	0.035584	0.025924	0.018940	0.013876	0.010194	0.007509	0.005546

TABLE 5 Present Value $(1 + i)^{-n}$ (continued)

n	10%	11%	12%	13%	14%	15%	16%
36	0.032349	0.023355	0.016910	0.012279	0.008942	0.006529	0.004781
37	0.029408	0.021040	0.015098	0.010867	0.007844	0.005678	0.004121
38	0.026735	0.018955	0.013481	0.009617	0.006880	0.004937	0.003553
39	0.024304	0.017077	0.012036	0.008510	0.006035	0.004293	0.003063
40	0.022095	0.015384	0.010747	0.007531	0.005294	0.003733	0.002640
41	0.020086	0.013860	0.009595	0.006665	0.004644	0.003246	0.002276
42	0.018260	0.012486	0.008567	0.005898	0.004074	0.002823	0.001962
43	0.016600	0.011249	0.007649	0.005219	0.003573	0.002455	0.001692
44	0.015091	0.010134	0.006830	0.004619	0.003135	0.002134	0.001458
45	0.013719	0.009130	0.006098	0.004088	0.002750	0.001856	0.001257
46	0.012472	0.008225	0.005445	0.003617	0.002412	0.001614	0.001084
47	0.011338	0.007410	0.004861	0.003201	0.002116	0.001403	0.000934
48	0.010307	0.006676	0.004340	0.002833	0.001856	0.001220	0.000805
49	0.009370	0.006014	0.003875	0.002507	0.001628	0.001061	0.000694
50	0.008519	0.005418	0.003460	0.002219	0.001428	0.000923	0.000599

TABLE 6 Amount of an Annuity $s_{\overline{n}|i} = \dfrac{(1 + i)^n - 1}{i}$

n	$\frac{1}{2}$%	$\frac{3}{4}$%	1%	$1\frac{1}{4}$%	$1\frac{1}{2}$%	$1\frac{3}{4}$%	2%
1	1.000000	1.000000	1.000000	1.000000	1.000000	1.000000	1.000000
2	2.005000	2.007500	2.010000	2.012500	2.015000	2.017500	2.020000
3	3.015025	3.022556	3.030100	3.037656	3.045225	3.052806	3.060400
4	4.030100	4.045225	4.060401	4.075627	4.090903	4.106230	4.121608
5	5.050251	5.075565	5.101005	5.126572	5.152267	5.178089	5.204040
6	6.075502	6.113631	6.152015	6.190654	6.229551	6.268706	6.308121
7	7.105879	7.159484	7.213535	7.268038	7.322994	7.378408	7.434283
8	8.141409	8.213180	8.285671	8.358888	8.432839	8.507530	8.582969
9	9.182116	9.274779	9.368527	9.463374	9.559332	9.656412	9.754628
10	10.228026	10.344339	10.462213	10.581666	10.702722	10.825399	10.949721
11	11.279167	11.421922	11.566835	11.713937	11.863262	12.014844	12.168715
12	12.335562	12.507586	12.682503	12.860361	13.041211	13.225104	13.412090
13	13.397240	13.601393	13.809328	14.021116	14.236830	14.456543	14.680332
14	14.464226	14.703404	14.947421	15.196380	15.450382	15.709533	15.973938
15	15.536548	15.813679	16.096896	16.386335	16.682138	16.984449	17.293417
16	16.614230	16.932282	17.257864	17.591164	17.932370	18.281677	18.639285
17	17.697301	18.059274	18.430443	18.811053	19.201355	19.601607	20.012071
18	18.785788	19.194718	19.614748	20.046192	20.489376	20.944635	21.412312
19	19.879717	20.338679	20.810895	21.296769	21.796716	22.311166	22.840559
20	20.979115	21.491219	22.019004	22.562979	23.123667	23.701611	24.297370
21	22.084011	22.652403	23.239194	23.845016	24.470522	25.116389	25.783317
22	23.194431	23.822296	24.471586	25.143078	25.837580	26.555926	27.298984
23	24.310403	25.000963	25.716302	26.457367	27.225144	28.020655	28.844963

A-18 APPENDIX

TABLE 6 Amount of an Annuity $s_{\overline{n}|i} = \dfrac{(1+i)^n - 1}{i}$ (continued)

n	$\frac{1}{2}$%	$\frac{3}{4}$%	1%	$1\frac{1}{4}$%	$1\frac{1}{2}$%	$1\frac{3}{4}$%	2%
24	25.431955	26.188471	26.973465	27.788084	28.633521	29.511016	30.421862
25	26.559115	27.384884	28.243200	29.135435	30.063024	31.027459	32.030300
26	27.691911	28.590271	29.525631	30.499628	31.513969	32.570440	33.670906
27	28.830370	29.804698	30.820888	31.880873	32.986678	34.140422	35.344324
28	29.974522	31.028233	32.129097	33.279384	34.481479	35.737880	37.051210
29	31.124395	32.260945	33.450388	34.695377	35.998701	37.363293	38.792235
30	32.280017	33.502902	34.784892	36.129069	37.538681	39.017150	40.568079
31	33.441417	34.754174	36.132740	37.580682	39.101762	40.699950	42.379441
32	34.608624	36.014830	37.494068	39.050441	40.688288	42.412200	44.227030
33	35.781667	37.284941	38.869009	40.538571	42.298612	44.154413	46.111570
34	36.960575	38.564578	40.257699	42.045303	43.933092	45.927115	48.033802
35	38.145378	39.853813	41.660276	43.570870	45.592088	47.730840	49.994478
36	39.336105	41.152716	43.076878	45.115505	47.275969	49.566129	51.994367
37	40.532785	42.461361	44.507647	46.679449	48.985109	51.433537	54.034255
38	41.735449	43.779822	45.952724	48.262942	50.719885	53.333624	56.114940
39	42.944127	45.108170	47.412251	49.866229	52.480684	55.266962	58.237238
40	44.158847	46.446482	48.886373	51.489557	54.267894	57.234134	60.401983
41	45.379642	47.794830	50.375237	53.133177	56.081912	59.235731	62.610023
42	46.606540	49.153291	51.878989	54.797341	57.923141	61.272357	64.862223
43	47.839572	50.521941	53.397779	56.482308	59.791988	63.344623	67.159468
44	49.078770	51.900856	54.931757	58.188337	61.688868	65.453154	69.502657
45	50.324164	53.290112	56.481075	59.915691	63.614201	67.598584	71.892710
46	51.575785	54.689788	58.045885	61.664637	65.568414	69.781559	74.330564
47	52.833664	56.099961	59.626344	63.435445	67.551940	72.002736	76.817176
48	54.097832	57.520711	61.222608	65.228388	69.565219	74.262784	79.353519
49	55.368321	58.952116	62.834834	67.043743	71.608698	76.562383	81.940590
50	56.645163	60.394257	64.463182	68.881790	73.682828	78.902225	84.579401
51	57.928389	61.847214	66.107814	70.742812	75.788070	81.283014	87.270989
52	59.218031	63.311068	67.768892	72.627097	77.924892	83.705466	90.016409
53	60.514121	64.785901	69.446581	74.534936	80.093765	86.170312	92.816737
54	61.816692	66.271796	71.141047	76.466623	82.295171	88.678292	95.673072
55	63.125775	67.768834	72.852457	78.422456	84.529599	91.230163	98.586534
56	64.441404	69.277100	74.580982	80.402736	86.797543	93.826690	101.558264
57	65.763611	70.796679	76.326792	82.407771	89.099506	96.468658	104.589430
58	67.092429	72.327654	78.090060	84.437868	91.435999	99.156859	107.681218
59	68.427891	73.870111	79.870960	86.493341	93.807539	101.892104	110.834843
60	69.770031	75.424137	81.669670	88.574508	96.214652	104.675216	114.051539
61	71.118881	76.989818	83.486367	90.681689	98.657871	107.507032	117.332570
62	72.474475	78.567242	85.321230	92.815210	101.137740	110.388405	120.679222
63	73.836847	80.156496	87.174443	94.975400	103.654806	113.320202	124.092806
64	75.206032	81.757670	89.046187	97.162593	106.209628	116.303306	127.574662
65	76.582062	83.370852	90.936649	99.377125	108.802772	119.338614	131.126155
66	77.964972	84.996134	92.846015	101.619339	111.434814	122.427039	134.748679
67	79.354797	86.633605	94.774475	103.889581	114.106336	125.569513	138.443652

TABLE 6 Amount of an Annuity $s_{\overline{n}|i} = \dfrac{(1+i)^n - 1}{i}$ (continued)

n	$\frac{1}{2}$%	$\frac{3}{4}$%	1%	$1\frac{1}{4}$%	$1\frac{1}{2}$%	$1\frac{3}{4}$%	2%
68	80.751571	88.283357	96.722220	106.188201	116.817931	128.766979	142.212525
69	82.155329	89.945482	98.689442	108.515553	119.570200	132.020401	146.056776
70	83.566105	91.620073	100.676337	110.871998	122.363753	135.330758	149.977911
71	84.983936	93.307223	102.683100	113.257898	125.199209	138.699047	153.977469
72	86.408856	95.007028	104.709931	115.673621	128.077197	142.126280	158.057019
73	87.840900	96.719580	106.757031	118.119542	130.998355	145.613490	162.218159
74	89.280104	98.444977	108.824601	120.596036	133.963331	149.161726	166.462522
75	90.726505	100.183314	110.912847	123.103486	136.972781	152.772056	170.791773
76	92.180138	101.934689	113.021975	125.642280	140.027372	156.445567	175.207608
77	93.641038	103.699199	115.152195	128.212809	143.127783	160.183364	179.711760
78	95.109243	105.476943	117.303717	130.815469	146.274700	163.986573	184.305996
79	96.584790	107.268021	119.476754	133.450662	149.468820	167.856338	188.992115
80	98.067714	109.072531	121.671522	136.118795	152.710852	171.793824	193.771958
81	99.558052	110.890575	123.888237	138.820280	156.001515	175.800216	198.647397
82	101.055842	112.722254	126.127119	141.555534	159.341538	179.876720	203.620345
83	102.561122	114.567671	128.388390	144.324978	162.731661	184.024563	208.692752
84	104.073927	116.426928	130.672274	147.129040	166.172636	188.244992	213.866607
85	105.594297	118.300130	132.978997	149.968153	169.665226	192.539280	219.143939
86	107.122268	120.187381	135.308787	152.842755	173.210204	196.908717	224.526818
87	108.657880	122.088787	137.661875	155.753289	176.808357	201.354620	230.017354
88	110.201169	124.004453	140.038494	158.700206	180.460482	205.878326	235.617701
89	111.752175	125.934486	142.438879	161.683958	184.167390	210.481196	241.330055
90	113.310936	127.878995	144.863267	164.705008	187.929900	215.164617	247.156656
91	114.877490	129.838087	147.311900	167.763820	191.748849	219.929998	253.099789
92	116.451878	131.811873	149.785019	170.860868	195.625082	224.778773	259.161785
93	118.034137	133.800462	152.282869	173.996629	199.559458	229.712401	265.345021
94	119.624308	135.803965	154.805698	177.171587	203.552850	234.732369	271.651921
95	121.222430	137.822495	157.353755	180.386232	207.606142	239.840185	278.084960
96	122.828542	139.856164	159.927293	183.641059	211.720235	245.037388	284.646659
97	124.442684	141.905085	162.526565	186.936573	215.896038	250.325542	291.339592
98	126.064898	143.969373	165.151831	190.273280	220.134479	255.706239	298.166384
99	127.695222	146.049143	167.803349	193.651696	224.436496	261.181099	305.129712
100	129.333698	148.144512	170.481383	197.072342	228.803043	266.751768	312.232306

TABLE 6 Amount of an Annuity $s_{\overline{n}|i} = \dfrac{(1+i)^n - 1}{i}$

n	3%	4%	5%	6%	7%	8%	9%
1	1.000000	1.000000	1.000000	1.000000	1.000000	1.000000	1.000000
2	2.030000	2.040000	2.050000	2.060000	2.070000	2.080000	2.090000
3	3.090900	3.121600	3.152500	3.183600	3.214900	3.246400	3.278100
4	4.183627	4.246464	4.310125	4.374616	4.439943	4.506112	4.573129
5	5.309136	5.416323	5.525631	5.637093	5.750739	5.866601	5.984711
6	6.468410	6.632975	6.801913	6.975319	7.153291	7.335929	7.523335
7	7.662462	7.898294	8.142008	8.393838	8.654021	8.922803	9.200435
8	8.892336	9.214226	9.549109	9.897468	10.259803	10.636628	11.028474
9	10.159106	10.582795	11.026564	11.491316	11.977989	12.487558	13.021036
10	11.463879	12.006107	12.577893	13.180795	13.816448	14.486562	15.192930
11	12.807796	13.486351	14.206787	14.971643	15.783599	16.645487	17.560293
12	14.192030	15.025805	15.917127	16.869941	17.888451	18.977126	20.140720
13	15.617790	16.626838	17.712983	18.882138	20.140643	21.495297	22.953385
14	17.086324	18.291911	19.598632	21.015066	22.550488	24.214920	26.019189
15	18.598914	20.023588	21.578564	23.275970	25.129022	27.152114	29.360916
16	20.156881	21.824531	23.657492	25.672528	27.888054	30.324283	33.003399
17	21.761588	23.697512	25.840366	28.212880	30.840217	33.750226	36.973705
18	23.414435	25.645413	28.132385	30.905653	33.999033	37.450244	41.301338
19	25.116868	27.671229	30.539004	33.759992	37.378965	41.446263	46.018458
20	26.870374	29.778079	33.065954	36.785591	40.995492	45.761964	51.160120
21	28.676486	31.969202	35.719252	39.992727	44.865177	50.422921	56.764530
22	30.536780	34.247970	38.505214	43.392290	49.005739	55.456755	62.873338
23	32.452884	36.617889	41.430475	46.995828	53.436141	60.893296	69.531939
24	34.426470	39.082604	44.501999	50.815577	58.176671	66.764759	76.789813
25	36.459264	41.645908	47.727099	54.864512	63.249038	73.105940	84.700896
26	38.553042	44.311745	51.113454	59.156383	68.676470	79.954415	93.323977
27	40.709634	47.084214	54.669126	63.705766	74.483823	87.350768	102.723135
28	42.930923	49.967583	58.402583	68.528112	80.697691	95.338830	112.968217
29	45.218850	52.966286	62.322712	73.639798	87.346529	103.965936	124.135356
30	47.575416	56.084938	66.438848	79.058186	94.460786	113.283211	136.307539
31	50.002678	59.328335	70.760790	84.801677	102.073041	123.345868	149.575217
32	52.502759	62.701469	75.298829	90.889778	110.218154	134.213537	164.036987
33	55.077841	66.209527	80.063771	97.343165	118.933425	145.950620	179.800315
34	57.730177	69.857909	85.066959	104.183755	128.258765	158.626670	196.982344
35	60.462082	73.652225	90.320307	111.434780	138.236878	172.316804	215.710755
36	63.275944	77.598314	95.836323	119.120867	148.913460	187.102148	236.124723
37	66.174223	81.702246	101.628139	127.268119	160.337402	203.070320	258.375948
38	69.159449	85.970336	107.709546	135.904206	172.561020	220.315945	282.629783
39	72.234233	90.409150	114.095023	145.058458	185.640292	238.941221	309.066463
40	75.401260	95.025516	120.799774	154.761966	199.635112	259.056519	337.882445
41	78.663298	99.826536	127.839763	165.047684	214.609570	280.781040	369.291865
42	82.023196	104.819598	135.231751	175.950545	230.632240	304.243523	403.528133
43	85.483892	110.012382	142.993339	187.507577	247.776496	329.583005	440.845665
44	89.048409	115.412877	151.143006	199.758032	266.120851	356.949646	481.521775

TABLE 6 Amount of an Annuity $s_{\overline{n}|i} = \dfrac{(1+i)^n - 1}{i}$ (continued)

n	3%	4%	5%	6%	7%	8%	9%
45	92.719861	121.029392	159.700156	212.743514	285.749311	386.505617	525.858734
46	96.501457	126.870568	168.685164	226.508125	306.751763	418.426067	574.186021
47	100.396501	132.945390	178.119422	241.098612	329.224386	452.900152	626.862762
48	104.408396	139.263206	188.025393	256.564529	353.270093	490.132164	684.280411
49	108.540648	145.833734	198.426663	272.958401	378.999000	530.342737	746.865648
50	112.796867	152.667084	209.347996	290.335905	406.528929	573.770156	815.083556

TABLE 6 Amount of an Annuity $s_{\overline{n}|i} = \dfrac{(1+i)^n - 1}{i}$

n	10%	11%	12%	13%	14%	15%
1	1.000000	1.000000	1.000000	1.000000	1.000000	1.000000
2	2.100000	2.110000	2.120000	2.130000	2.140000	2.150000
3	3.310000	3.342100	3.374400	3.406900	3.439600	3.472500
4	4.641000	4.709731	4.779328	4.849797	4.921144	4.993375
5	6.105100	6.227801	6.352847	6.480271	6.610104	6.742381
6	7.715610	7.912860	8.115189	8.322706	8.535519	8.753738
7	9.487171	9.783274	10.089012	10.404658	10.730491	11.066799
8	11.435888	11.859434	12.299693	12.757263	13.232760	13.726819
9	13.579477	14.163972	14.775656	15.415707	16.085347	16.785842
10	15.937425	16.722009	17.548735	18.419749	19.337295	20.303718
11	18.531167	19.561430	20.654583	21.814317	23.044516	24.349276
12	21.384284	22.713187	24.133133	25.650178	27.270749	29.001667
13	24.522712	26.211638	28.029109	29.984701	32.088654	34.351917
14	27.974983	30.094918	32.392602	34.882712	37.581065	40.504705
15	31.772482	34.405359	37.279715	40.417464	43.842414	47.580411
16	35.949730	39.189948	42.753280	46.671735	50.980352	55.717472
17	40.544703	44.500843	48.883674	53.739060	59.117601	65.075093
18	45.599173	50.395936	55.749715	61.725138	68.394066	75.836357
19	51.159090	56.939488	63.439681	70.749406	78.969235	88.211811
20	57.274999	64.202832	72.052442	80.946829	91.024928	102.443583
21	64.002499	72.265144	81.698736	92.469917	104.768418	118.810120
22	71.402749	81.214309	92.502584	105.491006	120.435996	137.631638
23	79.543024	91.147884	104.602894	120.204837	138.297035	159.276384
24	88.497327	102.174151	118.155241	136.831465	158.658620	184.167841
25	98.347059	114.413307	133.333870	155.619556	181.870827	212.793017
26	109.181765	127.998771	150.333934	176.850098	208.332743	245.711970
27	121.099942	143.078636	169.374007	200.840611	238.499327	283.568766
28	134.209936	159.817286	190.698887	227.949890	272.889233	327.104080
29	148.630930	178.397187	214.582754	258.583376	312.093725	377.169693
30	164.494023	199.020878	241.332684	293.199215	356.786847	434.745146
31	181.943425	221.913174	271.292606	332.315113	407.737006	500.956918

TABLE 6 Amount of an Annuity $s_{\overline{n}|i} = \dfrac{(1+i)^n - 1}{i}$ (continued)

n	10%	11%	12%	13%	14%	15%
32	201.137767	247.323624	304.847719	376.516078	465.820186	577.100456
33	222.251544	275.529222	342.429446	426.463168	532.035012	664.665524
34	245.476699	306.837437	384.520979	482.903380	607.519914	765.365353
35	271.024368	341.589555	431.663496	546.680819	693.572702	881.170156
36	299.126805	380.164406	484.463116	618.749325	791.672881	1014.345680
37	330.039486	422.982490	543.598690	700.186738	903.507084	1167.497532
38	364.043434	470.510564	609.830533	792.211014	1030.998076	1343.622161
39	401.447778	523.266726	684.010197	896.198445	1176.337806	1546.165485
40	442.592556	581.826066	767.091420	1013.704243	1342.025099	1779.090308
41	487.851811	646.826934	860.142391	1146.485795	1530.908613	2046.953854
42	537.636992	718.977896	964.359478	1296.528948	1746.235819	2354.996933
43	592.400692	799.065465	1081.082615	1466.077712	1991.708833	2709.246473
44	652.640761	887.962666	1211.812529	1657.667814	2271.548070	3116.633443
45	718.904837	986.638559	1358.230032	1874.164630	2590.564800	3585.128460
46	791.795321	1096.168801	1522.217636	2118.806032	2954.243872	4123.897729
47	871.974853	1217.747369	1705.883752	2395.250816	3368.838014	4743.482388
48	960.172338	1352.699580	1911.589803	2707.633422	3841.475336	5456.004746
49	1057.189572	1502.496533	2141.980579	3060.625767	4380.281883	6275.405458
50	1163.908529	1668.771152	2400.018249	3459.507117	4994.521346	7217.716277

TABLE 7 Amount of an Annuity for Fractional Interest Periods $s_{\overline{1/m}|i} = \dfrac{(1+i)^{1/m} - 1}{i}$

m	$\frac{1}{2}$%	$\frac{3}{4}$%	1%	$1\frac{1}{4}$%	$1\frac{1}{2}$%	$1\frac{3}{4}$%	2%
2	0.499377	0.499066	0.498756	0.498447	0.498139	0.497831	0.497525
3	0.332779	0.332503	0.332228	0.331954	0.331680	0.331408	0.331135
4	0.249533	0.249300	0.249068	0.248837	0.248606	0.248376	0.248147
6	0.166321	0.166148	0.165976	0.165805	0.165634	0.165464	0.165295
12	0.083143	0.083048	0.082954	0.082860	0.082766	0.082672	0.082579

m	3%	4%	5%	6%	7%	8%	9%
2	0.496305	0.495098	0.493902	0.492717	0.491543	0.490381	0.489229
3	0.330054	0.328985	0.327927	0.326880	0.325845	0.324820	0.323805
4	0.247236	0.246335	0.245445	0.244564	0.243693	0.242832	0.241980
6	0.164621	0.163955	0.163297	0.162647	0.162004	0.161368	0.160740
12	0.082209	0.081843	0.081482	0.081126	0.080774	0.080425	0.080081

A-23 APPENDIX

TABLE 7 Amount of an Annuity for Fractional Interest Periods $s_{\overline{1/m}|i} = \dfrac{(1+i)^{1/m} - 1}{i}$ (continued)

m	10%	11%	12%	13%	14%	15%	16%
2	0.488088	0.486958	0.485838	0.484728	0.483627	0.482537	0.481456
3	0.322801	0.321807	0.320824	0.319850	0.318885	0.317930	0.316985
4	0.241137	0.240303	0.239478	0.238661	0.237853	0.237054	0.236262
6	0.160119	0.159504	0.158897	0.158296	0.157702	0.157114	0.156532
12	0.079741	0.079405	0.079073	0.078745	0.078420	0.078099	0.077782

TABLE 8 Present Value of an Annuity $a_{\overline{n}|i} = \dfrac{1 - (1+i)^{-n}}{i}$

n	$\frac{1}{2}$%	$\frac{3}{4}$%	1%	$1\frac{1}{4}$%	$1\frac{1}{2}$%	$1\frac{3}{4}$%	2%
1	0.995025	0.992556	0.990099	0.987654	0.985222	0.982801	0.980392
2	1.985099	1.977723	1.970395	1.963115	1.955883	1.948699	1.941561
3	2.970248	2.955556	2.940985	2.926534	2.912200	2.897984	2.883883
4	3.950496	3.926110	3.901966	3.878058	3.854385	3.830943	3.807729
5	4.925866	4.889440	4.853431	4.817835	4.782645	4.747855	4.713460
6	5.896384	5.845598	5.795476	5.746010	5.697187	5.648998	5.601431
7	6.862074	6.794638	6.728195	6.662726	6.598214	6.534641	6.471991
8	7.822959	7.736613	7.651678	7.568124	7.485925	7.405053	7.325481
9	8.779064	8.671576	8.566018	8.462345	8.360517	8.260494	8.162237
10	9.730412	9.599580	9.471305	9.345526	9.222185	9.101223	8.982585
11	10.677027	10.520675	10.367628	10.217803	10.071118	9.927492	9.786848
12	11.618932	11.434913	11.255077	11.079312	10.907505	10.739550	10.575341
13	12.556151	12.342345	12.133740	11.930185	11.731532	11.537641	11.348374
14	13.488708	13.243022	13.003703	12.770553	12.543382	12.322006	12.106249
15	14.416625	14.136995	13.865053	13.600546	13.343233	13.092880	12.849264
16	15.339925	15.024313	14.717874	14.420292	14.131264	13.850497	13.577709
17	16.258632	15.905025	15.562251	15.229918	14.907649	14.595083	14.291872
18	17.172768	16.779181	16.398269	16.029549	15.672561	15.326863	14.992031
19	18.082356	17.646830	17.226008	16.819308	16.426168	16.046057	15.678462
20	18.987419	18.508020	18.045553	17.599316	17.168639	16.752881	16.351433
21	19.887979	19.362799	18.856983	18.369695	17.900137	17.447549	17.011209
22	20.784059	20.211215	19.660379	19.130563	18.620824	18.130269	17.658048
23	21.675681	21.053315	20.455821	19.882037	19.330861	18.801248	18.292204
24	22.562866	21.889146	21.243387	20.624235	20.030405	19.460686	18.913926
25	23.445638	22.718755	22.023156	21.357269	20.719611	20.108782	19.523456
26	24.324018	23.542189	22.795204	22.081253	21.398632	20.745732	20.121036
27	25.198028	24.359493	23.559608	22.796299	22.067617	21.371726	20.706898
28	26.067689	25.170713	24.316443	23.502518	22.726717	21.986955	21.281272
29	26.933024	25.975893	25.065785	24.200018	23.376076	22.591602	21.844385

TABLE 8 Present Value of an Annuity $a_{\overline{n}|i} = \dfrac{1-(1+i)^{-n}}{i}$ (continued)

n	$\frac{1}{2}$%	$\frac{3}{4}$%	1%	$1\frac{1}{4}$%	$1\frac{1}{2}$%	$1\frac{3}{4}$%	2%
30	27.794054	26.775080	25.807708	24.888906	24.015838	23.185849	22.396456
31	28.650800	27.568318	26.542285	25.569290	24.646146	23.769877	22.937702
32	29.503284	28.355650	27.269589	26.241274	25.267139	24.343859	23.468335
33	30.351526	29.137122	27.989693	26.904962	25.878954	24.907970	23.988564
34	31.195548	29.912776	28.702666	27.560456	26.481728	25.462378	24.498592
35	32.035371	30.682656	29.408580	28.207858	27.075595	26.007251	24.998619
36	32.871016	31.446805	30.107505	28.847267	27.660684	26.542753	25.488842
37	33,702504	32.205266	30.799510	29.478783	28.237127	27.069045	25.969453
38	34.529854	32.958080	31.484663	30.102501	28.805052	27.586285	26.440641
39	35.353089	33.705290	32.163033	30.718520	29.364583	28.094629	26.902589
40	36.172228	34.446938	32.834686	31.326933	29.915845	28.594230	27.355479
41	36.987291	35.183065	33.499689	31.927835	30.458961	29.085238	27.799489
42	37.798300	35.913713	34.158108	32.521319	30.994050	29.567801	28.234794
43	38.605274	36.638921	34.810008	33.107475	31.521232	30.042065	28.661562
44	39.408232	37.358730	35.455454	33.686395	32.040622	30.508172	29.079963
45	40.207196	38.073181	36.094508	34.258168	32.552337	30.966263	29.490160
46	41.002185	38.782314	36.727236	34.822882	33.056490	31.416474	29.892314
47	41.793219	39.486168	37.353699	35.380624	33.553192	31.858943	30.286582
48	42.580318	40.184782	37.973959	35.931481	34.042554	32.293801	30.673120
49	43.363500	40.878195	38.588079	36.475537	34.524683	32.721181	31.052078
50	44.142786	41.566447	39.196118	37.012876	34.999688	33.141209	31.423606
51	44.918195	42.249575	39.798136	37.543581	35.467673	33.554014	31.787849
52	45.689747	42.927618	40.394194	38.067734	35.928742	33.959719	32.144950
53	46.457459	43.600614	40.984351	38.585417	36.382997	34.358446	32.495049
54	47.221353	44.268599	41.568664	39.096708	36.830539	34.750316	32.838283
55	47.981445	44.931612	42.147192	39.601687	37.271467	35.135445	33.174788
56	48.737757	45.589689	42.719992	40.100431	37.705879	35.513951	33.504694
57	49.490305	46.242868	43.287121	40.593019	38.133871	35.885947	33.828131
58	50.239109	46.891184	43.848635	41.079524	38.555538	36.251545	34.145226
59	50.984189	47.534674	44.404589	41.560024	38.970973	36.610855	34.456104
60	51.725561	48.173374	44.955038	42.034592	39.380269	36.963986	34.760887
61	52.463245	48.807319	45.500038	42.503301	39.783516	37.311042	35.059693
62	53.197258	49.436545	46.039642	42.966223	40.180804	37.652130	35.352640
63	53.927620	50.061086	46.573903	43.423430	40.572221	37.987351	35.639843
64	54.654348	50.680979	47.102874	43.874992	40.957853	38.316807	35.921415
65	55.377461	51.296257	47.626608	44.320980	41.337786	38.640597	36.197466
66	56.096976	51.906955	48.145156	44.761462	41.712105	38.958817	36.468103
67	56.812912	52.513107	48.658571	45.196506	42.080891	39.271565	36.733435
68	57.525285	53.114746	49.166901	45.626178	42.444228	39.578934	36.993564
69	58.234115	53.711907	49.670199	46.050547	42.802195	39.881016	37.248592
70	58.939418	54.304622	50.168514	46.469676	43.154872	40.177903	37.498619

TABLE 8 Present Value of an Annuity $a_{\overline{n}|i} = \dfrac{1-(1+i)^{-n}}{i}$ (continued)

n	$\tfrac{1}{2}\%$	$\tfrac{3}{4}\%$	1%	$1\tfrac{1}{4}\%$	$1\tfrac{1}{2}\%$	$1\tfrac{3}{4}\%$	2%
71	59.641212	54.892925	50.661895	46.883630	43.502337	40.469683	37.743744
72	60.339514	55.476849	51.150391	47.292474	43.844667	40.756445	37.984063
73	61.034342	56.056426	51.634051	47.696271	44.181938	41.038276	38.219670
74	61.725714	56.631688	52.112922	48.095082	44.514224	41.315259	38.450657
75	62.413645	57.202668	52.587051	48.488970	44.841600	41.587478	38.677114
76	63.098155	57.769397	53.056486	48.877995	45.164138	41.855015	38.899132
77	63.779258	58.331908	53.521274	49.262218	45.481910	42.117951	39.116796
78	64.456973	58.890231	53.981459	49.641696	45.794985	42.376364	39.330192
79	65.131317	59.444398	54.437088	50.016490	46.103433	42.630334	39.539404
80	65.802305	59.994440	54.888206	50.386657	46.407323	42.879935	39.744514
81	66.469956	60.540387	55.334858	50.752254	46.706723	43.125243	39.945602
82	67.134284	61.082270	55.777087	51.113337	47.001697	43.366332	40.142747
83	67.795308	61.620119	56.214937	51.469963	47.292313	43.603275	40.336026
84	68.453042	62.153965	56.648453	51.822185	47.578633	43.836142	40.525516
85	69.107505	62.683836	57.077676	52.170060	47.860722	44.065005	40.711290
86	69.758711	63.209763	57.502650	52.513639	48.138643	44.289931	40.893422
87	70.406678	63.731774	57.923415	52.852977	48.412456	44.510989	41.071982
88	71.051421	64.249900	58.340015	53.188125	48.682222	44.728244	41.247041
89	71.692956	64.764169	58.752490	53.519136	48.948002	44.941764	41.418668
90	72.331300	65.274609	59.160881	53.846060	49.209855	45.151610	41.586929
91	72.966467	65.781250	59.565229	54.168948	49.467837	45.357848	41.751891
92	73.598475	66.284119	59.965573	54.487850	49.722007	45.560539	41.913619
93	74.227338	66.783245	60.361954	54.802815	49.972421	45.759743	42.072175
94	74.853073	67.278655	60.754410	55.113892	50.219134	45.955521	42.227623
95	75.475694	67.770377	61.142980	55.421127	50.462201	46.147933	42.380023
96	76.095218	68.258439	61.527703	55.724570	50.701675	46.337035	42.529434
97	76.711660	68.742867	61.908617	56.024267	50.937611	46.522884	42.675916
98	77.325035	69.223689	62.285759	56.320264	51.170060	46.705537	42.819525
99	77.935358	69.700932	62.659168	56.612606	51.399074	46.885049	42.960319
100	78.542645	70.174623	63.028879	56.901339	51.624704	47.061473	43.098352

TABLE 8 Present Value of an Annuity $a_{\overline{n}|i} = \dfrac{1-(1+i)^{-n}}{i}$

n	3%	4%	5%	6%	7%	8%	9%
1	0.970874	0.961538	0.952381	0.943396	0.934579	0.925926	0.917431
2	1.913470	1.886095	1.859410	1.833393	1.808018	1.783265	1.759111
3	2.828611	2.775091	2.723248	2.673012	2.624316	2.577097	2.531295
4	3.717098	3.629895	3.545951	3.465106	3.387211	3.312127	3.239720

A-26 APPENDIX

TABLE 8 Present Value of an Annuity $a_{\overline{n}|i} = \dfrac{1-(1+i)^{-n}}{i}$ (continued)

n	3%	4%	5%	6%	7%	8%	9%
5	4.579707	4.451822	4.329477	4.212364	4.100197	3.992710	3.889651
6	5.417191	5.242137	5.075692	4.917324	4.766540	4.622880	4.485919
7	6.230283	6.002055	5.786373	5.582381	5.389289	5.206370	5.032953
8	7.019692	6.732745	6.463213	6.209794	5.971299	5.746639	5.534819
9	7.786109	7.435332	7.107822	6.801692	6.515232	6.246888	5.995247
10	8.530203	8.110896	7.721735	7.360087	7.023582	6.710081	6.417658
11	9.252624	8.760477	8.306414	7.886875	7.498674	7.138964	6.805191
12	9.954004	9.385074	8.863252	8.383844	7.942686	7.536078	7.160725
13	10.634955	9.985648	9.393573	8.852683	8.357651	7.903776	7.486904
14	11.296073	10.563123	9.898641	9.294984	8.745468	8.244237	7.786150
15	11.937935	11.118387	10.379658	9.712249	9.107914	8.559479	8.060688
16	12.561102	11.652296	10.837770	10.105895	9.446649	8.851369	8.312558
17	13.166118	12.165669	11.274066	10.477260	9.763223	9.121638	8.543631
18	13.753513	12.659297	11.689587	10.827603	10.059087	9.371887	8.755625
19	14.323799	13.133939	12.085321	11.158116	10.335595	9.603599	8.950115
20	14.877475	13.590326	12.462210	11.469921	10.594014	9.818147	9.128546
21	15.415024	14.029160	12.821153	11.764077	10.835527	10.016803	9.292244
22	15.936917	14.451115	13.163003	12.041582	11.061240	10.200744	9.442425
23	16.443608	14.856842	13.488574	12.303379	11.272187	10.371059	9.580207
24	16.935542	15.246963	13.798642	12.550358	11.469334	10.528758	9.706612
25	17.413148	15.622080	14.093945	12.783356	11.653583	10.674776	9.822580
26	17.876842	15.982769	14.375185	13.003166	11.825779	10.809978	9.928972
27	18.327031	16.329586	14.643034	13.210534	11.986709	10.935165	10.026580
28	18.764108	16.663063	14.898127	13.406164	12.137111	11.051078	10.116128
29	19.188455	16.983715	15.141074	13.590721	12.277674	11.158406	10.198283
30	19.600441	17.292033	15.372451	13.764831	12.409041	11.257783	10.273654
31	20.000428	17.588494	15.592811	13.929086	12.531814	11.349799	10.342802
32	20.388766	17.873551	15.802677	14.084043	12.646555	11.434999	10.406240
33	20.765792	18.147646	16.002549	14.230230	12.753790	11.513888	10.464441
34	21.131837	18.411198	16.192904	14.368141	12.854009	11.586934	10.517835
35	21.487220	18.664613	16.374194	14.498246	12.947672	11.654568	10.566821
36	21.832252	18.908282	16.546852	14.620987	13.035208	11.717193	10.611763
37	22.167235	19.142579	16.711287	14.736780	13.117017	11.775179	10.652993
38	22.492462	19.367864	16.867893	14.846019	13.193473	11.828869	10.690820
39	22.808215	19.584485	17.017041	14.949075	13.264928	11.878582	10.725523
40	23.114772	19.792774	17.159086	15.046297	13.331709	11.924613	10.757360
41	23.412400	19.993052	17.294368	15.138016	13.394120	11.967235	10.786569
42	23.701359	20.185627	17.423208	15.224543	13.452449	12.006699	10.813366
43	23.981902	20.370795	17.545912	15.306173	13.506962	12.043240	10.837950
44	24.254274	20.548841	17.662773	15.383182	13.557908	12.077074	10.860505
45	24.518713	20.720040	17.774070	15.455832	13.605522	12.108402	10.881197
46	24.775449	20.884654	17.880066	15.524370	13.650020	12.137409	10.900181
47	25.024708	21.042936	17.981016	15.589028	13.691608	12.164267	10.917597
48	25.266707	21.195131	18.077158	15.650027	13.730474	12.189136	10.933575

TABLE 8 Present Value of an Annuity $a_{\overline{n}|i} = \dfrac{1 - (1+i)^{-n}}{i}$ (continued)

n	3%	4%	5%	6%	7%	8%	9%
49	25.501657	21.341472	18.168722	15.707572	13.766799	12.212163	10.948234
50	25.729764	21.482185	18.255925	15.761861	13.800746	12.233485	10.961683
51	25.951227	21.617485	18.338977	15.813076	13.832473	12.253227	10.974021
52	26.166240	21.747582	18.418073	15.861393	13.862124	12.271506	10.985340
53	26.374990	21.872675	18.493403	15.906974	13.889836	12.288432	10.995725
54	26.577660	21.992957	18.565146	15.949976	13.915735	12.304103	11.005252
55	26.774428	22.108612	18.633472	15.990543	13.939939	12.318614	11.013993
56	26.965464	22.219819	18.698545	16.028814	13.962560	12.332050	11.022012
57	27.150936	22.326749	18.760519	16.064919	13.983701	12.344491	11.029369
58	27.331005	22.429567	18.819542	16.098980	14.003458	12.356010	11.036118
59	27.505831	22.528430	18.875754	16.131113	14.021924	12.366676	11.042310
60	27.675564	22.623490	18.929290	16.161428	14.039181	12.376552	11.047991
61	27.840353	22.714894	18.980276	16.190026	14.055309	12.385696	11.053203
62	28.000343	22.802783	19.028834	16.217006	14.070383	12.394163	11.057984
63	28.155673	22.887291	19.075080	16.242458	14.084470	12.402003	11.062371
64	28.306478	22.968549	19.119124	16.266470	14.097635	12.409262	11.066395
65	28.452892	23.046682	19.161070	16.289123	14.109940	12.415983	11.070087
66	28.595040	23.121810	19.201019	16.310493	14.121439	12.422207	11.073475
67	28.733049	23.194048	19.239066	16.330654	14.132186	12.427969	11.076582
68	28.867038	23.263507	19.275301	16.349673	14.142230	12.433305	11.079433
69	28.997124	23.330296	19.309810	16.367617	14.151617	12.438245	11.082049
70	29.123421	23.394515	19.342677	16.384544	14.160389	12.442820	11.084449
71	29.246040	23.456264	19.373978	16.400513	14.168588	12.447055	11.086650
72	29.365088	23.515639	19.403788	16.415578	14.176251	12.450977	11.088670
73	29.480667	23.572730	19.432179	16.429791	14.183412	12.454608	11.090523
74	29.592881	23.627625	19.459218	16.443199	14.190104	12.457971	11.092223
75	29.701826	23.680408	19.484970	16.455848	14.196359	12.461084	11.093782
76	29.807598	23.731162	19.509495	16.467781	14.202205	12.463967	11.095213
77	29.910290	23.779963	19.532853	16.479039	14.207668	12.466636	11.096526
78	30.009990	23.826888	19.555098	16.489659	14.212774	12.469107	11.097730
79	30.106786	23.872008	19.576284	16.499679	14.217546	12.471396	11.098835
80	30.200763	23.915392	19.596460	16.509131	14.222005	12.473514	11.099849
81	30.292003	23.957108	19.615677	16.518048	14.226173	12.475476	11.100778
82	30.380586	23.997219	19.633978	16.526460	14.230069	12.477293	11.101632
83	30.466588	24.035787	19.651407	16.534396	14.233709	12.478975	11.102414
84	30.550086	24.072872	19.668007	16.541883	14.237111	12.480532	11.103132
85	30.631151	24.108531	19.683816	16.548947	14.240291	12.481974	11.103791
86	30.709855	24.142818	19.698873	16.555610	14.243262	12.483310	11.104396
87	30.786267	24.175787	19.713212	16.561896	14.246040	12.484546	11.104950
88	30.860454	24.207487	19.726869	16.567827	14.248635	12.485691	11.105459
89	30.932479	24.237969	19.739875	16.573421	14.251061	12.486751	11.105926
90	31.002407	24.267278	19.752262	16.578699	14.253328	12.487732	11.106354
91	31.070298	24.295459	19.764059	16.583679	14.255447	12.488641	11.106746
92	31.136212	24.322557	19.775294	16.588376	14.257427	12.489482	11.107107

TABLE 8 Present Value of an Annuity $a_{\overline{n}|i} = \dfrac{1-(1+i)^{-n}}{i}$ (continued)

n	3%	4%	5%	6%	7%	8%	9%
93	31.200206	24.348612	19.785994	16.592808	14.259277	12.490261	11.107437
94	31.262336	24.373666	19.796185	16.596988	14.261007	12.490983	11.107741
95	31.322656	24.397756	19.805891	16.600932	14.262623	12.491651	11.108019
96	31.381219	24.420919	19.815134	16.604653	14.264134	12.492269	11.108274
97	31.438077	24.443191	19.823937	16.608163	14.265546	12.492842	11.108509
98	31.493279	24.464607	19.832321	16.611475	14.266865	12.493372	11.108724
99	31.546872	24.485199	19.840306	16.614599	14.268098	12.493863	11.108921
100	31.598905	24.504999	19.847910	16.617546	14.269251	12.494318	11.109102

TABLE 8 Present Value of an Annuity $a_{\overline{n}|i} = \dfrac{1-(1+i)^{-n}}{i}$

n	10%	11%	12%	13%	14%	15%	16%
1	0.909091	0.900901	0.892857	0.884956	0.877193	0.869565	0.862069
2	1.735537	1.712523	1.690051	1.668102	1.646661	1.625709	1.605232
3	2.486852	2.443715	2.401831	2.361153	2.321632	2.283225	2.245890
4	3.169865	3.102446	3.037349	2.974471	2.913712	2.854978	2.798181
5	3.790787	3.695897	3.604776	3.517231	3.433081	3.352155	3.274294
6	4.355261	4.230538	4.111407	3.997550	3.888668	3.784483	3.684736
7	4.868419	4.712196	4.563757	4.422610	4.288305	4.160420	4.038565
8	5.334926	5.146123	4.967640	4.798770	4.638864	4.487322	4.343591
9	5.759024	5.537048	5.328250	5.131655	4.946372	4.771584	4.606544
10	6.144567	5.889232	5.650223	5.426243	5.216116	5.018769	4.833227
11	6.495061	6.206515	5.937699	5.686941	5.452733	5.233712	5.028644
12	6.813692	6.492356	6.194374	5.917647	5.660292	5.420619	5.197107
13	7.103356	6.749870	6.423548	6.121812	5.842362	5.583147	5.342334
14	7.366687	6.981865	6.628168	6.302488	6.002072	5.724476	5.467529
15	7.606080	7.190870	6.810864	6.462379	6.142168	5.847370	5.575456
16	7.823709	7.379162	6.973986	6.603875	6.265060	5.954235	5.668497
17	8.021553	7.548794	7.119630	6.729093	6.372859	6.047161	5.748704
18	8.201412	7.701617	7.249670	6.839905	6.467420	6.127966	5.817848
19	8.364920	7.839294	7.365777	6.937969	6.550369	6.198231	5.877455
20	8.513564	7.963328	7.469444	7.024752	6.623131	6.259331	5.928841
21	8.648694	8.075070	7.562003	7.101550	6.686957	6.312462	5.973139
22	8.771540	8.175739	7.644646	7.169513	6.742944	6.358663	6.011326
23	8.883218	8.266432	7.718434	7.229658	6.792056	6.398837	6.044247
24	8.984744	8.348137	7.784316	7.282883	6.835137	6.433771	6.072627
25	9.077040	8.421745	7.843139	7.329985	6.872927	6.464149	6.097092
26	9.160945	8.488058	7.895660	7.371668	6.906077	6.490564	6.118183
27	9.237223	8.547800	7.942554	7.408556	6.935155	6.513534	6.136364
28	9.306567	8.601622	7.984423	7.441200	6.960662	6.533508	6.152038

TABLE 8 Present Value of an Annuity $a_{\overline{n}|i} = \dfrac{1-(1+i)^{-n}}{i}$ (continued)

n	10%	11%	12%	13%	14%	15%	16%
29	9.369606	8.650110	8.021806	7.470088	6.983037	6.550877	6.165550
30	9.426914	8.693793	8.055184	7.495653	7.002664	6.565980	6.177198
31	9.479013	8.733146	8.084986	7.518277	7.019881	6.579113	6.187240
32	9.526376	8.768600	8.111594	7.538299	7.034983	6.590533	6.195897
33	9.569432	8.800541	8.135352	7.556016	7.048231	6.600463	6.203359
34	9.608575	8.829316	8.156564	7.571696	7.059852	6.609099	6.209792
35	9.644159	8.855240	8.175504	7.585572	7.070045	6.616607	6.215338
36	9.676508	8.878594	8.192414	7.597851	7.078987	6.623137	6.220119
37	9.705917	8.899635	8.207513	7.608718	7.086831	6.628815	6.224241
38	9.732651	8.918590	8.220993	7.618334	7.093711	6.633752	6.227794
39	9.756956	8.935666	8.233030	7.626844	7.099747	6.638045	6.230857
40	9.779051	8.951051	8.243777	7.634376	7.105041	6.641778	6.233497
41	9.799137	8.964911	8.253372	7.641040	7.109685	6.645025	6.235773
42	9.817397	8.977397	8.261939	7.646938	7.113759	6.647848	6.237736
43	9.833998	8.988646	8.269589	7.652158	7.117332	6.650302	6.239427
44	9.849089	8.998780	8.276418	7.656777	7.120467	6.652437	6.240886
45	9.862808	9.007910	8.282516	7.660864	7.123217	6.654293	6.242143
46	9.875280	9.016135	8.287961	7.664482	7.125629	6.655907	6.243227
47	9.886618	9.023545	8.292822	7.667683	7.127744	6.657310	6.244161
48	9.896926	9.030221	8.297163	7.670516	7.129600	6.658531	6.244966
49	9.906296	9.036235	8.301038	7.673023	7.131228	6.659592	6.245661
50	9.914814	9.041653	8.304498	7.675242	7.132656	6.660515	6.246259

TABLE 9 Present Value of an Annuity for Fractional Interest Periods $a_{\overline{1/m}|i} = \dfrac{1-(1+i)^{-1/m}}{i}$

m	$\tfrac{1}{2}$%	$\tfrac{3}{4}$%	1%	$1\tfrac{1}{4}$%	$1\tfrac{1}{2}$%	$1\tfrac{3}{4}$%	2%
2	0.498133	0.497205	0.496281	0.495361	0.494444	0.493532	0.492623
3	0.332227	0.331676	0.331128	0.330582	0.330038	0.329497	0.328957
4	0.249222	0.248835	0.248449	0.248065	0.247682	0.247301	0.246921
6	0.166182	0.165941	0.165701	0.165462	0.165224	0.164986	0.164750
12	0.083108	0.082997	0.082885	0.082774	0.082663	0.082553	0.082443

m	3%	4%	5%	6%	7%	8%	9%
2	0.489024	0.485483	0.481999	0.478569	0.475193	0.471869	0.468597
3	0.326818	0.324712	0.322637	0.320593	0.318578	0.316593	0.314636
4	0.245415	0.243932	0.242469	0.241027	0.239606	0.238204	0.236822
6	0.163812	0.162887	0.161974	0.161075	0.160187	0.159312	0.158448
12	0.082007	0.081576	0.081152	0.080733	0.080319	0.079911	0.079508

TABLE 9 Present Value of an Annuity for Fractional Interest Periods $a_{\overline{1/m}|i} = \dfrac{1 - (1 + i)^{-1/m}}{i}$ (continued)

m	10%	11%	12%	13%	14%	15%	16%
2	0.465374	0.462200	0.459073	0.455993	0.452958	0.449968	0.447021
3	0.312707	0.310805	0.308930	0.307081	0.305257	0.303459	0.301684
4	0.235459	0.234115	0.232788	0.231480	0.230188	0.228914	0.227657
6	0.157595	0.156754	0.155924	0.155104	0.154295	0.153496	0.152708
12	0.079111	0.078718	0.078330	0.077947	0.077569	0.077195	0.076826

TABLE 10

A Brief Table of Integrals

1. $\displaystyle\int u^n \, du = \dfrac{1}{n+1} u^{n+1} + c$

2. $\displaystyle\int e^u \, du = e^u + c$

3. $\displaystyle\int \dfrac{du}{u} = \ln u + c$

4. $\displaystyle\int u \, dv = uv - \int v \, du$

5. $\displaystyle\int \dfrac{du}{\sqrt{a^2 + u^2}} = \ln(u + \sqrt{a^2 + u^2}) + c$

6. $\displaystyle\int \dfrac{du}{a^2 - u^2} = \dfrac{1}{2a} \ln\left(\dfrac{a + u}{a - u}\right) + c$

7. $\displaystyle\int u e^u \, du = e^u(u - 1) + c$

8. $\displaystyle\int u^n e^u \, du = u^n e^u - n \int u^{n-1} e^u \, du$

9. $\displaystyle\int \dfrac{du}{u^2(a + bu)} = -\dfrac{1}{au} + \dfrac{b}{a^2} \ln\left(\dfrac{a + bu}{u}\right) + c$

10. $\displaystyle\int \ln u \, du = u \ln u - u + c$

11. $\displaystyle\int \dfrac{du}{\sqrt{u^2 - a^2}} = \ln(u + \sqrt{u^2 - a^2}) + c$

12. $\displaystyle\int \frac{du}{u^2 - a^2} = \frac{1}{2a} \ln\left(\frac{u-a}{u+a}\right) + c$

13. $\displaystyle\int \frac{du}{u\sqrt{a^2 - u^2}} = -\frac{1}{a} \ln\left(\frac{a + \sqrt{a^2 + u^2}}{u}\right) + c$

14. $\displaystyle\int \frac{du}{u\sqrt{a^2 + u^2}} = -\frac{1}{a} \ln\left(\frac{a + \sqrt{a^2 + u^2}}{u}\right) + c$

15. $\displaystyle\int \frac{du}{u\sqrt{a + bu}} = \frac{1}{\sqrt{a}} \ln\left(\frac{\sqrt{a + bu} - \sqrt{a}}{\sqrt{a + bu} + \sqrt{a}}\right) + c \quad (a > 0)$

16. $\displaystyle\int \frac{\sqrt{a^2 - u^2}}{u} du = -\sqrt{a^2 - u^2} - a \ln\left(\frac{a + \sqrt{a^2 - u^2}}{u}\right) + c$

17. $\displaystyle\int \frac{du}{u^2\sqrt{a^2 - u^2}} = -\frac{\sqrt{a^2 - u^2}}{a^2 u} + c$

18. $\displaystyle\int \frac{du}{(a^2 - u^2)^{3/2}} = \frac{u}{a^2\sqrt{a^2 - u^2}} + c$

19. $\displaystyle\int \frac{u\, du}{(a + bu)^2} = \frac{1}{b^2}\left[\frac{a}{a + bu} + \ln(a + bu)\right] + c$

20. $\displaystyle\int \sqrt{a^2 + u^2}\, du = \frac{u}{2}\sqrt{a^2 + u^2} + \frac{a^2}{2} \ln(u + \sqrt{a^2 + u^2}) + c$

21. $\displaystyle\int \frac{du}{(a^2 + u^2)^{3/2}} = \frac{u}{a^2\sqrt{a^2 + u^2}} + c$

22. $\displaystyle\int \frac{du}{u(a + bu)} = -\frac{1}{a} \ln\left(\frac{a + bu}{u}\right) + c$

ANSWERS TO SELECTED EXERCISES

CHAPTER 1

Section 1–1 Exercises

1. (A) true   (B) false   (C) false   (D) true
   (E) true   (F) true   (G) true   (H) true
   (I) true   (J) true   (K) false   (L) true
3. (A) true   (B) true   (C) true   (D) false
   (E) true   (F) true   (G) false   (H) true
5. (A) 0   (B) 1   (C) 1   (D) 21
   (E) 2   (F) 15 (G) 15 (H) 20
   (I) 20

Section 1–2 Exercises

1. (A) -2 (B) 10
3. (A) 7   (B) 5   (C) -3 (D) 13
5. (A) 5/7   (B) 5/8   (C) 5   (D) 1/3
7. (A) $C(x) = 0.25x + 10,000$
   (B) $C(0) = 10,000$; $10,000 is the fixed cost.
       $C(10) = 10,002.5$; it costs $10,002.50 to produce 10 bars of soap.
       $C(1000) = 10,250$; $10,250 is the cost of producing 1000 bars of soap.
9. (A) $R(x) = 10x$
   (B) $R(0) = 0$; $0 is the revenue from not selling any clocks.
       $R(10) = 100$; $100 is the total sales revenue gained from selling 10 clocks.
       $R(20) = 200$; $200 is the total sales revenue from selling 20 clocks.
11. (A) 42 pounds, 30 pounds.
    (B) $D(2) = 36$; the demand for EZ Chew Peanuts is 36 pounds, at the price of $2 per pound.
        $D(4) = 24$; at a price of $4 per pound, 24 pounds of EZ Chew Peanuts are demanded.
13. $A: (2, 3)$ $B: (0, 2)$ $C: (-2, 3)$ $D: (-2, 0)$ $E: (-1, 0)$
    $F: (2, 0)$ $G: (3, -1)$ $H: (-2, -2)$ $I: (0, -2)$ $J: (0, -3)$
17.

x	$S(x)$
1/4	4
1/2	2
1	1
2	1/2

19. All real numbers x such that $x \neq 5$.
21. A, B, C, D
23. No. $(2, 3)$ and $(2, -3)$ are both defined by this equation.
25. (A) $-3x^2 - 6hx - 3h^2 + 5x + 5h - 2$ (B) $-6hx - 3h^2 + 5h$ (C) $-6x - 3h + 5$
27. $-4x - 2h + 3$
29. $6x^2 + 6xh + 2h^2 - 3$

A-33

31.

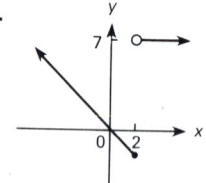

35. (A) 8/25   (B) 13,720

Section 1–3 Exercises

1. (A) 11/3   (B) 9/5   (C) 1/4   (D) −2/9
   (E) 10   (F) −2/3   (G) −7/5   (H) −14/3
3. (C)   (D)

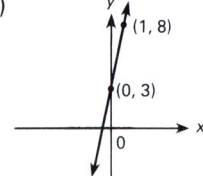

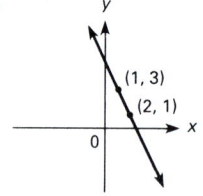

5. $50   **7.** (A) −5   (B) 5   (C) 5
9. (A) 0.80   (B) $0.80   (C) 0.20   (D) $0.20

Section 1–4 Exercises

1. (A) $y - 4 = 2(x - 1)$ or $y - 8 = 2(x - 3)$, $y = 2x + 2$
   (B) $y + 1 = -\frac{7}{5}(x - 4)$ or $y + 8 = -\frac{7}{5}(x - 9)$, $y = -\frac{7}{5}x + \frac{23}{5}$
   (C) $y + 10 = -2(x - 5)$ or $y + 14 = -2(x - 7)$, $y = -2x$
   (D) $y - 2 = 2(x - 3)$ or $y - 10 = 2(x - 7)$, $y = 2x - 4$
   (E) $y + 3 = -4(x + 2)$ or $y - 9 = -4(x + 5)$, $y = -4x - 11$

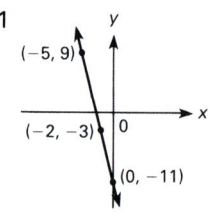

   (F) $y - 5 = x - 5$ or $y - 7 = x - 7$, $y = x$
   (G) $y - 12 = 3(x - 4)$ or $y - 18 = 3(x - 6)$, $y = 3x$

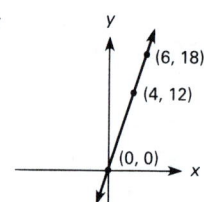

   (H) $y + 3 = -5(x - 5)$ or $y - 2 = -5(x - 4)$, $y = -5x + 22$
3. (A) $x = 4$ (B) $x = -3$
   (C) $y = -9$ (D) $y = 1$

A-35 ANSWERS TO SELECTED EXERCISES

5. $y = 4x - 13$

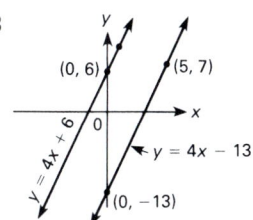

7. $y = \dfrac{-x}{3} + \dfrac{23}{3}$

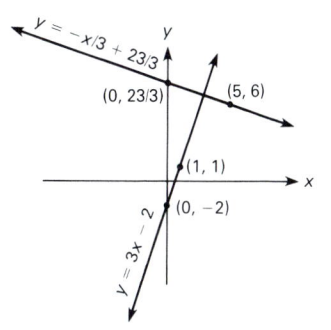

9. (A) and (B) are parallel.

11. (A) $C(x) = 50x + 2000$ (B) $2000; $50

13. $y = -5x + 100$

15. (A) $y = \dfrac{3x}{5} - \dfrac{11}{5}$ (B) $y = \dfrac{-x}{2} + \dfrac{15}{2}$

   (C) $y = \dfrac{3x}{2}$ (D) $y = \dfrac{-2x}{5} - \dfrac{16}{5}$

17. (A) $(0, -6); (10, 0); (15, 3)$ (B) $(7, 1); (14, -1)$
   (C) $(0, 0); (-2, 8/3)$ (D) $(1, 9); (-1/2, 0); (-1, -3)$
   (E) $(0, 4); (-1, 2); (-2, 0)$ (F) $(0, 0); (1, -3); (2, -6)$

19. (A) $y = 0.8x + 7$ (B) 75 billion dollars
   (C) For each dollar increase in disposable income, consumption increases by $0.80.

Section 1–5 Exercises

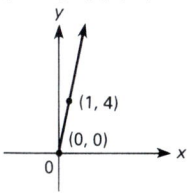

1. (A) $(6, 0) (0, -4)$ (B) $(9, 0) (0, 15)$

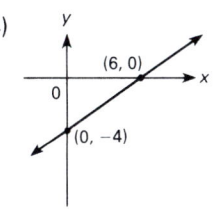

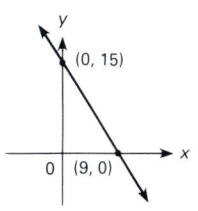

   (C) $(-8/3, 0) (0, 4)$ (D) $(2, 0) (0, -5/3)$
   (E) $(11/2, 0) (0, 11/7)$ (F) $(-3, 0) (0, -9/5)$
   (G) $(13/4, 0) (0, 13)$ (H) $(15/2, 0) (0, 15)$
   (I) $(-2/3, 0) (0, 4)$ (J) $(-26/5, 0) (0, 13)$

3. (B)

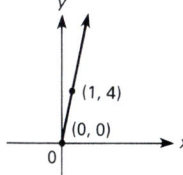

5. (B) $5000; $10   (C) $25,000   (D) $55,000
7. (B) 80   (C) $100   (D) 5

Section 1-6 Exercises

1. (A) $C(x) = 5x + 2000$ (B) $R(x) = 15x$ (C)

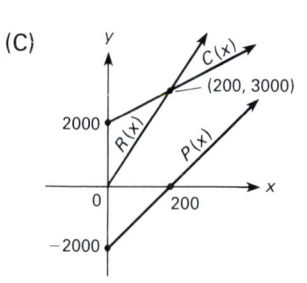

(D) (200, 3000) (E) $P(x) = 10x - 2000$ (F) See Answer 1C.
(G) $1000 (H) -$1000 (I) 4200

3. (A) $1300 (B) $C(x) = 1300x + 260,000$ (C) $R(x) = 1800x$
(E) (520, 936,000) (F) $P(x) = 500x - 260,000$
(H) $240,000 (I) 728

5. (A) (B) (9, 12) (C) $9

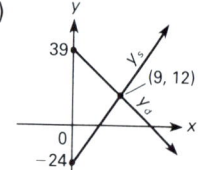

7. (A) $y = 30,000 - 2900x$ (C) $12,600
9. 50,000

CHAPTER 2

Section 2-1 Exercises

1. (A) 9 (B) 16/81 (C) 25
(D) -125 (E) 4096 (F) 1/125
(G) 1/16 (H) 1/9 (I) 64
(J) 9 (K) 78,125 (L) 1024
(M) 9/25 (N) 1 (O) 1/64

5. (A) $9x^2$ (B) $16y^4$ (C) $125 x^3 y^3$
(D) $x^9 y^9 z^9$ (E) 18 (F) 24

7. (A) 4096 (B) 27
(C) 1/3 (D) 729

3. (A) 3^{-2} (B) 5^{-6} (C) x^{-7}
(D) $(-5)^{-3}$ (E) x^{-n} (F) x^{-8}

Section 2-2 Exercises

1. (A–D)

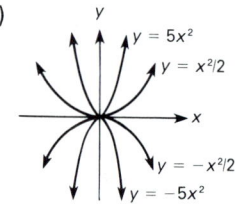

3.

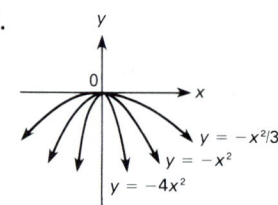

A-37 ANSWERS TO SELECTED EXERCISES

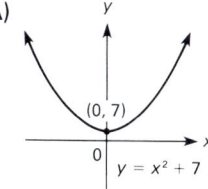

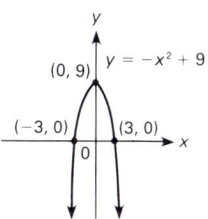

5. (A) $y = x^2 + 7$; vertex $(0, 7)$
(D) $y = -x^2 + 9$; $(0, 9)$, $(-3, 0)$, $(3, 0)$
(L) $(-2\sqrt{2}, 0)$, $(2\sqrt{2}, 0)$, $(0, -4)$

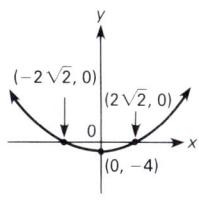

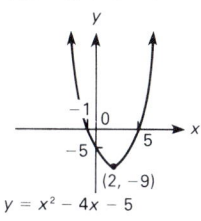

7. (B) 25 million dollars

9. (A, B) $y = (x + 3)^2$, $y = (x - 3)^2$; intersect at $(0, 9)$; x-intercepts $-3, 3$
(E, F) $f(x) = -2(x + 5)^2$, vertex $(-5, 0)$; $f(x) = -2(x - 3)^2$, vertex $(3, 0)$; y-intercepts $-50, -18$

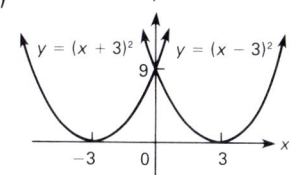

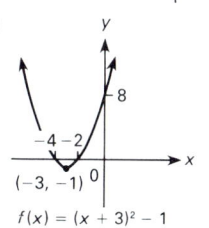

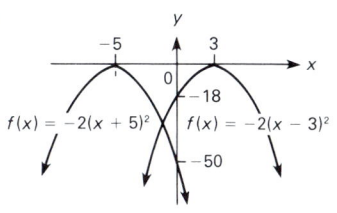

11. (A) $f(x) = (x + 3)^2 - 1$; vertex $(-3, -1)$; x-intercepts $-4, -2$; y-intercept 8
(I) $f(x) = -2(x - 4)^2 + 3$; vertex $(4, 3)$; x-intercepts $5/2, 11/2$; y-intercept -29

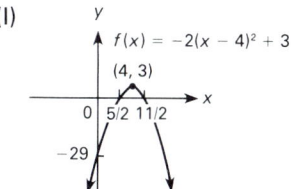

13. (A) $y = x^2 - 4x - 5$; vertex $(2, -9)$; x-intercepts $-1, 5$; y-intercept -5
(E) vertex $(0.6, 5.8)$; x-intercepts $-0.477, 1.68$; y-intercept 4

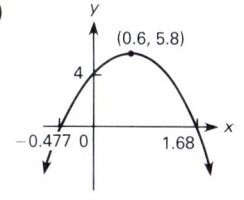

Section 2–3 Exercises

1. (A) $R(x) = -2x^2 + 40x$ (C) $200 (D) 10
3. (A) $R(p) = -3p^2 + 48p$ (C) $192 (D) $8
5. (A) $C(x) = x^2 - 100x + 2900$; vertex $(50, 400)$; y-intercept 2900
(B) 50 (C) $400 (D) $21
7. (B) (7, 16) (C) 7, 16
9. (A) $A(x) = -x^2 + 1000x$
(C) 500 ft. by 500 ft.; 250,000 sq. ft
11. (B) 6 seconds, 576 feet (C) 0 seconds and 12 seconds
13. (A) $R(x) = -0.3x^2 + 1.2x + 57.6$ (C) 2 weeks, $58.80
15. (A) 8.4 utiles (B) 2 (C) 7 million dollars, 10 utiles

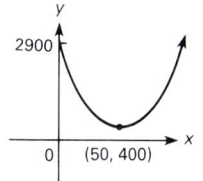

A-38 ANSWERS TO SELECTED EXERCISES

Section 2-4 Exercises

1. (A) $f(x) = x^3 - 7x + 6$

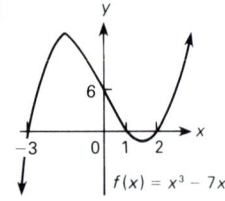

(C) 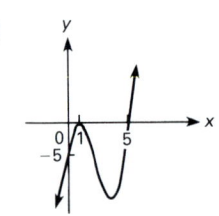 $f(x) = x^3 - 7x^2 + 11x - 5$

3. (B) $y = -\frac{1}{4}(x-3)^2(x+1)$

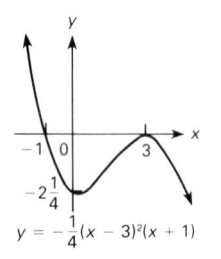

(C) 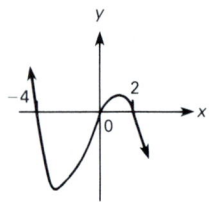 $f(x) = -3x^5(x-2)^3(x+4)$

Section 2-5 Exercises

1. (A) $P(x) = 8x - \dfrac{x^3}{2,000,000}$ (C) $x \leq 4000$ or $0 \leq x \leq 4000$

(D) $x \approx 2309$

3. (A) $P(x) = \dfrac{1}{10,000}x^2(x-20)(x-30)^2$ (B) 0, 20, 30   (C) no

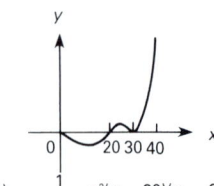

(D) yes   (E) yes   (F) $p(25) = \$7.81$; $p(35) = \$45.94$; $p(40) = \$320$

5. (B) -275 (C) 1600   (D) 0 seconds and 16 seconds

Section 2-6 Exercises

1. (A) $y = \dfrac{(x-3)^2(x+8)}{(x-1)^2}$

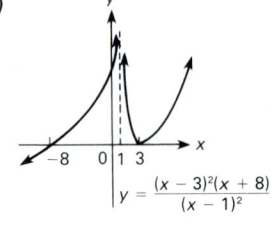

(C) $f(x) = \dfrac{(x-1)(x+3)^2}{(x+1)^3(x+4)}$

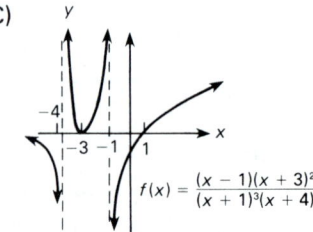

A-39 ANSWERS TO SELECTED EXERCISES

(E) $y = \dfrac{1}{x-4}$

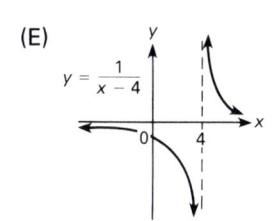

(L) $f(x) = \dfrac{x^3 - 8x^2}{x^2 - 36}$

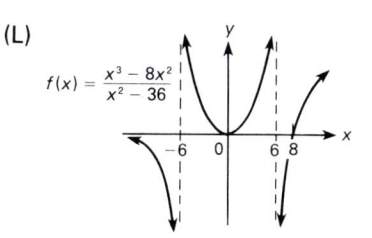

3. (B) 400   (C) 200   (D) 390
5. (B) $3050   (C) $80   (D) $50
7. (A) $\overline{C}(x) = (-x^2 + 20x + 125)/2x$
   (C) $40.25   (D) $20   (E) $11.25
9. (B) 10%: $2222
   20%: $5000
   50%: $20,000
   80%: $80,000
   90%: $180,000
   (C) no

CHAPTER 3

Section 3–1 Exercises

1. (A) $f(x) = 5^x$

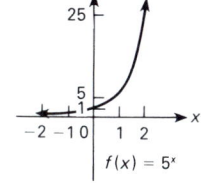

(J) $y = 3e^x$

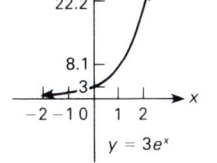

(M) $y = -3e^x$

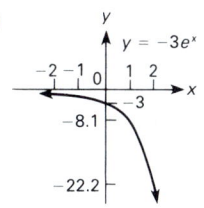

3. (K) $y = 7e^{-x}$

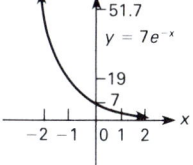

(N) $f(x) = -7e^{-x}$

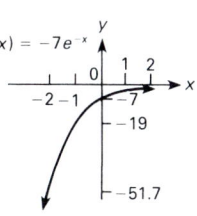

5. (E) $y = 5 \cdot \left(\dfrac{1}{5}\right)^x + 1$

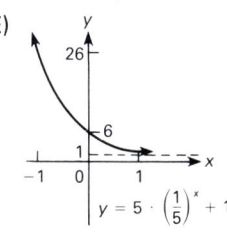

(F) $f(x) = -4 \cdot \left(\dfrac{1}{6}\right)^{-x} + 2$

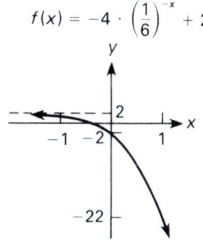

7. (A) $y = 500{,}000 \cdot 3^t$ (C) 40,500,000

A-40 ANSWERS TO SELECTED EXERCISES

Section 3-2 Exercises

1. (B) 3, 6, 12
3. (A) [graph showing $y = 1000e^{0.05t}$, y-axis marked 1000, 2000; x-axis marked 5, 10, 15]   (B) $1105.17
5. (B) 2000 grams   (C) 1481.636 grams   (D) 330.598 grams
7. (B) 215°   (C) 200.7°   (D) 65°   (E) 65°
9. (B) $3,978,615,600

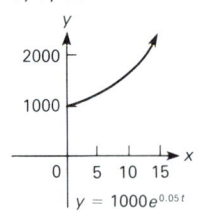

Section 3-3 Exercises

1. (A) $2 = \log_5 25$ (B) $2 = \log_4 16$ (C) $6 = \log_2 64$
   (D) $5 = \log_{10} 100{,}000$ (E) $-2 = \log_{10} .01$ (F) $1 = \log_{10} 10$
   (G) $w = \log_t S$ (H) $3 = \log_4 64$ (I) $x + y = \log_b N$

3. (A) 4   (B) 5   (C) 2
   (D) 0   (E) 0   (F) 0
   (G) 0   (H) 1   (I) 2
   (J) 3   (K) 4   (L) 5

5. 1

7. (A) 1.5694   (B) 2.5694   (C) 3.5694
   (D) 4.5694   (E) 5.5694   (F) −.4306
   (G) −1.4306   (H) −2.4306   (I) −3.4306

9. (A) .677607   (B) .941014   (C) 1.964260
   (D) 3.677607   (E) −1.171985   (F) −.096910

11. (A) 5.940185   (B) 8.029712
    (C) 594.0185   (D) 8029.712
    (E) 29.49850   (F) 256980.4

13. (A) 1.945910   (B) 1.223775   (C) 2.639057
    (D) 1.704748   (E) 6.620073

15. (A) $206,931.47; $216,094.38; $223,025.85; $234,011.97
    (B) $800.42 are the additional sales engendered by increasing advertising expenditures from $12,000 to $13,000.
    (C) $512.93

17. (A) $C(0) = 9000$; $9000 is the fixed production cost.
    $C(1) = 9006.93$; $9006.93 is the cost of producing 1 unit.
    $C(5) = 9017.92$; $9017.92 is the cost of producing 5 units.
    $C(10) = 9023.98$; $9023.98 is the cost of producing 10 units.
    (B) At a production level of 18 units, the increase in costs due to producing one more unit is $0.51.
    (C) $0.74

Section 3-4 Exercises

1. (A) −0.709541   (B) −0.645975   (C) 1.331911
   (D) 2.104120   (E) .111260   (F) 1.475772

A-41 ANSWERS TO SELECTED EXERCISES

3. (A) $x = \dfrac{10}{3} \cdot (8 - \ln p)$ (B) 19,342,583,000

5. (1, 1/2); (2, 1); (3, 2); (4, 4); (5, 8)

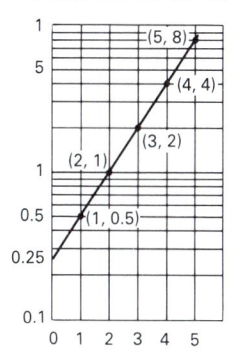

CHAPTER 4

Section 4–1 Exercises

1. (A) $I = \$210$, $S = \$1210$ (B) $I = \$1200$, $S = \$11,200$
(C) $I = \$200$, $S = \$5200$ (D) $I = \$240$, $S = \$8240$
(E) $I = \$60$, $S = \$2060$ (F) $I = \$900$, $S = \$9900$
3. $5813.95   **5.** $6666.67   **7.** (A) $2700   (B) $3300   (C) 16.36%
9. (A) $6944.44   (B) 9.72%   **11.** $612

Section 4–2 Exercises

1. (A) $S = \$2208.04$, $I = \$1208.04$ (B) $S = \$16,216.99$, $I = \$11,216.99$
(C) $S = \$14,326.78$, $I = \$6326.78$ (D) $S = \$12,311.80$, $I = \$9311.80$
3. (A) $22,080.40   (B) $12,080.40   **5.** $1326.65
7. $6245.97   **9.** $4396.46   **11.** (A) 6.09%   (B) 4.06%
(C) 12.68%   (D) 8.16%   **13.** (A) 13.87 years   (B) 18.64 years
15. $24,405.83   **17.** $1879.81
19. $2452.46

Section 4–3 Exercises

1. (A) $1197.22   (B) $7459.12   (C) $22,255.41   (D) $8099.15
3. (A) 6.18%   (B) 7.25%   (C) 8.33%   (D) 9.42%
5. (A) 18.31 years   (B) 15.69 years   **7.** 2.054068

Section 4–4 Exercises

1. (A) 312   (B) 189   (C) 38,227   **3.** (A) $6122.26   (B) $1322.26
5. (A) $47,575.42   (B) $63,937.37   **7.** $6183.48   **9.** $6122.26

A-42 ANSWERS TO SELECTED EXERCISES

11. $16,770.90   **13.** (A) 295.46971   (B) 212,661.11   (C) 3.028440

Section 4–5 Exercises

1. (A) $56,741.78   (B) $35,459.51   (C) $29,754.95   (D) $180,645.03
3. $7438.74   **5.** $53,349.26   **7.** $15,204.29
9. $19,327.03   **11.** $6671.38

Section 4–6 Exercises

1. $411.57   **3.** $352.31   **5.** $1956.11
7. (A) $2940.35   (B) $2357.90
    (C)

Payment Number	Payment	Interest	Total
1	$2940.35	$ 0	$ 2940.35
2	2940.35	147.02	6027.72
3	2940.35	301.39	9269.46
4	2940.35	463.47	12,673.28
5	2940.35	633.66	16,247.29
6	2940.35	812.36	20,000.00
		$2357.90	

9. (A) $1388.98   (B) $333,355.20, I = $243,355.20   (C) $60,899.37   (D) $29,100.63

Section 4–7 Exercises

1. $5127.98   **3.** $2191.04   **5.** $8265.97   **7.** $29,264.17
9. $3656.26

Section 4–8 Exercises

1. $8797.85   **3.** $8417.48   **5.** $5120.61

Section 4–9 Exercises

1. (A) $n = 21, c = 1/3, nc = 7$ (B) $n = 20, c = 3, nc = 60$
    (C) $n = 16, c = 2, nc = 32$ (D) $n = 10, c = 12, nc = 120$
3. S = $163,033.97; A = $33,958.19   **5.** S = $111,345.24; A = $27,522.84
7. $9981.43   **9.** $11,852.27   **11.** $591.98   **13.** $213.99

Section 4–10 Exercises

1. S = $44,594.69; A = $10,679.31   **3.** S = $24,714.85; A = $13,604.28
5. $955.12   **7.** $109.04

End of Chapter 4 Exercises

1. (A) $600   (B) $9400   **3.** (A) $540   (B) $1540
5. (A) $8000   (B) $18,000   **7.** $3954.59   **9.** $18,358.98

A-43 ANSWERS TO SELECTED EXERCISES

11. $53,429.04   **13.** $4307.69   **15.** $18,764.11
17. $836.28   **19.** $S = \$7935.35$; $I = \$3135.35$ **21.** $652.04
23. (A) $228.21   (B) $10,954.08   (C) $3954.08   **25.** $286.51
27. $198.67   **29.** $5648.11   **31.** $69,029.49   **33.** $2509.82
35. $S = \$3537.07$; $A = \$2380.35$ **37.** $S = \$60,852.45$; $A = \$38,179.58$
39. $3094.59

CHAPTER 5

Section 5–1 Exercises

1. (A) $x = -1, y = 1$ (B) $x = -2, y = 3$

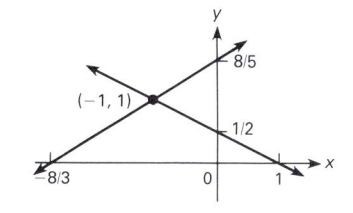

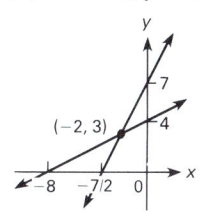

(C) $x = 97/28, y = -19/28$ (D) $x = 6, y = 5$

3. When expressed in slope-intercept form, both equations become $y = (5/7)x - 10$. Thus, the corresponding straight lines coincide and there are infinitely many solutions to the linear system (i.e., each ordered pair (x, y) corresponding to a point on the line is a solution to the linear system).

5. (A) $x = -3, y = -4$ (B) $x = -3, y = -2$ (C) $x = 1/2, y = 6$ (D) $x = 1/2, y = 2/3$
(E) $x = 5, y = 2$ (F) $x = 4, y = 3$ (G) $x = 6, y = 4$ (H) $x = 4, y = 7$

7. 2 pounds of meat and 1 pound of spinach.   **9.** 4000 wagons and 1000 cars.

11. $x = 400$ units; $y = \$10,000$

Section 5–2 Exercises

1. (A) 2×3 (B) 3×2 (C) 3×3 (D) 2×2 (E) 1×3 (F) 2×1 (G) 4×1
(H) 1×4 **3.** (B) and (C)   **5.** $x = 4, y = -1$

7. $x = 1, y = -4, z = 5, w = -7$ **9.** (A) $[5, -4, -2]$ (B) $[1, -4, 4]$
(C) $[-1, 4, -4]$ (D) $[9, -12, 3]$ (E) $[-4, 0, 6]$ (F) $[-1, -4, 7]$
(G) $[11, -12, 0]$ (H) $[-7, 12, -6]$ (I) $[7, -4, -5]$

11. $X = \begin{bmatrix} 2 \\ 5 \end{bmatrix}$

15. (A) Monday + Wednesday = $\begin{bmatrix} \$350 & \$580 & \$400 \\ \$350 & \$700 & \$360 \\ \$570 & \$330 & \$170 \\ \$270 & \$565 & \$410 \end{bmatrix}$

ANSWERS TO SELECTED EXERCISES

(B) Wednesday + Friday = $\begin{bmatrix} \$359 & \$260 & \$220 \\ \$560 & \$440 & \$380 \\ \$150 & \$125 & \$145 \\ \$215 & \$305 & \$330 \end{bmatrix}$

(C) Monday + Wednesday + Friday = $\begin{bmatrix} \$559 & \$760 & \$520 \\ \$660 & \$840 & \$590 \\ \$650 & \$405 & \$225 \\ \$365 & \$655 & \$580 \end{bmatrix}$

Section 5–3 Exercises

1. 9 **3.** 11 **5.** (A) $\begin{bmatrix} -14 & 16 \\ -11 & 18 \end{bmatrix}$ (B) $\begin{bmatrix} 14 & 2 \\ -32 & -10 \end{bmatrix}$ (C) no

9. (A) 2×4 (B) 3×5 (C) 2×3 (D) 4×4 (E) 4×4
11. (A) possible (B) not possible (C) possible (D) not possible
13. (A) $A = \begin{bmatrix} 2 & 3 \\ 1 & 1.5 \end{bmatrix}$ (B) $B = \begin{bmatrix} 100 \\ 200 \end{bmatrix}$ (C) $AB = \begin{bmatrix} 800 \\ 400 \end{bmatrix}$

15. (A) $\begin{bmatrix} 2 & 3 \\ -4 & 5 \end{bmatrix} \begin{bmatrix} x \\ y \end{bmatrix} = \begin{bmatrix} 7 \\ 9 \end{bmatrix}$ (B) $\begin{bmatrix} 1 & 5 \\ 4 & 8 \end{bmatrix} \begin{bmatrix} x_1 \\ x_2 \end{bmatrix} = \begin{bmatrix} 6 \\ 11 \end{bmatrix}$

(C) $\begin{bmatrix} 3 & -7 & -5 \\ 1 & 4 & -2 \\ 5 & 9 & 8 \end{bmatrix} \begin{bmatrix} x_1 \\ x_2 \\ x_3 \end{bmatrix} = \begin{bmatrix} 11 \\ 4 \\ 16 \end{bmatrix}$ (D) $\begin{bmatrix} 2 & 3 & 1 \\ 1 & 0 & 2 \\ 0 & 4 & 5 \end{bmatrix} \begin{bmatrix} x \\ y \\ z \end{bmatrix} = \begin{bmatrix} 11 \\ 9 \\ 17 \end{bmatrix}$

19. (A) $\begin{bmatrix} -16 & 75 \\ -25 & 34 \end{bmatrix}$ (B) $\begin{bmatrix} -512 & 0 \\ 52 & 8 \end{bmatrix}$ (C) $\begin{bmatrix} 17 & 22 & 17 \\ 44 & -9 & 8 \\ 34 & 8 & 18 \end{bmatrix}$

(D) $\begin{bmatrix} -117 & 86 & 51 \\ 113 & -94 & -23 \\ 136 & -112 & -56 \end{bmatrix}$ **21.** $A^n = \underbrace{A \cdot A \cdots A}_{nAs}$

Section 5–4 Exercises

1. (A) $x = -3, y = -4$ (B) $x_1 = 1/2, x_2 = 2/3$
(C) $x = 2, y = -4, z = 1$ (D) $x_1 = 2,$
$x_2 = 1/3, x_3 = -6$ (E) $x = 1/2, y = 1,$
$z = 3$ (F) $x_1 = 2, x_2 = -4, x_3 = 1/5$
3. $x_1 = 368/199, x_2 = 528/199, x_3 = -192/199,$
$x_4 = 38/199$

5. (A) no solution (B) no solution
(C) $x_1 = (1/4)x_2 + 5/4, x_2 = $ arbitrary
parameter (D) no solution (E) no
solution (F) $x_1 = -x_3 + 22, x_2 = -4,$
$x_3 = $ arbitrary parameter (G) no solution
(H) $x_1 = (2/5)x_3 + 4, x_2 = (-4/5)x_3,$
$x_3 = $ arbitrary parameter

7. (A) $x_1 = 3, x_2 = -2$ (B) no solution
(C) $x_1 = -3x_2 + 4, x_2 = $ arbitrary parameter
(D) no solution (E) $x_1 = 1/2, x_2 = 1,$
$x_3 = 3$ (F) $x_1 = -5x_3 + 4, x_2 = 3x_3, x_3 = $
arbitrary parameter

9. (A) $x = 31/2, y = -7/2$ (B) $x_1 = -1,$
$x_2 = 2$ (C) $x = 5, y = -1, z = -3$
(D) $x_1 = 2, x_2 = 1/3, x_3 = -6$
(E) $x_1 = 2, x_2 = -4, x_3 = 1/5$
(F) $x_1 = -2x_3 + 16/3, x_2 = x_3 + 23/3,$
$x_3 = $ arbitrary parameter

Section 5–5 Exercises

1. (A) yes (B) no (C) yes (D) no

3. (A) $A^{-1} = \begin{bmatrix} 11/62 & -5/62 \\ 4/31 & 1/31 \end{bmatrix}$

5. (A) $\begin{bmatrix} 7/32 & 1/32 \\ 3/32 & 5/32 \end{bmatrix}$ (B) $\begin{bmatrix} 0.4 & 0.3 \\ 0.2 & -0.1 \end{bmatrix}$ (C) $\begin{bmatrix} -1 & 1 \\ 1 & 0 \end{bmatrix}$ (D) $\begin{bmatrix} 2/9 & -1/9 & 2/9 \\ 1/9 & 22/9 & -17/9 \\ 2/9 & 8/9 & -7/9 \end{bmatrix}$

(E) $\begin{bmatrix} -1/7 & 2/7 & 1/7 \\ 4/7 & -1/7 & -4/7 \\ 0 & 0 & 1 \end{bmatrix}$ (F) $\begin{bmatrix} -2 & 3.5 & 1 \\ 1 & -0.5 & 0 \\ 1 & -2 & -0.5 \end{bmatrix}$

7. K^{-1} does not exist.

9. (A) $\begin{bmatrix} -1/3 & 0 & 0 & 1/6 \\ 1/3 & 0 & 1/3 & -1/6 \\ 1/3 & 0 & -1/3 & 1/3 \\ 0 & 1/2 & 0 & 0 \end{bmatrix}$

(B) Inverse does not exist.

Section 5–6 Exercises

1. (A) $\begin{bmatrix} x \\ y \end{bmatrix} = \begin{bmatrix} -3 \\ -4 \end{bmatrix}$ (B) $\begin{bmatrix} x_1 \\ x_2 \end{bmatrix} = \begin{bmatrix} -3 \\ -2 \end{bmatrix}$

(C) $\begin{bmatrix} x \\ y \\ z \end{bmatrix} = \begin{bmatrix} 2 \\ -4 \\ 1 \end{bmatrix}$ (D) $\begin{bmatrix} x_1 \\ x_2 \\ x_3 \end{bmatrix} = \begin{bmatrix} 23/11 \\ -146/11 \\ -9 \end{bmatrix}$

(E) $\begin{bmatrix} x_1 \\ x_2 \\ x_3 \end{bmatrix} = \begin{bmatrix} -2 \\ 1 \\ 4 \end{bmatrix}$ (F) $\begin{bmatrix} x \\ y \\ z \end{bmatrix} = \begin{bmatrix} 5 \\ -1 \\ -3 \end{bmatrix}$

3. $\begin{bmatrix} x_1 \\ x_2 \\ x_3 \\ x_4 \end{bmatrix} = \begin{bmatrix} 368/199 \\ 528/199 \\ -192/199 \\ 38/199 \end{bmatrix}$

Section 5–7 Exercises

1. (A) $A = \begin{bmatrix} 500 & 200 \\ 100 & 800 \end{bmatrix}$ (B) $B = \begin{bmatrix} 1200 \\ 100 \end{bmatrix}$ (C) $AX = B$
$\begin{bmatrix} 500 & 200 \\ 100 & 800 \end{bmatrix} \begin{bmatrix} x_1 \\ x_2 \end{bmatrix} = \begin{bmatrix} 1200 \\ 100 \end{bmatrix}$

(D) $\begin{bmatrix} x_1 \\ x_2 \end{bmatrix} = \begin{bmatrix} 2.47 \\ -0.18 \end{bmatrix}$

3. (A) $A = \begin{bmatrix} 0 & 1/2 \\ 1/4 & 0 \end{bmatrix}$ (B) 380 barrels of oil, 680 tons of coal.

5. $y = 2x^2 - 3x + 5$

7. $y = -2x^2 + 20x - 40$

CHAPTER 6

Section 6–1 Exercises

1. 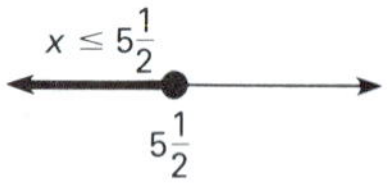 $x \leq 5\frac{1}{2}$

3. 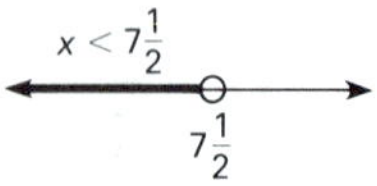 $x < 7\frac{1}{2}$

5. $x \leq 10\frac{1}{3}$

7. $x \leq 3$

9. $x \geq 11$

11. $x \geq 124$

13. $x < 7\frac{1}{2}$

Section 6–2 Exercises

1. (F) 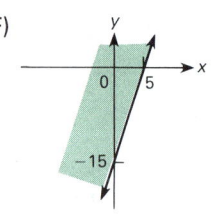 **2.** (A) 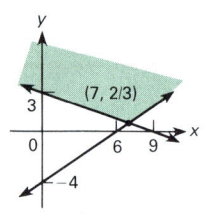 (D)

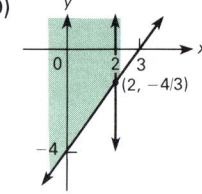

(F) 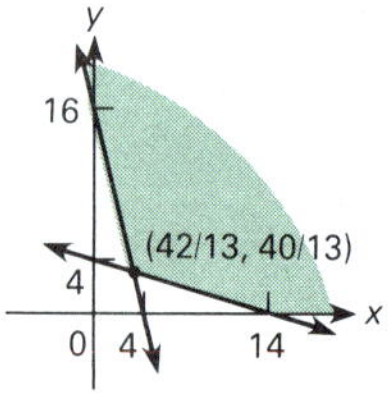 (I) 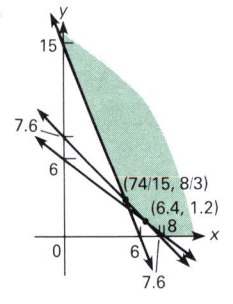 (J)

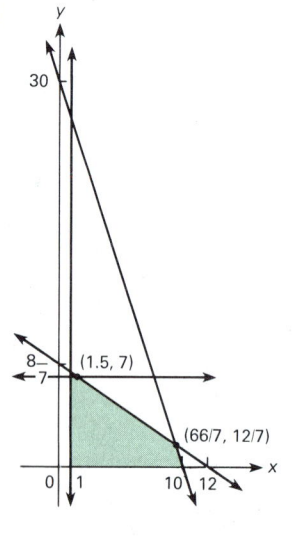

Section 6–3 Exercises

1. (A) 30 motorcycles, 10 mopeds; maximum profit is $4200. (B) 0 motorcycles, 30 mopeds; maximum profit is $2700.

3. 0 T140s, 10 T240s; maximum profit is $300.

5. 1960 gallons of F10, 1840 gallons of F20; minimum cost is $4964.

7. 170 units of product X and 80 units of product Y or 185 units of product X and 50 units of product Y; maximum profit is $840.

Section 6–4 Exercises

1. (A) Maximize $z = 120x_1 + 60x_2 + 0x_3 + 0x_4$
subject to $2x_1 + 3x_2 + x_3 = 90$
$5x_1 + x_2 + x_4 = 160$
$x_1 \geq 0, x_2 \geq 0, x_3 \geq 0, x_4 \geq 0$

Basic C_j	Basis	x_1 120	x_2 60	x_3 0	x_4 0	
0	x_3	2	3	1	0	90
0	x_4	5	1	0	1	160

(C) basic variables: $x_3 = 90, x_4 = 160$
nonbasic variables: $x_1 = 0, x_2 = 0$

3. (A) Maximize $z = 400x_1 + 500x_2 + 0x_3 + 0x_4$
subject to $x_1 + x_2 + x_3 = 100$
$60x_1 + 80x_2 + x_4 = 6600$
where $x_1 \geq 0, x_2 \geq 0, x_3 \geq 0, x_4 \geq 0$

Basic C_j	Basis	x_1 400	x_2 500	x_3 0	x_4 0	
0	x_3	1	1	1	0	100
0	x_4	60	80	0	1	6600

(C) basic variables: $x_3 = 100, x_4 = 6600$
nonbasic variables: $x_1 = 0, x_2 = 0$

5. (A) Maximize $z = 20x_1 + 42x_2 + 56x_3 + 0x_4 + 0x_5 + 0x_6$
subject to $2x_1 + 3x_2 + x_3 + x_4 = 6$
$4x_1 + 2x_2 + 3x_3 + x_5 = 12$
$4x_1 + 2x_2 + x_3 + x_6 = 8$
where $x_1 \geq 0, x_2 \geq 0, x_3 \geq 0, x_4 \geq 0, x_5 \geq 0, x_6 \geq 0$

(B)

Basic C_j	Basis	x_1 20	x_2 42	x_3 56	x_4 0	x_5 0	x_6 0	
0	x_4	2	3	1	1	0	0	6
0	x_5	4	2	3	0	1	0	12
0	x_6	4	2	1	0	0	1	8

(C) basic variables: $x_4 = 6, x_5 = 12, x_6 = 8$
nonbasic variables: $x_1 = 0, x_2 = 0, x_3 = 0$

Section 6–5 Exercises

1. Max $z = 4200$ at $x_1 = 30, x_2 = 10, x_3 = 0, x_4 = 0$.

3. Max $z = 43,000$ at $x_1 = 70, x_2 = 30, x_3 = 0, x_4 = 0$.

5. Max $z = 228$ at $x_2 = 6/7, x_3 = 24/7, x_6 = 20/7, x_1 = 0, x_4 = 0, x_5 = 0$.

A-48 ANSWERS TO SELECTED EXERCISES

7. Max profit = $1931.25 at 0 widgets, 562.5 gadgets, 1093.75 trinkets.
9. Max $z = 255$ at $x_6 = 8$, $x_1 = 25$, $x_5 = 15$, $x_2 = 30$, $x_3 = 0$, $x_4 = 0$.

Section 6–6 Exercises

1. (A) Shadow price of 2 is associated with the constraint $x_1 + 3x_2 + 2x_3 \leq 60$. (B) For each unit increase in the value of the b coefficient, $b_1 = 60$, of the constraint $x_1 + 3x_2 + 2x_3 \leq 60$, the value of the objective function increases by 2 units. For each unit decrease in the value of the b-coefficient, $b_1 = 60$, the value of the objective function decreases by 2 units.

3. (A) Shadow price of 5 is associated with the constraint $3x_1 + 6x_2 \leq 60$. (B) For each unit increase in the value of the b-coefficient, $b_1 = 60$, the value of the objective function increases by 5 units. For each unit decrease in the value of the b-coefficient, $b_1 = 60$, the value of the objective function decreases by 5 units.

5. (A) Shadow price of 13/7 is associated with the constraint $2x_1 + 3x_2 + 5x_3 \leq 20$; shadow price of 3/14 is associated with the constraint $6x_1 + 2x_2 + 3x_3 \leq 50$. (B) For each unit increase in the value of the b-coefficient, $b_1 = 20$, of the constraint $2x_1 + 3x_2 + 5x_3 \leq 20$, the value of the objective function increases by 13/7 units. For each unit decrease in the value of the b-coefficient, $b_1 = 20$, the value of the objective function decreases by 13/7 units. For each unit increase in the value of the b-coefficient, $b_2 = 50$, of the constraint $6x_1 + 2x_2 + 3x_3 \leq 50$, the value of the objective function increases by 3/14 unit. For each unit decrease in the value of the b-coefficient, $b_2 = 50$, the value of the objective function decreases by 3/14 unit.

7. (A) Shadow price of 80 is associated with the constraint $x_1 + x_2 + x_3 + x_4 \leq 40$.
(B) For each unit increase in the value of the b-coefficient, $b_1 = 40$, of the constraint $x_1 + x_2 + x_3 + x_4 \leq 40$, the value of the objective function increases by 80 units. For each unit decrease in the value of the b-coefficient, $b_1 = 40$, the value of the objective function decreases by 80 units.

Section 6–7 Exercises

1. (A) Min $z = 7.5$ at $x_1 = 2$, $x_2 = 1$, $x_3 = 0$, $x_4 = 0$, $x_5 = 0$, $x_6 = 0$. (B) Shadow price of 45/7600 is associated with the constraint $500x_1 + 200x_2 \geq 1200$; shadow price of 3/7600 is associated with the constraint $100x_1 + 800x_2 \geq 1000$. (C) For each unit increase in the value of the b-coefficient, $b_1 = 1200$, of the constraint $500x_1 + 200x_2 \geq 1200$, the value of the objective function increases by 45/7600 unit. For each unit decrease in the value of the b-coefficient, $b_1 = 1200$, the value of the objective function decreases by 45/7600. For each unit increase in the value of the b-coefficient, $b_2 = 1000$, of the constraint $100x_1 + 800x_2 \geq 1000$, the value of the objective function increases by 3/7600 unit. For each unit decrease in the value of the b-coefficient, $b_2 = 1000$, the value of the objective function decreases by 3/7600 unit.

3. (A) Min $z = 20$ at $x_2 = 10$, $x_5 = 6$, $x_1 = 0$, $x_3 = 0$, $x_4 = 0$, $x_6 = 0$, $x_7 = 0$. (B) Shadow price of 1 is associated with the constraint $x_1 + 2x_2 + x_3 \geq 20$.
(C) For each unit increase in the value of the b-coefficient, $b_1 = 20$, of the constraint $x_1 + 2x_2 + x_3 \geq 20$, the value of the objective function increases by 1 unit. For each unit decrease in the value of the b-coefficient, $b_1 = 20$, the value of the objective function decreases by 1 unit.

Section 6–8 Exercises

1. Let x_1 = amount invested in treasury bills, x_2 = amount invested in municipal bonds, x_3 = amount invested in real estate, x_4 = amount invested in a mutual fund, x_5 = amount invested in energy stocks, z = yield, then maximize $z = 0.083x_1 + 0.098x_2 + 0.159x_3 + 0.163x_4 + 0.184x_5$ subject to $-5.7x_1 - 4.7x_2 - 2.7x_3 - 1.7x_4 - 0.3x_5 \leq 0$, $x_1 + x_2 + x_3 + x_4 + x_5 \leq 800{,}000$,

$x_5 \leq 200{,}000$, and x_1, x_2, x_3, x_4, x_5 are nonnegative.

3. Let x_1 = number of gallons of Raw Material 1 used for Type A fertilizer, x_2 = number of gallons of Raw Material 1 used for Type B fertilizer, x_3 = number of gallons of Raw Material 2 used for Type A fertilizer, x_4 = number of gallons of Raw Material 2 used for Type B fertilizer, z = total cost, then minimize $z = 0.80x_1 + 0.80x_2 + 0.55x_3 + 0.55x_4$ subject to $0.2x_1 - 0.1x_3 \geq 0$, $0.02x_2 + 0.22x_4 \geq 0$, $x_1 + x_3 \geq 100{,}000$, $x_2 + x_4 \geq 70{,}000$ and x_1, x_2, x_3, x_4 are nonnegative.

5. $0 invented in treasury bills, $0 invested in municipal bonds, $0 invested in real estate, $600,000 invested in a mutual fund, $200,000 invested in energy stocks for a maximum yield of $134,600.

7. 33,333 gallons of Raw Material 1 used for Type A fertilizer, 0 gallons of Raw Material 1 used for Type B fertilizer, 66,667 gallons of Raw Material 2 used for Type A fertilizer, and 70,000 gallons of Raw Material 2 used for Type B fertilizer for a minimum cost of $101,833.25.

CHAPTER 7

Section 7–1 Exercises

1. (A) true (B) true (C) true (D) true
   (E) true (F) false (G) false (H) false

3. (A) yes (B) yes (C) yes

5. (A) $\{3, 5, 11, 12, a\}$ (B) $\{3, 11, 12, a\}$ (C) 5
   (D) $\{3, 5, 12, a\}$ (E) 4 (F) $\{3, 5, 11, 12, a\}$
   (G) $\{3, 12\}$ (H) 2 (I) $\{12\}$
   (J) $\{12\}$ (K) $\{3, 4, 5, 10, 11, a, b, c\}$ (L) $\{4, 10, b, c\}$
   (M) $\{4, 10, 11, a, b, c\}$ (N) $\{a\}$ (O) $\{3, 12\}$
   (P) $\varnothing$ (Q) $\{3, 4, 5, 10, 11, b, c\}$ (R) $\{3, 4, 5, 10, 11, 12, b, c\}$
   (S) $\{3, 5\}$ (T) $\{3, 12, a\}$ (U) $\{4, 10, b, c\}$

7. (A), (B), 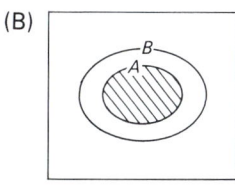 (C), 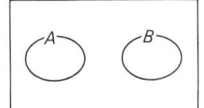 (D)

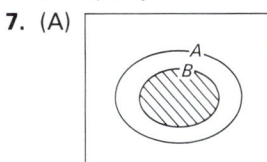

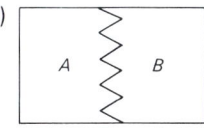

9. (A), 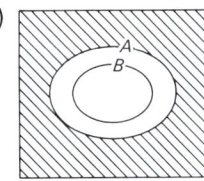 (B), 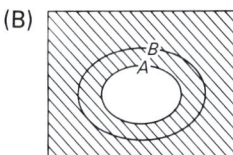 (C), 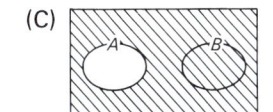 (D)

11. (G), 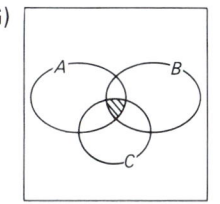 (H), 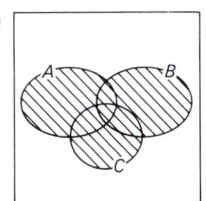 (N), 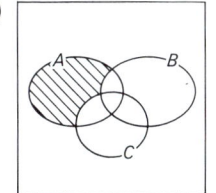 (P)

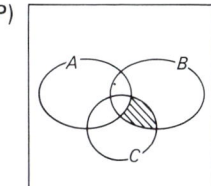

A-50 ANSWERS TO SELECTED EXERCISES

15. (A) 11 (B) 5 (C) 7 (D) 12
(E) 8 (F) 30 (G) 17

17. The number of elements in the set $A \cup B$ is the number of elements in A plus the number of elements in B, less the number of elements which appear in both sets A and B.

Section 7–2 Exercises

1. (A) {(M, O), (M, N), (M, E), (M, Y), (O, M), (O, N), (O, E), (O, Y), (N, M), (N, O), (N, E), (N, Y), (E, M), (E, O), (E, N), (E, Y), (Y, M), (Y, O), (Y, N), (Y, E)}
(B) 1/5 (C) 1/5 (D) 0

3. (A) {(P, N), (P, D), (P, Q), (N, P), (N, D), (N, Q), (D, P), (D, N), (D, Q), (Q, P), (Q, N), (Q, D)}
(B) 1/6 (C) 0

5. (A) {HHHH, HHHT, HHTH, HHTT, HTHH, HTHT, HTTH, HTTT, THHH, THHT, THTH, THTT, TTHH, TTHT, TTTH, TTTT}
(B) 1/4 (C) 15/16 (D) 5/16 (E) 1/8

7. 70%

9. (A) 1/13 (B) 1/13 (C) 12/13
(D) 1/2 (E) 1/13 (F) 12/13
(G) 1/26 (H) 1/52 (I) 25/26
(J) 2/13

Section 7–3 Exercises

1. (A) 1/2 (B) 2/3 (C) 1/3 (D) 5/6
3. (A) 1/5 (B) 1/10 (C) 3/10 (D) 7/10
5. (A) 12/25 (B) 1/5 (C) 3/25 (D) 14/25 (E) 17/25
7. .85
9. $n(A \cup B \cup C) = n(A) - n(A \cap B) + n(B) - n(B \cap C) + n(C) - n(A \cap C) + n(A \cap B \cap C)$
11. (A) .25 (B) .25 (C) .5

Section 7–4 Exercises

1. (A) 1/6 (B) 2/5 (C) 1/10 (D) 3/4
3. (A) .30 (B) .60 (C) .70
(D) 1/3 (E) 2/3 (F) .20
(G) .70 (H) .40 (I) .80
5. (A) 1/2 (B) 2/3
7. (A) 2/5 (B) 13/33 (C) 20/33
9. (A) 1/10 (B) 19/199 (C) 1/11 (D) 10/99
11. $P(A \mid B) = 0, P(B \mid A) = 0$

Section 7–5 Exercises

1. (A) 39/995 (B) 49/796 (C) 22/199 (D) 22/199 (E) 44/199
3. (A) .10 (B) 1/3
5. .56
7. (A) .81 (B) .99 (C) .01
9. (A) .9 (B) 4/7 (C) 2/3 (D) No. $P(A \mid B) \neq P(A)$

ANSWERS TO SELECTED EXERCISES

Section 7–6 Exercises

1. (A) 53/80 (B) 23/80 (C) 1/20 (D) 25/53 (E) 15/23
3. (A) 8/19 (B) 35/190 (C) 75/190
5. (A) .3635 (B) 240/727
7. 35/44

Section 7–7 Exercises

1. (A) 24 (B) 120 (C) 1
   (D) 1 (E) 720
3. (A) 35 (B) 1 (C) 6
   (D) 10 (E) 10 (F) 6
5. 120   7. 60   9. 10
11. 10   13. 105
15. (A) 4446/20285 ≈ .213 (B) 11/4165 ≈ .003 (C) 5359/20825 ≈ .257
17. (A) 134,596 (B) 27,132 (C) 51/253 ≈ .202
    (D) 765/1771 ≈ .432 (E) 510/1771 ≈ .288 (F) 1632/1771 ≈ .922
    (G) 139/1771 ≈ .078
19. (A) $(365)^6$ (B) 363; 362; 361; 360; (365)(364)(363)(362)(361)(360)
    (C) $P(E') = .96$, $P(E) = .04$, .04

Section 7–8 Exercises

1. (B) .90 is the probability that a player wins less than $5.
   (C) .96 is the probability that a player wins at most $5.
   (D) $E(x) = 2.1$; a player's long-run earnings would be approximately $2.10 per play.
   (E) $0.40
3. (A)

x	P(x)
+6,000,000	.70
−2,000,000	.30

   (B) $3,600,000
5. (A)

x	P(x)
200	.10
250	.20
300	.40
350	.20
400	.10

   (B) .20 (C) .30 (D) 350 (E) 300

Section 7–9 Exercises

1. (A) .2646 (B) .4116 (C) .2401 (D) .9163
   (E) .6517
3. (A) .0440 (B) .0010 (C) .0547 (D) .0107
   (E) .0097 (F) .0010 (G) .0107 (H) .0010
5. (A) .2668 (B) .1556 (C) .0404 (D) .4628
7. (A) .4096 (B) .1536 (C) .1808 (D) .0272

ANSWERS TO SELECTED EXERCISES

9. (A)

x	P(x)
0	.01024
1	.07680
2	.23040
3	.34560
4	.25920
5	.07776

(C) 3

11. (A)

x	P(x)
0	.0001
1	.0016
2	.0106
3	.0425
4	.1114
5	.2007
6	.2508
7	.2150
8	.1209
9	.0404
10	.0060

(B) P(x) graph with values: .0001, .0016, .0106, .0425, .1114, .2007, .2508, .2150, .1209, .0404, .0060 at x = 0, 1, 2, 3, 4, 5, 6, 7, 8, 9, 10

(C) 6

(D) $P(x \leq 2) = .0123$. Therefore, the probability of getting as few as 2 successes or less, is about 1.2%. We might question the claim that as many as 60% of the viewers buy the product.

13. (A)

x	P(x)
0	.3874
1	.3874
2	.1722
3	.0447
4	.0074
5	.0008
6	.0001
7	.0000
8	.0000
9	.0000

(C) .9

Section 7–10 Exercises

1. (A) 42% Democrats, 39% Republicans, 19% Independents.
   (B) 34.8% Democrats, 45.3% Republicans, 19.9% Independents.
   (C) 31.38% Democrats, 49.17% Republicans, 19.45% Independents.

3. (A)

From \ To	0	1
0	.6	.4
1	.4	.6

(B) 50% of digits are 0s and 50% are 1s.

A-53 ANSWERS TO SELECTED EXERCISES

Section 7–11 Exercises

1. (A) and (B)
3. (A) [.75 .25]   (B) [1/3 2/3]
5. [3/11 6/11 2/11]
In the long run, there will be approximately 27.27% Democrats, 54.55% Republicans, 18.18% Independents.
7. [.5 .5]
In the long run, 50% of the digits will be 0s and 50% will be 1s.

CHAPTER 8

Section 8–1 Exercises

1. (A) 4,   (B) 1   (C) 4

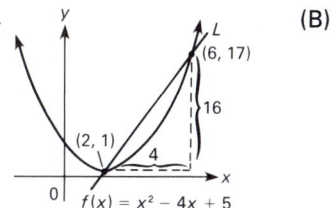

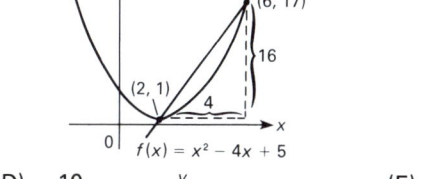

$f(x) = x^2 - 4x + 5$; points (2, 1), (6, 17), 16, 4

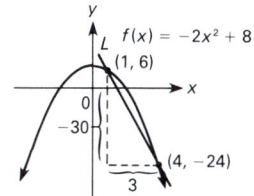

(D) -10,   (E) -23 (F) -12

$f(x) = -2x^2 + 8$; points (1, 6), (4, -24), -30, 3

3. (A) 0, $y = 1$
(B) -1, $y = -x - 1$,

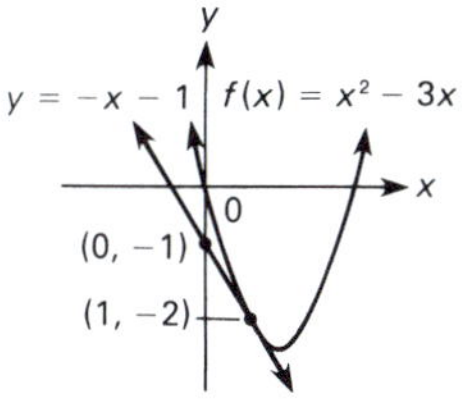

$y = -x - 1$ | $f(x) = x^2 - 3x$; points (0, -1), (1, -2)

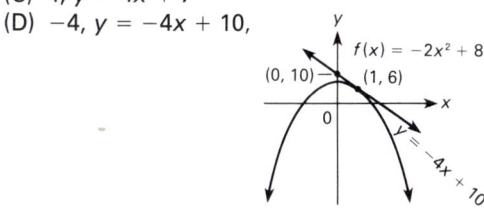

(C) 4, $y = 4x + 7$ (E) -13, $y = -13x - 2$
(D) -4, $y = -4x + 10$,   (F) -108, $y = -108x + 81$

$f(x) = -2x^2 + 8$; points (0, 10), (1, 6); $y = -4x + 10$

5. $6x - 2$ **7.** $3x^2 - 16$
9. (A) $2x - 3$ (B) -1 (C) 1

Section 8-2 Exercises

1. (A) Limit exists, $\lim_{x \to 5} f(x) = 6$. (B) Limit does not exist.
   (C) Limit exists, $\lim_{x \to 5} f(x) = 8$. (D) Limit does not exist.

3. (A) Limit exists, $\lim_{x \to \infty} f(x) = 0$. (B) Limit exists, $\lim_{x \to \infty} f(x) = 3$.
   (C) Limit does not exist. (D) Limit exists, $\lim_{x \to \infty} f(x) = 2$.

5. (A) 15 (B) 2 (C) does not exist (D) 4
   (E) −5 (F) 10 (G) 5 (H) does not exist
   (I) 0 (J) 5/2 (K) 0 (L) does not exist

7. (A) $4x^3$ (B) $5x^4$ (C) $6x - 2$ (D) 2 (E) $15x^2 - 2$
   (F) $-4x + 3$

Section 8-3 Exercises

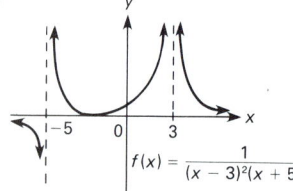

1. (A) and (D) 3. (A) and (B)
5. (A) (B) 3, −5 (C) 3, −5

$f(x) = \dfrac{1}{(x-3)^2(x+5)}$

Section 8-4 Exercises

1. (A) $3x^2$ (B) $10x^9$ (C) $20x^{19}$ (D) $\dfrac{1}{5}x^{-4/5}$
   (E) 4 (F) $-3x^{-4}$ (G) $-b/x^{b+1}$ (H) $-1/x^2$
   (I) $5x^4$ (J) $21(3x)^6$ (K) $-3/x^4$ (L) $-\dfrac{5}{2} \cdot \dfrac{1}{\sqrt{x^7}}$
   (M) $-\dfrac{1}{4} \cdot x^{-5/4}$ (N) 0 (O) 0

3. (A) $15x^2 - 16x + 16$ (B) $-4x + 4$ (C) $35x^6 - 24x^2 + 14x + 3/2$
   (D) $6x^5 - 5x^4 - \dfrac{8}{x^3} - \dfrac{6}{5}$ (E) $4x^3 - 12x^2 + 5$ (F) $x^2(5x^2 - 12x + 21)$
   (G) $2x + 3$ (H) $x(4x^2 - 3x - 24)$
   (I) $3x^2 - 12x + 5$ (J) $-24x^5 + 32x^3 - 6x + 2$
   (K) $10x^4 - 24x^2 + 14x - 3$

5. (A) $\dfrac{-x^4 + 8x^3 - 18x^2 + 18x - 21}{(x^3 - 3x + 9)^2}$ (B) $\dfrac{2x^3 + 15x^2}{(x+5)^2}$
   (C) $\dfrac{-x^3 - 28}{x^5}$ (D) $\dfrac{-2x^7 + 37x^6 - 160x^5 + 150x^4 - 244x^3 - 40x^2 + 36}{(x^5 - 7x^4 + 4x)^2}$
   (E) $\dfrac{x^4 - 27x^2 - 18x}{(x^2 - 9)^2}$

7. (A) $f'(x) = (-4)(-3x^{-4}) = 12x^{-4}$ (B) $f'(x) = \dfrac{x^3(0) - (-4)(3x^2)}{x^6} = \dfrac{12}{x^4}$

A-55 ANSWERS TO SELECTED EXERCISES

9. 4/25 **11.** 2
13. $y = -8x + 48$
15. (A) 32 feet per second, increasing. (B) 0 feet per second, neither increasing nor decreasing.
(C) 32 feet per second, decreasing. (D) [graph showing $S(x) = -16x^2 + 64x$ with $S'(1)=32$, $S'(2)=0$, $S'(3)=-32$]

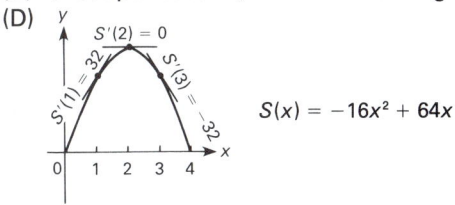

Section 8–5 Exercises

1. $5t^4 - 12t^2 + 7$ **3.** $5u^4$ **5.** $18u^5 - 40u^4 + 4$ **7.** $-21u^2 - 16u$
9. (A) $33x^2(x^3 + 5)^{10}$ (B) $(3x^5 - 2)(x^6 - 4x + 9)^{-1/2}$
(C) $3x^2 - 6$ (D) $3x^2 + 2x + 9$
(E) $60x^3(x^4 - 9)^{14}$ (F) $\dfrac{8 - 3x^2}{2\sqrt{(x^3 - 8x)^3}}$
(G) $\dfrac{16 - 12x^5}{3\sqrt[3]{(x^6 - 8x)^5}}$ (H) $\dfrac{36x - 12x^5}{5} \cdot (x^6 - 9x^2)^{-7/5}$
(I) $20(3x^2 - 8x)(x^3 - 4x^2 + 5)^{19}$ (J) $\dfrac{40x^2(6 - x)}{(x^4 - 8x^3 + 5)^{11}}$

11. $y = -3584x + 6656$
13. (A) $C'(t) = 15(5t + 4)^2$
(B) $C'(6) = 17{,}340$
At a 6-day production run, the rate of change of cost with respect to the number of days of the production run is $17,340 per day.

Section 8–6 Exercises

1. (A) $C'(x) = 10x$ (B) $30
(C) At a production level of 3 units, one more unit costs approximately $30.
3. (A) $R(x) = -2x^2 + 40x$ (B) $R'(x) = -4x + 40$
(C) $20, increasing   (D) $-$8, decreasing
5. (A) $P'(x) = 4 - (3x^2/1{,}000{,}000)$ (B) $3.97, increasing
(C) $-$6.83, decreasing
7. (A) $E = p/(p - 65)$ (B) $-10/3$
(C) A 3% increase in price is associated with a 10% decrease in demand.
9. (A) A 1% increase in price is associated with a 1.5% decrease in demand.
(B) elastic
11. $f(2) = -11 < 0$, $f(3) = 1 > 0$
$x_3 = x_4 = 2.9488284$
13. $x_3 = x_4 = 2.236068$

A-56 ANSWERS TO SELECTED EXERCISES

Section 8–7 Exercises

1. (A) $4e^{4x}$ (B) $-e^{-x}$
   (C) $-8e^{-2x}$ (D) e^{-x}
   (E) $0.2e^{-0.1x}$ (F) e^x
3. (A) $4^{x^2-3x}(2x-3)(1.386294)$ (B) $2^{3x^2-5x}(6x-5)(0.693147)$
   (C) $-4^{-0.02x}(0.027725)$ (D) $3^{-x}(2.197224)$
5. (A) $1/x$ (B) $4/x$ (C) $(2x-4)/(x^2-4x)$ (D) $-4x/(x^2+1)$ (E) $(2x-4)/(x^2-4x+1)$
   (F) $(-6x^5+3)/(-x^6+3x)$
7. (A) $1/(2.302585)x$ (B) $\dfrac{3}{(1.791759)(3x-4)}$
   (C) $\dfrac{3x^2+6}{(2.302585)(x^3+6x)}$ (D) $\dfrac{18x^8}{1.098612(2x^9-1)}$
9. (A) $1000e^{0.10x}$
   (B) 1349.86. At the end of the third year, the total amount is increasing by $1349.86 per year.
   (C) 1491.82. At the end of the fourth year, the total amount is increasing by $1491.82 per year.
   (D) $1648.72, increasing
   (E) $23,082.09
      Under normal conditions, amount would be $24,596.03.
11. $y = -0.367879x + 0.735758$
13. (A) $-1200e^{-0.60t}$ (B) -888.9816 (C) $y'(2) = -361.4328$
15. Decreasing at rate of 2.714511 degrees per minute.

CHAPTER 9

Section 9–1 Exercises

1. (A) $(-1, -16)$: relative minimum (B) $(-6, 325)$: relative maximum;
                                          $(1, -18)$: relative minimum
   (C) $(6 + \sqrt{34}, -1589 - 272\sqrt{34})$: relative minimum; (D) $(1, 9)$: relative minimum
       $(6 - \sqrt{34}, -1589 + 272\sqrt{34})$: relative maximum;
3. $(-5, 101)$: relative maximum; absolute maximum: $(11, 1893)$;
   $(1, -7)$: relative minimum; absolute minimum: $(-10, -249)$

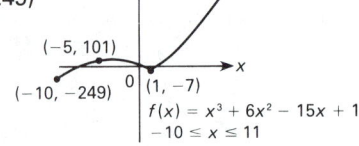

$f(x) = x^3 + 6x^2 - 15x + 1$
$-10 \le x \le 11$

5. $f'(x) = 2ax + b$
   $0 = 2ax + b$
   $x = \dfrac{-b}{2a}$
7. $y' = -e^{-x} < 0$ for all x

Section 9–2 Exercises

1. (A) $f'(x) = 3x^2 - 8x + 7$
 $f''(x) = 6x - 8$
 (C) $f'(x) = 7/(x + 6)^2$
 $f''(x) = -14/(x + 6)^3$
 (E) $f'(x) = (3x^2 + 5x^7)e^{x^5}$
 $f''(x) = (6x + 50x^6 + 25x^{11})e^{x^5}$

 (B) $f'(x) = 4x^3 - 21x^2 + 8$
 $f''(x) = 12x^2 - 42x$
 (D) $f'(x) = (1 + x)e^x$
 $f''(x) = (2 + x)e^x$
 (F) $f'(x) = 2x \ln(x + 1) + \dfrac{x^2}{(x + 1)}$
 $f''(x) = 2 \ln(x + 1) + \dfrac{(3x^2 + 4x)}{(x + 1)^2}$

 (G) $f'(x) = (x + 1)^3 (5x - 7)$
 $f''(x) = (x + 1)^2 (20x - 16)$

3. $f'(x) = 3x^2 - 8x + 7$ $f''(x) = 6x - 8$
 $f'(2) = 3$ $f''(4) = 16$

5. $y'(x) = (x + 3)(3x + 5)$ $y''(x) = 6x + 14$ $y''(2) = 26$

7. (A) (1, 3): relative minimum; concave up everywhere.
 (B) (−6, 224): relative maximum; (2, −32): relative minimum; (−2, 96): inflection point; concave up for $x > -2$; concave down for $x < -2$.
 (C) (−1, 13): relative maximum; (9, −487): relative minimum; (4, −237): inflection point; concave up for $x > 4$; concave down for $x < 4$.
 (D) (−2, 40/3): relative maximum; (2, 8/3): relative minimum; (0, 8): inflection point; concave up for $x > 0$; concave down for $x < 0$.
 (E) (3, −128): relative minimum; (0, 7): inflection point; (2, −73): inflection point; concave up for $x < 0$ and $x > 2$; concave down for $0 < x < 2$.
 (F) (0, 10): relative minimum; (1, 21): relative maximum; (8, −212,982): relative minimum; (6.5, −142,348.7): inflection point; (0.7, 15.9): inflection point; concave up for $x < 0.7$ and $x > 6.5$; concave down for $0.7 < x < 6.5$.

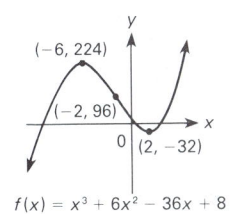

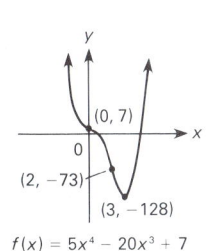

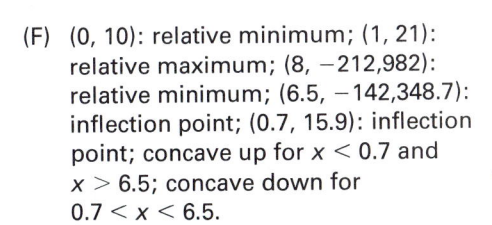

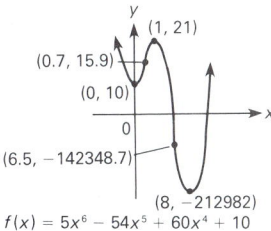

(G) $(-1, 4)$: relative minimum; $(1, 0)$: relative maximum; $(0, 2)$: inflection point; concave up for $x > 0$; concave down for $x < 0$.

(H) $(-3, 56)$: relative maximum; $(3, -52)$: relative minimum; $(0, 2)$: inflection point; concave up for $x > 0$; concave down for $x < 0$.

9. (A) $x < 0$ (B) $x > 0$
(C) $x = 0$; no.   (D)

11. (A) $(5, 0)$: inflection point; concave up for $x > 5$; concave down for $x < 5$.
(B) $(6, 0)$: relative minimum; concave up everywhere.
(C) $(-3, 2)$: relative minimum; concave up everywhere.
(D) $(8, 2)$: inflection point; concave up for $x > 8$; concave down for $x < 8$.
(E) $(-3/5, 1382.8)$: relative maximum; $(5, 6)$: relative minimum; $(4/5, 877.3)$: inflection point; concave up for $x > 4/5$; concave down for $x < 4/5$.
(F) $(-4.25, -252.2)$: relative minimum; $(-2.5, -149.1)$: inflection point; $(1, 1)$: inflection point; concave up for $x < -2.5$ and $x > 1$; concave down for $-2.5 < x < 1$.

13. (A)   (B)

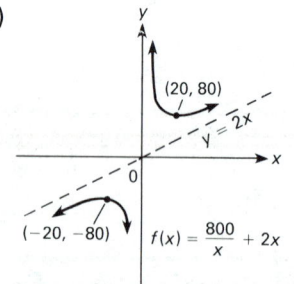

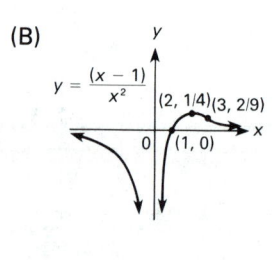

15. $(1/2, 2/e) = (0.5, 0.73)$: relative maximum; $(1, 4/e^2) = (1, 0.54)$: inflection point.

Section 9–3 Exercises

1. 6 million
3. 3 inches by 3 inches
5. $x = 8, y = 4$
7. $x = 500, y = 250$
9. (A) $C(Q) = \dfrac{7{,}350{,}000{,}000}{Q} + \dfrac{3Q}{2}$ (B) 70,000
   (C) $210,000
11. 6 seconds, 576 feet
13. 500 feet by 500 feet
15. 12, 12
17. a/b

Section 9–4 Exercises

1. (A) 2 (B) 82 (C) 8
3. (A) $3x^2 - 8y + 6x^2y^2$ (B) $5y^4 - 8x + 4x^3y$
5. (A) $24x^2 - 35x^4y^8 + 8y^2$ (B) $-4y - 56x^5y^7 + 16xy + 1$
   (C) -8904 (D) -7143
7. (A) $24x^5 - 24x^2 - 7 + 6y + 3x^2y^5$ (B) $6x + 8 + 5x^3y^4$
   (C) $120x^4 - 48x + 6xy^5$ (D) $20x^3y^3$
   (E) $6 + 15x^2y^4$ (F) $6 + 15x^2y^4$
9. (A) $-3x^2 + 12x^2y^6$ (B) $-2y + 24x^3y^5 + 8$
   (C) $-6x + 24xy^6$ (D) $-2 + 120x^3y^4$
   (E) $72x^2y^5$ (F) $72x^2y^5$
   (G) 36 (H) 9718
11. (A) $x^2 e^{x^2+y^2}(3 + 2x^2)$ (B) $2x^2 y e^{x^2+y^2}$
13. (A) $e^{2x+3y+5}(2x^3 + 3x^2 + 2y^2)$ (B) $e^{2x+3y+5}(3x^3 + 3y^2 + 2y)$
    (C) $2e^8 \approx 5962$
15. (A) $940 (B) $700
17. (A) Annual profits will increase by $18,002.
    (B) Annual profits will increase by $8005.

Section 9–5 Exercises

1. (A) relative minimum: (4, 5) (B) relative minimum: (7, 10)
   (C) relative minimum: $(-9/2, 5)$ (D) relative maximum: (10, 20)
   (E) relative maximum: (20, 50)
3. (9, 4) is the only critical point.
   $f_{xx}(9, 4) = 2 = A$, $f_{yy}(9, 4) = -4 = B$, $f_{xy}(9, 4) = 0 = C$
   $AB - C^2 = -8 < 0$
   Therefore, (9, 4) is a saddle point and there are no relative maxima or minima.
5. (A) $x = 200, y = 500$ (B) $1,660,000
7. (A) $R(x, y) = 200x - 20x^2 + 3xy + 300y - 15y^2$

ANSWERS TO SELECTED EXERCISES

(B) $x = 5 + \dfrac{315}{397} \approx \$5.79$

$y = 10 + \dfrac{230}{397} \approx \$10.58$

(C) $2166 + \dfrac{28{,}906}{397^2} \approx \$2166.25$

(D) $D_c = 94 + \dfrac{282}{397} \approx 95$

$D_m = 152 + \dfrac{356}{397} \approx 153$

CHAPTER 10

Section 10–1 Exercises

1. (A) $F'(x) = 7x^6 = f(x)$ (B) $F'(x) = 7x^6 = f(x)$
   (C) $F'(x) = x^2 - 8x + 5 = f(x)$ (D) $F'(x) = \dfrac{1}{\sqrt[5]{x}} = f(x)$

3. (A) $x^3 - 4x^2 + 5x + c$ (B) $x^4 - 16x + c$
   (C) $\dfrac{x^8}{4} - \dfrac{6x^5}{5} - x + c$ (D) $\dfrac{x^6}{6} - \dfrac{7x^2}{2} + c$
   (E) $x^5 - \dfrac{2\sqrt{x^3}}{3} + 14\sqrt{x} + c$ (F) $\dfrac{x^4}{4} + \dfrac{4}{x} + 6x + c$

5. $C(x) = 4x^2 + 2x + 1000$ 7. $C(x) = 3x^2 + x + 886$
9. (A) $S(x) = 0.5x - 0.24x^{1/2} - 38.34$ (B) $\$30{,}780{,}000{,}000$

Section 10–2 Exercises

1. (A) 80   (B) $-9/5$
   (C) 3   (D) 348
   (E) 15/8   (F) 300

3. $S = \dfrac{1}{4} + \dfrac{1}{2n} + \dfrac{1}{4n^2}$, as $n \to \infty$, $S \to \dfrac{1}{4} = F(1) - F(0)$

5. (A) upper: 3.78125   (B) upper: 3.715
        lower: 3.53125   lower: 3.615

7. $3\dfrac{2}{3} \approx 3.66667$

9. 132   11. 256

13. 40

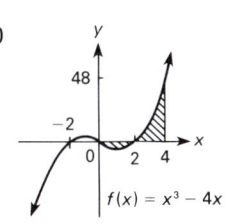

15. $4\dfrac{1}{4}$

17. 40

19. $588\dfrac{8}{15}$

Section 10–3 Exercises

1. $29\dfrac{1}{3}$

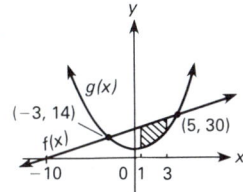

3. 77

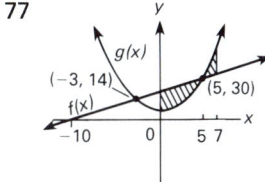

5. 1/4

7. 12

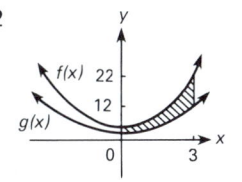

Section 10–4 Exercises

1. $D(x)$ is a decreasing linear function with intercepts (18, 0) and (0, 27). $S(x)$ is a parabola which opens upward with vertex at the origin. To find the equilibrium point, set $S(x) = D(x)$.

$\dfrac{1}{2}x^2 = -\dfrac{3}{2}x + 27$

$\dfrac{1}{2}x^2 + \dfrac{3}{2}x - 27 = 0$

$x^2 + 3x - 54 = 0$

$(x - 6)(x + 9) = 0$

If we require $x \geq 0$, the two functions intersect at $x = 6$. So the equilibrium point is (6, 18).

3. (A)

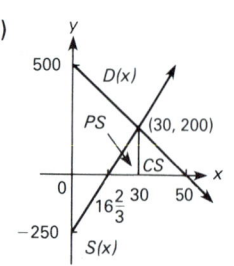

(B) (30, 200) (C) $2000 (D) $1333.33

Graph showing $D(x)$ starting at 500, $S(x)$ starting at -250, intersecting at $(30, 200)$, with PS above and CS below, x-intercepts at $16\frac{2}{3}$ and 50.

Section 10–5 Exercises

1. (A) $\dfrac{(x^3 - 7)^{11}}{11} + c$ (B) $\dfrac{(x^2 - 3)^5}{5} + c$

(C) $\dfrac{(x^3 - 4x)^6}{6} + c$ (D) $\dfrac{(5x + 6)^{13}}{13} + c$

3. (A) $\dfrac{-1}{27(x^3 - 5)^9} + c$ (B) $\dfrac{1}{12}\sqrt{(4x^2 + 5)^3} + c$

(C) $\dfrac{1}{2}\sqrt{x^4 - 6} + c$ (D) $\dfrac{1}{7}(\sqrt[3]{x^3 - 9})^7 + c$

(E) $\dfrac{-1}{9(3x - 5)^3} + c$ (F) $\dfrac{25}{9}\sqrt[5]{(x^3 - 9)^3} + c$

5. 4   **7.** 9012.125

9. Using the substitution principle, $u = x^3 - 5$, but then $du = 3x^2\,dx$

11. 232,800

Section 10–6 Exercises

1. (A) $e^x + c$ (B) $-e^{-x} + c$

(C) $\dfrac{1}{4}e^{4x} + c$ (D) $\dfrac{1}{3}[1 - e^{-3}] \approx 0.317$

(E) $2e^{x/2} + c$ (F) $-20e^{-0.3x} + c$

3. (A) $5 \ln x + c$ (B) $\ln 2 \approx 0.693$

(C) $\dfrac{1}{2}\ln(x^2 + 1) + c$ (D) $\dfrac{1}{4}\ln(x^4 - 1) + c$

(E) $2 \ln(x^3 + 4) + c$ (F) $\dfrac{5}{2}\ln(x^2 - 4) + c$

(G) $-\dfrac{1}{10}\ln(5x^2 - 6) + c$ (H) $-\dfrac{4}{3}\ln(x^3 + 5) + c$

5. $\ln 4 \approx 1.386$ **7.** $10{,}000{,}000\,(\ln 9 - \ln 5) \approx \$5{,}877{,}870$

9. $20(e^{10} - 1) \approx 440509$

Section 10–7 Exercises

1. (A) $\ln(x + \sqrt{81 + x^2}) + c$ (B) $\ln(x + \sqrt{x^2 + 64}) + c$

(C) $\ln 2 \approx 0.693$ (D) $\dfrac{1}{4}\ln\left(\dfrac{2 + x}{2 - x}\right) + c$

(E) $\dfrac{-7}{18}\ln\left(\dfrac{9 + x}{9 - x}\right) + c$ (F) $\dfrac{-1}{5x} + \dfrac{3}{25}\ln\left(\dfrac{5 + 3x}{x}\right) + c$

(G) $\dfrac{-3}{4x} - \dfrac{21}{16}\ln\left(\dfrac{4-7x}{x}\right) + c$ (H) $\dfrac{e^{5x}}{125}(25x^2 - 10x + 2) + c$

(I) $6 - 2e \approx 0.563$ (J) $\ln(2 + \sqrt{29}) - \ln 5 \approx 0.390$

Section 10–8 Exercises

1. (A) $\dfrac{10{,}000{,}000}{(0.12)^2}[1 - 0.4e^{-6}] \approx \$188{,}300{,}000$

   (B) $\dfrac{10{,}000{,}000}{(0.12)^2}[1 - 1.6e^{-0.6}] \approx \$84{,}653{,}000$

3. (A) $\dfrac{9{,}500{,}000}{(0.08)^2}[.84e^{0.16} - 0.6e^{0.4}] \approx \$134{,}564{,}890$

   (B) $\dfrac{9{,}500{,}000}{(0.08)^2}[1.16e^{-0.16} - 1.4e^{-0.4}] \approx \$74{,}276{,}700$

Section 10–9 Exercises

1. 1

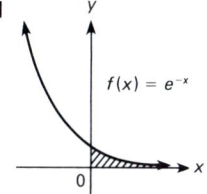

   [graph of $f(x) = e^{-x}$]

3. indeterminate

5. indeterminate

7. $40{,}000/0.09 \approx \$444{,}444.44$

Section 10–10 Exercises

1. (A) 0.4   (B) 0.25   (C) 0.25

3. (A) $f(x) = \begin{cases} 0.2 & \text{for } 2 \le x \le 7 \\ 0 & \text{otherwise} \end{cases}$ (B) 0.2   (C) 0.4

5. (A) $1 - e^{-0.5} \approx 0.3935$ (B) $e^{-1} - e^{-3} \approx 0.3181$ (C) $e^{-2} \approx 0.1353$

7. (A) $e^{-1} - e^{-2} \approx 0.2325$ (B) $1 - e^{-3} \approx 0.9502$ (C) $e^{-6} \approx 0.0025$

9. (A) $R(t) = e^{-0.5t}$

   (B) $R(1) = e^{-0.5} \approx 0.6065$; $R(1)$ is the probability that the life of the circuitry exceeds 1 year.

   (C) $R(4) = e^{-2} \approx 0.1353$, $R(4)$ is the probability that the life of the circuitry exceeds 4 years.

ANSWERS TO CASES

Case A

1. EPS = $97.96; price per share = $685.72.
2. EPS = $176.32; price per share = $1234.24.
3. 1985: EPS = $54.42; price per share = $435.36.
   1986: EPS = $97.96; price per share = $783.68.
   1987: EPS = $176.32; price per share = $1410.56.

Case B

1. NPV = $10,734.71. Since NPV > 0, the investment is earning a rate of return greater than 10% compounded annually.   2. $386,145.70

Case C

1.

	Well in Saudi Arabia	Well in Kuwait	Well in Egypt	Demand Requirements
Regular	0.2	0.3	0.4	19
Unleaded	0.1	0.2	0.1	10
Kerosene	0.3	0.1	0.4	20

2. x_1 = number of days the well in Saudi Arabia is operated, x_2 = number of days the well in Kuwait is operated, x_3 = number of days the well in Egypt is operated.

$$0.2x_1 + 0.3x_2 + 0.4x_3 = 19$$
$$0.1x_1 + 0.2x_2 + 0.1x_3 = 10$$
$$0.3x_1 + 0.1x_2 + 0.4x_3 = 20$$

3. $$\begin{bmatrix} 0.2 & 0.3 & 0.4 \\ 0.1 & 0.2 & 0.1 \\ 0.3 & 0.1 & 0.4 \end{bmatrix} \begin{bmatrix} x_1 \\ x_2 \\ x_3 \end{bmatrix} = \begin{bmatrix} 19 \\ 10 \\ 20 \end{bmatrix}$$

4. $X = A^{-1}B$
$$\begin{bmatrix} x_1 \\ x_2 \\ x_3 \end{bmatrix} = \begin{bmatrix} -7.78 & 8.89 & 5.56 \\ 1.11 & 4.44 & -2.22 \\ 5.56 & -7.78 & -1.11 \end{bmatrix} \begin{bmatrix} 19 \\ 10 \\ 20 \end{bmatrix} = \begin{bmatrix} 52.22 \\ 21.11 \\ 5.56 \end{bmatrix}$$

5. $X = A^{-1}B$
$$\begin{bmatrix} x_1 \\ x_2 \\ x_3 \end{bmatrix} = \begin{bmatrix} -7.78 & 8.89 & 5.56 \\ 1.11 & 4.44 & -2.22 \\ 5.56 & -7.78 & -1.11 \end{bmatrix} \begin{bmatrix} 20 \\ 10 \\ 15 \end{bmatrix} = \begin{bmatrix} 16.67 \\ 33.33 \\ 16.67 \end{bmatrix}$$

Case D

1. x_1 = number of chains produced, x_2 = number of desks produced, x_3 = number of tables produced.   2. Maximize $z = 15x_1 + 50x_2 + 80x_3$ subject to $30x_1 + 20x_2 + 10x_3 \leq 240$, $20x_1 + 60x_2 + 20x_3 \leq 320$, $10x_1 + 40x_2 + 90x_3 \leq 260$, $x_1 \geq 0$, $x_2 \geq 0$, $x_3 \geq 0$.   3. Max z = $311.30 at $x_1 = 5.5926$, $x_2 = 3.1852$, $x_3 = 0.85185$.

ANSWERS TO CASES

Case E

1. Binomial distribution with $n = 12$ and $p = 0.40$.

x	P(x)
0	0.002
1	0.017
2	0.064
3	0.142
4	0.213
5	0.227
6	0.177
7	0.101
8	0.042
9	0.012
10	0.002
11	0.000
12	0.000

2. P(x) bar chart over x = 0, 1, 2, ..., 12.

3. 0.064 **4.** 0.646 **5.** 4.8 **6.** Since there is very little likelihood (only 1.9%) of getting as few as one customer or less with reduced usage if the claim, $p = 0.40$, is true, we might question whether the claim, $p = 0.40$ is true.

Case F

1. $C(Q, S) = 5\left(\dfrac{1200}{Q}\right) + 1.15\dfrac{(Q - S)^2}{2Q} + 2.40\left(\dfrac{S^2}{2Q}\right)$

2. $Q = 124$, $S = 40$ **3.** $96.59 **4.** 40 **5.** 124 are ordered, 84 are available for sale.

INDEX

Absolute maximum, 533
Absolute minimum, 533
Absolute value, 5
Absolute value function, 16
Amortization, 203–5
   mortgage, 204–5
   schedule, 204
Annual maintenance cost, 137
Annuity, 186
   annuity due, 187
   complex (*see* Complex annuity)
   future value, 187–92
   ordinary annuity, 187, 190–92
   present value, 193–99
   term, 187
   total amount, 187–92
Antiderivative, 576
Antidifferentiation, 576
Antilogarithm, 145
A posteriori probability, 413
Apple growing, 91
A priori probability, 413
Area, 84–86, 90, 91
   between two curves, 599–603
   under a curve, 586–97
Asset depreciation, 126, 132
Augmented matrix, 270
Average cost function, 116–17
Average rate of change, 460–63
Average speed, 461
Axis of symmetry, 68

Back-order inventory model, 572
   back-order cost, 572
   carrying cost, 572
   stockout cost, 572
   total cost, 572
Bacteria growth, 131
Batch process, 103
Bayes' formula, 409–13
Blending problem, 363
Break-even point, 47–51, 80
Binomial distribution, 439–42
   expected value, 441–42
Binomial experiments, 435

Binomial probability, 435
Binomial theorem, 495
Birthday problem, 429

Calculus, 460
Capital expenditure analysis, 197
Capital formation, 582, 584
Capital investment decision, 236
Celsius temperature, 8
Combinations, 422–23
Complement law, 387
Complex annuity, 219
   future value, 222
   present value, 223
   total amount, 222
Complex annuity due, 227
   present value, 229
   total amount, 229
Compound interest, 169–78
   compound amount, 173
   compounding, 169–72
   conversion period, 169
   interest rate per conversion period, 172
   nominal rate, 172
   present value, 176
Concavity, 541–48
Conditional probability, 397
Consumers' surplus, 604–7
Consumption function, 44
Continuity, 484
Continuous cash flows, 624
   future value, 624
   present value, 624
Continuous compounding, 180–84
   compound amount, 182
   effective annual interest rate, 182
   nominal rate, 182
   present value, 183–84
Continuous probability distributions, 630–38
Continuous random variable, 630
Conversion period, 169
Cost-benefit curve, 112
Cost equation, 47–51

I-1

INDEX

Cost function, 18, 80–82
Cost minimization, 323
Counting number, 2
CRC Standard Mathematical Tables, 621
Critical values, 535
Curvilinear graph, 60

Defectives, 87–88
Deferred annuity, 211, 214–19
Demand function, 18, 83
De Morgan's laws, 379
Dependent event, 405
Dependent variable, 8, 560
Depreciation, 52–54
   economic life, 52
   salvage value, 52
Derivative, 460, 464
Derivative rules:
   chain rule, 503
   constant function rule, 489
   constant multiplier rule, 489
   exponential function, 518–19, 522
   logarithmic function, 523–24
   power rule, 487
   product rule, 492
   quotient rule, 493
   sum (or difference) rule, 490
Diet problem, 323
Difference quotient, 15
Differentiability, 482
Differential, 607
Differential calculus, 460
Discount, 165–67
   discount note, 165
   discount rate, 166
   maturity value, 166
   proceeds, 166
Discrete probability distribution, 630
Discrete random variable, 630
Discriminant, 76
Dot product, 255
Drug testing, 440

Earnings per share, 132, 156
Economic order quantity, 555, 572
Effective annual interest rate, 177
Elastic demand, 511
Elasticity of demand, 509–11
Employee growth, 133
Equalities
   rules of, 304
Equations of value, 209
   comparison point, 209
Equilibrium demand, 52
Equilibrium point, 51–52, 83–84, 246
Equilibrium price, 52
Expected value, 431–33
Exponent, 60
   base, 60

   laws of exponents, 60–62, 64–65
   rational exponent, 63
Exponential distribution, 634–37
Exponential equation, 149
Exponential function, 120–31

Factorial, 418
Fahrenheit temperature, 8
First-degree polynomial, 92
First-derivative test, 537
Fitting a parabola, 298
Fixed cost, 18, 34
Frequency distribution, 432
Fuel residue, 325
Function, 7
   dependent variable, 8
   domain, 10
   functional notation, 8–9
   independent variable, 8
   range, 10–11
   vertical line test, 11–13
Fundamental theorem of calculus, 591

Gamma distribution, 550
Gaussian elimination with back-substitution, 282
Gauss-Jordan method, 269–81
   more equations than variables, 278
   more variables than equations, 277
   $n \times n$ systems, 271
   parameter, 276
General annuity (*see* Complex annuity)
General law of addition (*see* Probability)
General law of multiplication (*see* Probability)
Geometric series, 185–86

Heating system installation, 365
Horizontal asymptote, 107, 113–14
Horizontal line, 27, 35–36, 38
Housing development, 245

Independent event, 405
Independent variable, 8, 560
Inelastic demand, 511
Inequalities
   rules of, 305
Inequality, 3
Inflection point, 542
Improper integral, 627–29
Instantaneous rate of change, 463
Instantaneous speed, 463
Integer, 3
Integral, 577
   definite, 585
   improper, 626–29
   indefinite, 577

I-3 INDEX

Integral calculus, 460
Integrals
  of exponential functions, 615
  of logarithmic functions, 617
Integration:
  constant multiplier rule, 579
  power rule, 577–78
  substitution, 607–11, 615–23
  sum (or difference) rule, 580
Interarrival time, 635, 637
Interest, 160
  future value, 160–61
  interest rate, 160
  maturity value, 163
  note, 163
  present value, 161
  principal, 160
Interval, 4
Inventory control, 555–57
  carrying cost, 555–56
  ordering cost, 555
Investment allocation, 362
Investment contract, 172, 179, 211
Irrational number, 3

Learning curve, 135, 529
Leases, 196
Leontief's input-output model, 296
  consumption matrix, 297
  net production matrix, 297
  technological matrix, 296
Limit, 467–78
  definition, 469
  infinity, 476–78
  theorems, 471–73
Limits of integration, 585
Linear equation, 30–38
  general form, 37–38
  point-slope form, 31–33
  slope-intercept form, 31–35, 38
Linear inequality:
  in one variable, 304–6
    graphing, 304–6
  in two variables, 307
    graphing, 307–17
Linear programming, 317
  constraints, 319
  objective function, 318
  region of feasible solutions, 319
  vertex points, 319–22
Linear systems, 240
  equivalent systems, 240
  inconsistent, 242–43
  intersection point, 241
  solution, 240
  solution by matrix inverse, 290–93
Logarithm, 130
  antilogarithm, 145
  base, 139
  common, 140

Napierian, 140
natural, 140
properties, 142
Logarithmic functions, 141

Maintenance expenses, 41
Marginal cost, 507–9
Marginal propensity to consume, 28
  disposable income, 28, 44–45
  personal consumption
    expenditures, 28, 44–45
Marginal propensity to save, 30, 584
Marginal revenue, 509
Marketing research, 375
Markov chain, 445
  equilibrium, 451
  regular transition matrix, 452
  steady-state, 451
  steady-state probabilities,
    451–55
  transition matrix, 446
  transition probabilities, 446
Matrix, 248
  column matrix, 249
  column vector, 249
  dimension, 249
  dot product, 255
  equality, 250
  equation, 262–63
  identity, 263–64
  inverse, 283–88
  product matrix, 257
  row matrix, 249
  row vector, 249
  square matrix, 249
  subtraction, 250–51
  sum, 250–51
Maximum value, 78
Minimizing surface area, 558
Minimizing surface area cost, 558
Minimum value, 78
Multivariate functions, 560–70
  critical point, 567
  maxima and minima, 566–70
  saddle point, 568
  second-derivative test, 568
Mutually exclusive events, 392

Natural number, 2
Net present value, 236–37
Newton's law of cooling, 136
Newton's method, 512–16
Nominal rate, 172
Normal probability density function,
  549
Note, 163–66
nth-degree polynomial, 93

Oil refinery scheduling, 302
Ordered pair, 9

INDEX

Parabola, 68–76
  axis of symmetry, 68
  vertex, 68
Parallel lines, 39
Parcel delivery charges, 16
Park management, 84
Partial derivative, 561–64
  first partial, 561–62
  instantaneous rate of change, 561
  second partial, 563–64
Particle movement, 106
Permutations, 419
Perpendicular lines, 39
Perpetual bond, 630
Perpetual cash flows, 629
  present value, 629
Pollution control, 117
Polynomial function, 92–104
  degree, 92–93
Population growth, 138, 155
Price earnings ratio, 156
Principal nth root, 62
Principal square root, 62
Probability, 382
  a posteriori probability, 413
  a priori probability, 413
  Bayes' formula, 409–13
  chance experiments, 382
  combinations, 422–23
  complement law, 387
  conditional probability, 397
  event, 382
  law of large numbers, 388
  laws of addition, 390–93
  laws of multiplication, 403–7
  multiplication rule of counting, 417
  mutually exclusive events, 392
  permutations, 419
  sample space, 382
  subjective probabilities, 388
  tree diagram, 385
Probability density function, 632
Probability distribution, 429
Proceeds, 100
Producers' surplus, 604–7
Production possibility curve, 116
Product mix, 244, 294
Product transformation curve, 77
Profit function, 49, 81–82
Projectile, 91
Project scheduling, 364

Quadrant, 10
Quadratic equation, 67
  degree, 67
Quadratic formula, 75
Quadratic function, 67–76
Quality control, 99

Radical, 62
Radioactive decay, 134
Random variable, 430
  continuous, 430
  discrete, 430
  expected value, 431
  mean, 431
Rate of change, 25–26
Rational function, 106–15
  horizontal asymptote, 107, 113–14
  vertical asymptote, 106–15
Rational number, 3
Rectangular coordinate system, 9–10
  origin, 9
  x-axis, 9
  x-coordinate, 9
  y-axis, 9
  y-coordinate, 9
Real estate development, 614
Real number, 2–3
Real number line, 2
Relative maximum, 532
Relative minimum, 532
Reliability, 637
Reliability function, 637–38
Restrictive sample space, 397
Revenue flow, 612, 614
Revenue function, 79–83
Reynolds number, 560
Riemann sum, 589
Root, 62
  cube root, 62
  nth root, 62–63
  principal nth root, 62
  principal square root, 62
  square root, 62
Row operations, 269
Rules of equalities, 304
Rules of inequalities, 304

Sales revenue equation, 47–51, 79
Sales revenue function, 18
Sample space (*see* Probability)
Scientific notation, 65–66
Secant line, 461–63
Second-degree polynomial, 92
Second derivative, 541
Second-derivative rule, 543
Second-derivative test, 544
Semilogarithmic graph paper, 152
Sets, 370–77
  complement, 374
  De Morgan's laws, 379
  disjoint sets, 372
  elements, 370
  empty set, 371
  equality, 371
  intersection, 372
  null set, 371

I-5 INDEX

Sets (continued)
  union, 373
  universal set, 372
  Venn diagram, 371
Simple annuity (see Annuity)
Simple interest (see Interest)
Simplex method, 326
  a-coefficients, 331
  alternate optima, 348
  artificial variable, 353
  basic feasible solution, 327
  basic variable, 328
  basis, 328
  b-coefficients, 331
  c-coefficients, 331
  degeneracy, 348–49
  degenerate solution, 349
  indicators, 337
  initial tableau, 329
  maximization, 329
  minimization, 352
  nonbasic variable, 328
  optimum solution, 330
  pivot element, 338
  shadow price, 349–51
  shadow price indicator, 350
  slack variable, 327
  surplus variable, 327, 353
Sinking fund, 200–202
  schedule, 203
Shifting a graph, 70
Slope, 23–28
Special law of addition (see Probability)
Square root, 62
Stock price forecasting, 156

Supply function, 19, 83
System of linear equations (see Linear systems)

Tableau, 270
Tables of integrals, 621
Third-degree polynomial, 93
Time series, 154
Time series data, 154
Tree diagram, 385

Uniform distribution, 630
Unit pricing, 76, 102
Utile, 86
Utility, 86, 92
Utility function, 86–87, 92

Variable annuity, 212
Variable cost per unit, 18, 26
Vector, 249
  column, 249
  row, 249
Venn diagram, 371
Vertex, 68
Vertical asymptote, 106–15
Vertical line, 27, 36, 38
Vitamin requirements, 323
Volume, 100–102

Waiting time, 634, 636
Whole number, 3
Windmill design, 22

x-intercept, 41–44

y-intercept, 26

MANAGEMENT

Apple growing, 91, 559
Average cost function, 116
Break-even analysis, 47–51, 54, 55, 56–57, 79–81, 248
Cost analysis, 17, 18, 26, 34, 40, 46, 116, 117, 149, 507, 516, 546–48, 550–53, 566, 571
Cost-benefit curve, 112, 117
Cost minimization, 323, 325
Decision-making:
    capital investment decision, 236–37
    utility, 86, 92
Employee growth, 133, 529
Maintenance expenses, 41, 137
Minimizing average cost, 550–51, 557
Profit analysis, 77, 81–82, 89, 501–2, 549, 566
Profit analysis and defectives, 87–89
Profit maximization, 318, 324, 325, 326, 368
Revenue analysis, 18, 44, 79, 82, 89, 148, 465, 467, 509, 516, 517, 561, 566, 569, 571
Sales, 137
Sales and advertising expenditures, 139, 148
Sales decay, 529
Toll booth revenue, 626
Unit pricing, 76
Unit pricing and revenue, 102, 105
Wage analysis, 396

MARKETING

Advertising, 444
Brand switching, 449, 450, 455
Consumer behavior, 415
Customer classification, 408
Marketing reseach, 375, 394, 395, 396, 400, 401
Preference analysis:
    taste test, 444
Survey analysis, 457

MISCELLANEOUS

Approximating roots, 518
Birthday problem, 429
Communication systems, 451
Cross-classification table, 380, 394, 395, 397, 400
Education:
    learning curve, 135, 138, 529
    testing, 438, 442, 443
Enclosure, 84–86, 90, 91, 553, 558, 559
Law enforcement:
    radar traps, 443
Population growth, 138, 155
Scientific notation, 65–67

NATURAL SCIENCE

Cooling, 530
Newton's law of cooling, 136, 138
Projectile, 91, 501, 559
Temperature conversion, 8
Tomato seed germination, 444
Trajectory, 501
Weather forecasting, 392
Windmill design, 22

POLITICAL SCIENCE

Party switching, 450
Sampling voters, 451
Voting analysis, 408, 450, 451

PRODUCTION MANAGEMENT

Batch process and unit profits, 103, 105
Blending problem, 363, 367
Manufacturing, 465, 571
Maximizing volume, 100–102, 105, 557
Minimizing surface area, 558
Minimizing surface area cost, 558
Oil refinery scheduling, 302
Production possibility curve, 116
Production time management, 260–62, 267
Product mix, 244, 245, 294–96, 299, 300
Product transformation curve, 77
Project scheduling, 364–66, 367
Quality control and profits, 99, 104–5

QUALITY CONTROL

Acceptance sampling, 445
Battery life, 637
Brand defective determination, 413, 415
Calculator circuitry length of life, 638
Production lots, 444
Product reliability, 637
Sampling without replacement, 402, 424, 428
Sampling with replacement, 402
Service reliability, 409
Warranty life, 637

REAL ESTATE

Minimizing energy costs, 551–53, 557
Real estate development, 614
Rental income management, 91

RISK ANALYSIS

Credit customer analysis, 416
Gambling machine, 429, 433
Insurance driver classification, 451

TRANSPORTATION

Mail-order operations, 416
Parcel service charges, 16, 22
Time shipment data, 251, 254

COLLEGE MATHEMATICS